The Enzymes

KINETICS AND MECHANISM

VOLUME II

Third Edition

CONTRIBUTORS

THOMAS C. BRUICE
W. W. CLELAND
SAMUEL J. DI MARI
GORDON G. HAMMES
ALBERT S. MILDVAN
G. POPJÁK
J. H. RICHARDS
IRWIN A. ROSE
PAUL R. SCHIMMEL
ESMOND E. SNELL
SERGE N. TIMASHEFF

ADVISORY BOARD

THE ENZYMES

Kinetics and Mechanism

Edited by PAUL D. BOYER

Molecular Biology Institute
University of California
Los Angeles, California

Volume II

THIRD EDITION

ACADEMIC PRESS New York and London 1970

ACADEMIC PRESS, INC.
111 Fifth Avenue, New York, New York 10003

United Kingdom Edition published by
ACADEMIC PRESS, INC. (LONDON) LTD.
Berkeley Square House, London W1X 6BA

LIBRARY OF CONGRESS CATALOG CARD NUMBER: 75-117107

PRINTED IN THE UNITED STATES OF AMERICA

Contents

8. Some Physical Probes of Enzyme Structure in Solution

SERGE N. TIMASHEFF

9. Metals in Enzyme Catalysis

ALBERT S. MILDVAN

List of Contributors

Numbers in parentheses indicate the pages on which the authors' contributions begin.

THOMAS C. BRUICE (217), Department of Chemistry, University of California, Santa Barbara, California

W. W. CLELAND (1), Department of Biochemistry, University of Wisconsin, Madison, Wisconsin

SAMUEL J. DI MARI (335), Department of Biochemistry, University of California, Berkeley, California

GORDON G. HAMMES (67), Department of Chemistry, Cornell University, Ithaca, New York

ALBERT S. MILDVAN (445), The Institute for Cancer Research, Philadelphia, Pennsylvania

G. POPJÁK (115), Departments of Biological Chemistry and Psychiatry, University of California School of Medicine, Los Angeles, California

J. H. RICHARDS (321), Division of Chemistry and Chemical Engineering, California Institute of Technology, Pasadena, California

IRWIN A. ROSE (281), The Institute for Cancer Research, Philadelphia, Pennsylvania

PAUL R. SCHIMMEL (67), Departments of Biology and Chemistry, Massachusetts Institute of Technology, Cambridge, Massachusetts

ESMOND E. SNELL (335), Department of Biochemistry, University of California, Berkeley, California

SERGE N. TIMASHEFF (371), Graduate Department of Biochemistry, Brandeis University, and Pioneering Research Laboratory for Physical Biochemistry, United States Department of Agriculture, Eastern Utilization Research and Development Division, Waltham, Massachusetts

Preface

The remarkable expansion of information on enzymes since 1959 when Volume I of the Second Edition of "The Enzymes" appeared is so widely recognized that justification for a third edition is unnecessary. The growth rate of information is akin to that of a bacterial culture in early log phase; further expansion of knowledge leading to a deeper understanding of enzymes at the molecular level shows promise of continuing at this rate.

During the last decade, the primary sequence of an increasing number of enzymes has been unraveled, and, even more important, the elegant beauty of the three-dimensional structure for some has been determined. Today, a student of enzymology and metabolism has a high base for biocatalytic insight and experimentation. Enzyme subunits and their interplay have been recognized and a prominent beginning has been made into the understanding of mechanisms of control whose very existence was only hazily forecast ten years ago. Correlated with advances in molecular genetics has come a realization of the dictation of the tertiary and quaternary structure by the primary amino acid sequence. These and related achievements have implications for biological development and evolution considerably beyond understanding the function of single protein molecules.

Somewhat overshadowed by these dramatic thrusts into new intellectual regions has been a continued progress in enzyme isolation, physical characterization, and chemical modification, and in further understanding kinetics, specificity, and mechanism. The goal is to acquire sufficient information to allow generalization and to achieve the satisfaction and power of understanding. In these steady advances lies the hope for a rational molecular pharmacology.

For the present edition, it seemed that any editor, or group of editors, would lack the perspective and competence to cover the entire field. Also, Professor Henry Lardy, who contributed so much to the second edition, was not available. Thus, the approach evolved for this edition was to plan each volume with the help of a highly competent Advisory Board composed of researchers in a particular field. Invaluable service has been given in organizing Volumes I and II by Advisory Board members John M. Buchanan, W. W. Cleland, William P. Jencks, Daniel E. Koshland, Jr., and Emil L. Smith.

The objective of "The Enzymes" is to present fully the information available on enzymes and enzyme catalysis at a molecular level. The scope of the third edition is similar to that of the second edition, except that separate volumes are not planned on cofactors. The chapters will stress contributions of the past ten years, depending on previous editions for much of the earlier material. The treatise does not attempt to give information on all enzymes that have been described. Indeed, there are hints that among enzymes it is considered something of an honor to achieve sufficient status to be presented as a separate chapter in "The Enzymes."

The first two volumes cover basic properties of enzymes in general. Volume I contains chapters on structure and control, Volume II on kinetics and mechanism. Subsequent volumes will cover specific areas of catalysis. Those in preparation cover hydrolysis of peptides and other C—N linkages; cleavage of glycoside bonds and of carboxyl, phosphate, and sulfate esters; phosphorolysis; hydration, dehydration, and elimination; carboxylation and decarboxylation; aldol cleavage and condensation and other C—C cleavage; and isomerization. Volumes for which Advisory Boards are still to be chosen will cover group transfer, energy-requiring syntheses, oxidation–reduction, and special processes.

Timeliness is difficult to achieve in multiauthored treatises. I am pleased that the present volumes excell in this respect. Manuscripts included in Volumes I and II were received within a three-month period. Through the splendid cooperation of the authors and the publisher, these volumes are published in less than a year after receipt of the manuscripts! The users of the volume owe their appreciation to the authors and to the staff of Academic Press. Any credit to me must be shared by my capable and charming wife, Lyda, for many organizational details, manuscript handling, and editorial assistance.

Paul D. Boyer

Los Angeles, California
August, 1970

Contents of Other Volumes

Volume I: Structure and Control

Volume III: Peptide Bond Hydrolysis

1

Steady State Kinetics

W. W. CLELAND

I. Introduction

Since the publication of the second edition of this work in 1959, enzyme kinetics has developed to a powerful experimental tool for studying enzyme mechanisms that is routinely applied by the modern enzymologist and should be understood by all biochemists. The traditional method for introducing this subject is to lead the reader through a tortuous maze of algebra to try to convince him that the resulting rate equations have some basis in reality, and can be applied to experimental data. Certainly this is how the field developed, but we shall try in this chapter instead to give the reader first an intuitive understanding of what kinetics is all about and then show how the equations whose algebraic form we have been considering are derived and manipulated. In this way, the nonmathematical person can learn enough of the subject to enable him to use the tools in his work without having to bother with the actual detailed algebra.

Kinetics is the study of rates of reactions and can be carried out on any time scale. However, this chapter is restricted to the kinetics of the steady state, that is, to conditions where the reactant concentrations and their Michaelis constants greatly exceed that of the catalyst (the usual laboratory situation) or where there is a dynamic steady state with substrates being continually supplied and products continually removed (the usual situation in the cell) (*1*). Transient state and nonsteady state kinetics will be discussed in the next chapter. While much can be learned from rapid reaction studies, the steady state kinetic mechanism for the enzyme must normally be known first in order properly to interpret the observed effects.

We will not discuss the history of enzyme kinetics as such. Chapters 1 to 3 of the second edition cover this well.

II. Nomenclature and Notation

This chapter uses the nomenclature and notation of Cleland (*1a, 2*); the reader is warned that he will find other systems in the literature and will have to examine the definitions given for the kinetic constants in order to relate them to those used here.

Substrates are designated by the letters A, B, C, and D in the order in which they add to the enzyme, and products by the letters P, Q, R, and S in the order in which they leave the enzyme. Inhibitors are designated I and J. There are two types of enzyme forms—those which are stable on the time scale of the reaction and combine with reactants in bimolecular steps, and those which are basically enzyme–reactant complexes and will dissociate in a reasonable amount of time. Stable enzyme forms are designated E, F, and G, with E being the least complex or "free" enzyme if such a distinction is possible. The stable forms other than free enzyme (phosphoenzyme, carboxybiotin enzyme, or aminoacyladenylate enzyme, for instance) can usually be isolated by chemical methods and later shown to transfer the group they are carrying to one of the reactants,

1. More precisely, the steady state assumption requires that the change in concentration with time of all enzyme forms is very much smaller than the change in concentration with time of substrate or product concentrations. As long as initial velocity appears constant at short time intervals, one normally assumes that steady state conditions exist.

1a. W. W. Cleland, *BBA* **67**, 104 (1963).

2. W. W. Cleland, *BBA* **67**, 104 (1963).

although any such complex with a half-life of minutes would still appear kinetically to be a stable enzyme form.

Enzyme–reactant complexes which readily break down are called transitory complexes and consist of two types: those in which the active site is not completely filled, so that they can still bind another reactant (noncentral transitory complexes), and those in which the active site is completely filled with reactants, so that only dissociation of substrates or products is possible. The latter are called central complexes and correspond to the "Michaelis complex" of classic enzyme theory. For alcohol dehydrogenase, for example, E-DPN and E-DPNH are noncentral transitory complexes, while E-DPN-ethanol and E-DPNH-acetaldehyde are central complexes. Transitory complexes are designated by letters indicating their components such as EA, EAB, FB, and EQ. Central complexes are enclosed in parentheses—(EAB) or (EPQ)—and, since steady state kinetic studies cannot normally distinguish the number of central complexes, we usually assume the presence of only one while indicating the possibilities for isomerization—(EAB–EPQ) or (EA–FP). It should be remembered, however, that the actual catalytic reaction normally involves conversion of one central complex of enzyme and substrates into another of enzyme and products (*3a*).

The number of kinetically important reactants in a given direction is designated the reactancy in this direction and indicated by the syllables Uni, Bi, Ter, and Quad. Thus a reaction with one substrate and two products is Uni Bi and is unireactant in the forward direction and bireactant in the reverse. Since kinetic studies are usually run at constant pH, hydrogen ion is not considered a reactant in this analysis. The variation of kinetic constants with pH is a subject of considerable interest by itself and is treated separately below.

Kinetic mechanisms fall into two major groups. Those in which all reactants must combine with the enzyme before reaction can take place and any products be released are called sequential. Mechanisms in which one or more products are released before all substrates have added are called *ping pong;* in such mechanisms the enzyme oscillates between two or more stable enzyme forms, while in sequential mechanisms there is only one stable enzyme form (free enzyme). Sequential mechanisms are

3. C. P. Henson and W. W. Cleland, *JBC* **242,** 3833 (1967).

3a. The justification for lumping all central complexes into one is that for nonbranched mechanisms the rate equation will be the same when written in coefficient form (*1a*) regardless of how many central complexes are included in the mechanism. Steady state kinetics can only distinguish between mechanisms with rate equations which are algebraically different when expressed in coefficient form. Branched mechanisms such as that shown by aconitase are exceptions, and here one can get some information on the number of central complexes (*3*).

called ordered, if reactants combine with the enzyme and dissociate in obligatory order, or random, if alternate pathways exist and the order of combination or release is not obligatory. The term ordered refers to the apparent kinetic mechanism; certain mechanisms where the kinetics of the chemical reaction appear ordered on the basis of product or dead end inhibition studies can be shown by sensitive isotopic exchange techniques to have small contributions from the alternate paths. The term rapid equilibrium is applied to random or ordered mechanisms in which certain steps involving the formation of enzyme–substrate complexes have unimolecular rate constants for dissociation much larger than the overall maximum velocity, so that in the steady state these steps are essentially at thermodynamic equilibrium.

In many cases, a description such as Ordered Uni Bi, Random Ter Ter, or Ping Pong Bi Bi is sufficient to describe the kinetic mechanism. In more complex ping pong mechanisms, however, the groups of substrate additions and product releases are designated by Uni, Bi, Ter. Thus Bi Bi Uni Uni Ping Pong refers to a mechanism where two substrates add, two products are released, the third substrate adds, and the third product then leaves. In this system of nomenclature water and hydrogen ion are not considered as reactants.

Isomerization of transitory complexes never changes the algebraic form of the rate equation for a kinetic mechanism, and thus the mechanism is still of the same fundamental type. Hydrolytic reactions are frequently Ordered Uni Bi, for instance, in which dissociation of the first product P gives a covalent EQ complex which isomerizes to an ionic EQ′ complex before Q dissociates. Such isomerization of noncentral transitory complexes can often be demonstrated by looking at quantitative relationships between kinetic constants [a classic example is the study of potato acid phosphatase by Hsu *et al.* (*4*)]. However, if a stable enzyme form isomerizes so that the form generated by the release of a product must undergo a conformation change before binding the next substrate by a step which is at least partially rate limiting when all substrates are saturating, then the rate equation has new terms in the denominator. Cleland called these *Iso mechanisms* (*1a*), but since clear-cut examples of such mechanisms are not yet known we do not discuss them further here.

Kinetic constants corresponding to the various reactants are designated by K with an identifying lower case subscript: K_a, K_b, K_p, and K_q indicate the limiting Michaelis constants (the concentrations of the indicated reactants which give half of the maximum velocity when all other substrates are at infinite concentrations, and no products or other inhibitors are present); K_{ia}, K_{ib}, etc., are used for inhibition or dissociation con-

4. R. Y. Hsu, W. W. Cleland, and L. Anderson, *Biochemistry* **5**, 799 (1966).

stants, and other special constants may be defined as needed. All such constants have dimensions of concentration. The equilibrium constant is K_{eq}. The maximum velocities in forward and reverse directions are designated V_1 and V_2; these have dimensions of concentration per unit time such as M/sec. If E_t is the concentration of independent active sites (this will normally exceed the concentration of enzyme molecules since most enzymes are oligomers and contain two, four, or more identical active sites per molecule), the expressions V_1/E_t and V_2/E_t are the turnover numbers or catalytic center activities and have dimensions of reciprocal time. They can be evaluated only when the molecular weight and number of catalytic sites per molecule are known so that E_t can be expressed in molar units. In the absence of such information, the maximum velocity per unit weight of protein is all that can be determined.

In order to describe kinetic mechanisms, the shorthand notation of Cleland (*1a*) will be used. The enzyme is denoted by a horizontal line, and the successive additions of substrates and release of products are designated by vertical arrows. Enzyme forms are shown under the line and rate constants are indicated for each step, those on the left side of the arrow or top of the line being for the forward reaction (*4a*). The following examples illustrate the notation:

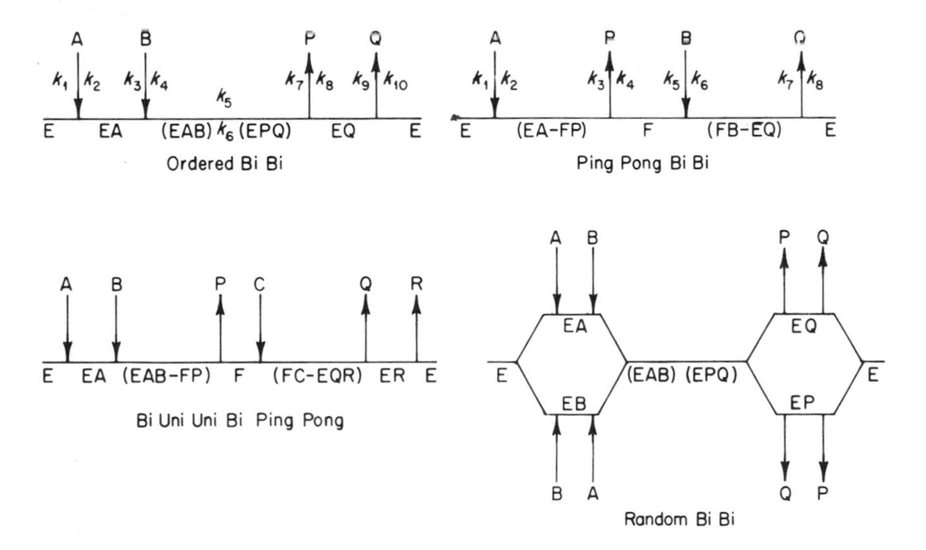

4a. Some kineticists use rate constants with plus and minus subscripts, but our experience is that this multiplies errors tremendously; thus, rate constants are numbered consecutively, with the odd ones normally being for the forward reaction and the even ones for the reverse reaction.

III. Basic Theory

The rate equation which describes the behavior of most enzymes when only one substrate concentration is varied is

$$v = \frac{VA}{K_a + A} \tag{1}$$

Clearly at very high A, $v = V$, while at low A, $v = (V/K_a)$ A; V and V/K_a (or their reciprocals) are thus the two fundamental kinetic parameters varying independently with the concentrations of other substrates, inhibitors, activators, or such variables as pH, temperature, or ionic strength. The ratio of these two independent parameters K_a is called the Michaelis constant and represents the level of A giving $v = V/2$. While it is not an independent constant, it is a very useful one for the biochemist, since it is also a clue to the physiological level of the substrate. (A substrate concentration around K_a utilizes most of the catalytic potential of the enzyme, while still maintaining proportional control; at high substrate levels the rate does not vary with substrate concentration, and one has no control.) The Michaelis constant is not usually identical with the dissociation constant of the enzyme–substrate complex, but rather it represents the apparent dynamic dissociation constant under steady state conditions; that is, the dissociation constant equals (E)(A)/EA at thermodynamic equilibrium, while the Michaelis constant equals the same expression under steady state conditions (*4b*). The Michaelis constant thus measures the apparent affinity under steady state as opposed to equilibrium conditions. Michaelis constants may be larger, smaller, or the same as dissociation constants, depending on the mechanism and the relative size of certain rate constants.

IV. The Tools of the Kineticist

One of the major reasons for carrying out kinetic studies on enzymes is to determine the kinetic mechanism. To do this kinetic data are com-

4b. More rigorously, this expression is

$$K = \frac{\text{(A) (}\Sigma\text{ concentrations of enzyme forms that would be present when A = 0)}}{\text{(}\Sigma\text{ concentrations of enzyme forms that would be present when A is infinite)}}$$

The truth of this expression can be demonstrated by substituting in the appropriate concentrations of the enzyme forms from the distribution equations derived below. If all other substrate concentrations are saturating, this expression gives the limiting Michaelis constant K_a; if not, it gives an apparent Michaelis constant.

pared with the predicted kinetics of all possible mechanisms, and those which do not fit are eliminated. This may seem a nearly impossible job, but the number of possible (or at least reasonably plausible) mechanisms is not large; the possibilities can rapidly be narrowed and then tested, one by one, by kinetic techniques designed specifically to distinguish them. If all of the techniques described here are applied, one can in most cases deduce the actual kinetic mechanism with little or no ambiguity. The types of experiments that the kineticist has at his disposal, and the types of knowledge he gets from each, are described below.

A. Initial Velocity Studies

1. *Bireactant Initial Velocity Studies*

The rate equations describing the steady state kinetics of enzyme-catalyzed reactions have the form:

$$v = \frac{dP}{dt} = \frac{f(A,B\ldots)}{g(A,B\ldots)} \qquad (2)$$

where $f(A,B\ldots)$ and $g(A,B\ldots)$ are linear polynomials involving the concentrations of the reactants. If these equations are integrated in order to follow the time course of the reaction, the expressions are quite complex except in the simplest cases. [Darvey has done this for a number of mechanisms (*5*).] Further, the algebraic form of the integrated equation is not very sensitive to differences in mechanism, and thus this approach cannot be used very profitably to identify kinetic mechanisms. The integrated equations can be compared with data in order to verify the rate equation once the mechanism and the kinetic constants have been determined by the more usual techniques; Schwert has recently done this for lactic dehydrogenase (*6*).

The rate equations are thus used in their differential form and the velocity is determined as a function of reactant concentrations. The normal method for determining initial velocity patterns is to use a separate reaction mixture for each substrate concentration, but it is possible in special cases where the equilibrium constant is high and the products do not inhibit (or their mode of inhibition is known) to determine velocities as a function of substrate concentration from the progress curve of the reaction (that is, one determines the slopes of the progress curve and the remaining substrate concentrations at different times). If the products do not inhibit, one such curve will provide enough data to determine

5. I. G. Darvey and J. F. Williams, *BBA* **85,** 1 (1964).
6. G. W. Schwert, *JBC* **244,** 1278 and 1285 (1969).

the maximum velocity and Michaelis constant, but if the products are inhibitory more than one curve is needed. This approach spares enzymes and reagents and is most useful for studying the effect of pH, temperature, or ionic strength on the values of the kinetic parameters.

A determination of initial velocity patterns usually involves variation of the concentration of one substrate at different fixed levels of the other one (or ones), and in the absence of products (or, in certain special cases, in the presence of a fixed level of a product). Bireactant mechanisms give one of three patterns. Most sequential mechanisms will show a rate equation of the form

$$v = \frac{VAB}{K_{ia}K_b + K_aB + K_bA + AB} \tag{3}$$

where v is the initial velocity, A and B are substrate concentrations, and V and the K's are constants. Plots of v vs. A at constant B, or v vs. B at constant A are hyperbolas and not very informative, but a plot of $1/v$ vs. $1/A$ or $1/B$ under these conditions is a straight line, and if one makes such reciprocal plots vs. $1/A$ at various B levels, or vice versa, one gets a series of lines intersecting to the left of the vertical axis. The crossover point has a horizontal coordinate of $-1/(K_{ia})$ when A is the substrate being varied, and $-K_a/(K_{ia}K_b)$ when B is varied, and a vertical coordinate of $1/V\ [1 - (K_a)/(K_{ia})]$ in either plot (*6a*). It is thus obvious that the pattern crosses above the horizontal axis when K_{ia} is greater than K_a, below the axis when K_{ia} is less than K_a, and on the axis when $K_{ia} = K_a$. The vertical position of the crossover point depends only on the ratio of K_{ia} to K_a, and it has no diagnostic value as far as distinguishing mechanisms. Equality of K_{ia} and K_a is purely fortuitous, regardless of mechanism, and should not be taken as evidence for a random as opposed to ordered mechanism. This initial velocity pattern is generally referred to as an intersecting or sequential one. Note that the equation is algebraically symmetrical with respect to A and B so that one cannot tell them apart. The numerical values of the constants are extracted graphically by replotting the slopes and intercepts of the reciprocal plots vs. the reciprocal concentrations of the nonvaried substrates (Fig. 1).

Ping pong bireactant mechanisms give an initial velocity rate equation similar to Eq. (3) except that there is no constant term in the denominator. The result is that the pattern of reciprocal plots becomes parallel. Again the equation is algebraically symmetrical with respect

6a. In a random mechanism $K_{ia}K_b = K_aK_{ib}$; thus, $-1/(K_{ib})$ is the horizontal coordinate of the crossover point when B is varied, but in an ordered mechanism this relationship does not hold and K_{ib} has a different definition.

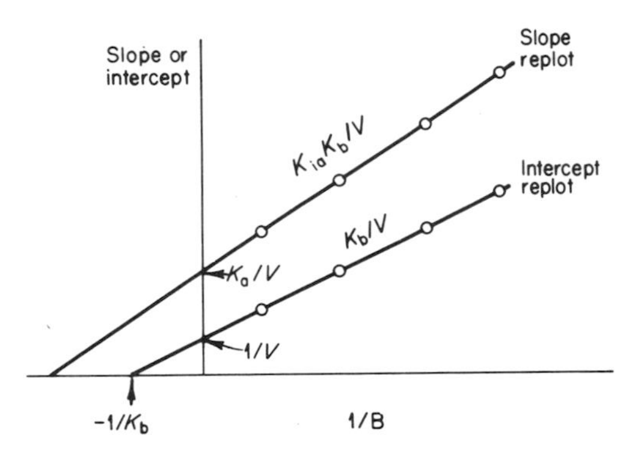

FIG. 1. Slope and intercept replots for an intersecting initial velocity pattern. The ratio of the slopes of the two replots determines K_{ia}, and the ratio of intercepts determines K_a; K_b and V are determined directly from the intercept replot.

to A and B, and the numerical values of the constants are obtained from replots of the intercepts just as with the intersecting pattern (the slope replot is, of course, horizontal). This is referred to as a parallel or ping pong initial velocity pattern. While an intersecting pattern definitely intersects, there is always some doubt about whether a parallel pattern is truly parallel or very slightly intersecting. One must therefore be cautious in interpreting an apparently parallel pattern as ping pong in the absence of corroborative data. Mammalian hexokinases were thought for a time to be ping pong on this basis ($K_{ia} \ll K_a$, and the lines intersect far below the horizontal axis and look nearly parallel), but have since been shown to be sequential (7). It cannot be overemphasized that a variety of different kinetic experiments must be run to establish the kinetic mechanism.

In certain special mechanisms, the K_aB term in Eq. (3) is missing. The equation is now algebraically unsymmetrical with respect to A and B, and different patterns are obtained when A or B are varied. When A is varied, the pattern intersects to the left of the vertical axis at $-1/(K_{ia})$ and above the horizontal axis at $1/V$, and a replot of the slopes of the reciprocal plots vs. 1/B goes through the origin (Fig. 2), instead of having a finite vertical intercept as is normally the case. The reciprocal plot when B is infinite is thus horizontal, or, in other words, the requirement for A appears to disappear at very high B concentration. When B is varied at different A levels, the pattern intersects on the vertical axis, again showing that infinite B removes dependence on A. This initial velocity pattern results from the ordered addition of two reactants with the addition of the first being at thermodynamic equi-

7. J. Ning, D. L. Purich, and H. J. Fromm, *JBC* **244**, 3840 (1969).

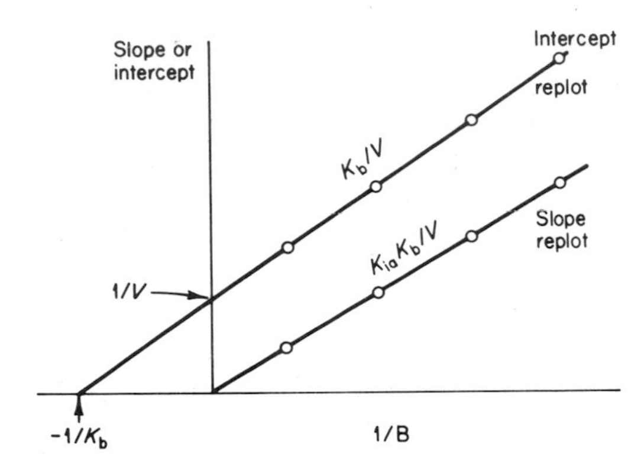

FIG. 2. Slope and intercept replots from the equilibrium ordered initial velocity pattern with A as variable substrate. The ratio of the slopes of the two replots determines K_{ia}.

librium. (Combination of EA with B then displaces the $E + A \rightleftharpoons EA$ equilibrium to the right, and an infinite concentration of B converts all E to EAB, regardless of the A level—remember E_t is much less than A, and thus a level of A equal to E_t is essentially zero on the scale of concentration we are using for A.) This situation arises in two ways. First, when addition of an activator (such as a metal cation) must take place before addition of the substrate, the activator cannot dissociate once the substrate has added, and the activator need not leave the enzyme during each catalytic cycle but may remain bound, the step involving addition of activator will be at equilibrium in the steady state, even if the rate constants for addition and release of activator are slow compared with the maximum velocity of reaction. This is a not uncommon situation with metal cations that activate enzymes by direct combination, as opposed to activation by prior complexing of a substrate (such as MgATP). A less common situation giving this kinetic pattern, which we can call the equilibrium ordered initial velocity pattern, is a rapid equilibrium ordered addition of substrates. In this case, the rate constant for dissociation of EA must be much larger than the maximum velocity, but this does occur in some cases. For this pattern, note that although the Michaelis constant for A(K_a) appears equal to zero, the dissociation constant of A(K_{ia}) is finite and easily determined.

2. *Predicting Initial Velocity Patterns*

An understanding of why patterns are observed for a given mechanism will aid the reader in predicting the patterns for other mechanisms.

To do this, the slope and intercept of a reciprocal plot must be observed to see how they are affected by the concentrations of other substrates.

The vertical intercept of a reciprocal plot is the reciprocal velocity when the variable substrate is at infinite concentration. Thus, to alter the intercept, a compound must alter the reaction rate when the variable substrate is saturating. This is normally always true for a second substrate, regardless of mechanism, since it is combining with an enzyme form other than that with which the variable substrate combines and will affect the rate even when the variable substrate is at infinite concentration. Thus the two common initial velocity patterns (the intersecting or sequential and the parallel or ping pong patterns) both show an effect on the intercepts by the substrate other than the variable one. The only exception is in the equilibrium ordered case, where saturation with B eliminates dependence on A by displacing the $E + A \rightleftharpoons EA$ equilibrium completely to the right as EA reacts with B. When B is varied there is no intercept effect by A, and the lines cross on the vertical axis since no A, other than an amount stoichiometric with enzyme, is needed to give the full maximum velocity as long as B is infinite.

The interpretation and prediction of slope effects is more subtle. The slope of a reciprocal plot is the reciprocal of the apparent first-order rate constant for reaction of substrate when it is at very low levels. [Remember that most enzyme-catalyzed reactions follow Eq. (1) and at low substrate levels $v = (V/K_a)A$; the slope of the reciprocal plot of Eq. (1) is K_a/V.] If A is the substrate we are considering, and it combines with the enzyme by the step $E' + A \rightleftharpoons E'A$, it is clear that in order to observe the most rapid net combination of E′ with A at low A levels (highest V/K_a or lowest K_a/V) we want the concentration of E′ to be as high as possible and the concentration of E′A as low as possible. Thus any factors which raise the level of E′ with respect to E′A or decrease E′A with respect to E′ will lower the slope of the reciprocal plot. Likewise, any factor which increases the ratio of E′A to E′, either by raising E′A, lowering E′, or both, will increase the slope.

Consider now the effect of a substrate other than the variable one (which we will call the changing fixed substrate). A substrate combining with E′A would obviously lower E′A with respect to E′; thus, increasing concentrations of this substrate would lower the slope of the reciprocal plot. Likewise, a substrate which combines with free enzyme to give E′ would raise E′ with respect to E′A and again lower the slope. In both of these cases, we would observe the intersecting or sequential initial velocity patterns. Then, under what conditions would one not see a slope effect? Whenever there is not a reversible connection between the points of addition in the reaction sequence of the variable and the changing fixed substrates. The classic cases are ping pong mechanisms such as:

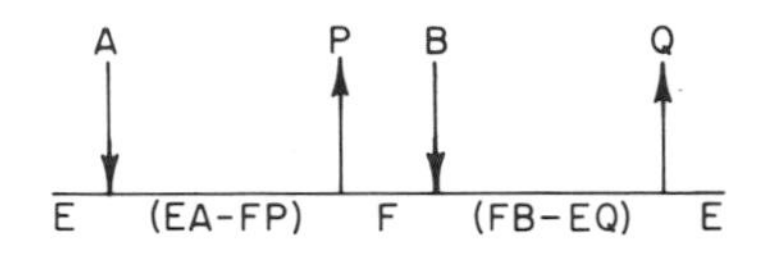

Here one or more products are released after the addition of each substrate; in the absence of initially added products, these steps are irreversible. (One can only reverse them by having products present, and the initial rate is determined at the start of reaction before appreciable products accumulate.) If A is the variable substrate, raising the B level will convert F to the second central complex, but the ratio of E to the first central complex is unchanged because of the lack of reversible connection between E and F. We thus have no slope effect and observe the parallel or ping pong initial velocity pattern.

However, if we add a finite amount of product P to the system initially we see a different picture: P now continually combines with F and converts it back to (EA–FP), thus, the steady state level of this central complex is higher than it would be in the absence of P. If we now raise the B level, keeping P constant, B will compete with P for F and by producing (FB–EQ) from F prevent P from returning it to (EA–FP). The steady state level of (EA–FP) thus drops specifically with respect to E as the level of B is increased, and we have a slope effect which produces an intersecting instead of a parallel pattern. The conversion of a parallel into an intersecting pattern by the addition of a fixed level of product is one diagnostic for a ping pong mechanism of this type.

The parallel initial velocity pattern is observed for a ping pong mechanism only when form F is kinetically stable on the time scale of the reaction. If F can spontaneously hydrolyze to give E plus the fragment of A that was transferred to the enzyme in producing P, the initial velocity pattern becomes intersecting when Q release is measured (if P release is measured the effect is different—see below under alternate product inhibition). This is because at low B levels most of the catalytic cycles do not result in formation of Q but rather in direct hydrolysis of substrate. Thus the apparent first-order rate constant for reaction of enzyme and substrate to give Q is nearly zero. Raising the level of B does not affect the true rate of combination of A with E, but since reaction flux is directed from hydrolysis to production of Q, it raises the apparent rate of reaction of E and A and thus decreases the slope of reciprocal plots so that the intersecting pattern is observed.

There are other irreversible steps besides product release which may interrupt reversible reaction sequences and thus prevent observation of a slope effect in an initial velocity pattern. One of these is saturation with a substrate—no step can be reversed in the face of an infinite concentration of reactants—but by saturation true infinity is meant, not just 10 times the Michaelis constant. Saturation with a substrate has a unique effect on the initial velocity pattern only when there are three or more substrates, and this is discussed further below. Finally, a step can be irreversible simply because the equilibrium constant of the reaction is extremely high, and the dissipation of free energy occurs during this step. This is not very common, since most enzymic reactions are freely reversible, but a classic case is the oxidation of 3-hydroxyanthranilate by oxygen, in which both atoms of oxygen are found in the product; thus, the mechanism has to be sequential, but the initial velocity pattern is the parallel one (*8*). Apparently combination of oxygen with the enzyme leads to a very large decrease in free energy before 3-hydroxyanthranilate adds; in the absence of substrate this oxygen–enzyme complex breaks down slowly with irreversible loss of enzymic activity.

The above discussion of the meaning and causes of slope and intercept effects allows us to formulate the following rules for prediction of initial velocity patterns.

The slopes of reciprocal plots will be affected by the changing fixed substrate whenever the points of combination of it and the variable substrate in the reaction sequence are connected by reversible steps. Saturation with substrates or release of products at zero concentration are considered irreversible steps. If no reversible connection exists, no slope effect is observed.

The intercepts of reciprocal plots will be affected by the changing fixed substrate unless (a) its combination with enzyme is at thermodynamic equilibrium and (b) the variable substrate combines with the complex containing the changing fixed substrate in such a way that it traps the changing fixed substrate on the enzyme so it cannot leave.

With these rules, we can now look at cases where there are three or more substrates and analyze the expected initial velocity patterns.

3. *Terreactant Initial Velocity Patterns*

In bireactant cases, ping pong mechanisms give the parallel initial velocity pattern and sequential mechanisms give the intersecting or occasionally the equilibrium ordered pattern. When there are three substrates, however, this is no longer always true, and parallel patterns may occur with sequential mechanisms and intersecting ones with ping pong mechanisms. Consider, for example, the ordered addition of three substrates:

8. N. Ogasawara, J. E. Gander, and L. M. Henderson, *JBC* **241,** 613 (1966).

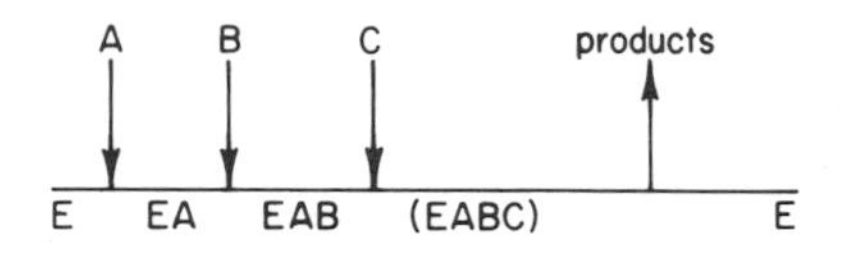

Unless rapid equilibrium conditions apply, there will always be intercept effects here, and if no substrate is saturating the reaction sequence from E to (EABC) is reversible, so there will be slope effects. Thus the initial velocity pattern will be intersecting regardless of whether A and B, A and C, or B and C are the variable and changing fixed substrates. In each case, the third substrate would be held at constant concentration for the entire pattern. If substrate B is truly saturating, however, the reversible sequence is broken and the A-C initial velocity pattern becomes a parallel one. In practice the slope effect becomes smaller and smaller as B is raised, but unless B is raised to over 100 times its Michaelis constant, a completely parallel pattern will not be seen. The A-B and B-C initial velocity patterns will always be intersecting, regardless of the level of the other substrate. This approach was first used by Frieden with glutamic dehydrogenase (*9*); however, he was unable to raise ammonia high enough to convert the TPNH-α-ketoglutarate initial velocity pattern to a really parallel one (ammonia was 10 times its Michaelis constant). Although this was considered an excellent paper when it first appeared, the modern kineticist 10 years later would not be satisfied with this and would obtain TPNH-α-ketoglutarate initial velocity patterns at several levels of ammonia and extrapolate the observed slope effect to infinite ammonia to see whether it really disappeared. One must also be cautious about jumping to conclusions about mechanisms on the basis of initial velocity data alone. A classic case occurred with tyrosine transaminase, which was first postulated to be ordered terreactant on the basis of two intersecting and one parallel initial velocity patterns (*10*). Later work showed, however, that the mechanism was really ping pong, in agreement with the parallel initial velocity pattern for tyrosine and α-ketoglutarate at high pyridoxal phosphate levels (*11*). The intersecting initial velocity patterns seen for tyrosine and pyridoxal phosphate as well as α-ketoglutarate and pyridoxal phosphate resulted from the loose binding of pyridoxal phosphate to the enzyme. Analysis in this case was complicated by the nonenzymic formation of Schiff bases between tyrosine and pyridoxal phosphate.

While an ordered terreactant mechanism shows a parallel initial veloc-

9. C. Frieden, *JBC* **234**, 2891 (1959).
10. T. I. Diamondstone and G. Litwack, *JBC* **238**, 3859 (1963).
11. G. Litwack and W. W. Cleland, *Biochemistry* **7**, 2072 (1968).

ity pattern when substrate B is saturating, a completely random mechanism shows intersecting patterns at all times *(11a)*. If one substrate must add first, but the other two can add randomly, parallel initial velocity patterns will be obtained when either B or C is saturating. This is easily understood if one remembers that saturation with B leads to addition in the order A, B, and C, while saturation with C causes the order to be A, C, and B (that is, saturation at the branch point diverts all reaction flux through one path or the other). This is the situation found for citrate cleavage enzyme [MgATP adds first; CoA and citrate then add randomly *(12)*].

Terreactant ping pong mechanisms can be of two types. If there are three stable enzyme forms and a product is released after each substrate is added (Hexa-Uni ping pong), all initial velocity patterns will be parallel unless products are added. Mechanisms of this sort normally involve an enzyme bound prosthetic group which moves from one catalytic site to another (lipoic acid in pyruvic dehydrogenase, for example), but since none has had a thorough kinetic study, we will not discuss these further. The common ping pong terreactant mechanisms are either Bi Bi Uni Uni (biotin-containing carboxylases) or Bi Uni Uni Bi (many pyrophosphorylytic synthetases). These mechanisms are similar as far as initial velocity analysis goes:

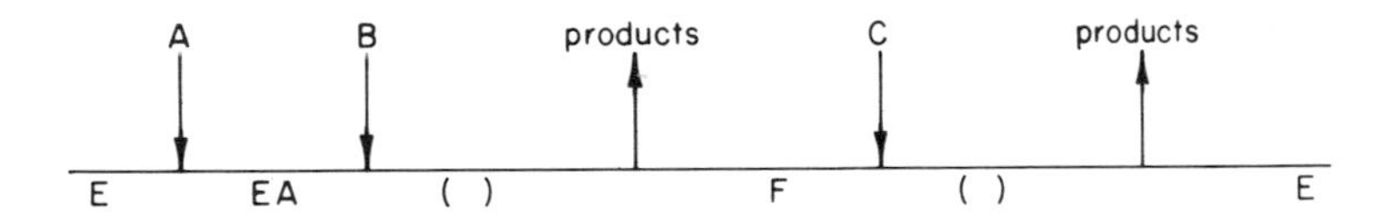

It is clear that A and B are reversibly connected and will always give an intersecting initial velocity pattern, although some preliminary unpublished work suggests that pyruvic carboxylase may show the rapid equilibrium ordered pattern, MgATP adding first, bicarbonate second. The A-C or B-C initial velocity patterns will be parallel, however, since the release of products interrupts the reversible connection between F and the E, EA sequence. The presence of fixed levels of all products released between B and C or between C and A would convert these patterns to intersecting ones, since the reversible connection would be reestablished. The resulting initial velocity patterns could be reconverted to parallel ones in certain cases by saturation with the third sub-

11a. In theory random mechanisms give curved reciprocal plots, but in practice one usually cannot see the curvature, which occurs near the vertical axis. The initial velocity patterns thus resemble those expected for a rapid equilibrium random mechanism.

12. K. M. Plowman and W. W. Cleland, *JBC* **242,** 4239 (1967).

Initial velocity experiments do determine the Michaelis constants and, in some cases, the dissociation constants of the reactants, and the resulting numerical information is of great interest to the biochemist. It must be remembered, however, that in multireactant cases any individual reciprocal plot determines only an apparent maximum velocity and Michaelis constant, and the true maximum velocity and limiting Michaelis constants are obtained only by analysis of the full initial velocity patterns.

B. Inhibition Studies

1. *Nomenclature and Basics*

The second major tool of the kineticist is inhibition studies. Rarely is a kinetic mechanism worked out without extensive resort to such experiments. Inhibitors may be products, substrates, or other molecules which resemble these and play musical chairs with the intended substrates for their absorption pockets in the enzyme. We will not consider in this discussion such materials as heavy metal ions, reagents that react covalently with the enzyme, or other reagents that lead to denaturation. Such compounds generally cause irreversible inhibition or denaturation and are not much help in elucidating kinetic mechanisms. We will also not discuss molecules which inhibit by combination at sites remote from the catalytic site (allosteric inhibition); this phenomenon is discussed by other authors of this treatise particularly by Koshland, Chapter 7, Volume I.

First we should define some terms. A total inhibitor causes the velocity to become zero when its concentration is infinite. Conversely, a finite rate is still observed when a partial inhibitor is present at infinite concentration. Partial inhibitions by allosteric inhibitors are often observed, but in our present discussion partial inhibitors are limited to those that divert the reaction flux to an alternate (and slower) reaction pathway or molecules which are too small to prevent completely the absorption of the substrate in its pocket but whose presence raises the dissociation constants of the substrates or distorts the geometry so that the maximum velocity is reduced, or both. Examples of the first are alternate product inhibitors (see below), while an example of the second is the partial inhibition by methanol of horse liver alcohol dehydrogenase [it partially blocks the absorption of other alcohols (*15*)].

Product inhibition results from the formation of the same enzyme–

15. C. C. Wratten and W. W. Cleland, *Biochemistry* 4, 2442 (1965).

reactant complexes that form when this product is a substrate in the reverse reaction, and inhibition by products is always expected (although not always observed; the inhibitory level may be greater than the solubility of the inhibtor). An alternate product is a product which would be observed if a different substrate were used; we will have more to say about the use of alternate product inhibitors below. A molecule that plays musical chairs with the other reactants for the absorption site but is not chemically altered is a dead end inhibitor, since the formation of enzyme–inhibitor complexes gives a dead end step in the mechanism. When a substrate acts as a dead end inhibitor as well as a substrate, we observe substrate inhibition. Products may also act as dead end inhibitors and give mixed product and dead end inhibition.

There are three basic types of inhibition, depending on the effect the inhibitor has on the slope and intercept of reciprocal plots. Initial velocities are determined in the usual manner except that different fixed levels of inhibitor (including zero) are present in the reaction mixtures. If the slope of the reciprocal plot is affected, but the intercept is not, the result is competitive inhibition. Reciprocal plots at different inhibitor levels cross on the vertical axis, and saturation with the variable substrate eliminates the inhibition. If the intercepts are affected, and the slopes are not, the inhibition pattern is a series of parallel lines or uncompetitive inhibition. If both slopes and intercepts are increased by the inhibitor, reciprocal plots will converge to the left of the vertical axis and the inhibition is called noncompetitive. In noncompetitive cases, the lines may cross above, below, or on the horizontal axis or they may not all cross at the same point. Some workers have limited the use of the term noncompetitive to those cases where the reciprocal plots cross on the horizontal axis and called other noncompetitive patterns mixed, but there is no good theoretical reason for doing this and we consider any inhibition where both slopes and intercepts are affected as noncompetitive.

As is done with initial velocity patterns, inhibition patterns are further analyzed by replotting slopes and/or intercepts vs. inhibitor concentration. For many inhibitions such replots are linear, and the inhibition can be called linear competitive, linear noncompetitive, etc. The inhibition constant for the slopes (K_{is}) or intercepts (K_{ii}) is the horizontal intercept of such a replot (see Fig. 3). This is because the expression $(1 + I/K_i)$ is a factor of the term in the rate equation which represents the slope or intercept of the reciprocal plot, and the horizontal intercept of the replot represents the point where slopes or intercepts appear to be zero, and thus where $I = -K_i$. In linear noncompetitive inhibition, reciprocal plots alway intersect at a single point to the left of the vertical axis.

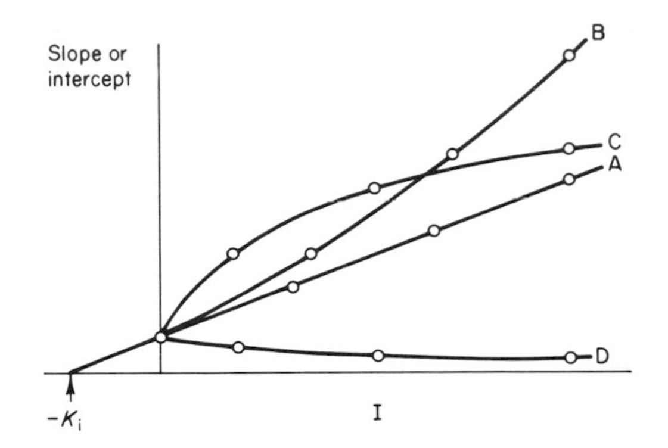

FIG. 3. Characteristic shapes of slope and intercept replots from inhibition experiments: A, linear; B, parabolic; C, hyperbolic inhibition; and D, hyperbolic activation. The horizontal intercept determines K_i for a linear inhibition.

While many inhibitions are linear, it is not unusual for the slope or intercept replot, or both, to be curved. Where the replot is concave up and has the form

$$\text{slope or intercept} = a(1 + b\text{I} + c\text{I}^2) \tag{4}$$

we call the inhibition parabolic (see Fig. 3); such inhibitions are caused by the combination of at least two molecules of inhibitor in such a way that an I^2 term results (see below for the rules for predicting when this will occur). In a noncompetitive inhibition, just the slope or intercept may be parabolic, in which case we will call it S-parabolic I-linear noncompetitive, or S-linear I-parabolic noncompetitive. Parabolic noncompetitive means both replots are parabolas. If the slope and intercept replot are different functions, the reciprocal plots will not cross in a single point. If both are parabolas, the lines cross at a single point only if the ratio of b to c in Eq. (4) is the same for both replots.

Where the replot is convex upward with a horizontal asymptote at high inhibitor levels and has the form

$$\text{slope or intercept} = \frac{\text{a}(1 + \text{I}/K_{\text{i num}})}{(1 + \text{I}/K_{\text{i denom}})} \tag{5}$$

we call the inhibition hyperbolic (see Fig. 3). The inhibition is a partial one, and at infinite inhibitor concentration the slope or intercept reaches the value $\text{a}(K_{\text{i denom}})/(K_{\text{i num}})$, It is obvious that this will be inhibition only if $K_{\text{i denom}}$ is larger than $K_{\text{i num}}$; if the reverse is true, the effect will be activation, and this often happens in these cases. These inhibitions (or activations) occur when the inhibitor (or activator) causes the reaction flux to be diverted to an alternate pathway that is slower (or faster) than the normal one. The reciprocal plots of a hyperbolic noncompetitive pattern cross at a single point only if the $K_{i\ denom}$'s are the same in the slope and intercept terms; if they are not, or if only slopes or intercepts are hyperbolic, the pattern does not show a single crossover point.

Replots may have more complex algebraic forms, but at this point we only want to stress the importance of making these plots of slopes and intercepts vs. inhibitor to determine the form of the inhibition. Often inhibition data are tacitly assumed to represent linear inhibition, but a close analysis indicates nonlinear replots would be obtained. If a nonlinear replot is obtained the nonlinearity can be confirmed directly by maintaining the substrate at a fixed level and varying only the inhibitor concentration and then plotting $1/v$ vs. inhibitor concentration. Linear inhibition gives straight lines, parabolic ones, parabolas, etc. This plot is useful because one can determine many more points on the line than are feasible for a replot where a full reciprocal plot must be constructed to get each point. This $1/v$ vs. I plot must only be used after the type of inhibition (competitive, uncompetitive, or noncompetitive) is known, however, since it gives no indication of whether the observed inhibition results from changes in slopes or intercepts of standard reciprocal plots.

2. *Prediction of Inhibition Patterns*

Inhibition patterns may be predicted in very much the same way as initial velocity patterns. Basically, the expected effects of the inhibitor on the slopes and intercepts of reciprocal plots are determined separately, and then the results are combined to predict the pattern. The intercepts of the plots will be increased by the presence of an inhibitor whenever the variable substrate and inhibitor combine with different enzyme forms, and saturation with the variable substrate cannot eliminate the inhibition. In practice, this usually means in a steady state mechanism that *the intercept effect is seen whenever the inhibitor and the variable substrate do not combine with the same enzyme form,* but the exceptions are important and demonstrate the power of the prediction technique. In a mechanism such as the Theorell–Chance mechanism

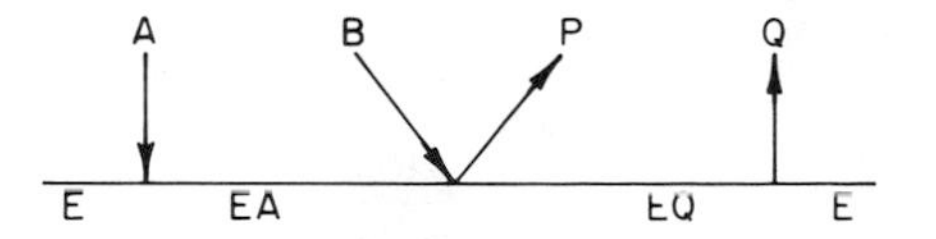

saturation with B eliminates any inhibition by P, because the only effect of P is to convert EQ back to EA and it cannot do this in the face of an infinite concentration of B (the reaction is $EA + B \rightleftharpoons EQ + P$). Thus there is no intercept effect and the inhibition is competitive. This mechanism is really a special case of an ordered mechanism where the steady state level of the (EAB–EPQ) central complexes is very low, and in practice if one puts in very high levels of both B and P, one can usually see the small intercept effect caused by formation of the central complexes.

Another situation where intercept effects are not seen even though the inhibitor and variable substrate react with different enzyme forms is where rapid equilibrium conditions prevail so that the steps in the reaction sequence between the points of addition of substrates and inhibitor are at thermodynamic equilibrium. In a rapid equilibrium ordered sequence, for instance, an inhibitor reacting with E will not affect the intercepts of the reciprocal plots when B is varied, since saturation with B displaces the reactions

$$\mathrm{EI} \rightleftharpoons \mathrm{E} + \mathrm{I} \qquad \mathrm{E} + \mathrm{A} \rightleftharpoons \mathrm{EA} \qquad \mathrm{EA} + \mathrm{B} \rightleftharpoons (\mathrm{EAB})$$

entirely toward formation of (EAB). This can happen only if these steps are at equilibrium; in the usual steady state mechanism, infinite B would decrease the concentration of EA but not of E to zero, and the inhibition would show a finite intercept effect. The same sort of reasoning predicts that all product inhibitions in a rapid equilibrium random mechanism will be competitive, since all reactants combine with free enzyme, and saturation with any substrate will decrease the level of free enzyme to zero and thus prevent enzyme–product complexes from forming (remember that steps such as $\mathrm{E} + \mathrm{P} \rightleftharpoons \mathrm{EP}$ are at thermodynamic equilibrium in such a mechanism). But the very nature of random mechanisms also dictates that if reactants can come and go independently of each other, one will get dead end complex formation, at least between the smaller pair of reactants (see Fig. 4). For example, with creatine kinase, which has a rapid equilibrium random mechanism, an enzyme–creatine–MgADP dead end complex is observed (*16*). Since saturation with creatine cannot prevent MgADP from combining to form this complex, an intercept effect will be seen and MgADP gives noncompetitive inhibition vs. creatine. MgATP, which occupies the same location as MgADP, can prevent the combination of MgADP; thus, MgADP is a competitive inhibitor vs. MgATP. The situation with the larger pair of reactants (MgATP, creatine phosphate here) depends on whether they can both be present at once. If they can, then they give noncompetitive inhibition versus each other; if they are truly mutually exclusive in their combination, they give competitive inhibition. The whole matter of predicting intercept effects by inhibitors thus is merely a matter of deciding

16. J. F. Morrison and E. James, *BJ* **97**, 37 (1965).

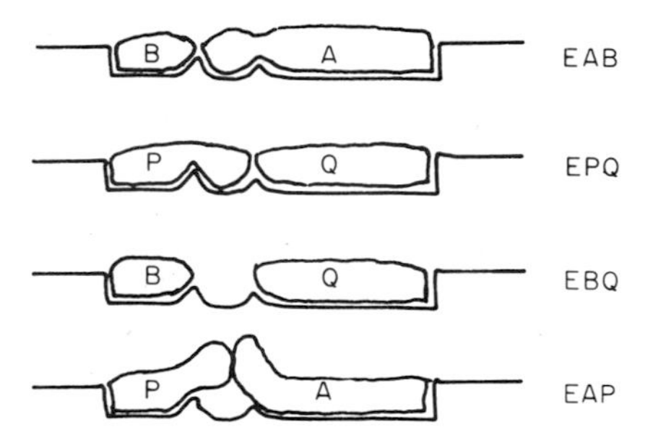

FIG. 4. Possible ternary complexes in a random sequential bireactant mechanism. In addition to these, binary EA, EB, EP, EQ complexes and free enzyme can exist. The EAP complex does not always form and when it does the dissociation constants are usually elevated.

whether the variable substrate at infinite concentration can prevent the combination of the inhibitor.

The effects of an inhibitor on the slopes of reciprocal plots is predicted in similar fashion for that used for initial velocity patterns. The key point is whether the inhibitor can specifically change the ratio of enzyme–variable substrate complex to uncomplexed enzyme; that is, if the variable substrate reacts according to $E' + A \rightleftharpoons E'A$, the inhibitor must raise E′A with respect to E′, or lower E′ with respect to E′A, in order to affect the slope. If the variable substrate combines downstream in the reaction sequence from the point of combination of inhibitor, and the steps in the sequence are reversible, a slope effect will be present since raising the inhibitor level will specifically lower E′ by drawing enzyme off into enzyme–inhibitor complexes before it gets to E′. This will be true regardless of whether the inhibitor is a product or dead end inhibitor. However, when the point of combination of inhibitor is downstream from the point of combination of the variable substrate and reversibly connected to it (but the sequence downstream from the inhibitor to the substrate is not reversible), a product inhibitor will cause a slope effect, but a dead end inhibitor will not. This is because a product can partially reverse the reaction sequence since the enzyme–product complex is a normal intermediate; thus, increasing the product concentration will back the reaction up so that E′A is increased with respect to E′. However, a dead end inhibitor combining at the same downstream point cannot back up the reaction sequence but can only draw the enzyme off into dead end complexes. As a result a dead end inhibitor will give slope effects only when its point of addition is upstream from and reversibly connected to the point of combination of the variable substrate. This situation is analogous to water flow in a river. Both closing the gates in the dam (product inhibition) and breaking the levee (dead end inhibition) lower water levels downstream, but only the dam produces a higher water level back upstream.

When an inhibitor combines more than once in the reaction sequence with the enzyme, the expected effects are first determined separately at each point of combination by means of the rules just discussed. For example, combination at one point may produce both slope and intercept effects, while combination at the other produces only an intercept effect. In this case, the inhibition is clearly noncompetitive and the slope will be a linear function of inhibitor. We must still determine, however, whether the intercept effect will be linear or parabolic. The effects will be parabolic only if combination of I at one point specifically increases the steady state concentration of the enzyme form with which the second molecule of I combines. For dead end inhibitors this only occurs when a second molecule of inhibitor reacts with the EI complex. Otherwise all effects are linear. For products, parabolic effects occur when two molecules of the inhibiting product are released with a reversible sequence between the points of release, or when the product also acts as a dead end inhibitor with the combination as dead end inhibitor preceding the combination as product, with a reversible sequence in between. In this case the combination as product backs up the reaction sequence sufficiently to raise the level of the enzyme form to which the product adds in dead end fashion. An example of this sort of thing is the combination of P in dead end fashion with EA in an Ordered Bi Bi mechanism:

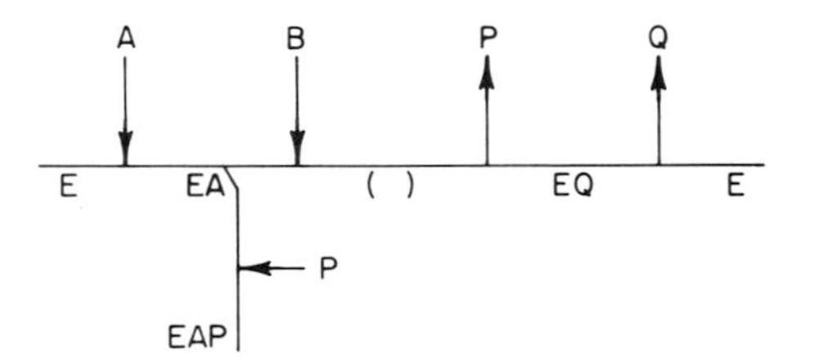

If A is variable substrate, the combination of P with EQ will give both slope and intercept effects, while the combination with EA gives only an intercept effect (P is combining as a dead end inhibitor downstream from the combination of A and thus cannot cause a slope effect). The intercept effect is parabolic, since combination of P with EQ increases the level of the central complexes, and because of their more rapid breakdown, also increases the level of EA. Since the concentrations of both EA and P are thus functions of the level of P, a P^2 term is generated when P combines with EA. If B is variable substrate, combination of P with EQ predicts both slope and intercept effects, while combination with EA now predicts only a slope effect. The slope effect is parabolic for the same reason the intercept effect was when A was varied.

In this analysis one must keep in mind that the intercept represents saturation with the variable substrate. For instance, if P reacts with E in dead end fashion in the following mechanism, and B is the variable substrate

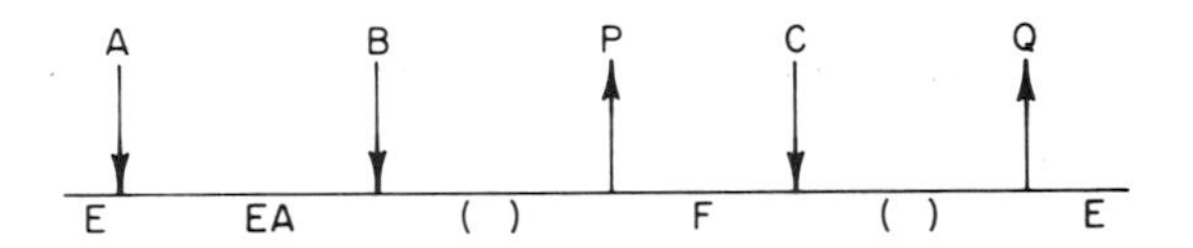

combination of P in each location predicts both slope and intercept effects. The two points of combination appear properly connected by a reversible sequence in order to get parabolic effects, but saturation with B (the variable substrate) interrupts this reversible sequence; thus, only the slopes will show the parabolic effect, not the intercepts. Saturation with a nonvaried substrate may change these patterns, too. In the present example, for instance, saturation with A will eliminate the dead end combination of P and leave only the normal linear noncompetitive product inhibition.

3. *Types of Inhibition Experiments*

a. Product Inhibition. With the rules for predicting inhibition patterns in hand, let us now look at the types of inhibition experiments the kineticist uses to deduce mechanisms. First of all, inhibition by the products is a basic part of any kinetic investigation. The products are the substrates for the reverse reaction and thus are expected to form complexes with the enzyme. Since these complexes tie up enzymes in nonuseful form, the effect of the products is always to inhibit the reaction. The type of inhibition expected depends on which substrate is varied and which product is the inhibitor, and a full product inhibition study involves using each product in turn as an inhibitor versus each substrate as the variable one. For example, consider an Ordered Bi Bi mechanism such as that shown by many pyridine nucleotide-linked dehydrogenases:

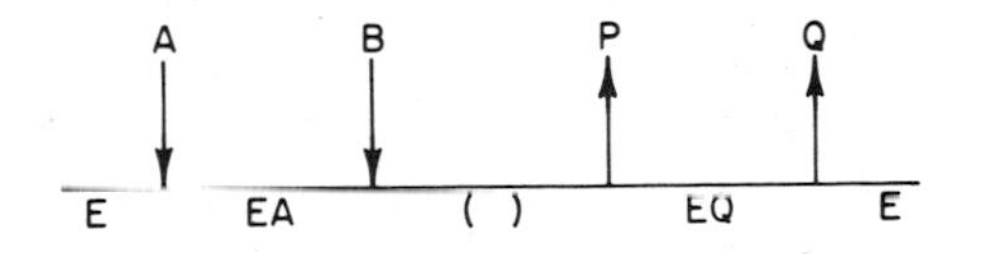

The only reactants which combine with the same enzyme form are A and Q, which both combine with E. These give competitive inhibition when one is the variable substrate and the other the inhibitor. In every other case (Q vs. B, or P vs. A or B) the variable substrate cannot prevent combination of the inhibitor, and since reversible connections exist in every case, the inhibitions are noncompetitive at nonsaturating levels of the nonvaried substrates. When B is the variable substrate, saturation with A will eliminate all inhibition by Q but not by P. In practice, this effect is shown by elevated apparent inhibition constants for Q as the A level is raised; however, unless the ratio of solubility to dissociation constant is much lower for Q than for A, one cannot hope to raise A sufficiently to make the apparent inhibition constant for Q sufficiently high so that inhibition by Q is not seen at all. When A is the variable substrate and B is saturating, the inhibition by P becomes uncompetitive, since the reversible connection from A to P is broken and the slope no longer shows the effect of the inhibitor. This is an easier phenomenon to observe than the elimination of inhibition by Q by saturation with A, since the effect of P on the intercepts will remain, and once the apparent inhibition constant for the slope is 100 times that for the intercept, one cannot tell the pattern from an uncompetitive one. A level of B 100 times the Michaelis constant will probably be sufficient to make inhibition by P appear uncompetitive.

The Ordered Bi Bi mechanism predicts three noncompetitive product inhibitions and one competitive product inhibition, with one pattern becoming uncompetitive at high levels of the other substrate; this permits one to identify the order of combination and release of substrates. It must be remembered that when one starts the product inhibition study, it is not usually known which substrate is A or B and which product is P or Q. If the rate limiting step is solely the breakdown of EQ, and the steady state level of the (EAB–EPQ) central complexes is very low, this mechanism reduces to the Theorell–Chance mechanism described earlier. In that case the B–P pattern changes to competitive and there will be two competitive and two noncompetitive patterns. All inhibition by P will be eliminated by saturation with B, and all inhibition by Q eliminated by saturation with A. In such a mechanism the order of combination and release of substrates cannot be deduced from product inhibition studies. In practice, if very high levels of both B and P are used, the intercept effect resulting from formation of the central complexes can usually be seen and identification made of A, B, P, and Q.

If combination of reactants is random (*16a*), all product inhibition patterns are competitive except between molecules that can form dead end complexes with the enzyme, and these are noncompetitive. The two reactants lacking the piece transferred during the reaction will always

16a. Rigorously, this discussion applies only to rapid equilibrium mechanisms which in pure form are rare (creatine kinase is one). However, simulation studies

form a dead end complex with the enzyme, while the two reactants both possessing this piece may or may not form such a complex [or if it forms the dissociation constants may be greatly elevated (see Fig. 4)]. Thus one either has two competitive and two noncompetitive patterns (if both dead end complexes can form) or three competitive patterns and one noncompetitive pattern. The only possible confusion is between a Theorell–Chance mechanism and a random mechanism with two dead end complexes, but these are easily distinguished by isotopic exchange or dead end inhibition studies.

In the classic Ping Pong Bi Bi mechanism

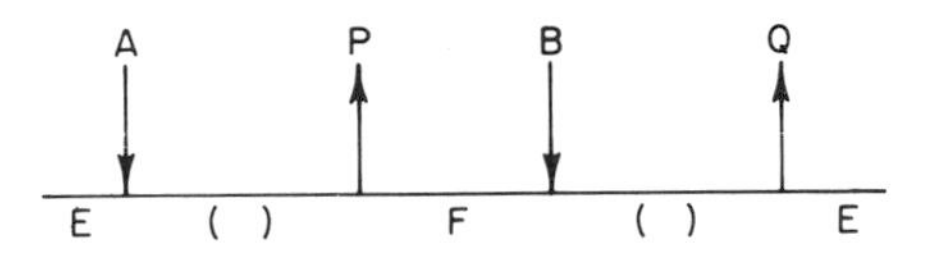

A and Q combine with E, and B and P combine with F. Both of these pairs thus give competitive product inhibition. The A and P, or B and Q pairs give noncompetitive inhibition, and in each case the inhibition is eliminated by saturation with the nonvaried substrate. These are the patterns shown by a number of transaminases, and by nucleoside diphosphate kinase (*14*). Presumably for mechanisms like these there is only one absorption pocket in the enzyme, and all reactants fit into it.

Not all ping pong mechanisms involve similarly shaped reactants, however. For instance, although the transcarboxylase which catalyzes reaction between methylmalonyl CoA and pyruvate to give propionyl CoA and oxalacetate shows ping pong initial velocity kinetics, the product inhibition patterns are not those of the classic ping pong mechanism; rather the two keto acids are mutually competitive and so are the two CoA thioesters (*18*). In the classic mechanism these inhibitions would be noncompetitive. Pyruvate and propionyl CoA would be competitive in the classic mechanism but are actually noncompetitive, and the same is true for methylmalonyl CoA and oxalacetate. This enzyme contains bound biotin which serves to carry a carboxyl group from one catalytic center where the keto acids are absorbed to the other where the CoA thioesters combine (Fig. 5). Since the first product is released before the

---

have shown that random mechanisms, unless very unusual values are assumed for the rate constants, resemble rapid equilibrium mechanisms in their initial velocity and product inhibition patterns, even though the rate limiting step is not solely the interconversion of two central complexes. Certain kinases have random, but not rapid equilibrium mechanisms [galactokinase (*17*), hexokinase (*7*)].

17. J. S. Gulbinsky and W. W. Cleland, *Biochemistry* **7**, 566 (1968).
18. D. B. Northrop, *JBC* **244**, 5808 (1969).

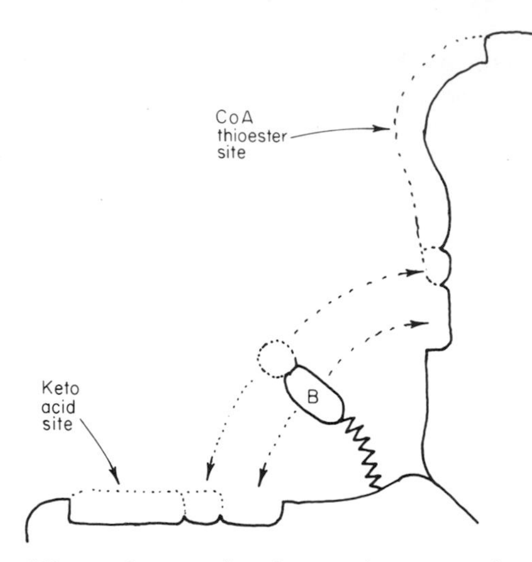

FIG. 5. Biotin (B) carrying a carboxyl group between active centers in a two-site ping pong mechanism [in this case, oxalacetate transcarboxylase (*18*)].

second substrate gets a chance to react with the carboxybiotin, the initial velocity pattern is ping pong, but the reversed product inhibition patterns show that each group of reactants combines only at its own site and that there seems to be independent combination at the two sites. It seems likely that the two catalytic sites are on different subunits, but this is as yet unproven; it also seems likely that other biotin enzymes will have similar two-site ping pong mechanisms.

When mechanisms are ter- or quadreactant, one can deduce the product inhibition patterns in the same fashion. When there are three products, some of the product inhibitions are likely to be uncompetitive, even at nonsaturating levels of the nonvaried substrates. For instance, the malic enzyme has an Ordered Bi Ter mechanism:

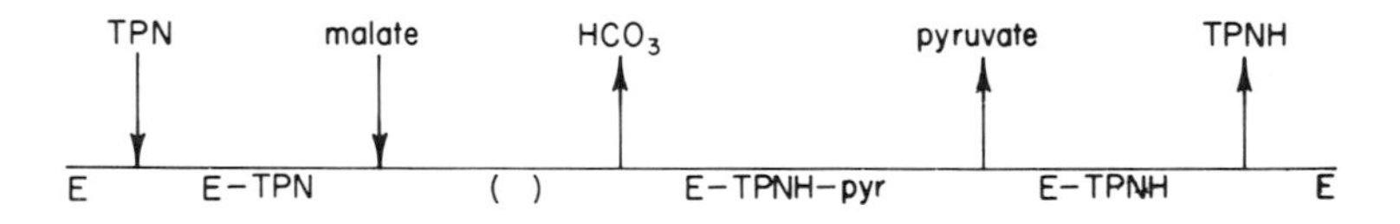

Since it is surrounded by product release steps which are irreversible, pyruvate should give only uncompetitive product inhibition, regardless of which substrate is varied. TPNH should be competitive vs. TPN and noncompetitive vs. malate, and bicarbonate should be noncompetitive vs. both substrates. These are the patterns actually observed (*19*).

Life is not always this simple, however, and with the closely related TPN isocitric dehydrogenase complications arise since α-ketoglutarate

19. R. Y. Hsu, H. A. Lardy, and W. W. Cleland, *JBC* **242,** 5315 (1967).

does not give solely uncompetitive inhibition because it combines very strongly in dead end fashion with E–TPN and thus gives noncompetitive inhibition vs. malate. It still gives uncompetitive inhibition vs. TPN, however. This propensity of products to form dead end complexes, in addition to combining as expected, frequently complicates analysis of product inhibition patterns, but the expected patterns for mixed product and dead end inhibition are easily deduced as described above. The additional dead end inhibition may not change the pattern (that is, linear noncompetitive may remain linear noncompetitive), but it will change the numerical values of the apparent inhibition constants; this becomes apparent when one tries to evaluate individual kinetic constants by comparison with the expected rate equation. In fact, this problem constitutes the major limitation on the usefulness of product inhibition studies, but at the same time it does provide more detail about the true characteristics of the kinetic mechanism.

It should be noted that the products of the reaction as it normally occurs *in vivo* are more likely to give mixed dead end and product inhibition than the products in the reverse direction. This is because product inhibition does not normally occur *in vivo,* and thus the product concentrations *in vivo* are low enough not to give dead end inhibition either. There has been no evolutionary pressure to eliminate undesirable kinetic properties of enzymes which are seen only at nonphysiological reactant concentrations, and thus where the product looks like a substrate, it is not surprising to see it absorbed in the pocket in place of this substrate. The phenomenon is rare, or at least not very prominent in the reverse reaction, since dead end inhibition by products in this direction is equivalent to substrate inhibition in the physiological direction; this again is deleterious and does not occur at physiological substrate concentrations.

Uncompetitive product inhibitions may also be seen in terreactant ping pong mechanisms such as

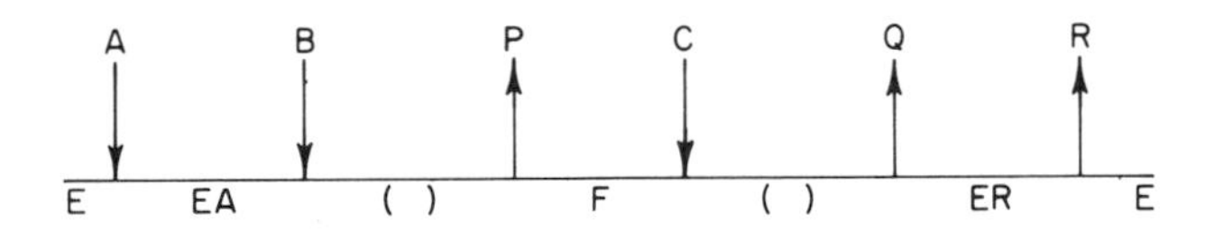

If release of Q and R is ordered, R will be an uncompetitive inhibitor vs. C, competitive vs. A, and noncompetitive vs. B; Q will be noncompetitive vs. C and uncompetitive vs. A or B. If release of Q and R is random, however, the patterns will be different. Q and R will be competitive against whichever occupies the same part of the absorption pocket—A or B—and noncompetitive vs. whichever of these they can form a dead end complex with. Asparagine synthetase has this mechanism where A and B are MgATP and aspartate, Q and R are asparagine and AMP, and P and C are MgPP and ammonia. Asparagine is competitive vs. aspartate and noncompetitive vs. MgATP, while AMP is competitive vs. MgATP and noncompetitive vs. aspartate, demonstrating that in this case product release is random, and not ordered (*20*).

One can play tricks with product inhibition patterns to extract information seemingly unavailable at first glance. For instance, an Ordered Bi Quad mechanism

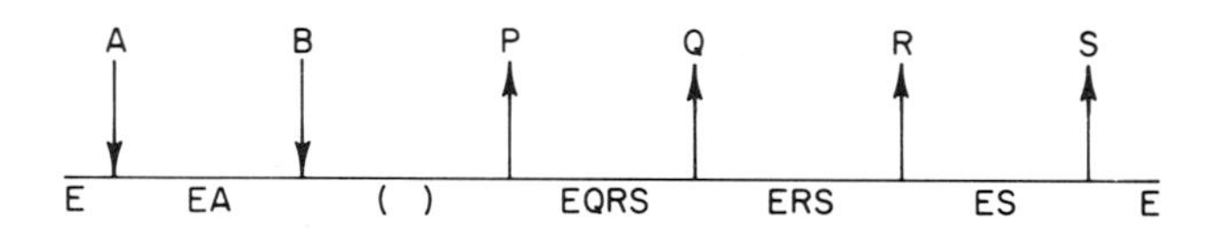

predicts uncompetitive product inhibition by both Q and R; thus, they appear not to be distinguishable. However, a fixed level of S will make the sequence from R to A or B reversible and convert the pattern to noncompetitive when R is the product inhibitor, while a fixed level of P will likewise make Q noncompetitive vs. either A or B. The change of a pattern by addition of a fixed level of product in this manner is often very diagnostic, and it is a technique which should be used more often to confirm kinetic mechanisms.

b. Alternate Product Inhibition. In certain cases, alternate products may be used to observe quite distinctive kinetic patterns and thus to confirm or identify the kinetic mechanism. For example, an enzyme catalyzing the Ordered Uni Bi hydrolysis of various phosphate esters, where the final product (phosphate) is the same, but the first product released depends on which substrate is used, has a kinetic mechanism which can be diagrammed as

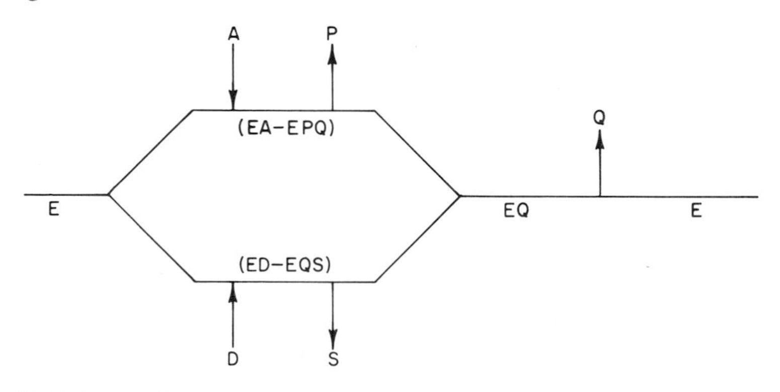

20. H. Cedar and J. H. Schwartz, *JBC* **244,** 4122 (1969).

If the rate of Q release is considered as the reaction rate and A is the variable substrate, P will act as a product inhibitor giving linear noncompetitive inhibition, and likewise S will act as a linear noncompetitive inhibitor. In the latter case the intercepts are affected since A and S combine with different enzyme forms. To understand why the slopes are affected, we must remember that the slope of a reciprocal plot is the reciprocal of the apparent first-order rate constant at low substrate concentration for reaction of E with A to give the measured product. In the present case, only the direct breakdown of EQ to E and Q yields the measured product, and increasing the level of S causes many of the catalytic cycles to produce D rather than Q as a product. Not every combination of E with A at low substrate concentration results in the formation of Q, and as a result the presence of S lowers the apparent first-order rate constant and thus raises the slope of the reciprocal plot.

If release of D is taken as the reaction rate, we have a ping pong mechanism (A and S as substrates) with an unstable F form (EQ here) which can break down spontaneously to give E. As mentioned earlier, S will act as a substrate, and the initial velocity pattern will be an intersecting one because of the $EQ \rightarrow E + Q$ step.

The unique pattern occurs when release of P is taken as the reaction rate and S is added. The rate of release of P measures both the hydrolysis of A to P plus Q and also reaction of A with S to give P plus D; this rate is in fact the sum of the rates of release of Q and D. The effect of S in this system is to divert the reaction flux from one path to another, but since each catalytic cycle produces P, regardless of whether Q or D is also produced, the apparent first-order rate constant for reaction of E with A at very low A is not affected. No slope effect by S is thus predicted. The effect of S on the intercept depends on whether the unimolecular step for release of D from (ED–EQS) is faster or slower than the breakdown of EQ to give Q. If EQ breakdown is faster, S will increase the intercept and cause inhibition; if breakdown of EQ is slower, it will cause activation. If the two rates are the same or in the event that the isomerization of (EA) to (EPQ) or the release of P from (EPQ) is very slow compared to both rates, no effect by S on the intercept will be observed. Further, the effect of S on the intercept will be hyperbolic; that is, saturation with S will result in a new finite rate, either slower or faster than in its absence. The resulting pattern is thus hyperbolic uncompetitive inhibition or activation.

This unique pattern can be used not only to confirm the kinetic mechanism but also to yield additional information. Thus, if S can combine with E to give a dead end complex (not an unexpected phenomenon since the portion of the absorption pocket designed for S is unoccupied in the free enzyme), a linear slope inhibition will be observed, and the pattern will now be S-linear inhibition, I-hyperbolic inhibition or activation. If D were used as substrate and S as product inhibitor, such a situation would give S-parabolic I-linear noncompetitive inhibition, and one would have to try to extract the dissociation constant of the dead end complex from the parabolic slope replot. With A as substrate, the linear slope replot determines this value directly and cleanly with no interference. The absence of any slope effect would also show that the dead end complex did not form and that all of the slope effect in the noncompetitive inhibition of S vs. D truly resulted from product inhibition [this has been done with potato acid phosphatase (*4*)].

Such experiments are not limited to Uni Bi hydrolyses but can be applied to other enzymes such as dehydrogenases where an alternate second substrate is oxidized to an alternate first product. In this case the experiment also can be used to prove directly the presence of central complexes (and thus the invalidity of the Theorell–Chance mechanism). If there were no central complexes

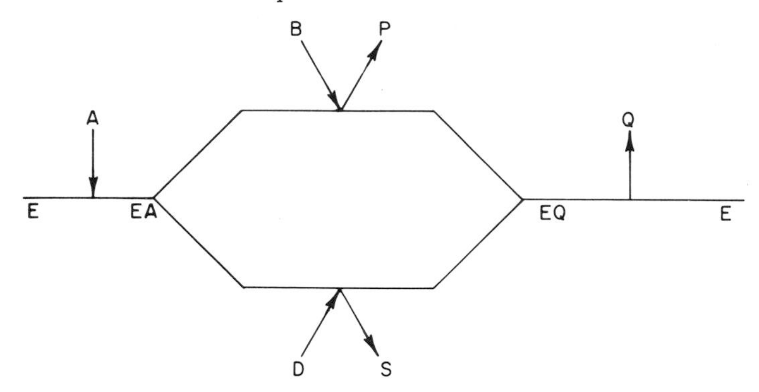

and one measured the release of P, it is obvious that at infinite B and S the reaction cycle consists solely of bimolecular steps, and thus should have an infinite rate. The observation that when S is increased to high levels the intercepts of reciprocal plots of $1/v$ vs. 1/B do not decrease to zero shows that this situation does not prevail and that central complexes do exist. This approach has been used with liver alcohol dehydrogenase, where acetaldehyde gave S-linear inhibition I-hyperbolic activation when *p*-hydroxybenzyl alcohol was the varied substrate and *p*-hydroxybenzaldehyde the measured product (*15*).

c. Dead End Inhibition. The kinetics of substrates, products, and alternate products define the form of the rate equation and are certainly necessary to deduce the kinetic mechanism. However, they are often not sufficient to do this uniquely and other kinetic experiments are necessary, especially when a reaction can be studied in only one direction. One of the most useful procedures in such cases involves the use of dead end inhibitors. These nonreactants play musical chairs for vacant portions of the absorption pocket and produce very specific inhibition effects in addition to providing evidence concerning the geometry of the absorption pocket. The effects of an inhibitor that combines with any enzyme form are easy to deduce (see above): One has merely to remember that slope effects arise only when the inhibitor combines before the variable substrate in the reaction sequence with reversible steps in between. Further, the inhibitions are always linear, except where two or more molecules of the inhibitor combine with the same enzyme form.

One of the most useful functions of dead end inhibitors is in verifying ordered addition of substrates in cases where product inhibition studies cannot or do not give an unequivocal answer. Consider, for example, formic dehydrogenase, which catalyzes the oxidation of formate to bicarbonate by DPN (*21*). The equilibrium constant is very large and neither the reverse reaction nor isotopic exchange between products and substrates can be demonstrated. DPNH is competitive vs. DPN, with the inhibition constant independent of formate concentration, and noncompetitive vs. formate. Bicarbonate is competitive vs. formate and noncompetitive vs. DPN, which might suggest a random mechanism, except that the levels of bicarbonate used (up to 700 mM) were so high that one hesitates to consider the observed effects as specific product inhibition rather than a combination of product inhibition, dead end inhibition, and perhaps other effects as well.

When ADPR was used as a dead end inhibitor, it was competitive vs. DPN ($K_i = 12\ \mu M$) and noncompetitive vs. formate. Nitrate, on the other hand, was competitive vs. formate (K_i around 1 mM) and uncompetitive vs. DPN. These patterns are consistent only with ordered addition of DPN followed by formate, unless one wishes to assume that DPN must be present to permit nitrate binding but not to permit formate binding.

Whenever several compounds give competitive inhibition vs. one substrate, and particularly when they are rather dissimilar, the question arises whether the inhibitors are combining in the same part of the absorption pocket or whether they are combining in different places. If they absorb in the same place their binding should be mutually exclusive, as it is with the substrate; but if they absorb in different parts of the pocket, they may both be able to bind at the same time. The degree to which they antagonize or synergize each other's combination can thus yield data

21. D. Peacock, in preparation.

on the geometry of the absorption pocket and the conformation changes induced by absorption of molecules in it.

If there are two inhibitors, I and J, the kinetic question is whether an EIJ complex forms, and if so, whether the dissociation constant of I from EIJ is the same as from EI (and similarly for J from EIJ and EJ). To answer this, all substrate concentrations are held constant, the concentrations of the two inhibitors are varied, and initial velocities are determined (*22*). A plot of $1/v$ vs. I at the different J levels (including zero) gives lines that are parallel or that cross somewhere to the left of the vertical axis (Fig. 6). The vertical intercepts of these plots will always vary with J (they represent the effect of J when $I = 0$), but the slopes will vary only if an EIJ complex forms. The horizontal intercept of a $1/v$ vs. I plot is the dissociation constant (K_i) of the inhibitor in a simple situation where the inhibitor combines with only one enzyme form, while the vertical intercept is the uninhibited rate. If J combines to give an EJ complex which cannot combine with I, the rates in the absence of I would be decreased by the factor $(1 + J/K_j)$ or the vertical intercepts of $1/v$ vs. I plots will be increased by this factor. But the removal of enzyme to form EJ also raises the apparent dissociation constant of the EI complex by the same factor, so the slopes of the plots, which are the ratio of vertical to horizontal intercepts, remain unchanged.

If I can combine with EJ as readily as with E, then while J still in-

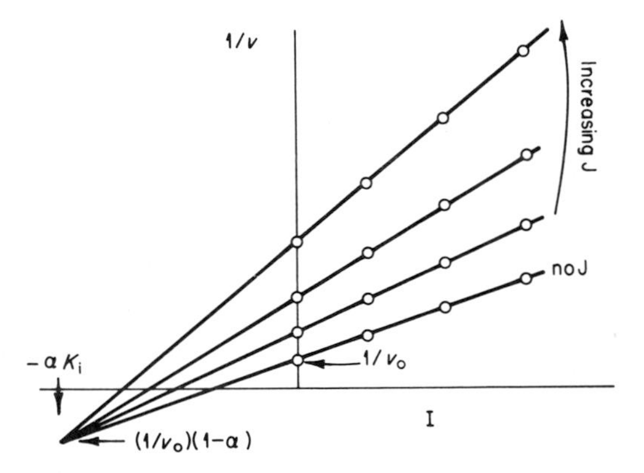

FIG. 6. Determination of whether inhibitors are mutually exclusive. In the case shown $\alpha > 1$, and the inhibitors interfere with but do not prevent each other's binding. Parallel lines would indicate mutually exclusive binding, while lines crossing on the horizontal axis would indicate that the presence of one inhibitor did not affect the binding of the other. In all cases binding of inhibitor and substrate is mutually exclusive.

22. T. Yonetani and H. Theorell, *ABB* **106,** 243 (1964).

hibits the reaction in the absence of I, the apparent K_i for I is unchanged by the presence of J, since both E and EJ are available for reaction. Thus the vertical intercepts vary with J, but the horizontal intercept is constant and the pattern crosses on the horizontal axis. If the presence of J raises the dissociation constant of I, but not to infinity, the pattern will now cross below the horizontal axis, while if J actually lowers the dissociation constant of I, the pattern will cross above the horizontal axis. The equation for this situation is

$$\frac{1}{v} = \frac{1}{v_0}\left(1 + \frac{I}{K_i} + \frac{J}{K_j} + \frac{IJ}{\alpha K_i K_j}\right) \tag{6}$$

where v_0 is the uninhibited rate, K_i and K_j are dissociation constants for EI and EJ, and α is an interaction coefficient which tells the extent to which the binding of I and J is mutually exclusive. If α is infinity, binding is exclusive, and the pattern is parallel. If α is above 1, the presence of one inhibitor hinders but does not prevent binding of the other, and the lines cross below the horizontal axis. If $\alpha = 1$, binding is independent, and the pattern crosses on the horizontal axis. When α is less than 1, binding is synergistic, and the crossover point is above the axis. The coordinates of the crossover point are $I = -\alpha K_i$, $1/v = (1/v_0)(1 - \alpha)$.

This approach has been used with liver alcohol dehydrogenase to show that *o*-phenanthroline combines at the nicotinamide part of the absorption pocket (*22*). Values of α for *o*-phenanthroline and other inhibitors were ADPR, 1.0; ADP, 0.5; and AMP, 0.3. ADPR, ADP, and AMP gave α values of infinity among themselves as expected. The synergism between AMP or ADP, but not ADPR and *o*-phenanthroline, suggests that conformation changes result from binding that facilitate combination in the rest of the absorption pocket, with the effect being maximal for the smallest nucleotide which has the highest dissociation constant.

d. Substrate Inhibition. At high concentrations the substrates will often act as dead end inhibitors, particularly when a reaction is being studied in the nonphysiological direction (see above). Substrate inhibition does not normally occur at physiological substrate concentrations and thus is of interest to some biochemists only in that it sets upper limits to the *in vivo* levels. Thus the differing sensitivities of muscle and heart lactate dehydrogenase isozymes to substrate inhibition by pyruvate reflect the different pyruvate levels in muscle and heart. To the kineticist, however, substrate inhibitions arc onc of the best diagnostic tools for studying mechanisms, and their importance cannot be overemphasized. The assays are easy to run, and interpretation is usually very straightforward.

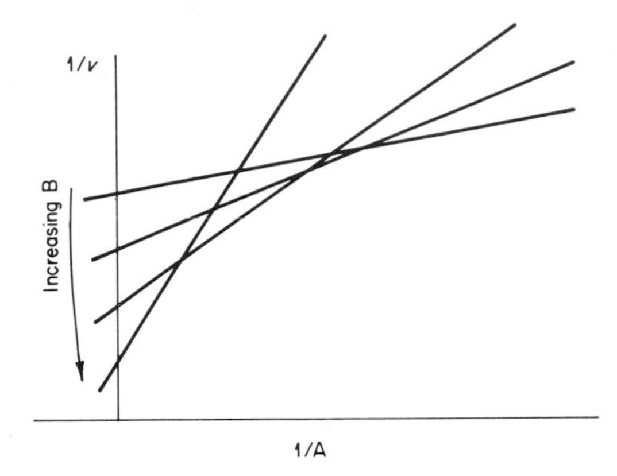

FIG. 7. Competitive substrate inhibition in a ping pong mechanism. A replot of slopes will be linear and similar to curve A in Fig. 3. Data points are not shown for clarity.

The proper way to study a substrate inhibition is to vary a noninhibitory substrate at differing high levels of the inhibitory one and see whether the slopes, intercepts, or both of reciprocal plots show the inhibitory effect. These cases are called competitive, uncompetitive, and noncompetitive substrate inhibition, respectively. While most substrate inhibitions result from dead end combination of the substrate with an enzyme form that it is not supposed to react with, and thus can be called linear (that is, a $1/v$ vs. substrate inhibitor plot will become linear at very high concentrations), some substrate inhibitions arise because reaction flux is diverted into slower alternate pathways. These are hyperbolic or partial substrate inhibitions. Parabolic inhibition could of course be observed if the substrate combined twice with the same enzyme form in dead end fashion.

Competitive substrate inhibition (Fig. 7) is characteristic of classic ping pong mechanisms, and observation of such inhibition in an initial velocity study makes one feel more secure about assigning a ping pong mechanism to an observed parallel initial velocity pattern. This type of substrate inhibition results when substrate B combines with form E as well as with F:

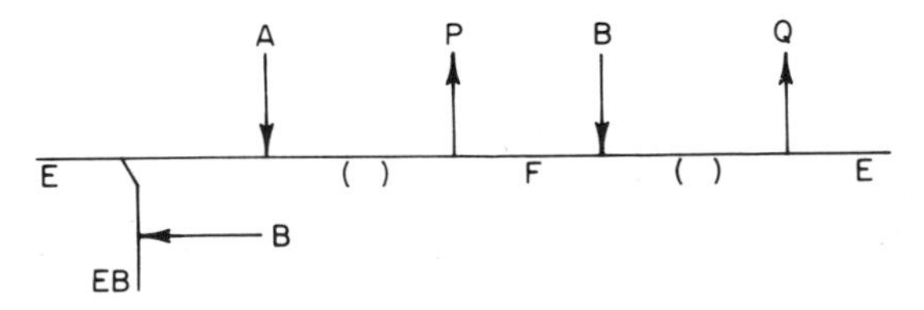

The effect is competitive since both substrates now combine with the same enzyme form (E). Saturation with the variable substrate eliminates the inhibition, and thus the intercepts of reciprocal plots are unaffected. Since the only variation in the slopes of the reciprocal plots comes from the substrate inhibition, the dissociation constant for the dead end combination is easily obtained by replotting the slopes against the inhibitory substrate concentration (the horizontal intercept of this replot with sign changed is the dissociation constant—see curve A, Fig. 3).

Substrate A can sometimes combine with form F in dead end fashion so that it gives competitive substrate inhibition as well as (or in place of) B. Nucleoside diphosphate kinase has a ping pong mechanism in which all substrates can combine with both free and phosphoenzyme, although only for MgADP (*in vivo* always a product) is there less than a factor of 10 between the dissociation constants for the proper and improper form (*14*). Competitive substrate inhibition by both substrates is a wild looking but very characteristic pattern (Fig. 8), but since the data can be fitted to the rate equation by least squares methods to extract the values of all kinetic constants there is no special problem in analysis. Strong double competitive substrate inhibition is expected only in the nonphysiological direction. For example, the reaction of CoA with acetoacetyl CoA to give two molecules of acetyl CoA catalyzed by yeast β-ketothiolase is ping pong and shows strong competitive substrate inhibition by both substrates (*23*). Although this is the thermodynamically

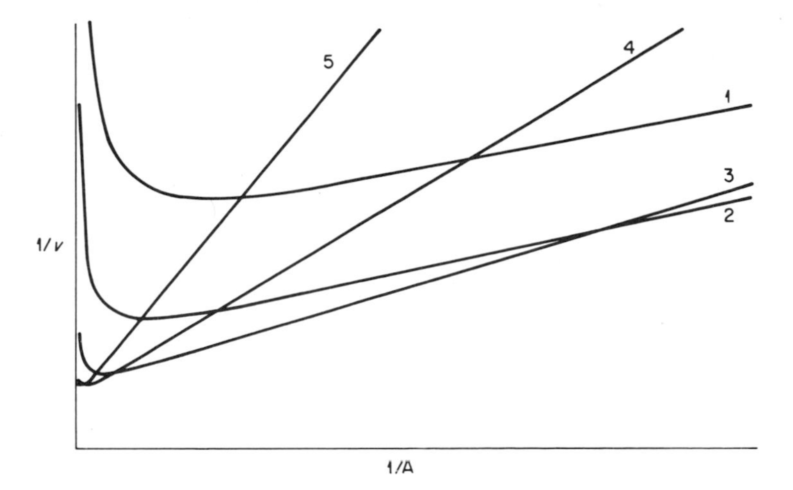

FIG. 8. Double competitive substrate inhibition in a ping pong mechanism. The concentration of B increases from 1 to 5. Data points are not shown for clarity.

23. P. R. Stewart and H. Rudney, *JBC* **241,** 1212 (1966).

favorable direction (and the easiest one to assay), the reaction *in vivo* goes the other way with acetoacetyl CoA drawn off by subsequent reaction with another molecule of acetyl CoA. This second condensation is also ping pong, but it shows competitive substrate inhibition only by acetoacetyl CoA and only at levels above 15 μM. It is thus very likely that the *in vivo* level of acetoacetyl CoA in this pathway is less than 15 μM.

While competitive substrate inhibition is characteristic of ping pong mechanisms, uncompetitive substrate inhibition is the expected pattern for ordered sequential mechanisms. This occurs when the second substrate to add can combine in dead end fashion with the EQ as well as the EA complex, and the EBQ complex can break down to give only EQ and B.

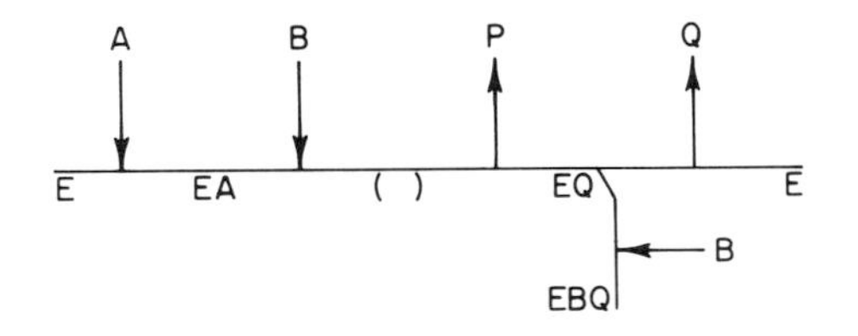

The inhibition only affects the intercepts of reciprocal plots when A is varied, since in the absence of Q there is no reversible connection downstream from EQ to the point where A adds. The resulting initial velocity pattern (Fig. 9) thus shows that the slopes of reciprocal plots behave as expected as B is raised but that the intercepts go through a minimum and then increase again at high B levels. The replot of intercepts vs.

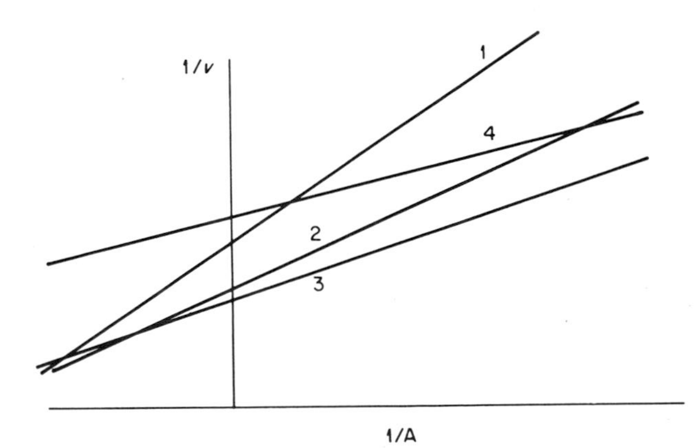

FIG. 9. Uncompetitive substrate inhibition in an ordered sequential mechanism. The slope replot will appear normal (see Fig. 1), but the intercept replot will be a hyperbola (Fig. 10). Substrate concentration increases from 1 to 4. Data points are not shown for clarity.

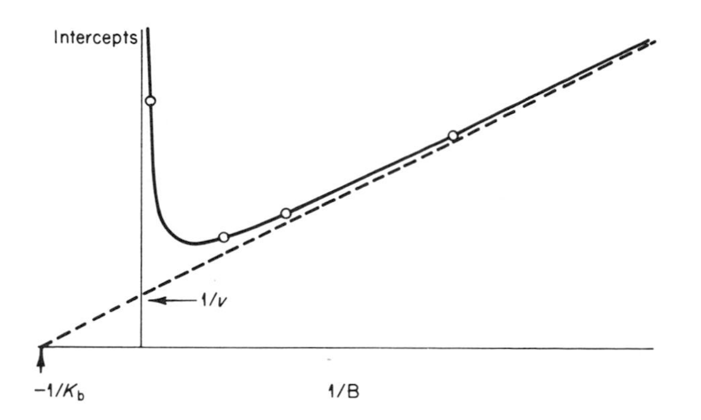

FIG. 10. Intercept replot for data such as those in Fig. 9. If the intercepts are plotted vs. B, the shape is the same, but the horizontal intercept gives the apparent substrate inhibition constant.

either B or 1/B is thus a hyperbola (Fig. 10), and the exact dissociation constant for the dead end combination is unknown since one is never sure how much enzyme is in the form of the EQ complex as opposed to central complexes. This type of substrate inhibition is seen in several pyridine nucleotide-linked dehydrogenases. With pig heart TPN isocitric dehydrogenase, for example, α-ketoglutarate gives very prominent uncompetitive substrate inhibition against either bicarbonate or TPNH as a result of dead end combination with E–TPN. The inhibition is linear up to 60 mM α-ketoglutarate (over 100 times the apparent inhibition constant and 2000 times the Michaelis constant), indicating that TPN cannot leave the enzyme while α-ketoglutarate remains absorbed. Bicarbonate also gives some substrate inhibition in this reaction at very high concentrations, but it is competitive vs. TPNH and noncompetitive vs. α-ketoglutarate, as would be expected for dead end combination of bicarbonate with free enzyme at the TPNH site. The inhibition vs. α-ketoglutarate is noncompetitive, since bicarbonate combines as a dead end inhibitor prior to combination of α-ketoglutarate, and the steps between are reversible. The same rules used for predicting dead end inhibition patterns are thus directly applicable to substrate inhibition as long as the inhibitory substrate is the changing fixed one.

A truly rapid equilibrium random mechanism would not show substrate inhibition unless one substrate had some affinity for the part of the absorption pocket designed for the other. Such situations are uncommon, and the inhibition should be of the competitive type, so it should be readily recognized. If a predominantly ordered mechanism has some small degree of randomness, however, the substrate inhibition pattern may be quite complex and at the same time very informative. Liver alcohol dehydrogenase has a largely ordered kinetic mechanism, and the initial velocity, product inhibition, and alternate product inhibition patterns are those predicted for the ordered case (*15, 24*). Substrate inhibition is given only by primary alcohols, which are such good substrates that the rate limiting step is solely the rate of breakdown of the E–DPNH complex. Secondary alcohols, which are much poorer substrates, do not give substrate inhibition since the rate limiting step is now the catalytic reaction itself, and the steady state level of E–DPNH available for combination with the alcohol to give a dead end complex is very small (*25*).

The substrate inhibition by primary alcohols is not linear, however, and a plot of $1/v$ vs. alcohol concentration is hyperbolic and reaches a plateau. Further, the inhibition is not uncompetitive as is expected from strictly dead end inhibitors combining with EQ. Rather the inhibitions are noncompetitive, and the slope effects are hyperbolic as well. In fact, with cyclohexanol the effect on the intercept is activation rather than inhibition, so that one observes S-hyperbolic substrate inhibition, I-hyperbolic substrate activation by cyclohexanol. Hyperbolic effects are normally the result of alternate pathways of reaction, and in this case the unusual pattern results from breakdown of E-alcohol-DPNH to give E-alcohol and DPNH, and the subsequent reaction of E-alcohol with DPN to give the usual ternary complex. Most primary alcohols hinder the release of DPNH from the E-alcohol-DPNH complex, but cyclohexanol apparently increases the rate constant for release of DPNH. These effects are seen on the intercepts when DPN is varied at high alcohol levels since the combination of alcohol to give the dead end complex is with an enzyme form other than free enzyme. The hyperbolic substrate inhibition of the slopes of reciprocal plots comes about because DPN reacts with E-alcohol to give central complexes much less readily than it reacts with free enzyme to give E–DPN, and thus the apparent first-order rate constant at very low DPN for combination of DPN with enzyme is decreased when much of the enzyme exists as E-alcohol instead of free E.

While substrate inhibitions are usually analyzed graphically by making reciprocal plots vs. a noninhibitory substrate, it is useful to be able to analyze data where the inhibitory substrate is the variable one. All linear substrate inhibitions resulting from dead end combination of the substrate with some improper enzyme form will give the following equation, where V, K_a, and K_I are apparent constants:

$$v = \frac{VA}{K_a + A + A^2/K_I} \tag{7}$$

24. C. C. Wratten and W. W. Cleland, *Biochemistry* **2**, 935 (1963).
25. K. Dalziel and F. M. Dickinson, *BJ* **100**, 34 and 491 (1966).

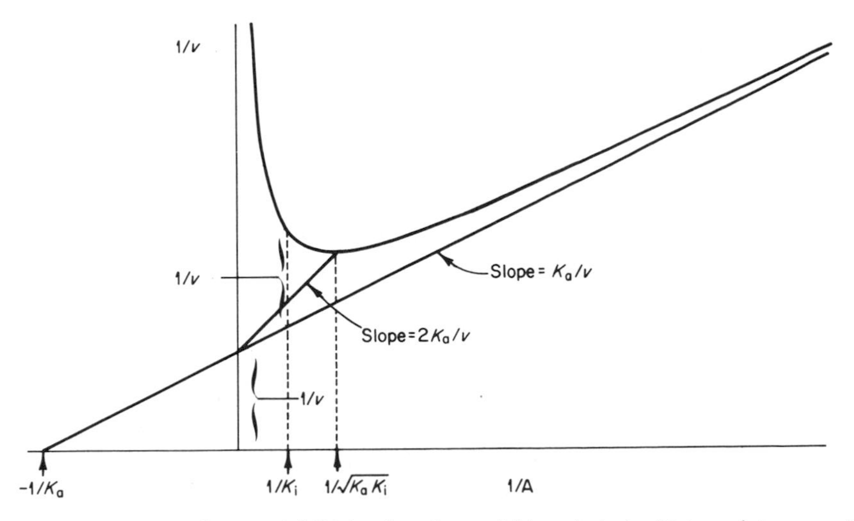

FIG. 11. Linear substrate inhibition by the variable substrate. Data points are not shown for clarity. Note that the most common error in graphical analysis is placing the asymptote too close to the curve; it is for this reason that the data should be fitted to Eq. (7) by the least squares method.

If $1/v$ is plotted vs. 1/A, the slopes and intercepts of the asymptote determine K_a/V and $1/V$ as usual (Fig. 11). The minimum point has coordinates of $1/A = 1/(K_aK_I)^{1/2}$ and $1/v = (1/V)[1 + 2(K_a/K_I)^{1/2}]$ and the line connecting the vertical intercept of the asymptote and the minimum point has a slope of $2K_a/V$ or double that of the asymptote. In order to determine the correct position of the asymptote, therefore, one adjusts it until the line from its vertical intercept to the minimum point has twice its slope. The point where $1/A = 1/K_I$ can then be located by finding where the vertical separation of the curve and the asymptote is equal to the asymptote intercept $(1/V)$; at this point the curve has a vertical coordinate of $(1/V)(2 + K_a/K_I)$ and the asymptote has a coordinate of $(1/V)(1 + K_a/K_I)$. The substrate inhibition curve can be linearized by letting $\alpha = A/A_{min}$, where A_{min} is the value of A at the minimum point of the curve, and plotting $1/v$ vs. $(\alpha + 1/\alpha)$ according to the equation *(25a)*:

$$1/v = (K_a/K_I)^{1/2}[\alpha + (1/\alpha)]/V + 1/V \tag{8}$$

Since $A_{min} = (K_aK_I)^{1/2}$, one can combine the horizontal intercept of this linearized plot $[(K_I/K_a)^{1/2}]$ with the value of A_{min} by multiplication to give K_I and by division to give K_a. Substrate inhibition data can also be

25a. C. Marmasse, *BBA* **77**, 530 (1963).

represented as a lovely bell-shaped curve by plotting v vs. log A, from which the constants can be obtained as follows:

$$K_I = A_1 + A_2 - 4A_m$$
$$K_a = A_m^2/K_I$$
$$V = A_m[1 + 2(K_a/K_I)^{1/2}]$$

where A_m, A_1, and A_2 are values of A at the maximum point of the bell-shaped curve, and at the two points where the velocity is half of that at the maximum. In practice, of course, the kineticist fits these data to Eq. (7) by a least squares method and then plots the results in whatever way meets his fancy.

The discussion above assumes that substrate inhibition results from the dead end combination of substrates with enzyme forms they are not supposed to react with. While available data certainly seem to support this view in most closely studied cases, many authors have taken the alternative viewpoint that substrate inhibition involves two molecules of substrate binding within a pocket designed for only one, with each only partially able to enter the absorption pocket. Kinetically this involves dead end combination of two molecules of substrate to give an EA_2 complex. If there are two or more substrates, one can easily distinguish this possibility from those discussed above, since the resulting substrate inhibition will appear to result from combination by two molecules of the inhibitory substrate with the same enzyme form that it combines with properly. Where there is only one substrate, as in hydrolytic reactions where one cannot vary the concentration of water, this approach does not work, and at first glance it appears one cannot tell the difference between the "classic" picture (two molecules of A forming a dead end complex with free enzyme) and the more likely mechanism where A combines in dead end fashion with EQ, since both give Eq. (7). If one varies the substrate in the inhibitory range in the presence of different levels of the first product P, however, the distinction becomes simple. In the presence of P, the rate equation for a Uni Bi mechanism is

$$v = VA/[K_a + A + K_{ia}P/K_{ip} + AP/K_{ip}] \tag{9}$$

If substrate combines twice with free enzyme to give an EA_2 dead end complex, the K_a and $K_{ia}P/K_{ip}$ terms are multiplied by a factor $(1 + aA + bA^2)$, while if A combines in dead end fashion with EQ, an additional term in A^2 occurs in the denominator. A little quick algebra will show that when the resulting equations are rearranged into the form of Eq. (7), in both cases apparent K/V and apparent $1/V$ will be linear functions of P, but in the A with EQ case apparent $1/(VK_I)$ does not vary with P, while when E combines with two molecules of A to give

EA_2, it is also a linear function of P, with the same apparent inhibition constant as for apparent K/V. As of the moment, this approach has not yet been applied to substrate inhibition in a Uni Bi mechanism.

C. Isotopic Exchange

1. *Basic Considerations*

All of the kinetic experiments we have discussed have involved measurement of the initial velocity of the net chemical reaction as a function of the concentrations of substrates, products, inhibitors, etc. We say "net chemical reaction" because certain steps in the mechanism may proceed at rates considerably above the net rate of chemical reaction, but what one sees for these steps is the small difference between two large rates in the forward and reverse directions. There is no way to observe flux rates in the individual steps in a reaction by following the chemical reaction, but one can do this very nicely with isotopes, and study of the kinetics of isotopic exchange is the third major tool of the kineticist. As above, we defer derivation of the rate equations until later and try to develop an intuitive understanding of the kinetic patterns observed in such studies. Our discussion is limited to isotopic exchange between substrates and products and does not consider exchanges with water, although such experiments are very informative and the kinetics of proton or oxygen exchanges when they occur should be a part of any study of mechanism. We also do not discuss "isotope effects"; that is, altered rates when one isotope replaces another, which are discussed by Richards in Chapter 6 of this volume.

Isotopic exchange studies are always carried out in the presence of one or more products and frequently (but not necessarily) at chemical equilibrium. If no products are present initially, the rate of isotopic transfer from substrates to products simply equals the chemical reaction rate, and this is an excellent way to determine initial velocity patterns if chemical or optical assay methods are unavailable. One may follow the transfer of label from substrate to the pool of product present initially or from product back to a substrate. If the reaction is at thermodynamic equilibrium, this distinction disappears and the rate of transfer is equal in both directions (that is, ATP $\rightarrow$ ADP exchange must show the same rate as ADP $\rightarrow$ ATP exchange). The initial velocity of isotopic exchange can either be determined directly by measuring label incorporation over a short time period, or, if the reaction is at chemical equilibrium, by making measurements of incorporation when about half of the label has been transferred, and, finally, at isotopic equilibrium (if the reaction is at chemical equilibrium, the time course of isotopic transfer is a standard exponential curve). The former is more precise, while the latter does not require as high a specific activity for the labeled starting material.

2. *Exchange in Sequential Mechanisms*

With sequential mechanisms, isotopic exchange can be measured either at chemical equilibrium with all reactants present or between a product and a substrate while the chemical reaction is proceeding. An example of the latter is the observation of isotopic transfer from the first product back into the substrate in an Ordered Uni Bi hydrolysis. As long as there is a reasonable amount of the EQ complex present in the steady state this back exchange will be reasonably fast compared to the rate of the chemical reaction proceeding in the forward direction; in fact, the ratio of the $P^* \rightarrow A$ exchange rate to the forward chemical reaction rate is essentially the ratio of the concentration of P to its slope inhibition constant as a product inhibitor. The exchange of Q into products is much slower, and is dependent on the presence of P. One can thus use exchange studies to determine the order of product release: Whichever product shows exchange into substrate in the absence of the other is the first one released. Both products should be tested at levels equivalent to their product inhibition constants or above. If both products show exchange, product release is evidently random; moreover, the rate limiting steps are those of product release (otherwise there would not be reasonable levels of both EQ and EP in the steady state). This approach was used elegantly by Byrne and Hass with glucose-6-phosphatase to show ordered release of glucose before phosphate (*26*).

The most widely applied use of isotopic exchange experiments to date has been in distinguishing ordered from random Bi Bi mechanisms. Initial velocity studies make no distinction between these, and product inhibition patterns are often either complicated by additional dead end inhibitions or simply not decisive. For instance, a Theorell–Chance mechanism and a rapid equilibrium random mechanism with two dead end complexes give identical initial velocity and product inhibition patterns. Isotopic exchange studies readily distinguish between these; further, they are the only really good techniques available for deciding whether a random mechanism is truly a rapid equilibrium one or not.

In an Ordered Bi Bi mechanism

26. L. F. Hass and W. L. Byrne, *JACS* **82,** 947 (1960).

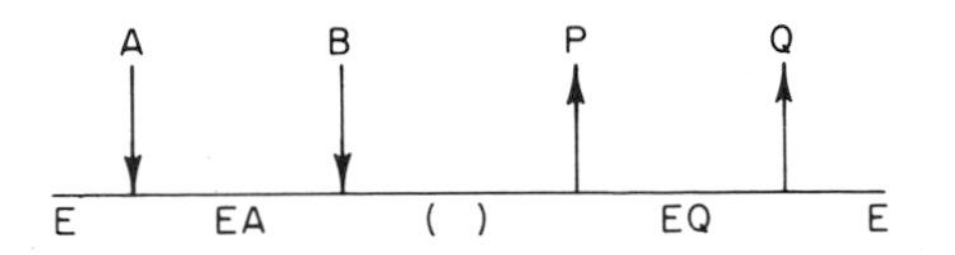

we can conceivably measure isotopic exchange between A and Q, B and P, and either between A and P or B and Q, but not both. Thus for alcohol dehydrogenase, one can measure DPN–DPNH, alcohol–aldehyde, and alcohol–DPNH exchanges, but no exchange takes place between aldehyde and DPN. These exchange studies are normally carried out at chemical equilibrium; thus, the rate of exchange from A to Q is the same as from Q to A, and which one we measure is a matter of convenience. To remain at chemical equilibrium, one substrate and one product are usually varied together in constant ratio and the form of the resulting reciprocal plots of initial velocity of isotopic exchange is determined. For the ordered mechanism, the reciprocal plot will be linear when A and Q are the variable reactants, regardless of which exchange is measured. Raising the concentration of B to infinity, however, decreases the concentration of EA to zero and thus eliminates any exchange involving A, since A can only exchange with EA when there is something to exchange with. Since we are at chemical equilibrium, saturation with B also draws all of E through EA into the central complexes. Thus, when B is one of the variable reactants, the reciprocal plot of exchange velocity will show total substrate inhibition of A–Q or A–P exchanges. A similar argument shows that when P is a variable reactant, it will cause substrate inhibition of any exchange involving Q. The usual experimental protocol thus is to vary either A and Q or B and P in constant ratio and determine the effects on the rate of A–Q and B–P exchanges. Varying A and Q gives linear reciprocal plots for both exchanges, while varying B and P gives a linear reciprocal plot for the B–P exchange (*26a*) and a total substrate inhibition pattern for the A–Q exchange [that is, $1/v$ goes to infinity as 1/B or 1/P goes to zero; see Fig. 11 and Eq. (7)]. It is important experimentally to test the effects of increased B and P on both exchanges since high concentrations of reactants may inhibit the exchange by other mechanisms such as dead end complex formation, increased ionic strength, or increased concentration of an inhibitory counterion (particularly a problem when raising the concentration of magnesium nucleotide if chloride is inhibitory—it may

26a. In theory, this plot is not completely linear and may curve near the vertical axis (it is a 2/1 function; see Section IV,D,2 below), but it may not have a minimum and thus one does not observe substrate inhibition. In practice, the plot will usually appear linear within experimental error.

be necessary to use other means of preparing the complex other than mixing $MgCl_2$ and nucleotide). In such cases both exchanges should show the effects, however.

The danger of dead end complex formation lessens the usefulness of raising A and P or B and Q in constant ratio, although this may be a good way to confirm the formation or nonformation of such complexes. Substrate inhibition occurs with both exchanges if there is dead end complex formation, but if only one exchange is inhibited the mechanism is ordered. With malate dehydrogenase, raising DPN and oxalacetate inhibits DPN–DPNH exchange, but not oxalacetate–malate exchange, showing ordered combination of DPNH and oxalacetate and no dead end E-DPN–oxalacetate complex. On the other hand, when malate and DPNH are raised together, both exchanges are inhibited, suggesting a dead end E-DPNH–malate complex (*27*).

One could also test the effects of these reactants on the B–Q (or A–P) exchange, but since the results should be the same as for the A–Q exchange, this has not often been done.

In ordered mechanisms the relative maximum rates of A–Q and B–P exchange given by the intercepts (or asymptote intercepts, if substrate inhibition occurs) of reciprocal plots will depend on where the rate limiting steps are in the mechanism. When the catalytic transfer that converts one central complex into another is rate limiting, the two exchanges may go at nearly the same rate, but if the rate limiting step, as is common for dehydrogenases, is the release of the second product, the B–P exchange is much faster than the A–Q exchange. For the Theorell–Chance mechanism, where the steady state level of the central ternary complexes is essentially zero (see above), the reciprocal plot for B–P exchange as B and P are varied together will go through the origin (that is, the maximum exchange velocity is infinite). Observation of a finite maximum exchange velocity in such an experiment shows that finite concentrations of central complexes exist, and to date no enzyme meets this criterion of a Theorell–Chance mechanism.

While completely ordered kinetic mechanisms such as malate or lactic dehydrogenases at pH 8.0 (*27, 28*) give total substrate inhibition of the A–Q exchange by infinite B and P, a largely ordered but partly random mechanism will show only partial substrate inhibition of the A–Q exchange; that is, the reciprocal plot may have a minimum but also a finite (rather than infinite) vertical intercept. This is the situation with liver and yeast alcohol dehydrogenases (*29*) and with malate and lactic

27. E. Silverstein and G. Sulebele, *Biochemistry* **8,** 2543 (1969).
28. E. Silverstein and P. D. Boyer, *JBC* **239,** 3901 (1964).
29. E. Silverstein and P. D. Boyer, *JBC* **239,** 3908 (1964).

dehydrogenases at very high pH (*27, 28*). The finite rates at saturating B and P result from slow dissociation of A and Q from the ternary complexes to give binary EB and EP complexes which then recombine with A and Q to regenerate the central complexes and thus give exchange. The partial inhibition shows that the rate constants for release of A and Q from the central complexes are quite small and the alternate pathway unfavorable. This is the same picture as that obtained for liver alcohol dehydrogenase from the substrate inhibition data discussed above. It is intriguing that a small molecule like ethanol or acetaldehyde cannot completely prevent the dissociation of the nucleotide from the central complexes but that larger ones such as malate and lactate can.

Random mechanisms give different isotopic exchange patterns. Since no reactant can prevent another from leaving the enzyme, no total substrate inhibition patterns are observed unless dead end complexes form, and then all exchanges are inhibited. The normal patterns are that linear reciprocal plots are obtained for all exchanges when the similar substrate and product are varied together, but total substrate inhibition is seen for all exchanges where unlike reactants that can form dead end complexes are varied.

For example, with galactokinase one gets linear plots for either ATP–ADP or galactose–galactose 1-phosphate exchange when MgATP and MgADP, or galactose and galactose 1-phosphate are varied together, but varying MgADP and galactose, or MgATP and galactose 1-phosphate gives substrate inhibition because of the formation of the dead end complexes E-MgADP–galactose or E-MgATP–galactose 1-phosphate (*17*). In no case should one exchange show substrate inhibition and another not as long as the mechanism is really random.

Isotopic exchange experiments can also tell whether a random mechanism is a rapid equilibrium one or not, and in fact this is really the only way to do this. If the rate limiting step is solely the interconversion of two central complexes, all isotopic exchanges at chemical equilibrium will proceed at identical rates regardless of the concentrations of reactants. This is simply because all exchanges are limited by the same step. So far only one enzyme, creatine kinase from rabbit muscle, has been shown to have such a mechanism (*30*). In all other cases studied nonidentical exchange rates have been found, indicating that the release of at least one reactant from the enzyme is one of the rate limiting steps. For yeast hexokinase (*29*) and *E. coli* galactokinase (*17*), the nucleotide exchange is faster than the sugar–sugar phosphate exchange by factors

30. J. F. Morrison and W. W. Cleland, *JBC* **241**, 673 (1966).

of 1.5 to 2.5, showing that the dissociation of sugar and/or sugar phosphate from the enzyme must partly limit the latter exchange.

3. *Exchange Patterns in Ping Pong Mechanisms*

In contrast to sequential mechanisms, ping pong mechanisms permit isotopic exchange studies to be made on isolated parts of the reaction sequence in the absence of the other reactants and the following discussion is restricted to such situations. Since overall chemical reaction cannot take place (assuming always that the enzyme concentration is much lower than reactant concentrations) the system is automatically at chemical equilibrium as soon as the steady state is reached, and one can vary reactants individually at fixed levels of others to obtain isotopic exchange velocity patterns entirely analogous to the initial velocity patterns obtained for the chemical reaction. Exchange studies are thus in one sense a more powerful tool for studying ping pong than sequential mechanisms, since the patterns can be interpreted more specifically when only one reactant is varied at a time, and numerical values can be extracted for kinetic constants, which is difficult when two reactants are varied simultaneously. For instance, the initial velocity of isotopic exchange between A and P in a classic ping pong Bi Bi mechanism is given by the equation:

$$v = \frac{[V_1(K_{ia}/K_a)]\text{AP}}{K_{ia}\text{P} + K_{ip}\text{A} + \text{AP}} \tag{10}$$

which has the same algebraic form as the equation describing the initial velocity of the chemical reaction of A and B. The denominator terms have dissociation constants rather than Michaelis constants as their coefficients [since we are at thermodynamic equilibrium the dynamic steady state dissociation constant (Michaelis constant) is identical with the equilibrium dissociation constant], and these dissociation constants are most easily measured by this type of isotopic exchange study (*14*). The apparent maximum exchange velocity is not necessarily the same as either maximum velocity for chemical reaction (the apparent maximum exchange velocity between A and P can also be expressed as V_2K_{ip}/K_p). The equation for B–Q exchange is analogous to Eq. (10) with B and Q replacing A and P.

If A is varied at different P levels, Eq. (10) predicts a parallel pattern of reciprocal plots, just as is seen for the initial velocity of the chemical reaction. This and other patterns for isotopic exchange velocity in partial ping pong sequences can be deduced by the sort of intuitive analysis we have used previously. As before, we predict the effect on slopes and intercepts of reciprocal plots separately and combine the results to get the expected pattern. The intercepts represent the apparent maximum velocity of exchange at infinite concentration of the varied reactant, and increasing the concentration of any reactant on the opposite side of the central complexes or any reactant along the exchange path on the same side of the central complexes will raise the apparent maximum velocity of exchange and thus give an intercept effect. Reactants inside the exchange points may give total substrate inhibition of exchange at very high levels, however, and the prediction of these effects will also be a part of the pattern. In the simple A–P case under consideration here, P will obviously cause an intercept effect when A is the variable reactant by increasing the rate of exchange of P with the central complex.

The slope of the reciprocal plot is again the reciprocal of the apparent first-order rate constant at very low varied reactant concentrations for reaction of enzyme and varied reactant to give exchange. In predicting effects here, we must remember that the system is at thermodynamic equilibrium and thus that at extremely low varied reactant concentration all of the enzyme will be in enzyme forms on the same side of the central complexes as the point of combination of the varied reactant as long as the other nonvaried reactant concentrations are finite. Thus any reactant combining on the opposite side of the central complexes from the variable one cannot increase the exchange rate at very low variable reactant concentration and cause a decrease in slopes. It can, however, *increase* the slope if it combines within the exchange path, and high concentrations will inhibit the exchange even at very low variable reactant concentrations. In the present simple A–P case, P causes no change in slopes since it combines on the other side of the central complexes; and at very low A, all of the enzyme is in form E regardless of the concentration of P. The overall pattern is thus the parallel one.

Before proceeding to more complex cases, it should be pointed out that this pattern may show competitive substrate inhibition by either reactant, just as in the case of initial velocity studies of this mechanism. Such inhibition will occur whenever P reacts with E, or A with F in dead end fashion. When the kinetics of isotopic exchange for nucleoside diphosphate kinase were investigated, the expected parallel pattern showed competitive substrate inhibition by both nucleoside diphosphates and nucleoside triphosphates, just as was observed in the initial velocity patterns (*14*).

The real power of isotopic exchange studies in elucidating details of ping pong mechanisms can be seen by looking at the predicted exchange patterns for the Bi Uni portion of a ping pong mechanism such as Bi Uni Uni Bi:

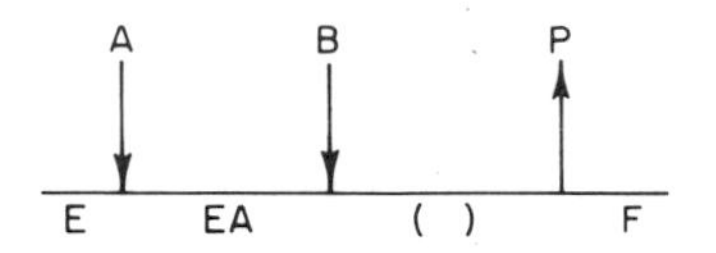

In a real situation (such as a pyrophosphorylytic split of MgATP where P is MgPP), one can measure only A–P or B–P exchange, but not both, depending on the nature of the reaction. This is not a handicap, however, since the pattern will indicate whether the molecule involved in exchange is A or B or whether combination of A and B is random (in which case the designation is arbitrary). Consider A–P exchange first. We have three possible patterns with A and B, A and P, and B and P as the variable reactants. The A–P pattern will be a parallel one since P is on the opposite side of the central complexes from A, and thus no slope effects are expected. Since neither A nor P is within the exchange path (they are at its ends) no substrate inhibition is predicted unless dead end combination of P with E or A with F occurs (and such inhibitions would be competitive).

The B–P pattern is also basically parallel when A–P exchange is observed, since B and P are on opposite sides of the central complexes, but now B should give total noncompetitive substrate inhibition at high levels. Both slopes and intercepts are affected, since B is on the opposite side of the central complexes from P, but within the exchange path so that saturation by B lowers EA to zero and eliminates A–P exchange regardless of the concentration of P. Dead end complex formation by B or P would give competitive substrate inhibition (even if P combined with E, since at equilibrium a high level of B would drive all EA and E into the central complexes). However, this substrate inhibition by B will not be seen if addition of A and B is random, since exchange of A can then take place by direct dissociation of the central complexes regardless of the concentration of B.

In contrast to the A–P and B–P patterns, the A–B pattern in this mechanism is intersecting. Both slopes and intercepts will vary with B when A is the variable reactant since B must convert EA into the central complexes in order for exchange to occur regardless of whether the concentration of A is high or low. Total uncompetitive substrate inhibition will be given by B, however, since at finite A levels infinite B lowers EA to zero and prevents exchange. There is no substrate inhibition of the slopes, however, since at very low A levels there is very little EA for B to react with, and most of the enzyme will be in the E form and available to react with A even at very high B. Dead end complex formation is unlikely to complicate this pattern, especially if a reasonably high level of P is present. Again, if reactant combination is random total substrate inhibition by B will not be seen.

If it turns out that the molecule exchanging with P is B, the patterns for B–P exchange will be as follows. Both B–P and A–P patterns will be parallel ones, and no substrate inhibitions should be observed as long as dead end complexes do not form. The A and B pattern will be the equilibrium ordered one; that is, reciprocal plots when B is varied will have a common vertical intercept, and when A is varied the slope replot will go through the origin. This is because the reaction is at chemical equilibrium, and A is only needed to form EA and does not participate in the exchange itself. Saturation with B will convert all EA to central complexes as long as A is finite, and thus the dependence of exchange on A appears to disappear at infinite B. Substrate inhibition should not occur.

It can be seen that by measuring exchange of P with its corresponding reactant and determining the form of the A–B pattern we can distinguish ordered from random addition of A and B and distinguish A from B if addition is ordered. If P exchanges with A and combination is ordered, the A–B pattern is intersecting with uncompetitive substrate inhibition by B; while if P exchanges with B and combination is ordered, the A–B pattern is an equilibrium ordered one. If A and B combine randomly, the A–B pattern is an intersecting one with no substrate inhibition. Cedar and Schwarz have used this approach to demonstrate that MgATP and aspartate add randomly to asparagine synthetase (*20*). In this case the pyrophosphate–ATP exchange also shows substrate inhibition by pyrophosphate as a result of dead end combination with free enzyme. The final release of AMP and asparagine was also shown by product inhibition to be random, so it appears that this enzyme has a Bi Uni Uni Bi ping pong mechanism with random combination in the bireactant parts of the sequence.

The other common partial ping pong sequence that merits consideration is the Bi Bi sequence that occurs as part of Bi Bi Uni Uni mechanisms:

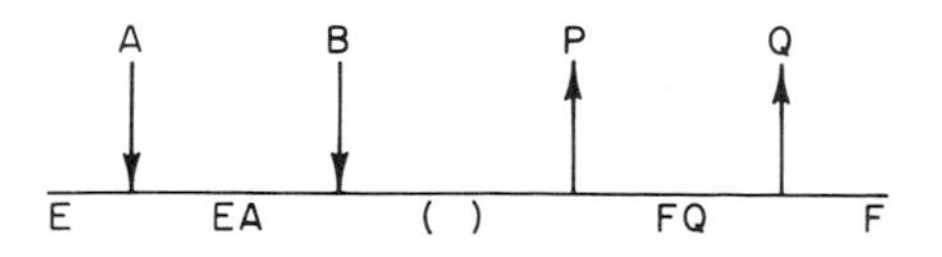

Here again we can vary any reactant at fixed levels of any other one and automatically achieve chemical equilibrium as soon as the steady state is set up. Patterns where the varied reactants are on opposite sides of central complexes will be basically parallel, with superimposed substrate inhibition by any reactant that is within the exchange path (assuming ordered addition of reactants). The A–B and P–Q patterns will be intersecting and will show the same characteristics as those described above for the Bi Uni case. It should be simple to distinguish ordered from random addition of reactants from these patterns, but as yet such experiments have not been carried out on this type of mechanism. The biotin-containing carboxylases probably have this mechanism and are natural targets for this type of experimental approach.

D. A Few Loose Ends

1. *Variation of Kinetic Parameters with pH*

Most kinetic experiments are conducted at constant pH, but considerable information about the groups present at or near the absorption pocket in the enzyme can be obtained by studying the variation of the kinetic parameters with pH. The entities which show independent variation with pH are V (the maximum velocity when all substrates are saturating), V/K (the apparent first-order rate constant for reaction of enzyme with substrate at very low substrate concentration but with all other substrates saturating), and those inhibition constants which are true dissociation constants. The pH variation of the Michaelis constant, in particular, is not a simple function but rather displays the effects of dividing the expression for V as a function of pH by the expression for V/K. Data are usually plotted in the form of log V or log V/K vs. pH. Let us see first what the log V vs. pH curve should look like for various situations.

The maximum velocity is dependent on those unimolecular steps involving actual catalytic reaction or the dissociation of reactants, and the pH variation of V will reflect the pH variation of those steps which are rate limiting. Changes in V with pH will normally reflect the protonization or deprotonization of a group in the transitory complex which must undergo the rate limiting step. While the ionization state of one of the bound reactants may also affect V, the usual effects seem to come from ionization of those groups on the enzyme responsible for the actual catalysis, such as the imidazole group of histidine, the carboxyl groups of glutamate or aspartate, the SH group of cysteine, the OH of tyrosine, the α-amino group at the end of the protein chain, or the ϵ-amino group of lysine. (The deprotonation of the guanidinium group of arginine occurs at too high a pH to be of any practical interest.) Ionization or protonation may cause total loss of activity or may result in lowered but still finite activity. The gradual change in the ionization state of the entire enzyme may also influence its conformation in such a way that the maximum velocity changes slowly with pH, but we will limit this discussion to cases where the ionization state of a single group produces a specific effect since most data can be explained on this basis.

If a single group must be protonated for reaction to occur, and the enzyme form in which that group is unprotonated is inactive, then the apparent rate constant is divided by a factor $(1 + K_h/\mathrm{H})$ where H is hydrogen ion concentration and K_h is the dissociation constant of this proton. At low pH, H exceeds K_h sufficiently so that this function is 1.0, but at high pH the term K_h/H becomes larger by a factor of 10 for each pH unit and rapidly reduces the apparent value of the rate constant involved. The plot log V vs. pH then follows the equation:

$$\log V = \log V_0 - \log(1 + K_h/\mathrm{H}) \tag{10a}$$

where V_0 is the V at low pH. At high pH, this reduces to $\log V = \log V_0 + \mathrm{p}K_h - \mathrm{pH}$, and thus the low pH asymptote of Eq. (10a) is horizontal, while the high pH asymptote is linear with a slope of −1. The two asymptotes intersect at $\mathrm{p}K_h$, and the value of log V at this point is $\log(V_0/2)$. If it is the unprotonated rather than the protonated form which is active, the factor involved here is $(1 + \mathrm{H}/K_h)$, and the curve is flat at high pH and drops off on the low pH side with a limiting slope of 1.

If a second group also required for activity ionizes at a higher pH than the first one, the curve will undergo a second change in slope to a value of −2, and the intersection of the asymptotes to this portion of the curve will define the second pK. The overall plot of log V vs. pH thus appears to be a curve with segments whose asymptotes or tangents have whole number slopes (or zero) and define the pK's involved by their intersection points. Several reservations must be made, however. First, if the pK's in question are not separated by more than two pH units, the portion of the curve between them will not have a linear segment, and the pK's cannot be accurately determined by inspection (one can of course use statistical techniques and make a least squares fit to extract these values), although the extrapolated linear segments will intersect at the average of the two pK's. This situation arises if, for example, two histidines are necessary for catalysis and one must be protonated and the other unprotonated; the curve of log V vs. pH has no flat portion but merely consists of sections with slopes of 1 and −1 connected by a curved portion about 2 pH units wide [the pH factor appearing in the rate equation will be $(1 + \mathrm{H}/K_1 + K_2/\mathrm{H})$, where K_1 and K_2 are dissociation constants of the groups that must be unprotonated and protonated, respectively].

A more serious reservation concerning interpretation of log V vs. pH curves is that V often depends on more than one rate constant, and these may not all have the same pH dependence. For instance, the ionization state of a histidine may be very important for catalytic transfer but have no effect or little effect on such steps as release of products. Let us look at what can happen in such a situation. We will assume that V is limited by two unimolecular constants k_1 and k_2 so that $V = k_2k_1/(k_1 + k_2)$. If k_2 involves a step where a group must be protonated to show activity, but k_1 does not change with pH, we can express the pH variation by

$$V = \frac{k_1\{k_2/[1 + (K_h/H)]\}}{k_1 + \{k_2/[1 + (K_h/H)]\}} = \frac{k_1k_2}{k_1(1 + K_h/H) + k_2} \tag{11}$$

Once again, log V is constant at low pH and decreases with a slope of —1 per pH unit at high pH, but now the apparent pK (intersection point of the extrapolated linear segments) is not pK_h but rather pK_h + $\log(1 + k_2/k_1)$. If k_2 were originally the rate limiting step ($k_2 \ll k_1$), the true pK will be observed, but to the extent that k_2 is originally greater than k_1 at low pH, the apparent pK will be displaced to higher pH; if $k_2 = 100\ k_1$, the displacement is 2 pH units! If the group involved must be unprotonated for the enzyme to be active, the pK will be displaced to lower pH's; thus, the pK's observed in a log V vs. pH curve must be taken as limiting values only, unless one can definitely assign the pH variation to the rate limiting steps. The true pK can only be higher than the observed one when protonation destroys activity or lower for one where deprotonation eliminates activity. Some of the differences in pK's for particular groups that have been observed in experiments of this type are certainly caused by these effects rather than by alteration of the pK by the environment on the protein surface.

A still further complication arises when an ionization change leads to finite rather than zero activity. One now has present enzyme molecules with different activities; if the Michaelis constants are affected, slightly nonlinear reciprocal plots (concave down) may be observed at pH's where the ionization is partial (see Section IV,D,2). If we limit ourselves for a moment to consideration of the true maximum velocity, however, the net effect of this situation is to introduce a sigmoidal wave into the log V vs. pH curve. Unless the activity after the ionization change is less than 1% of the original value, the sigmoid curve will not have a linear segment between the two reverse curvatures, but in any case the pK will be the point at which V drops to the average of its values before and after the change. Unless the change in V is less than a factor of 10, the pK occurs near the top of the wave and not near the middle (the midpoint occurs at the geometric rather than the arithmetic mean of the two V values). Waves of this kind are quite common in log V or log V/K vs. pH curves. It should be remembered that for log V vs. pH curves, if the step being looked at is not rate limiting on both sides of the curve, the apparent pK may be displaced by the phenomenon described earlier.

Thus while the theory of pH variation of V is fairly straightforward, interpretation is frequently not; a good knowledge of the kinetic mechanism and of which steps are rate limiting at various pH's is necessary before interpretation is possible.

The pH variation of V/K is often as informative as that of V. Since V/K is the apparent first-order rate constant for reaction of enzyme with substrate at very low substrate levels, it reflects the ability of enzyme (specifically the form that combines with the substrate) and substrate to form a complex. Thus any ionizations occurring on either the substrate, or the enzyme form which combines with the substrate, will show up in the log V/K vs. pH plot if they affect the ability of these two to combine; but they will not show up if they do not. One advantage of this analysis is that the observed pK's will be the true values since they reflect only one step, the combination of enzyme and substrate. The pK's of the substrates are normally known so they can be accounted for (although the ionization state of the substrate may not affect its ability to combine with the enzyme), and any other observed pK's must belong to groups on the enzyme (or at least on the enzyme form which reacts with the substrate; the pK's of bound reactants may enter into this). Thus the observation that log V/K vs. pH is constant at high pH but shows a break at a pK of 7.0, with a slope of 1 below this, might indicate that a histidine must be unprotonated in order for the substrate to combine with the enzyme.

If ionization causes only a change in the value of V/K rather than eliminating combination of enzyme and substrates, a wave in the log V/K vs. pH curve will occur, although some nonlinearity in the primary reciprocal plots in the region of the ionization may also be detected since whenever different enzyme forms with different apparent Michaelis constants are present reciprocal plots are hyperbolas, concave down, with the curvature near the vertical axis, and a linear asymptote at high reciprocal substrate concentrations (see below). If the Michaelis constants are not too different, the curvature will be slight; but if they differ by more than an order of magnitude, the effect will be quite prominent. The reciprocal plots will be linear, however, at pH's above and below the ionization region. More use should be made of log V/K vs. pH plots; many workers have only plotted V vs. pH or have plotted the Michaelis constant vs. pH, which is quite confusing since it merely shows the result of dividing the pH variation of V by the pH variation of V/K and the resulting curve is quite complex and difficult to interpret.

2. *Nonlinear Reciprocal Plots*

Most of the previous discussion has assumed that reciprocal plots of initial velocity vs. substrate concentration were linear and that if there were two or more active catalytic sites per molecule they were identical and independent. This may be true for many enzymes, but there are important exceptions. Let us now consider what may make a reciprocal plot nonlinear.

If the variable substrate adds twice with reversible connection between the points of combination, reciprocal plots are parabolas, concave up. A parabola has no linear asymptote, and at very low substrate concentrations the reaction rate is proportional to substrate concentration squared. A reciprocal plot vs. $(1/A)^2$ will normally not be linear either, but it will be concave down because there is usually a term in $1/A$ present as well as the $(1/A)^2$ term.

Nonlinear reciprocal plots will always be seen when two enzymes are present, both catalyzing the same reaction but with different apparent Michaelis constants. The initial velocity is thus

$$v = \frac{V_1 A}{K_{a1} + A} + \frac{V_2 A}{K_{a2} + A} \tag{12}$$

where the subscripts 1 and 2 refer to the kinetic constants for the two enzymes (*30a*). The ratio of the V_1 and V_2 values will, of course, depend on how much of each enzyme is present as well as on the turnover number of each, but K_{a1} and K_{a2} are intrinsic properties of each form and their values will not depend on the levels of each present. Inversion of this equation produces the double reciprocal form

$$\frac{1}{v} = \frac{1 + a(1/A) + b(1/A)^2}{d + c(1/A)} \tag{13}$$

where a is the sum of the Michaelis constants, b is their product, d is the sum of the maximum velocities, and c is $(V_1K_{a2} + V_2K_{a1})$. This curve is what we call a 2/1 function; that is, the numerator is a second-order polynomial in $1/A$ and the denominator a first-order one (a 4/3 function is a fourth-order polynomial divided by a third-order one, and

30a. This situation occurs whenever one has a true mixture of different isozymes or where some effect, such as ionization of a group on the enzyme, has produced two populations with different Michaelis constants. The same effect will be seen in hybrid isozymes when the different subunits present have different intrinsic Michaelis constants, although in such cases the subunits may appear to have different Michaelis constants in the hybrid than in the homogeneous oligomer.

so forth). When $1/v$ is plotted vs. 1/A, the curve is a hyperbola with a linear asymptote at high 1/A values. The portion of the curve near the vertical axis is concave down in this particular case, although 2/1 functions arising from other mechanisms may be concave up and even have a minimum point.

Care must be taken in interpreting data of this form. The asymptote and the initial tangent do not by themselves determine K_{a1} and K_{a2}; the apparent values determined this way are less than the larger of the two and larger than the smaller one. The data should be fitted to Eq. (13) and the coefficients a and b combined algebraically to solve for K_{a1} and K_{a2}.

The random bireactant addition of substrates should in theory result in reciprocal plots that are 2/1 functions unless one substrate is saturating (in which case the reaction becomes ordered, with that substrate adding first), or unless certain relationships apply among the rate constants, the simplest of which involves the conversion of (EAB) to products being much slower than the dissociation of substrates from the enzyme (the rapid equilibrium assumption). However, simulation studies have shown that for realistic values of rate constants the expected curvature is very slight and that random mechanisms appear to be rapid equilibrium ones, even when the rapid equilibrium assumption is far from being valid. To date, no one has clearly identified a nonlinear reciprocal plot caused by random addition of substrates, but such a situation might exist and workers should keep the possibility in mind. The expected 2/1 functions can be either concave up or down, and they may have minimum points if concave up. This latter situation would result whenever saturation with the variable substrate forces the enzyme to use a pathway that is slower than the one normally used.

If the combination and/or reaction of one or more substrates at one catalytic center is not independent of the presence of this reactant at the other catalytic center, reciprocal plots may be nonlinear. Such cooperativity may be negative (the presence of the reactant raises the apparent Michaelis constant), or positive (the reactant lowers the apparent Michaelis constant); negative cooperativity normally gives reciprocal plots concave down, and positive cooperativity gives plots concave up. Negative cooperativity permits the apparent Michaelis constant to be adjusted to the physiological concentration; thus, the enzyme always operates in the region of proportional control. As the Michaelis constant is raised, the maximum velocity will normally be increased also in order to make full use of catalytic potential. Citrate cleavage enzyme shows such reciprocal plots for citrate (*12*).

Positive cooperativity, on the other hand, serves to provide a sensitive means of controlling reaction rate. In such a situation, the binding of the substrate with the free enzyme (or its reaction once bound, or both) is very difficult but becomes easier as other substrate molecules are absorbed on the other subunits of the enzyme. Sigmoid curves of v vs. A are thus observed, and it is clear that varying the degree of cooperativity from zero (normal Michaelis kinetics with linear reciprocal plots) to very high varies the reaction rate at what are originally Michaelis constant levels of substrate from a large proportion of maximum velocity to nearly zero. In such a system, control is thus mediated not by variation in substrate concentration but normally by changes in cooperativity produced by absorption of modifiers at other locations on the enzyme (see Volume I, Chapter 7 by Koshland).

The question arises how to analyze reciprocal plots that show cooperativity, particularly those that show positive cooperativity where reciprocal plots are concave upward. From the usual reciprocal plots one can tell only whether the curve has an asymptote and a minimum point, but one cannot tell by inspection how complex the rate equation really is. The rate equations for enzymic reactions (see below) are the ratios of two polynomials in substrate concentrations, and in reciprocal form such equations may be 2/1, 3/2, 4/3 or more complex functions. Where positive cooperativity is present and there are n subunits, one expects the rate equation in reciprocal form to be at least a $n/(n-1)$ function; thus, some knowledge of n is helpful in deducing the number of cooperative subunits present in the enzyme. The most common method used for this is to plot $\log[v/(V-v)]$ vs. log A, where V is the apparent maximum velocity obtained by extrapolation on a reciprocal plot. If the rate equation is assumed to be $v = V\mathrm{N}/\mathrm{D}$ where N and D are polynomials in substrate concentration, we can derive the expression $v/(V - v) = \mathrm{N}/(\mathrm{D}-\mathrm{N})$. Since D and N are both polynomials in A, and their highest power terms in A are identical (otherwise v would not equal V at infinite A), (D — N) is now a polynomial of one less degree than N. If the reciprocal plot for this rate equation is a $n/(n-1)$ function, N will be a polynomial with highest power n, and D — N with highest power $n - 1$. We then have

$$\log[v/(V-v)] = \log \mathrm{N} - \log(\mathrm{D}-\mathrm{N}) = \log[\mathrm{A}^n + a\mathrm{A}^{n-1} + b\mathrm{A}^{n-2} + \cdots + i\mathrm{A}^2 + j\mathrm{A}] - \log[k\mathrm{A}^{n-1} + l\mathrm{A}^{n-2} + \cdots + x\mathrm{A} + y] \quad (14)$$

where $a, b \ldots i, j$, and y are positive coefficients, but $k, l \ldots x$ may be either positive or negative. If we differentiate with respect to log A to get the slope of this plot, we have

$$\frac{d\{\log[v/(V-v)]\}}{d(\log \mathrm{A})} = \frac{(n\mathrm{A}^n + (n-1)a\mathrm{A}^{n-1} + \cdots + 2i\mathrm{A}^2 + j\mathrm{A})}{(\mathrm{A}^n + a\mathrm{A}^{n-1} + \cdots + i\mathrm{A}^2 + j\mathrm{A})} - \frac{[(n-1)k\mathrm{A}^{n-1} + (n-2)l\mathrm{A}^{n-2} + \cdots x\mathrm{A}]}{(k\mathrm{A}^{n-1} + l\mathrm{A}^{n-2} + \cdots x\mathrm{A} + y)} \quad (15)$$

Several things can be deduced from examination of this equation. First, the slope at very low A will be 1 as long as the numerator of the original rate equation for v as a function of A contains an A term. If it contains only A^2 and higher terms, the limiting slope will be 2, and so forth. Stated in another way this slope at very low A values equals the difference between the two numbers when we describe the type of function in reciprocal form (that is, the slope is one for 2/1, 3/2, 4/3 functions but two for 3/1, 4/2, 5/3 functions, etc.). A limiting slope of two at low substrate concentrations shows that at least two molecules of substrate must be present to get reaction; this is usually not the case for cooperative situations, and limiting slopes of 1 are to be expected.

Second, the slope at extremely high substrate concentration should always be 1 as long as the original rate equation contains terms in numerator and denominator in A^{n-1}, which is normally always the case. This is of little practical importance, however, since this region of the curve cannot be determined because the value of $\log[v/(V-v)]$ becomes very inaccurate when v becomes an appreciable portion of V. Thus values of $\log[v/(V-v)]$ above $+0.6$ (or at the very best, 1.0) should not be plotted.

Finally, the maximum possible slope that can be observed is n, but the value actually observed will be considerably less than this unless the degree of cooperativity is very high. The maximum slope can thus be used as a lower limiting value of n, and values of n above 1 indicate that some cooperativity exists (or the reciprocal plot is concave upward for one of the other reasons discussed earlier). A maximum slope of 1.8 does not mean that n is 2, however; the true value may well be 4 or 6. If the true value of n is known, the maximum slope of the log $[v/(V-v)]$ vs. log A plot thus gives by comparison a convenient empirical measure of the degree of cooperativity which can be used to quantitate the effects of modifiers on the kinetics. It must be remembered, however, that such plots are not expected to be linear but to have limiting slopes of 1 at low log A values and that the maximum, not the average slope, must be taken as a minimum value of n. The curve should get flatter at high log A values, but values of $\log[v/(V-v)]$ above 0.6 will usually be unreliable because they depend on the small difference between v and V, and V is not accurately known.

3. *Metals*

We have discussed the kinetics of the organic reactants involved in enzymic reactions in considerable detail; however, we have not said much about the inorganic cations that are necessary for many reactions but are not altered by the reactions. These can be divided into two groups, those that complex the enzyme, and those that complex one or more reactants. The classic cases of the latter are the magnesium complexes of di- and triphosphates such as MgATP, MgADP, and MgPRPP. As far as we know, all of these compounds exist largely as their magnesium complexes in the cell, and the uncomplexed compounds are inactive as substrates and in fact will behave as inhibitors if they are absorbed on the enzyme (*30b*). Since the dissociation constant of MgATP and similar triphosphates is around 14 μM at pH 8, one can maintain nearly all added reactant as the magnesium complex by adding a fixed excess of magnesium over ATP (1 mM has frequently been used since this level of free magnesium is rarely inhibitory and is probably close to the *in vivo* level). With ADP this procedure will lead to errors since the dissociation constant of MgADP at pH 8 is 0.25 mM and at 1 mM free magnesium 20% of the ADP is not complexed. Calculation of the exact amount of ADP and magnesium to add to get the desired concentrations of free magnesium and of MgADP is therefore essential. One must also beware of buffers that bind either magnesium (imidazole, for instance) or the other components (ATP with tris, for example), although the use of a metal ion buffer might be very advantageous in some cases. What one must *not* do is vary the total amount of substrate added at a fixed level of total metal or vary metal and substrate together in constant ratio; in these cases, too many things vary at once to permit clear interpretations. Many older kinetic studies are spoiled by such lack of proper handling of metal ion concentrations.

With proper precautions, the concentrations of MgATP or of whatever magnesium complex one desires are known, and this concentration can be varied in the same way as other substrates but with the concentration of free metal kept constant. If there is doubt whether the free or complexed molecule is actually the substrate, the suspected substrate is varied at fixed levels of the other molecule. This technique has shown that MgATP and MgADP are always the substrates and that free ATP and ADP are either inhibitors or have no effect (*30c*). Likewise, free magnesium normally will appear to be inhibitory or to have no effect. With citrate cleavage enzyme, however, both magnesium citrate and free citrate act as substrates (*12*).

When the metal activates by complexing with the enzyme, the situation is not so simple. The kinetic mechanism of the organic reactants can be studied at a saturating metal concentration, but to determine the role of the metal its concentration must be varied. Since it is unrealistic in most cases to expect the metal to combine with and leave the enzyme

30b. An exception is myokinase which catalyzes phosphorylation of free ADP by MgADP (*13a*).

30c. J. F. Morrison and W. J. O'Sullivan, *BJ* **94**, 221 (1965).

during each catalytic cycle, realistic mechanisms to consider are ones where the metal must be present before substrates can combine or where metal and substrate combination can be random. The former situation will give the equilibrium ordered initial velocity pattern for the metal and the substrate which cannot add until after metal is present (that is, saturation with this substrate will appear to eliminate the requirement for the metal, and the reciprocal plots vs. this substrate at different metal concentrations will cross on the vertical axis), while random order of addition should give normal intersecting patterns between metal and substrate [this is the case with phosphoglucomutase (*31*)]. Parallel patterns would be observed only if metal stimulated reaction at high concentrations of the variable substrate but had no kinetic effect at very low levels of this substrate; such a mechanism is difficult to imagine.

V. Derivation of Rate Equations

A. Rate Equations for Chemical Reaction

Thus far we have largely avoided deriving rate equations and have attempted to show intuitively how the observed kinetic patterns can be explained. But there comes a time in any kinetic study when the appropriate rate equation is needed to compute the numerical values of the kinetic constants. Although one can depend on published equations in many cases, it is preferable to have at least some familiarity with the methods involved.

First, let us consider the rate equation for net chemical reaction. A diagram of the mechanism is made which shows the rate constants for each step and the concentrations of reactants where appropriate. Thus for an ordered Bi Bi mechanism, we have

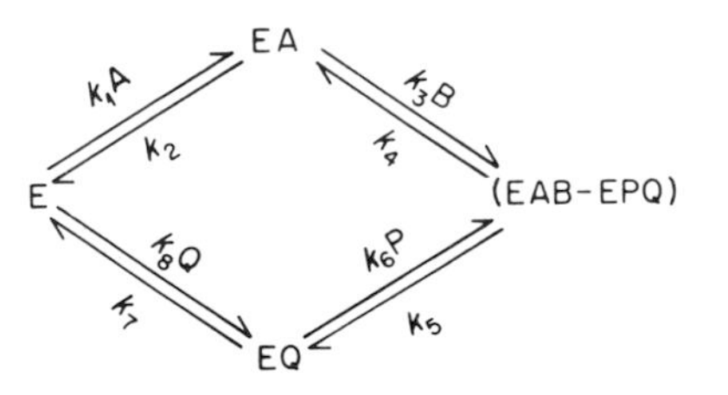

The differential equations which describe the rates of change with time of the concentration of each enzyme form are

31. W. J. Ray, Jr. and G. Roscelli, *JBC* **241**, 2596 (1966).

$$\frac{d(\text{E})}{dt} = -k_1\text{A}(\text{E}) - k_8\text{Q}(\text{E}) + k_2(\text{EA}) + k_7(\text{EQ})$$

$$\frac{d(\text{EA})}{dt} = k_1\text{A}(\text{E}) - k_2(\text{EA}) - k_3\text{B}(\text{EA}) + k_4(\text{EAB–EPQ}) \tag{16}$$

$$\frac{d(\text{EAB–EPQ})}{dt} = k_3\text{B}(\text{EA}) - (k_4 + k_5)(\text{EAB–EPQ}) + k_6\text{P}(\text{EQ})$$

$$\frac{d(\text{EQ})}{dt} = k_8\text{Q}(\text{E}) + k_5(\text{EAB–EPQ}) - k_6\text{P}(\text{EQ}) - k_7(\text{EQ})$$

If the concentration of enzyme is much lower than the concentrations of the reactants (normally true in laboratory experiments) or if we are going to assume a dynamic steady state (normally true *in vivo*) these rates of change will be negligible compared with the reaction rate ($d\text{P}/dt$ or $d\text{Q}/dt$) and can be set equal to zero (this is the steady state assumption). These four equations are not independent, however, and thus three of them can be combined with a conservation equation for enzyme

$$\text{E} + \text{EA} + (\text{EAB–EPQ}) + \text{EQ} = E_t$$

to give four equations which can be solved simultaneously for E, EA, (EAB–EPQ), and EQ as a function of the reactant concentrations, the different rate constants, and E_t. With the concentrations of the enzyme forms in hand, one can then substitute into an equation such as

$$v = d\text{Q}/dt = k_7(\text{EQ}) - k_8\text{Q}(\text{E}) \tag{17}$$

to get the desired rate equation.

Clearly, the only difficult part of this is solving four equations in four unknowns, algebraically. This is not difficult with three unknowns, but by the time one reaches five or six enzyme forms solution by standard algebraic methods becomes quite tedious. As a result, King and Altman developed a schematic method to simplify this solution and permit one to write the distribution equations for E/E_t, EA/E_t, etc., directly by inspection (*32*). This is a very useful method, and can save lots of labor, but since it is described in standard textbooks (*32a*) we will not describe it further here.

The above discussion centered on nonrandom mechanisms. When the rigorous steady state treatment is applied to random mechanisms, however, the resulting equations become the ratios of very complex polynomials in reactant concentrations, and even the King–Altman procedure

32. E. L. King and C. Altman, *J. Phys. Chem.* **60,** 1375 (1956).

32a. H. R. Mahler and E. H. Cordes, "Biological Chemistry." Harper, New York, 1966.

involves a tremendous amount of work. Further, simulation studies indicate that the resulting equations give kinetic patterns that nearly resemble those predicted by the much simpler equations for rapid equilibrium random mechanisms. It is therefore very useful to have equations for rapid equilibrium random cases available and for more complex mechanisms where only part of the sequence is random to have available methods for deriving rate equations that permit assumption of rapid equilibrium only for the random segments.

The rate equation for a rapid equilibrium random sequential mechanism is very easily derived by direct algebraic means. For example, let us derive the initial velocity equation for a rapid equilibrium random bireactant mechanism. We can write the equations:

$$K_{ia} = \frac{(E)A}{(EA)} \qquad K_{ib} = \frac{(E)B}{(EB)} \qquad K_a = \frac{(EB)A}{(EAB)} \qquad K_b = \frac{(EA)B}{(EAB)} \tag{18}$$

$$E + EA + EB + EAB = E_t$$

$$v = (V/E_t)(EAB)$$

which are very easily combined into

$$v = \frac{VAB}{K_{ia}K_b + K_bA + K_aB + AB} \tag{19}$$

where $K_{ia}K_b = K_aK_{ib}$. The extension of this procedure to full equations with products present is straightforward.

When a mechanism is basically steady state but contains random segments, one can obtain an equation that closely matches the real situation by assuming rapid equilibrium for only the random segments. This can be accomplished by the procedure of Cha (*33*). Once the rate equation has been derived, it is conveniently expressed by defining various kinetic constants in terms of groupings of rate constants; thus, the final equation looks like those dealt with in this chapter. Cleland describes this process in detail (*1a*), and it will not be further discussed here.

B. Rate Equations for Isotopic Exchange

When we derive the rate equations for chemical reaction, we are concerned with the chemical concentrations of each enzyme form. In order to derive rate equations for isotopic exchange, however, we are concerned only with the concentrations of those molecules of enzyme which contain bound labeled reactant rather than the total chemical concen-

33. S. Cha, *JBC* **243**, 820 (1968).

trations of these enzyme forms. Let us consider, for example, A–Q exchange in a Bi Bi Uni Uni ping pong mechanism:

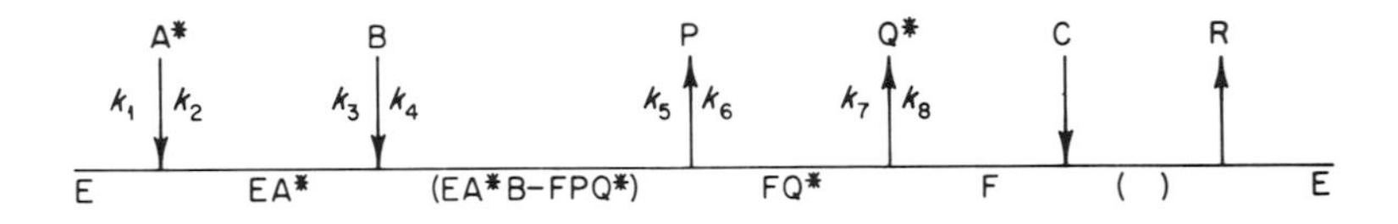

The enzyme forms that will become labeled during exchange will be EA*, (EA*B–FPQ*), and FQ*. The initial rate of exchange will be given by $v^* = dQ^*/dt = k_7(FQ^*)$, where (FQ*) refers to the concentration of *labeled* FQ* and not to the total chemical concentration which may be much greater. Since we are assuming that no labeled Q exists to start with, there is no negative term [that is, the term $-k_8Q^*(F)$ is zero]. Our job is thus to find out what the concentration of (FQ*) is in the steady state.

To do this we can write the differential equations for the rates of change with time of the concentrations of *labeled* enzyme forms (not total chemical concentrations). These will be

$$
\begin{aligned}
\frac{d(\mathrm{EA}^*)}{dt} &= k_1\mathrm{A}^*(\mathrm{E}) + k_4(\mathrm{EA}^*\mathrm{B{-}FPQ}^*) - k_2(\mathrm{EA}^*) - k_3\mathrm{B}(\mathrm{EA}^*) \\
\frac{d(\mathrm{EA}^*\mathrm{B{-}FPQ}^*)}{dt} &= k_3\mathrm{B}(\mathrm{EA}^*) + k_6\mathrm{P}(\mathrm{FQ}^*) - (k_4 + k_5)(\mathrm{EA}^*\mathrm{B{-}FPQ}^*) \qquad (20) \\
\frac{d(\mathrm{FQ}^*)}{dt} &= k_5(\mathrm{EA}^*\mathrm{B{-}FPQ}^*) - k_6\mathrm{P}(\mathrm{FQ}^*) - k_7(\mathrm{FQ}^*)
\end{aligned}
$$

As long as the enzyme concentration is very much lower than the concentrations of the reactants, these rates of change are much lower than the rate of exchange, and we can set all three of these equations equal to zero, giving us three equations which we can solve for the concentrations of EA*, (EA*B–FPQ*), and FQ* as a function of A*, B, P, E, and the various rate constants. Note that we do not need a conservation equation as we do for deriving the rate equation for chemical reaction but that the final equation will contain the chemical concentration of E, which must be evaluated by the standard methods described above. These equations can be solved simultaneously by any convenient method; Cleland has described a scheme similar to the King–Altman method to do this schematically, but it will not be presented here because of lack of space (*34*). The only solution we are interested in is that for FQ^* which is

$$FQ^* = \frac{k_1k_3k_5A^*B(E)}{k_2(k_4 + k_5)k_7 + k_3k_5k_7B + k_2k_4k_6P} \tag{21}$$

If we now substitute this into the equation $v^* = k_7(FQ^*)$ and put in the expression for E in this mechanism, we will have the rate equation for A–Q exchange. By comparing this equation with the definitions of the kinetic constants for this mechanism, we can express this and any other rate equation for isotopic exchange in terms of the same kinetic constants used for the chemical reaction (*34a*). Note that when we substitute the equation for E into Eq. (21) we will have A^* (which is the concentration of labeled substrate) in the numerator and A, the total chemical concentration of A, in the denominator. To simplify things, we normally consider A^* and A to be the same; this means A^* is now the concentration of material with a given specific activity rather than the concentration of actual labeled molecules. The equations would be perfectly valid, however, if v^* and A^* were expressed in terms of cpm/ml/min and cpm/ml and A in normal concentration units.

Acknowledgment

The author is indebted to Professor William Ray, Jr., who made many helpful comments on the manuscript.

34. W. W. Cleland, *Ann. Rev. Biochem.* **36,** 77 (1967).
34a. For the example given here this has been done by Cleland (*34*); however, in Eq. (13) of that reference, the K_cABPQ term was inadvertantly omitted from the denominator.

2

Rapid Reactions and Transient States

GORDON G. HAMMES • PAUL R. SCHIMMEL

I. Introduction

The elucidation of the mechanism of action of enzymes is a long-standing goal of biochemistry. In order to understand how an enzyme works on a molecular basis, dynamic information must be obtained for the entire time sequence of the reaction; that is, the discrete intermediates involved in the catalytic conversion of substrates to products must be detected and their lifetimes and modes of breakdown and formation determined. In order to examine the entire range of possible lifetimes, time constants approaching the period of molecular vibrations ($\sim 10^{-13}$ sec) must sometimes be measured. This often requires special experimental techniques and methods of data analysis. When detailed kinetic information is combined with known molecular structures of enzymes and substrates, an understanding of the molecular nature of the mechanism of action of enzymes may be attained.

The most conventional approach to enzyme kinetics is the use of the steady state method (*1–8*). Extremely low concentrations of enzyme (e.g., 10^{-7} to 10^{-10} M) are employed so that the overall rate of conversion of substrates to products is slow enough to be conveniently measured. Under these circumstances all enzyme species are in a steady state, and the rate of appearance of products (or disappearance of substrates) is the only parameter which is experimentally determined. An exact steady state rate law is relatively easy to derive, and it may usually be cast into a form which is convenient for the experimental determination of a number of characteristic steady state parameters (*1–8*). However, the kinetic constants obtained from steady state studies are complicated functions of the individual rate constants, and information about the rates of the individual steps cannot be obtained for any realistic mechanism. Nevertheless, lower bounds to rate constants of individual steps (*7*, *9*) and the order of addition of substrates in multiple substrate–multiple product reactions (*8*) can be determined by steady state methods (see Chapter 1 by Cleland).

The principal disadvantage to the steady state approach is that direct information about the states occurring between the uptake of substrates and release of products is not obtained; i.e., intermediate transient states are not directly observable. This difficulty may be surmounted by employing higher concentrations of enzyme (e.g., 10^{-5} to 10^{-3} M). The reaction then occurs very rapidly with typical half-times of milliseconds. In order to observe transient states with lifetimes of milliseconds or less, special experimental methods are required. Also, the steady state approximation is not valid for an appreciable time interval when the enzyme and substrate concentrations are comparable; thus, solution of the kinetic equations is often difficult.

The first instrument designed for the purpose of measuring rapid reactions was a rapid mixing device originally developed by Hartridge and

1. L. Michaelis and M. L. Menten, *Biochem. Z.* **49,** 333 (1913).
2. G. E. Briggs and J. B. S. Haldane, *BJ* **19,** 338 (1925).
3. J. B. S. Haldane, "Enzymes." Longmans, Green, New York, 1930.
4. R. A. Alberty, *Advan. Enzymol.* **17,** 1 (1956).
5. R. A. Alberty, "The Enzymes," 2nd ed., Vol. 1, p. 143, 1959.
6. J. Z. Hearon, S. A. Bernhard, S. L. Friess, D. J. Botts, and M. F. Morales, "The Enzymes," 2nd ed., Vol. 1, p. 49, 1959.
7. L. Peller and R. A. Alberty, *Progr. Reaction Kinetics* **1,** 235 (1961).
8. W. W. Cleland, *Ann. Rev. Biochem.* **36,** 73 (1967).
9. L. Peller and R. A. Alberty, *JACS* **81,** 5907 (1959).

Roughton (*10*). The most modern rapid mixing devices have typical mixing times of a few milliseconds (*11, 12*), and instruments with mixing times as short as tenths of a millisecond have been developed (*13*). This approach has two principal inherent limitations: (1) Transient states with lifetimes shorter than the apparatus mixing time cannot be studied; and (2) the mathematical treatment of the kinetic behavior of reactants and intermediates, in systems far from equilibrium, is generally very complicated. In fact, exact integrated rate laws can be obtained for very few complex reaction mechanisms. The second limitation has been surmounted in part, however, by the rapid development of high-speed analog and digital computers (see Section II,C).

About fifteen years ago an entirely new experimental approach to the study of rapid reactions was introduced by Eigen and co-workers (*14*). Relaxation methods were developed which permit the measurement of reaction rates with characteristic time constants as short as 10^{-10} sec (*15*). The underlying principle of these methods rests on the fact that the equilibrium position of a chemical reaction generally depends upon one or more intensive thermodynamic variables such as the temperature, pressure, and electric field intensity. A change in one of these variables will alter the thermodynamic state of the system, and consequently the equilibrium concentrations will be changed. The rates at which the concentrations approach the new equilibrium state is a measure of the rates of the chemical reactions occurring in the system. Thus this method is not limited by how fast reactants can be mixed but rather by how fast an appropriate intensive variable can be changed. By use of ultrasonic and electromagnetic radiation some intensive variables can be altered in times as short as 10^{-10} to 10^{-11} sec, which is considerably shorter than mixing times. A second advantage of relaxation methods is that the thermodynamic perturbations are small; thus, the corresponding deviations from chemical equilibrium are quite small. For such perturbations, the rate equations for any mechanism can be linearized; consequently, an exact solution to the rate equations governing the system can always be obtained, regardless of the complexity of the reaction mechanism.

The application of relaxation techniques to biological systems has just recently become widespread. However, a number of important biological

10. H. Hartridge and F. J. W. Roughton, *Proc. Roy. Soc.* **B104**, 376 (1923).
11. B. Chance, *Tech Org. Chem.* **8**, Part II, 1314 (1963).
12. Q. H. Gibson, *Ann. Rev. Biochem.* **35**, 435 (1966).
13. R. L. Berger, R. Balko, W. Borcherdt, and W. Friauf, *Rev. Sci. Instr.* **39**, 486 (1968).
14. M. Eigen, *Discussions Faraday Soc.* **17**, 194 (1954).
15. M. Eigen and L. de Maeyer, *Tech. Org. Chem.* **8**, Part II, 895 (1963).

reactions have been examined and some of the intermediate transient states in enzymic catalysis have been directly observed and characterized (*16–19*). Thus far, the chief limitation of these methods has been that they are only easily applicable to reactions which are readily reversible. However, the underlying principles are applicable to steady states as well as equilibrium states so that perturbations can be applied during the time course of an essentially irreversible reaction. This approach has been used somewhat by combination of rapid mixing and relaxation techniques (*20–22*). A summary of the available experimental techniques for studying the dynamics of enzymic reactions and their time range of application is given in Fig. 1.

This chapter emphasizes the kinetic treatment and analysis of rapid reactions and transient states. The experimental techniques and results

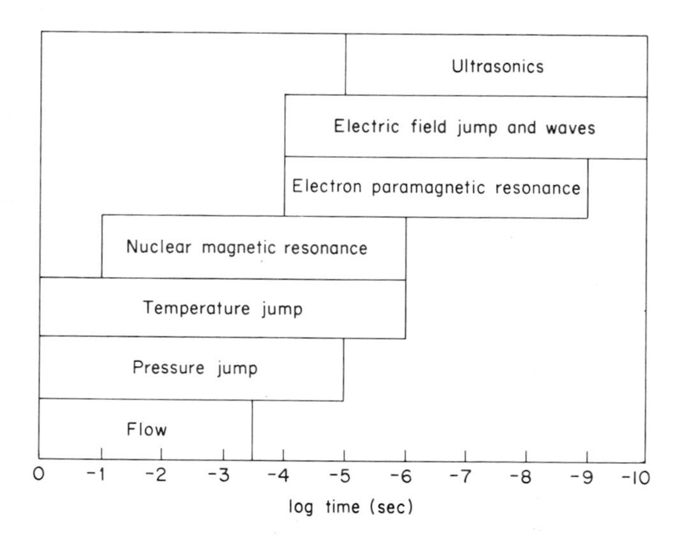

FIG. 1. Summary of the time ranges of fast reaction techniques which have been used to study enzymic reactions and related processes. Magnetic resonance methods are not discussed in this chapter.

16. M. Eigen, *in* "Fast Reactions and Primary Processes in Chemical Kinetics" (S. Claesson, ed.), p. 333. Wiley (Interscience), New York, 1967.
17. M. Eigen and G. G. Hammes, *Advan. Enzymol.* **25**, 1 (1963).
18. G. G. Hammes, *Advan. Protein Chem.* **23**, 1 (1968).
19. G. G. Hammes, *Accounts Chem. Res.* **1**, 321 (1968).
20. M. Eigen and L. de Maeyer, *in* "Rapid Mixing and Sampling Techniques in Biochemistry" (B. Chance *et al.*, eds.), p. 175. Academic Press, New York, 1964.
21. G. Czerlinski, *in* "Rapid Mixing and Sampling Techniques in Biochemistry" (B. Chance *et al.*, eds.), p. 183. Academic Press, New York, 1964.
22. J. E. Erman and G. G. Hammes, *Rev. Sci. Instr.* **37**, 746 (1966).

have been recently reviewed (*15–19*, *23*) and are not considered in detail. Instead an attempt has been made to give a comprehensive description of major approaches to the kinetic treatment of transient states in enzyme kinetics. By necessity and intention both the theoretical treatments and examples discussed are selective in nature. Section II treats kinetic systems far from equilibrium. Some exact and approximate solutions to rate equations are considered, and the use of computers for solving rate equations is discussed. In Section III, kinetic systems near equilibrium are treated. The general theory is developed along with consideration of some particularly useful cases. Throughout this chapter an attempt has been made to develop examples which are directly relevant to enzymic mechanisms. Finally, in Section IV, a few selected experimental results are presented to illustrate the type of information which can be obtained about enzymic reactions by use of rapid reaction techniques.

II. Kinetic Systems Far from Equilibrium

In this section some methods of treatment and analysis of kinetic systems which are not restricted to being near equilibrium are considered. Although integrated rate laws are most convenient to use, very few complex mechanisms exist for which integrated rate laws can be obtained; therefore, much of this section is devoted to various mathematical approximation techniques, as well as computer methods, for solving complex rate equations.

A. Integrated Rate Expressions

In general, reaction mechanisms involving only unimolecular processes are the only multistep mechanisms which can be analytically treated without introducing one or more approximations such as steady state assumptions and restrictions to the early or late phase of the reaction. Such mechanisms constitute an important class of biological reactions. Consider, for example, the sequential unimolecular mechanism

$$X_0 \underset{k_{-1}}{\overset{k_1}{\rightleftharpoons}} X_1 \underset{k_{-2}}{\overset{k_2}{\rightleftharpoons}} \cdots \underset{k_{-i}}{\overset{k_i}{\rightleftharpoons}} X_i \underset{k_{-(i+1)}}{\overset{k_{(i+1)}}{\rightleftharpoons}} \cdots \underset{k_{-n}}{\overset{k_n}{\rightleftharpoons}} X_n \tag{1}$$

23. K. Kustin, ed., "Methods in Enzymology," Vol. 16, Academic Press, New York, 1969.

The species X_i might represent different conformational states of a protein, or a polypeptide or polynucleotide chain. The differential equations which govern the kinetic behavior of this mechanism are

$$-\frac{d(X_0)}{dt} = k_1(X_0) - k_{-1}(X_1)$$

$$\vdots$$

$$-\frac{d(X_i)}{dt} = -k_i(X_{i-1}) + (k_{-i} + k_{(i+1)})(X_i) - k_{-(i+1)}(X_{i+1})$$

$$\vdots$$

$$-\frac{d(X_n)}{dt} = -k_n(X_{n-1}) + k_{-n}(X_n) \tag{2}$$

Equation (2) constitutes a set of $n + 1$ linear first-order differential equations in $n + 1$ unknowns; because of the condition of mass conservation [(X_T) = total concentration]

$$(X_T) = \sum_{i=0}^{n} (X_i) \tag{3}$$

one of the variables in Eq. (2) may be eliminated. The resulting n-linear differential equations may then be solved by standard methods of linear algebra, and the solutions are of the form

$$(X_i) = (\bar{X}_i) + \sum_{j=1}^{n} A_{ij}e^{-\lambda_j t} \quad (i = 0, 1, 2 \ldots n) \tag{4}$$

where $\bar{X}_i$ is the final (time-independent) equilibrium value of X_i, A_{ij} is a constant which depends upon initial conditions, and λ_j is a reciprocal time constant which is a known (in principle) function of the rate constants. In particular, for the case $n = 1$

$$(X_1) = (\bar{X}_1) + [(X_1)^0 - (\bar{X}_1)]e^{-\lambda t} \tag{5}$$

where

$$\lambda = (k_1 + k_{-1})$$

and $(X_1)^0$ is the value of X_1 at $t = 0$. For the cases where $n > 1$, the λ_j's are obtained by solving a secular equation. Experimental determination of all of the λ_j's, together with a knowledge of the equilibrium constants, permits evaluation of all $2n$ rate constants.

The mathematical techniques required to solve the above case are discussed in detail in Section III, where rate equations near equilibrium are considered. [See also standard textbook discussions of unimolecular reactions (*24, 25*).] The point to be emphasized here is that an analytical solution of the rate equations governing the mechanism of Eq. (1) is possible.

Consider now the bimolecular reaction between an enzyme E and a ligand S

$$\mathrm{E} + \mathrm{S} \underset{k_{-1}}{\overset{k_1}{\rightleftharpoons}} \mathrm{X} \tag{6}$$

where S may be a substrate, effector, inhibitor, etc. The rate law for this mechanism is

$$-\frac{d(\mathrm{S})}{dt} = k_1(\mathrm{E})(\mathrm{S}) - k_{-1}(\mathrm{X}) \tag{7}$$

Use of the mass conservation conditions

$$\begin{aligned} (\mathrm{E})_0 &= (\mathrm{E}) + (\mathrm{X}) \\ (\mathrm{S})_0 &= (\mathrm{S}) + (\mathrm{X}) \end{aligned} \tag{8}$$

permits Eq. (7) to be integrated. The form of the solution, however, is rather complex. For the case that $(\mathrm{E})_0 = (\mathrm{S})_0$ and $(\mathrm{X}) = 0$ at $t = 0$

$$\ln\left[\frac{2(\mathrm{S}) + (k_{-1}/k_1)(1+\alpha)}{2(\mathrm{S}) + (k_{-1}/k_1)(1-\alpha)}\right] = k_{-1}\alpha t + \text{const}, \quad \alpha = [1 + 4(\mathrm{S})_0 k_1/k_{-1}]^{1/2} \tag{9}$$

If the equilibrium constant is known, Eq. (9) can be readily used to obtain the rate constants from the experimental data. The integrated rate equations for more general boundary conditions are even more complex and are not extremely useful. However, Eq. (7) can be readily integrated numerically with a digital computer and both rate constants can be extracted from the data. This procedure has been used in a kinetic investigation of the interaction of cyanide and fluoride with peroxidase (*26*). Discussion of the general procedures for numerical integrations is deferred until later in this section.

Although integrated rate equations can be obtained for more complex reactions than those of Eqs. (1) and (6) in special situations, these mechanisms are essentially the only types involving reversible reactions

24. S. W. Benson, "The Foundations of Chemical Kinetics," pp. 39–42. McGraw-Hill, New York, 1960.
25. A. A. Frost and R. G. Pearson, "Kinetics and Mechanism," 2nd ed., pp. 173–177. Wiley, New York, 1960.
26. W. D. Ellis and H. B. Dunford, *Biochemistry* **7**, 2054 (1968).

which can be handled analytically. The kinetic equations governing more complex mechanisms generally must be solved by introduction of various approximations.

B. Single Substrate–Single Product Michaelis–Menten Mechanism

A general mechanism for an enzyme-catalyzed single substrate–single product reaction is (*9*)

$$\mathrm{E} + \mathrm{S} \underset{k_{-1}}{\overset{k_1}{\rightleftharpoons}} \mathrm{X}_1 \underset{k_{-2}}{\overset{k_2}{\rightleftharpoons}} \cdots \underset{k_{-n}}{\overset{k_n}{\rightleftharpoons}} \mathrm{X}_n \underset{k_{-(n+1)}}{\overset{k_{n+1}}{\rightleftharpoons}} \mathrm{E} + \mathrm{P} \tag{10}$$

where E is the enzyme, S is the substrate, P is the product, and the X_i are binary intermediate complexes. The kinetics of this reaction scheme is governed by the $n + 3$ differential equations

$$\begin{aligned}
-\frac{d(\mathrm{S})}{dt} &= k_1(\mathrm{E})(\mathrm{S}) - k_{-1}(\mathrm{X}_1) \\
-\frac{d(\mathrm{E})}{dt} &= k_1(\mathrm{E})(\mathrm{S}) + k_{-(n+1)}(\mathrm{E})(\mathrm{P}) - k_{-1}(\mathrm{X}_1) - k_{(n+1)}(\mathrm{X}_n) \\
-\frac{d(\mathrm{X}_1)}{dt} &= -k_1(\mathrm{E})(\mathrm{S}) + (k_{-1} + k_2)(\mathrm{X}_1) - k_{-2}(\mathrm{X}_2) \\
&\vdots \\
-\frac{d(\mathrm{X}_i)}{dt} &= -k_i(\mathrm{X}_{(i-1)}) + (k_{-i} + k_{(i+1)})(\mathrm{X}_i) - k_{-(i+1)}(\mathrm{X}_{i+1}) \\
&\vdots \\
-\frac{d(\mathrm{P})}{dt} &= -k_{(n+1)}(\mathrm{X}_n) + k_{-(n+1)}(\mathrm{E})(\mathrm{P})
\end{aligned} \tag{11}$$

Use of the mass conservation equations

$$(\mathrm{E})_0 = (\mathrm{E}) + \sum_{i=1}^{n} (\mathrm{X}_i), \qquad (\mathrm{S})_0 = (\mathrm{S}) + (\mathrm{P}) + \sum_{i=1}^{n} (\mathrm{X}_i) \tag{12}$$

permits elimination of two of the variables in Eqs. (11) with the result that $n + 1$ differential equations in $n + 1$ unknowns are obtained. Because the first three and last two equations of Eqs. (11) are nonlinear, there is no general solution to this set of differential equations, even when $n = 1$. However, solutions can be obtained under certain conditions: for example, when $n = 1$ and $k_1 = k_{-2}$; when the reaction is in the very early stages; when all of the binary enzyme intermediates are in a steady state; and when the system is near equilibrium. Each of these situations is considered below, the last being treated in Section III.

1. *General Solutions to the One Intermediate Mechanism*

The exact solution to Eqs. (11) can be obtained under the special circumstances that $n = 1$ and $k_1 = k_{-2}$. The latter stipulation is precisely the condition that the Michaelis constant for the forward reaction, $K_S = (k_{-1} + k_2)/k_1$, equals that for the reverse reaction, $K_P = (k_{-1} + k_2)/k_{-2}$. The solution to Eqs. (11) is then given by (*27*)

$$(X_1) = \frac{2k_1(E)_0(S)_0(1 - e^{-\lambda_1 t})}{(\lambda_1 - \lambda_2) + (\lambda_1 + \lambda_2)e^{-\lambda_1 t}} \tag{13a}$$

and

$$(P) = \frac{2k_2(S)_0}{(k_{-1} + k_2)[(\lambda_1 - \lambda_2) + (\lambda_1 + \lambda_2)e^{-\lambda_1 t}]}[\lambda_1(1 - e^{-\lambda_3 t}) - \lambda_3(1 - e^{-\lambda_1 t})] \tag{13b}$$

where

$$\lambda_1 = \{(k_1[(E)_0 + (S)_0] + k_{-1} + k_2)^2 - 4k_1^2(E)_0(S)_0\}^{1/2}$$

$$\lambda_2 = -\{k_1[(E)_0 + (S)_0] + k_{-1} + k_2\}$$

$$\lambda_3 = \frac{k_1(E)_0[2k_1(S)_0 + \lambda_2 - \lambda_1]}{\lambda_2 - \lambda_1}$$

The boundary conditions $(S) = (S)_0$ and $(P) = 0$ at $t = 0$ have been employed in arriving at Eqs. (13). The rates of approach to equilibrium of X and P for some arbitrarily selected values of $(E)_0$, $(S)_0$, and the rate constants are shown in Fig. 2. The concentration of X_1 rises rapidly and then approaches a constant value corresponding to its steady state level. The concentration of P increases continuously and the term $e^{-\lambda_3 t}$ corresponds to the steady state production of P from S. The steady state rate law can be derived if the assumption is made that $(E)_0 \ll (S)_0$. Equations (13) can then be written as

$$(X_1) = k_1(E)_0(S)_0(1 - e^{-\lambda_1 t})/\lambda_1 \tag{14a}$$

and

$$(P) = \frac{k_2(S)_0}{k_{-1} + k_2}[(1 - e^{-\lambda_3 t}) - (\lambda_3/\lambda_1)(1 - e^{-\lambda_1 t})] \tag{14b}$$

where

27. W. G. Miller and R. A. Alberty, *JACS* **80**, 5146 (1958).

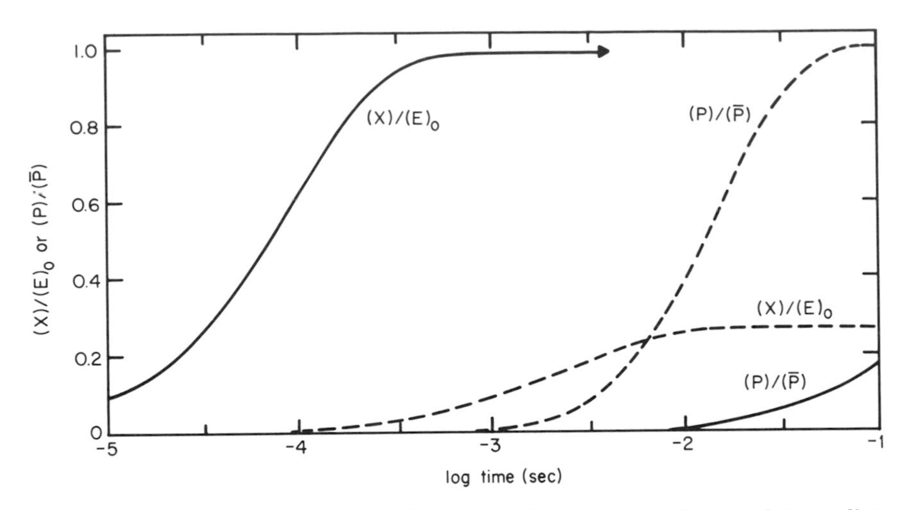

FIG. 2. Rate of approach to equilibrium of the species in the one intermediate Michaelis–Menten mechanism. The quantities $(X)/(E)_0$ and $(P)/(\bar{P})$ are plotted as functions of time, where $(\bar{P})$ is the final equilibrium concentration of P. These curves were calculated according to Eqs. (13a) and (13b) assuming that $k_1 = 10^8$ M^{-1} $\sec^{-1}$, $k_{-1} = k_2 = 10^2$ $\sec^{-1}$, $(E)_0 = 10^{-6}\,M$, and $(S)_0 = 10^{-4}\,M$ (solid curves) or $10^{-6}\,M$ (dashed curves).

$$\lambda_1 = [k_1(S)_0 + k_{-1} + k_2] \qquad \text{and} \qquad \lambda_3 = \frac{k_1(E)_0(k_{-1} + k_2)}{k_1(S)_0 + k_{-1} + k_2}$$

Since $(E)_0 \ll (S)_0$, $\lambda_1 \gg \lambda_3$, and when $t \gg \lambda_1^{-1}$, (X_1) is in a steady state and Eq. (14b) reduces to the usual Michaelis–Menten integrated steady state equation (*27*). Upon substitution of the minimum values of the rate constants given by Frieden *et al.* for the fumarase reaction (*28*) together with $(E)_0 = 5 \times 10^{-10}\,M$ and $(S)_0 = 10^{-4}\,M$, X_1 is found to reach 0.001% of its steady state value within 10^{-3} sec.

When $k_1 \neq k_{-2}$ an analytical solution cannot be obtained. However, a perturbation solution expressed as a power series in $(k_1 - k_{-2})$ may be obtained (*27*). The coefficients in this series are extremely complicated quantities, however, each of which involves sums of various complicated integrals. Although this solution is itself quite unwieldly, it can be used in conjunction with a computer to perform a comprehensive analysis of the one intermediate mechanism.

This rather detailed consideration of the simplest possible Michaelis–Menten mechanism serves to illustrate several points. First, a convenient general analytical solution cannot even be obtained in this case. Second, even the approximate solutions are quite complex and of little practical utility. Finally, the time constants describing the transient behavior of

28. C. Frieden, R. Wolfe, Jr., and R. A. Alberty, *JACS* **79**, 1524 (1957).

intermediates for typical enzymes are so short that they cannot be measured with conventional techniques.

2. *Early Phase Solutions*

A number of mathematical treatments of the early period (transient or pre-steady state stage) of simple enzymic mechanisms have been presented (*29–35*). Early phase solutions to Eqs. (11) have been obtained for the case $k_{-(n+1)} \approx 0$. If $n = 1$ and $(P) = d(P)/dt = 0$ at $t = 0$ (*31*),

$$(P) = (S)_0 \left(1 - \frac{\lambda_2 e^{-\lambda_1 t} - \lambda_1 e^{-\lambda_2 t}}{\lambda_2 - \lambda_1}\right) \tag{15}$$

where λ_1 and λ_2 are the two roots $\lambda_{1,2}$ of the quadratic

$$\lambda_{1,2} = \frac{[(E)_0 + (S)_0 + K_S] \pm \{[(E)_0 + (S)_0 + K_S]^2 - 4k_2(E)_0/k_1\}^{1/2}}{(2/k_1)} \tag{16}$$

This result is valid only at early times, but if λ_1 or λ_2 can be measured, the three rate constants can be determined since K_S and $k_2(E)_0$ (the maximal velocity) can be determined from conventional steady state measurements.

For cases where $n > 1$, expressions for $P(t)$ have only been obtained under quite restrictive conditions, i.e., $(E)_0 \ll (S)_0$ as well as $k_{-(n+1)} = 0$ (*30, 33, 35*). Even in these instances the results are so complex that they have not been extensively used. However, in all cases a very short transient period is indicated prior to the establishment of a steady state period.

3. *Steady State Solutions*

From the foregoing discussion it is apparent that if $(E)_0 \ll (S)_0$, the transient period is very short and is followed by a relatively long steady state period. This latter period, in fact, is equivalent to the situation usually treated where all of the enzyme species are assumed to be in a steady state. A single rate equation then governs the system (*4*):

$$\frac{d(P)}{dt} = \frac{V_S/K_S(S) - V_P/K_P(P)}{1 + (S)/K_S + (P)/K_P} \tag{17}$$

29. B. Chance, *JBC* **151,** 553 (1943).
30. H. Gutfreund, *Discussions Faraday Soc.* **20,** 167 (1955).
31. M. F. Morales and D. E. Goldman, *JACS* **77,** 6069 (1955).
32. K. J. Laidler, *Can. J. Chem.* **33,** 1614 (1955).
33. L. Ouellet and K. J. Laidler, *Can. J. Chem.* **34,** 146 (1956).
34. P. A. T. Swoboda, *BBA* **23,** 70 (1957).
35. L. Ouellet and J. A. Stewart, *Can. J. Chem.* **37,** 737 (1959).

where V_S and V_P are the maximal velocities in the forward and reverse directions, respectively, and K_S and K_P are the Michaelis constants for substrate and product, respectively. These parameters are complicated aggregates of rate constants (*9*):

$$V_S = \frac{(E)_0}{\Sigma_{i=1}^{N} \Sigma_{s=i}^{N} K_i^s/k_{(s+1)}} \tag{18a}$$

$$K_S = \frac{\Sigma_{s=0}^{N} K_0^s/k_{(s+1)}}{\Sigma_{i=1}^{N} \Sigma_{s=i}^{N} K_i^s/k_{(s+1)}} \tag{18b}$$

$$V_P = \frac{(E)_0}{\Sigma_{i=1}^{N} \Sigma_{s=0}^{i-1} K_i^s/k_{(s+1)}} \tag{18c}$$

$$K_P = \frac{\Sigma_{s=0}^{N} K_0^s/k_{(s+1)}}{K_0^{n+1} \Sigma_{i=1}^{N} \Sigma_{s=0}^{i-1} K_i^s/k_{(s+1)}} \tag{18d}$$

where

$$K_i^s = \prod_{r=i+1}^{s} \frac{k_{-r}}{k_r} = \frac{(\bar{X}_i)}{(\bar{X}_s)}, \qquad K_0^{n+1} = \prod_{r=1}^{n+1} \frac{k_{-r}}{k_r}$$

and the overbar denotes equilibrium concentration. Conventional steady state experiments using either initial velocity measurements or an integrated form of Eq. (17) can be used to obtain the four steady state parameters. Clearly, since the form of the rate law is independent of the number of intermediates in the mechanism, the number of elementary steps involved cannot be deduced from steady state rate measurements alone (*9*). Furthermore, individual rate constants cannot be ascertained except when $n = 1$; in this case the four rate constants can be deduced from the four kinetic parameters. When $n > 1$, lower bounds to the rate constants can be calculated (*9*):

$$k_1 \geq \frac{V_S + V_P}{K_S(E)_0} \tag{19a}$$

$$k_{-(n+1)} \geq \frac{V_S + V_P}{K_P(E)_0} \tag{19b}$$

$$k_{(i+1)} \geq V_S/(E)_0, \quad i \neq 0 \tag{19c}$$

$$k_{-i} \geq V_P/(E)_0, \quad i \neq n + 1 \tag{19d}$$

Thus, although the steady state solution to the rate equations is rather simple in form, detailed information about the elementary steps of the mechanism is obscured. This brief discussion of steady state kinetics is only intended to establish some perspective for the relative merits of steady state vs. transient measurements. For more comprehensive considerations of steady state enzyme kinetics, see references *4–9, 36–38*, and Chapter 1 of this volume by Cleland.

4. *Concluding Remarks*

The above discussion illustrates the difficulties encountered in obtaining integrated rate laws for complex reaction mechanisms and, in particular, even for the relatively simple Michaelis–Menten mechanism. Although only a single substrate–single product mechanism has been considered, it should be apparent that an analysis of multisubstrate reactions would be more difficult but similar in nature. Thus, the analysis of kinetic systems far from equilibrium is hampered by the inherent mathematical limitations of treating transient states. In fact the treatments that can be made frequently involve so many restrictions and approximations that detailed kinetic information is sacrificed: Often only aggregates of rate constants are obtainable, and the number of elementary steps in the mechanism cannot be deduced from the approximate rate expressions. These shortcomings can be considerably overcome by obtaining numerical solutions to the differential equations with an analog or digital computer.

C. Computer Solutions to Rate Equations

The rapid development of high-speed computers has had a considerable impact on the treatment and analysis of kinetic systems. Computer applications to kinetic systems fall into two broad categories: statistical analyses of experimental data and simulations and analyses of complex mechanisms. Only the latter application is considered here [cf. Cleland (*39*) for a discussion of statistical analysis of steady state data]. Simulations are used, for example, to investigate the kinetic behavior of intermediates in complex mechanisms, where the values of rate constants for the various steps are at least approximately known. The computer numerically integrates the differential equations which govern the kinetic behavior of the mechanism. Alternatively, a computer may also be used to find a mechanism and rate constants which will quantitatively account for a given set of experimental data.

Two types of computing machines are available for kinetic investiga-

36. V. Bloomfield, L. Peller, and R. A. Alberty, *JACS* **84,** 4367 (1962).
37. F. M. Huennekens, *Tech. Org. Chem.* **8,** Part II, 1231 (1963).
38. M. Dixon and E. C. Webb, "Enzymes," 2nd ed. Academic Press, New York, 1964.
39. W. W. Cleland, *Advan. Enzymol.* **29,** 1 (1967).

tions: the analog and digital computers. An analog computer is a device in which certain physical variables serve as analogs or models for the actual variables of interest. These variables satisfy a known set of equations which are analogous to the equations which are to be solved. In an electronic analog computer the variables are voltages (dependent variables) and time (independent variables). In a kinetic application, various voltages may correspond, for example, to concentrations of certain chemical species. The computer is constructed so that the voltages obey the same differential equations as those for the corresponding concentration variables in the reaction mechanism.

For illustrative purposes, consider the analog computer solution to the rate law for the one intermediate ($n = 1$) Michaelis–Menten equation [Eqs. (11)]. Concentrations (E), (S), (X_1), and (P) are replaced by variable voltages V_E, V_S, V_{X_1}, and V_P that are functions of computer time t, which is a constant multiple of the real time. The total substrate concentration and total enzyme concentration are assigned constant voltages V_{S_0} and V_{E_0}, respectively, and constant multipliers α_i replace rate constants k_i. The electric analog of the two independent rate equations is

$$-\frac{dV_S}{dt} = \alpha_1(V_{E_0} - V_{X_1})V_S - \alpha_{-1}V_{X_1}$$

$$-\frac{dX_1}{dt} = -\alpha_1 V_S(V_{E_0} - V_{X_1}) - \alpha_{-2}(V_{E_0} - V_{X_1})(V_{S_0} - V_{X_1} - V_S) + (\alpha_{-1} + \alpha_2)V_{X_1}$$

The computer is started with a given set of initial conditions. For example, at $t = 0$, V_S could be set equal to V_{S_0} and V_{X_1} to 0. The solution is then generated and voltages V_S and V_{X_1} are displayed as a function of time on an oscilloscope. Initial conditions and constant multipliers (rate constants) may readily be varied in order to investigate their effects on the transient behavior of the chemical species and to attempt to fit experimental data to a given mechanism. This procedure is readily extended to multiple intermediate mechanisms. For example, Fig. 3 displays some simulated concentration–time curves for a Michaelis–Menten mechanism with $n = 4$ and $k_{-i} = 0$ for all i. The concentrations of all of the intermediates approach approximately constant levels after a short time; at sufficiently long times they will go to zero although this is not shown on the figure. Analog computer analyses of experimental data have been extensively applied to the reactions catalyzed by peroxidase and catalase (cf. *40, 41*).

The principle of operation of a digital computer is quite distinct from

40. J. Higgins, *Tech. Org. Chem.* **8,** Part I, 285 (1961).
41. B. Chance, *Science* **92,** 455 (1940).

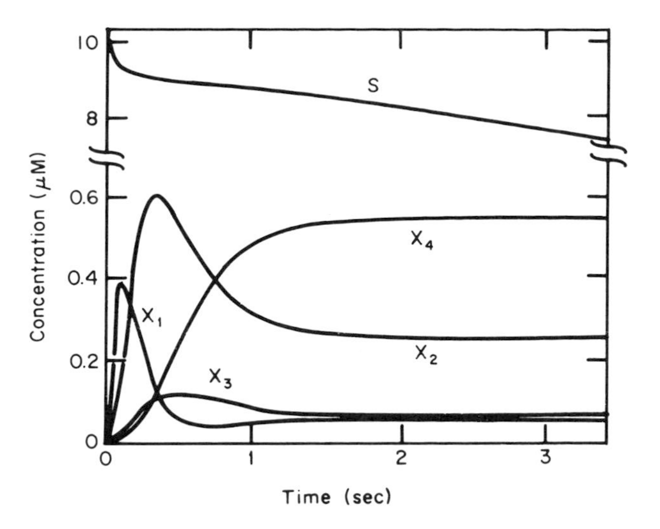

FIG. 3. Typical analog computer solutions for the four intermediate Michaelis–Menten mechanism [Eq. (10), $n = 4$]. The parameters used were $(S)_0 = 10\ \mu M$, $(E)_0 = 1\ \mu M$, $k_1 = 0.9 \times 10^6\ M^{-1}\ sec^{-1}$, $k_2 = 9\ sec^{-1}$, $k_3 = 1\ sec^{-1}$, $k_4 = 9\ sec^{-1}$, $k_5 = 2\ sec^{-1}$, and $k_{-i} = 0$ for all i. (Courtesy of Professor J. Higgins, cf. *40*.)

that of the analog computer. In addition to the ability to perform standard arithmetic operations, the digital computer possesses a storage or "memory" unit. Because of the memory capacity, the digital computer can solve a much wider range of complex problems than can the analog computer. The differential rate equations for mechanisms of virtually any degree of complexity can be numerically integrated to a high degree of accuracy. The results are usually typed on paper with a high-speed automatic printer, but many other output methods are also available [cf. Dickson (*42*) for a detailed discussion of digital computers].

The main problem confronting the user is the choice of the method of numerical integration. A large number of methods exist, for example, the Euler procedure, the Adams method, the Milne methods, and the various Runge-Kutta techniques (*43*). Each method possesses its own particular advantages and disadvantages, and the one of choice is usually dictated by the particular problem under investigation. Generally, the user seeks to maximize the precision without using an excessive amount of computer time. Relatively simple procedures such as the modified

42. T. R. Dickson, "The Computer and Chemistry." Freeman, San Francisco, California, 1968.

43. F. B. Hildebrand, "Introduction to Numerical Analysis." McGraw-Hill, New York, 1956.

(first-order) Euler method (*43*), or one described recently by DeTar and DeTar (*44*), appear to be sufficiently accurate and fast for most kinetic problems.

As an illustration, a procedure for numerical integration of Eqs. (11) for the case $n = 2$ is constructed. Integration of the three rate equations is accomplished most easily by computing the change Δ^m in the concentration of the mth species occurring in a small time interval Δt. The sum of all such changes occurring in all time intervals between $t = 0$ and $t = t$, together with the initial conditions, gives the concentration of the mth species at any time t. Starting at $t = 0$, with $(S) = (S)_0$ and $(E) = (E)_0$, the change $\Delta_1{}^m$ in the concentration of the mth species after the first time interval Δt is

$$\Delta_1{}^s = -k_1(E)_0(S)_0\Delta t, \qquad \Delta_1^{x_1} = k_1(E)_0(S)_0\Delta t = -\Delta_1{}^s, \qquad \Delta_1{}^p = 0$$

In the jth time interval, at $t = j\Delta t$, these changes are

$$\Delta_j{}^s = [-k_1(E)_{j-1}(S)_{j-1} + k_{-1}(X_1)_{j-1}]\Delta t, \quad j > 1$$
$$\Delta_j^{x_1} = [k_1(E)_{j-1}(S)_{j-1} - (k_{-1} + k_2)(X_1)_{j-1} + k_{-2}(X_2)_{j-1}]\Delta t, \quad j > 1$$
$$\Delta_j{}^p = [-k_{-3}(E)_{j-1}(S)_{j-1} + k_3(X_2)_{j-1}]\Delta t, \quad j > 1$$

where the subscript $j - 1$ on the concentration variables refers to concentrations at $t = (j - 1)\Delta t$. The concentrations of the various species at time $t = j\Delta t$ is then given by

$$(S)_j = (S)_0 + \sum_{k=1}^{j} \Delta_k{}^s, \qquad (X_1)_j = \sum_{k=1}^{j} \Delta_k^{x_1}, \qquad (P)_j = \sum_{k=1}^{j} \Delta_k{}^p \tag{20}$$

Concentrations of E and X_2 may be computed from the mass conservation equations. The time interval Δt must be short enough so that the solution becomes reasonably insensitive ($\pm 1\%$) to further decreases in its length. In practice, such a simple approach is not generally used because of inherent limitations in accuracy. Instead, the modified (first-order) Euler procedure is frequently employed (cf. *43*). This method of integration is fast and reasonably accurate.

Extremely complex mechanisms have been investigated with a digital computer. One of the most ambitious series of undertakings has been the studies of complex metabolic pathways. For example, Kerson *et al.* have conducted simulation studies of mammalian pyruvate kinase (*45*), and Garfinkel and Hess have simulated the glycolytic and respiratory pathway of Ehrlich ascite tumor cells (*46*). In the latter case, 89 reactions

44. D. F. DeTar and C. E. DeTar, *J. Phys. Chem.* **70,** 3842 (1966).
45. L. A. Kerson, D. Garfinkel, and A. S. Mildvan, *JBC* **242,** 2124 (1967).
46. D. Garfinkel and B. Hess, *JBC* **239,** 971 (1964).

involving 65 chemical species (enzymes, substrates, and intermediates) were considered. Because of the great speed and capacity of modern digital computers, very complex mechanisms can now be analyzed, and in fact digital computers have generally supplanted analog computers in the analysis of rate equations.

III. Kinetic Systems near Equilibrium

The analysis of kinetic systems near equilibrium is now considered. The time-dependent behavior of systems near equilibrium is generally described as chemical relaxation, and an exact treatment of such phenomena is possible. In Section III,A general solutions to the kinetic problem are presented and the calculation of kinetic constants (relaxation times) for simple and complex mechanisms are presented. Convenient methods for the analysis and interpretation of experimental results are also discussed. The amplitudes of relaxation processes are discussed in Section III,B. First a transformation to normal coordinates is used to obtain the general solution to coupled rate equations near equilibrium. A thermodynamic formulation for chemical relaxation is then given, and this is followed by explicit calculations of the amplitudes of relaxation processes.

A. Relaxation Spectra

1. *One-Step Mechanisms*

The simplest mechanism displaying the important features of rate processes near equilibrium is a bimolecular reaction such as that given by Eq. (6), e.g., the combination of enzyme and substrate. The rate law near equilibrium may be obtained by introducing new variables:

$$(\mathrm{E}) = (\mathrm{E}^0) + (\Delta\mathrm{E}), \qquad (\bar{\mathrm{E}}) = (\mathrm{E}^0) + (\Delta\bar{\mathrm{E}}), \quad \text{etc., for (S) and (X)} \tag{21}$$

where (E^0) is a time-independent reference concentration (e.g., the stoichiometric concentration), the Δ's denote deviations from reference concentrations, and the overbar denotes equilibrium values of the variables. In terms of these new variables, the rate law for this mechanism is

$$\frac{d(\mathrm{X})}{dt} = \frac{d\Delta(\mathrm{X})}{dt} = k_1\{(\bar{\mathrm{E}}) + [\Delta(\mathrm{E}) - \Delta(\bar{\mathrm{E}})]\}\{(\bar{\mathrm{S}}) + [\Delta(\mathrm{S}) - \Delta(\bar{\mathrm{S}})]\} - k_{-1}\{(\bar{\mathrm{X}}) + [\Delta(\mathrm{X}) - \Delta(\bar{\mathrm{X}})]\} \tag{22}$$

This equation may be simplified by making use of the relationships $k_1(\bar{E})(\bar{S}) = k_{-1}(\bar{X})$ (condition for equilibrium), and $\Delta(S) - \Delta(\bar{S}) = (S) - (\bar{S}) = \Delta(E) - \Delta(\bar{E}) = -[\Delta(X) - \Delta(\bar{X})]$ (mass conservation):

$$\frac{d\Delta(X)}{dt} = -\{k_1[(\bar{E}) + (\bar{S})] + k_{-1}\}[\Delta(X) - \Delta(\bar{X})] + k_1[\Delta(X) - \Delta(\bar{X})]^2 \tag{23}$$

Near equilibrium the quadratic term may be assumed to be negligible compared to the other terms in the rate law since $\Delta(X) - \Delta(\bar{X}) = (X) - (\bar{X})$ is very small compared to $(\bar{X})$. In fact this may be regarded as a definition of the term *near equilibrium*. Equation (23) then becomes

$$\frac{d\Delta(X)}{dt} = -\frac{\Delta(X)}{\tau} + \frac{\Delta(\bar{X})}{\tau} \tag{24}$$

where

$$\frac{1}{\tau} = k_1[(\bar{E}) + (\bar{S})] + k_{-1} \tag{25}$$

Thus near equilibrium the rate law for a second-order reaction can be described by a linear first-order differential equation. The relaxation time τ is a function of the equilibrium concentrations, and its variation with concentration can be used to determine the two rate constants.

2. *Transient and Stationary Solutions of Rate Equations*

Integration of Eq. (24) gives

$$\Delta[X(t)] = \Delta[X(0)]e^{-t/\tau} + \frac{e^{-t/\tau}}{\tau}\int_0^t e^{\theta/\tau}\Delta[\bar{X}(\theta)]d\theta \tag{26}$$

Two general types of solutions to this equation are of interest: the transient solution and the stationary solution. The transient solution is characteristic of when $\Delta(\bar{X})$ changes monotonically from an initial to a final value. The jump techniques—temperature jump, pressure jump, electric field jump—are typical examples of this. For example, if a temperature jump is applied to the system exponentially then $\Delta(\bar{X})$ would also change exponentially, i.e.,

$$\Delta[\bar{X}(t)] = \Delta[\bar{X}(\infty)](1 - e^{-kt}) \tag{27}$$

where k is a reciprocal time constant, typically about 10^6 sec^{-1}. Equation (26) can now be evaluated (with $\Delta[X(0)] = 0$) to give

$$\begin{aligned}\Delta[X(t)] &= \Delta[\bar{X}(\infty)]\left(1 - \frac{k\tau}{k\tau - 1}e^{-t/\tau} + \frac{e^{-kt}}{k\tau - 1}\right), \quad k\tau \neq 1 \\ &= \Delta[\bar{X}(\infty)][1 - (1 + t/\tau)e^{-t/\tau}], \quad k\tau = 1\end{aligned} \tag{28}$$

If the relaxation time is much shorter than the rise time of the perturbation, i.e., $k\tau \ll 1$, $\Delta(\mathrm{X})$ follows $\Delta(\bar{\mathrm{X}})$ exactly; whereas in the converse situations, $k\tau \gg 1$, $\Delta(\mathrm{X})$ changes exponentially to its final equilibrium value, i.e.,

$$\Delta[\mathrm{X}(t)] = \Delta[\bar{\mathrm{X}}(\infty)](1 - e^{-t/\tau}), \quad k\tau \gg 1 \tag{29}$$

Obviously, measurements of the relaxation time can only be made if $\tau > k^{-1}$. The behavior of $\Delta[\mathrm{X}(t)]/\Delta[\bar{\mathrm{X}}(\infty)]$ for various values of $k\tau$ is shown in Fig. 4.

The most useful stationary solution to Eq. (26) is obtained when a sinusoidal periodic perturbation is applied to the system, i.e.,

$$\Delta[\bar{\mathrm{X}}(t)] = \Delta(\bar{\mathrm{X}})_0 e^{i\omega t} \tag{30}$$

where ω is the radial frequency and $\Delta(\bar{\mathrm{X}})_0$ is a constant. Substitution of this equation into Eq. (26) and integration gives

$$\Delta[\mathrm{X}(t)] = \Delta[\mathrm{X}(0)]e^{-t/\tau} + \frac{\Delta(\bar{\mathrm{X}})_0}{1 + i\omega\tau}(e^{i\omega t} - e^{-t/\tau}) \tag{31}$$

For $t \gg \tau$ the transient term vanishes and

$$\Delta[\mathrm{X}(t)] = \frac{\Delta[\bar{\mathrm{X}}(t)]}{1 + i\omega t} = \frac{\Delta[\bar{\mathrm{X}}(t)]e^{-i\Phi}}{[1 + (\omega\tau)^2]^{1/2}} \tag{32}$$

where $\tan \Phi = \omega\tau$. This result indicates that $\Delta[\mathrm{X}(t)]$ oscillates with the

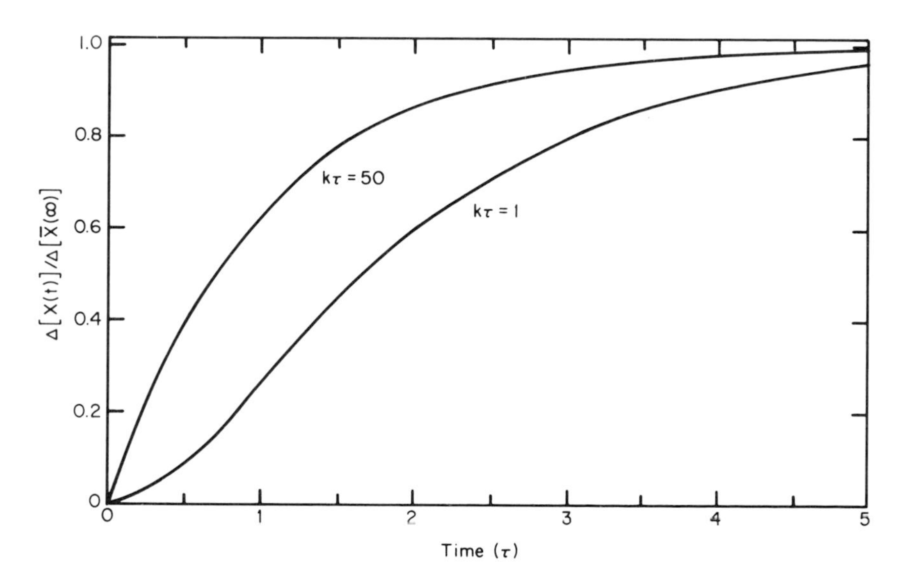

FIG. 4. Rate of approach to equilibrium according to Eqs. (28) and (29). $\Delta[\mathrm{X}(t)]/\Delta[\bar{\mathrm{X}}(\infty)]$ is plotted versus time (in units of τ) for $k\tau = 50$ and $k\tau = 1$.

same frequency as $\Delta[\bar{X}(t)]$ but lags behind by the phase angle Φ; moreover, the amplitude of $\Delta[X(t)]$ is decreased from that of $\Delta[\bar{X}(t)]$ by the factor $(1 + \omega^2\tau^2)^{-1/2}$. These features are illustrated in Fig. 5 for the following three limiting conditions. First, when the period of oscillation is much longer than the relaxation time ($\omega\tau \ll 1$), $\Phi \rightarrow 0$ and $\Delta(X)$ follows $\Delta(\bar{X})$ exactly. In the other extreme $\omega\tau \gg 1$, and the reaction is unperturbed by the wave $[\Delta(X) \rightarrow 0]$. Finally, if $\omega\tau = 1$, $\Phi = \pi/4 = 45°$, and $\Delta(X)$ lags 45° behind $\Delta(\bar{X})$, the amplitude of oscillation is decreased by $1/\sqrt{2}$. An extensive discussion of other types of periodic and transient functions of interest experimentally is given in Eigen and de Maeyer (*15*).

As a matter of convenience transient perturbations can be treated as step function perturbations with the final equilibrium concentrations being taken as reference concentrations, i.e., $(E^0) = (\bar{E})$, etc. Thus, after application of the step function perturbations $\Delta(\bar{X}) = 0$, and Eq. (24) be easily integrated to give

$$\Delta[X(t)] = \Delta[X(0)]e^{-t/\tau}$$

The results discussed thus far indicate the same relaxation time is applicable regardless of how the perturbation is applied. In fact, to a good approximation in dilute liquid solutions, the relaxation time is not dependent on the type of perturbation applied, although in principle the relaxation time depends slightly on the thermodynamic variables being held constant [cf. Section III,B and Eigen and de Maeyer (*15*) for further consideration of this point]. For the purpose of calculating relaxation times, it is most convenient to assume a step function perturbation and to set $\Delta(\bar{X}) = 0$. This will be done in the ensuing discussion, and it

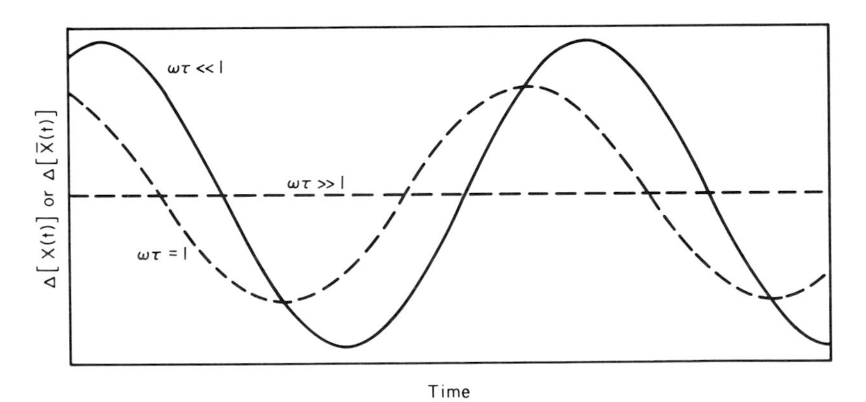

FIG. 5. $\Delta[X(t)]$ and $\Delta[\bar{X}(t)]$ versus time for different limits of $\omega\tau$, calculated according to Eq. (32). The solid curve represents $\Delta[\bar{X}(t)]$ and the dashed curves $\Delta[X(t)]$. When $\omega\tau \ll 1$ the curves coincide, and they are represented by the solid curve.

should not be construed as any limitation on the validity of the results obtained.

3. *Two-Step Mechanism*

The two-step mechanism

$$E + S \underset{k_{-1}}{\overset{k_1}{\rightleftharpoons}} X_1 \underset{k_{-2}}{\overset{k_2}{\rightleftharpoons}} X_2 \tag{33}$$

is frequently encountered; for example, an enzyme and substrate might form a complex X_1 which then undergoes a conformational change to X_2. The rate equations for this mechanism may be linearized near equilibrium as described above to give

$$\begin{aligned} -\frac{d\Delta(E)}{dt} &= \{k_1[(\bar{E}) + (\bar{S})] + k_{-1}\}\Delta(E) + k_{-1}\Delta(X_2) \\ &= a_{11}\Delta(E) + a_{12}\Delta(X_2) \end{aligned} \tag{34a}$$

$$\begin{aligned} -\frac{d\Delta(X_2)}{dt} &= k_2\Delta(E) + (k_{-2} + k_2)\Delta(X_2) \\ &= a_{21}\Delta(E) + a_{22}\Delta(X_2) \end{aligned} \tag{34b}$$

where the a_{ij}'s are defined by these equations. The solutions to a set of coupled first-order linear homogeneous differential equations is a sum of exponential terms, with the number of terms being equal to the number of independent rate equations. For the case under consideration the solutions can be written as

$$\begin{aligned} \Delta(E) &= A_{11}e^{-t/\tau_1} + A_{12}e^{-t/\tau_2} \\ \Delta(X_2) &= A_{21}e^{-t/\tau_1} + A_{22}e^{-t/\tau_2} \end{aligned} \tag{35}$$

The condition for a nontrivial solution can be shown to be

$$\begin{vmatrix} a_{11} - 1/\tau & a_{12} \\ a_{21} & a_{22} - 1/\tau \end{vmatrix} = 0 \tag{36}$$

or $$(1/\tau)^2 - (a_{11} + a_{22})(1/\tau) + (a_{11}a_{22} - a_{12}a_{21}) = 0.$$

The solutions to this quadratic equation are

$$\frac{1}{\tau_{1,2}} = \frac{(a_{11} + a_{22})}{2}\left\{1 \pm \left[1 - \frac{4(a_{11}a_{22} - a_{12}a_{21})}{(a_{11} + a_{22})^2}\right]^{1/2}\right\} \tag{37}$$

where τ_1 corresponds to the positive sign and τ_2 to the negative sign.

Thus a spectrum of two relaxation times describes the kinetic behavior near equilibrium of the mechanism of Eq. (33). The two relaxation times have different dependencies on concentration and consequently can readily be distinguished if a wide enough range of concentrations is investigated. All four rate constants can then be obtained.

Quite often the bimolecular step in Eq. (33) is much more rapid (nearly diffusion controlled) than the second step in the mechanism. In this case $k_1[(\bar{E}) + (\bar{S})] + k_{-1} \gg k_2 + k_{-2}$ or $a_{11} \gg a_{22}$ and the bracketed term in Eq. (37) may be expanded [recall $(1 - X)^{1/2} \approx 1 - X/2$, when $X \ll 1$] with the result

$$\frac{1}{\tau_1} = a_{11} = k_1[(\bar{E}) + (\bar{S})] + k_{-1} \tag{38a}$$

$$\frac{1}{\tau_2} = \frac{k_2}{1 + \{(k_{-1})/k_1[(\bar{E}) + (\bar{S})]\}} + k_{-2} \tag{38b}$$

The expression for $1/\tau_1$ is identical to that derived for the one-step mechanism [Eq. (25)]. At low concentrations of $(\bar{E})$ and $(\bar{S})$ both $1/\tau_1$ and $1/\tau_2$ are linear functions of $[(\bar{E}) + (\bar{S})]$, and consequently they are indistinguishable under such conditions. At high values of $(\bar{E}) + (\bar{S})$, however, $1/\tau_2$ becomes concentration independent and reaches the limiting value of $(k_2 + k_{-2})$ (which is the relaxation time for the isolated reaction $X_1 \rightleftarrows X_2$), while $1/\tau_1$ continues to increase. This type of concentration dependence is illustrated in Fig. 6.

The approximate expressions for the relaxation times given in Eqs. (38) can be derived without solving the secular determinant [Eq. (36)]. The expression for τ_1 is readily derived by assuming that the unimolecular step is sufficiently slow that the concentration of X_2 is fixed while

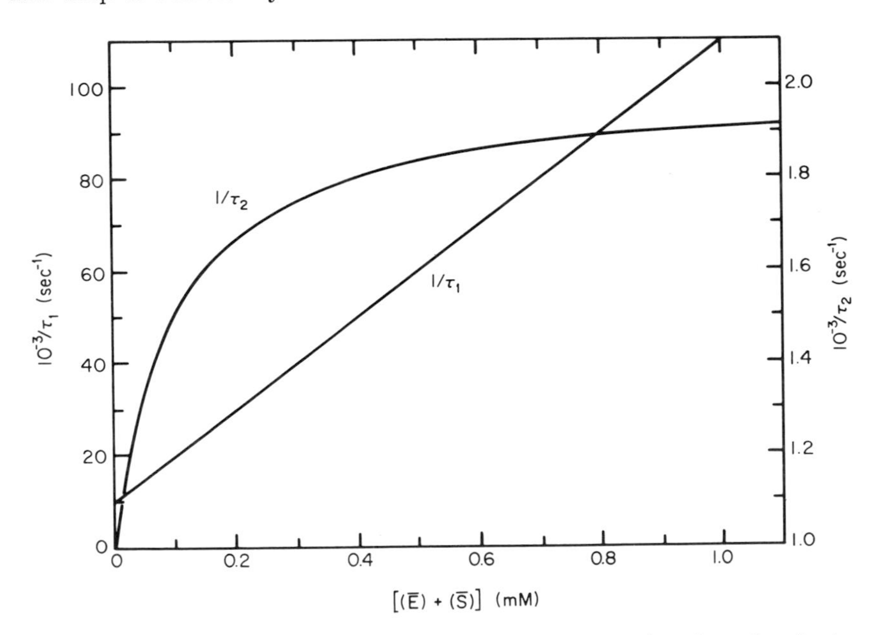

FIG. 6. Concentration dependence of the reciprocal relaxation times for the two-step mechanism of Eq. (33). Reciprocal relaxation times were calculated from Eqs. (38a) and (38b) assuming $k_1 = 10^8\ M^{-1}\ \text{sec}^{-1}$, $k_{-1} = 10^4\ \text{sec}^{-1}$, $k_2 = k_{-2} = 10^3\ \text{sec}^{-1}$.

the bimolecular step equilibrates. The second relaxation time may then be derived by assuming that the fast bimolecular step is always at equilibrium while the second step equilibrates; that is, the equilibrium constant relationship $k_1/k_{-1} = (\bar{X}_1)/[(\bar{E})(\bar{S})]$ can be differentiated to give $\Delta(X_1) = \{k_1[(\bar{E}) + (\bar{S})]/k_{-1}\}\ \Delta(E)$. Using this relationship with the mass conservation equation $\Delta(X_1) = -\Delta(E) - \Delta(X_2)$ enables Eq. (34b) to be written

$$-\frac{d\Delta(X_2)}{dt} = \frac{\Delta(X_2)}{\tau_2}$$

4. *Multistep Mechanisms*

The single substrate–single product Michaelis–Menten mechanism with an arbitrary number of intermediates [Eq. (10)] is a good example of a multistep mechanism of coupled reactions. The $n+1$ linearized rate equations for this mechanism can be written as

$$\begin{aligned}
-\frac{d\Delta(X_1)}{dt} &= a_{11}\Delta(X_1) + a_{12}\Delta(X_2) + \ldots + a_{1,n+1}\Delta(P) \\
-\frac{d\Delta(X_2)}{dt} &= a_{21}\Delta(X_1) + a_{22}\Delta(X_2) + \ldots + a_{2,n+1}\Delta(P) \\
&\ \ \vdots \\
-\frac{d\Delta(P)}{dt} &= a_{n+1,1}\Delta(X_1) + a_{n+1,2}\Delta(X_2) + \ldots + a_{n+1,n+1}\Delta(P) \qquad (39)
\end{aligned}$$

where the a_{ij}'s are functions of rate constants and equilibrium concentrations which may be obtained by methods outlined above. The solution to Eq. (39) is obtained by solving the secular equation

$$|\mathbf{a} - (1/\tau)\mathbf{I}| = 0 \qquad (40)$$

where $\mathbf{a}$ is the $n+1$ square matrix of the a_{ij}'s, $\mathbf{I}$ is the unit matrix, and $|\mathbf{a} - (1/\tau)\ \mathbf{I}|$ is the determinant of $\mathbf{a} - (1/\tau)\ \mathbf{I}$. Equation (40) possesses $n+1$ roots which correspond to the $n+1$ relaxation times that comprise the relaxation spectrum of the mechanism. The time dependence of the deviation from equilibrium of the various reactants is given by equations of the form

$$\Delta(X_i) = \sum_{j=1}^{n+1} A_{ij}e^{-t/\tau_j}, \quad i = 1, 2 \ldots n+1 \qquad (41)$$

where the A_{ij}'s are constants which depend on the initial conditions.

For the case $n > 1$, an exact analytical solution to Eq. (40) is not possible. However, in virtually all cases encountered in practice many of the individual steps equilibrate at substantially different rates (*18, 19*);

consequently, more than two steps will rarely equilibrate at comparable rates and will at the same time be kinetically coupled to each other. Therefore, it is almost always possible to introduce simplifying approximations such as in the two-step mechanism discussed above. As a further example consider the mechanism of Eq. (10) for $n = 2$:

$$E + S \underset{k_{-1}}{\overset{k_1}{\rightleftharpoons}} X_1 \underset{k_{-2}}{\overset{k_2}{\rightleftharpoons}} X_2 \underset{k_{-3}}{\overset{k_3}{\rightleftharpoons}} E + P \tag{42}$$

The three relaxation times which characterize this mechanism may be obtained by solving the third-order secular equation arising from the three linearized rate equations. Although the resulting cubic equation cannot be solved analytically, an approximate solution can be obtained if the unimolecular step, $X_1 \rightleftarrows X_2$, equilibrates more slowly (a factor of 10 is sufficient) than the two bimolecular steps. This can always be achieved at sufficiently high concentrations of S and P. The two short relaxation times are computed by assuming that the unimolecular step is "frozen" while the bimolecular steps relax. The linearized rate equations describing the relaxation of the kinetically coupled bimolecular steps are

$$\begin{aligned}
-\frac{d\Delta(S)}{dt} &= \{k_1[(\bar{E}) + (\bar{S})] + k_{-1}\}\Delta(S) + k_1(\bar{S})\Delta(P) \\
&= a_{11}\Delta(S) + a_{12}\Delta(P) \\
-\frac{d\Delta(P)}{dt} &= k_{-3}(\bar{P})\Delta(S) + \{k_{-3}[(\bar{E}) + (\bar{P})] + k_3\}\Delta(P) \\
&= a_{21}\Delta(S) + a_{22}\Delta(P)
\end{aligned} \tag{43}$$

where the mass conservation relationships $-\Delta(E) = \Delta(X_1) + \Delta(X_2)$, $\Delta(S) = -\Delta(X_1)$, and $\Delta(P) = -\Delta(X_2)$ have been used. The relaxation times associated with Eqs. (43) are given by Eqs. (37). The long relaxation time may now be computed from the rate equation

$$\begin{aligned}
-\frac{d[\Delta(S) + \Delta(X_1)]}{dt} &= \frac{d[\Delta(P) + \Delta(X_2)]}{dt} \\
&= k_2\Delta(X)_1 - k_{-2}\Delta(X)_2
\end{aligned} \tag{44}$$

The mass conservation relationships are $\Delta(E) + \Delta(X_1) + \Delta(X_2) = 0$ and $\Delta(S) + \Delta(P) + \Delta(X_1) + \Delta(X_2) = 0$, and the assumption that the bimolecular steps are always at equilibrium while the slow step relaxes implies that the equations $(\bar{E})(\bar{P})/(\bar{X}_2) = k_3/k_{-3}$ and $(\bar{E})(\bar{S})/(\bar{X}_1) = k_{-1}/k_1$ can be differentiated to give $k_{-3}(\bar{E})\Delta(P) + k_{-3}(\bar{P})\Delta(E) = k_3\Delta(X_2)$ and $k_1(\bar{E})\Delta(S) + k_1(\bar{S})\Delta(E) = k_{-1}\Delta(X_1)$. Insertion of these relationships into Eq. (44) yields

$$\frac{-d[\Delta(\mathrm{S}) + \Delta(\mathrm{X}_1)]}{dt} = [\Delta(\mathrm{S}) + \Delta(\mathrm{X}_1)]/\tau_3 \tag{45a}$$

where

$$\frac{1}{\tau_3} = \frac{k_2}{1 + \dfrac{k_1k_3(\bar{\mathrm{S}}) + k_{-1}k_{-3}(\bar{\mathrm{E}}) + k_{-1}k_{-3}(\bar{\mathrm{P}}) + k_{-1}k_3}{k_1k_3(\bar{\mathrm{E}}) + k_1k_{-3}(\bar{\mathrm{E}})[(\bar{\mathrm{E}}) + (\bar{\mathrm{S}}) + (\bar{\mathrm{P}})]}} + \frac{k_{-2}}{1 + \dfrac{k_1k_3(\bar{\mathrm{S}}) + k_{-1}k_{-3}(\bar{\mathrm{P}}) + k_1k_3(\bar{\mathrm{E}}) + k_{-1}k_3}{k_{-1}k_{-3}(\bar{\mathrm{E}}) + k_1k_{-3}(\bar{\mathrm{E}})[(\bar{\mathrm{E}}) + (\bar{\mathrm{S}}) + (\bar{\mathrm{P}})]}} \tag{45b}$$

At high concentrations of substrate and product $1/\tau_3 \approx k_2 + k_{-2}$ as expected. Thus, the entire spectrum of relaxation times for the three-step mechanism has been explicitly evaluated. If all three relaxation times can be determined experimentally over a range of concentrations all six rate constants can be readily evaluated.

The assumptions which may be invoked to simplify the relaxation spectrum of complex enzymic mechanisms vary for each particular case; references *47–58* should be consulted for the treatment of various mechanisms required to describe actual experimental data.

5. *Alternative Treatment of Multistep Mechanisms*

For mechanisms characterized by more than two relaxation times, the secular equation [Eq. (40)] can be most easily obtained and solved by an alternative method. The relaxation times can be shown to be given as the roots of the secular determinant

$$|\mathbf{b} - (1/\tau)\mathbf{I}| = 0 \tag{46}$$

where $\mathbf{b} \neq \mathbf{a}$, but the eigenvalues (reciprocal relaxation times) of $\mathbf{b}$ are the same as those of $\mathbf{a}$. The alternative secular equation [Eq. (46)] is obtained by casting the kinetic problem into the framework of non-

47. G. G. Hammes and P. Fasella, *JACS* **84,** 4644 (1962).
48. R. E. Cathou and G. G. Hammes, *JACS* **86,** 3240 (1964).
49. T. C. French and G. G. Hammes, *JACS* **87,** 4669 (1965).
50. R. E. Cathou and G. G. Hammes, *JACS* **87,** 4674 (1965).
51. J. E. Erman and G. G. Hammes, *JACS* **88,** 5607 (1966).
52. J. E. Erman and G. G. Hammes, *JACS* **88,** 5614 (1966).
53. G. G. Hammes and F. G. Walz, Jr., *JACS* **91,** 7179 (1969).
54. E. J. del Rosario and G. G. Hammes, *JACS* **92,** 1750 (1970).
55. K. Kirschner, M. Eigen, R. Bittman, and B. Voight, *Proc. Natl. Acad. Sci. U. S.* **56,** 1661 (1966).
56. G. G. Hammes and J. L. Haslam, *Biochemistry* **7,** 1519 (1968).
57. G. G. Hammes and J. L. Haslam, *Biochemistry* **8,** 1591 (1969).
58. G. G. Hammes and J. K. Hurst, *Biochemistry* **8,** 1083 (1969).

equilibrium thermodynamics (*59*). This alternative approach has two principal advantages: First the secular equation may be written merely by inspecting the mechanism (that is, it is not necessary to go through the computational labor of writing out and linearizing all of the rate equations); and, second, the resulting secular determinant is much easier to manipulate and evaluate than is the other equivalent equation (*59, 60*).

The matrix **b** may be written as the product of two matrices **r** and **g** (i.e., **b** = **r g**), each of which may be computed according to rules given by Castellan (*59*). The matrix **r** is a diagonal matrix whose typical diagonal element r_i is the equilibrium exchange rate of the ith reaction, i.e.,

$$r_i = k_i \prod_{\alpha} \bar{C}_{\alpha}^{\;-\nu_{\alpha i}} \tag{47}$$

where k_i, $\bar{C}_\alpha$, and $\nu_{\alpha i}$ are the forward rate constant of the ith reaction, equilibrium concentration of the species α, and stoichiometric coefficient of species α in the ith reaction. The product in Eq. (47) is taken over reactants of the ith reaction. (Clearly r_i can also be written in terms of reverse rate constants and concentrations of products.) The typical element g_{ij} of **g** is

$$g_{ij} = \sum_{\alpha=1}^{n} \frac{\nu_{\alpha i}\nu_{\alpha j}}{\bar{C}_\alpha} = g_{ji} \tag{48}$$

The element g_{ij} arises from the "coupling" between the ith and jth reaction. Consider for illustration the mechanism

$$\mathrm{E} + \mathrm{S} \underset{k_{-1}}{\overset{k_1}{\rightleftharpoons}} \mathrm{X}_1 \underset{k_{-2}}{\overset{k_2}{\rightleftharpoons}} \mathrm{X}_2 \underset{k_{-3}}{\overset{k_3}{\rightleftharpoons}} \cdots \underset{k_{-n}}{\overset{k_n}{\rightleftharpoons}} \mathrm{X}_n \tag{49}$$

where, for example, the initial enzyme–substrate complex undergoes a series of sequential isomerizations. Utilization of Eqs. (47) and (48) gives

$$\mathbf{b} = \begin{vmatrix} k_1[(\bar{\mathrm{E}}) + (\bar{\mathrm{S}})] + k_{-1} & -k_{-1} & 0 & \cdot & \cdot & 0 \\ -k_2 & k_2 + k_{-2} & -k_{-2} & & & \\ 0 & -k_3 & k_3 + k_{-3} & -k_{-3} & 0 \;\cdots & 0 \\ \cdot & & & & & \cdot \\ \cdot & & & & & \cdot \\ \cdot & & & & & \\ 0 & \cdot & \cdot & \cdot & -k_n & k_n + k_{-n} \end{vmatrix} \tag{50}$$

59. G. W. Castellan, *Ber. Bunsenges. Physik. Chem.* **67,** 898 (1963).
60. G. G. Hammes and P. R. Schimmel, *J. Phys. Chem.* **70,** 2319 (1966).

This matrix is in simple triple diagonal form, and in this form it is particularly easy to manipulate *(60)*. Note that the jth equilibria gives rise to the jth row and column of **b**, a fact which enables one to reduce the order of the matrix for various special cases.

As an example of the utility of the **b** matrix, consider the case where the jth reaction equilibrates slowly compared to all of the remaining equilibria so that

$$k_j + k_{-j} \ll k_i + k_{-i} \quad i \neq j \tag{51}$$

(For the first step k_i should be replaced by $k_1[(\bar{\mathrm{E}}) + (\bar{\mathrm{S}})]$.) Under this condition,

$$\frac{1}{\tau_j} = \frac{|\mathbf{b}|}{|\mathbf{b}_{jj}|} \tag{52}$$

where $|\mathbf{b}_{jj}|$ is the cofactor of $\mathbf{b}_{jj}$ and is obtained by deleting the jth row and column of $|\mathbf{b}|$ *(60)*. The remaining $n-1$ eigenvalues are now obtained from the secular equation

$$|\mathbf{b}_{jj} - (1/\tau)\mathbf{I}| = 0 \tag{53}$$

However, $\mathbf{b}_{jj}$ is a block diagonal form, the two blocks $\mathbf{b}_1$ and $\mathbf{b}_2$ consisting of all of the matrix elements above and to the left of the element b_{jj} in **b** and below and to the right of b_{jj}, respectively. Since the eigenvalues of the block diagonal form $\mathbf{b}_{jj}$ are the eigenvalues of the individual blocks, Eq. (53) may be further decomposed into two secular equations

$$|\mathbf{b}_1 - (1/\tau)\mathbf{I}| = 0 \tag{54a}$$
$$|\mathbf{b}_2 - (1/\tau)\mathbf{I}| = 0 \tag{54b}$$

where the order of the first equation is $j-1$, and that of the second is $n-j$. The longest relaxation time of each submatrix $\mathbf{b}_1$ and $\mathbf{b}_2$ can now be computed by exactly the same procedure used for the original **b** matrix. This procedure can be continued until the shortest relaxation time of each submatrix is obtained; this is the diagonal element associated with the fastest reaction in each submatrix.

Even more complex mechanisms can be treated in a similar fashion using this formalism; Castellan *(59)* and Hammes and Schimmel *(60, 61)* should be consulted for more details. Thus for complex mechanisms the **b** matrix is frequently more convenient to use than alternate formulations.

6. *Thermodynamically Dependent Reactions*

The preceding examples involved cases where the reaction mechanisms consisted of thermodynamically independent reactions, i.e., no cyclic

61. G. G. Hammes and P. R. Schimmel, *J. Phys. Chem.* **71,** 917 (1967).

processes or parallel pathways were involved. In these cases, therefore, the number of relaxation times characterizing the mechanism equals the number of equilibria. However, when thermodynamically dependent reactions occur, the number of equilibria exceeds the number of relaxation times. In general, the number of relaxation times is equal to the number of independent concentration variables.

A simple example of a mechanism containing thermodynamically dependent reactions is

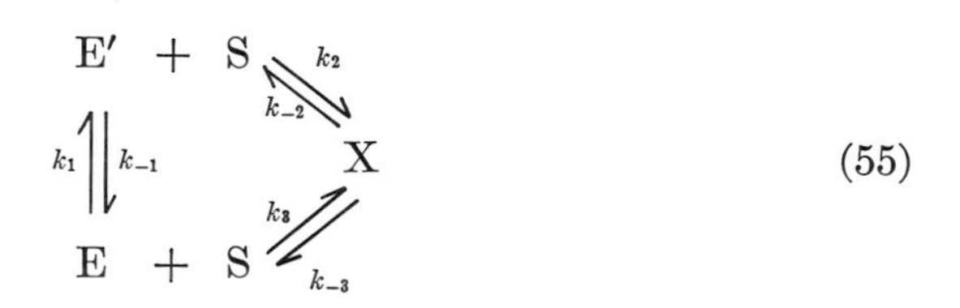

(55)

where E and E′ are, for example, different conformational states of the free enzyme which have different affinities for the substrate S. The mass conservation equations for this system are

$$\Delta(\mathrm{S}) + \Delta(\mathrm{X}) = 0, \qquad \Delta(\mathrm{E}) + \Delta(\mathrm{E}') + \Delta(\mathrm{X}) = 0$$

so that only two independent rate equations are needed to completely specify the kinetic behavior. For example, two possible linearized rate equations are

$$-\frac{d\Delta(\mathrm{E})}{dt} = \{k_1 + k_{-3} + k_3[(\bar{\mathrm{E}}) + (\bar{\mathrm{S}})]\}\Delta(\mathrm{E}) + [k_{-3} - k_{-1} + k_3(\bar{\mathrm{E}})]\Delta(\mathrm{E}') \tag{56a}$$

$$-\frac{d\Delta(\mathrm{E}')}{dt} = [k_{-2} - k_1 + k_2(\bar{\mathrm{E}}')]\Delta(\mathrm{E}) + \{k_{-1} + k_{-2} + k_2[(\bar{\mathrm{E}}') + (\bar{\mathrm{S}})]\}\Delta(\mathrm{E}') \tag{56b}$$

Thus, the mechanism of Eq. (55) possesses two relaxation times, although three equilibria exist.

An important example of thermodynamically dependent reactions arises in connection with the effect of hydrogen ion concentration on enzymic reactions. The pH dependence of enzymic reactions is often attributed to the fact that the enzyme can exist in different ionization states which have different catalytic properties. The effect of pH on a single substrate–single product Michaelis–Menten mechanism [Eq. (10)] is often accounted for by the modified mechanism (*61, 62*)

62. R. A. Alberty and V. Bloomfield, *JBC* **238**, 2804 (1963).

$$
\begin{array}{ccccccccc}
S + E & \underset{k'_{-1}}{\overset{k_1'}{\rightleftharpoons}} & X_1 & \rightleftharpoons \cdots & X_i & \underset{k'_{-(i+1)}}{\overset{k'_{(i+1)}}{\rightleftharpoons}} & \cdots X_n & \underset{k'_{-(n+1)}}{\overset{k'_{(n+1)}}{\rightleftharpoons}} & E + P \\
\uparrow\downarrow K_{0b} & & \uparrow\downarrow K_{1b} & & \uparrow\downarrow K_{ib} & & \uparrow\downarrow K_{nb} & & \uparrow\downarrow K_{0b} \\
S + EH & \underset{k_{-1}}{\overset{k_1}{\rightleftharpoons}} & HX_1 & \rightleftharpoons \cdots & HX_i & \underset{k_{-(i+1)}}{\overset{k_{(i+1)}}{\rightleftharpoons}} & \cdots HX_n & \underset{k_{-(n+1)}}{\overset{k_{(n+1)}}{\rightleftharpoons}} & EH + P \\
\uparrow\downarrow K_{0a} & & \uparrow\downarrow K_{1a} & & \uparrow\downarrow K_{ia} & & \uparrow\downarrow K_{na} & & \uparrow\downarrow K_{0a} \\
S + EH_2 & \underset{k''_{-1}}{\overset{k_1''}{\rightleftharpoons}} & H_2X_1 & \rightleftharpoons \cdots & H_2X_i & \underset{k''_{-(i+1)}}{\overset{k''_{(i+1)}}{\rightleftharpoons}} & \cdots H_2X_n & \underset{k''_{-(n+1)}}{\overset{k''_{(n+1)}}{\rightleftharpoons}} & EH_2 + P
\end{array}
\qquad (57)
$$

where the K_{ia}'s and K_{ib}'s are ionization constants and protons have been omitted for the sake of simplicity. The vertical protolytic steps generally equilibrate more rapidly than the horizontal steps (especially in buffered solutions). Hence, it may be assumed that the protolytic steps are always at equilibrium while the horizontal steps relax. If the solution is buffered, i.e., $\Delta(H^+) = 0$ while the horizontal steps relax, then the relaxation times for Eq. (57) are given by the same expressions as for the mechanism in Eq. (10) except for a redefinition of terms for the former mechanism: The rate constants k_i and k_{-i} must be replaced by $k_i{}^*$ and k^*_{-i}; in addition, the free enzyme concentration must be written as $(\bar{E}) + (\overline{EH}) + (\overline{EH_2})$ rather than as just $(\bar{E})$. The asterisked rate constants are defined in terms of the rate constants in the mechanism of Eq. (57). For example, the rate of conversion of X_1 to X_2 is given by

$$k_2''(X_1H_2) + k_2(X_1H) + k_2'(X) = k_2{}^*[(X_1H_2) + (X_1H) + (X_1)]$$

where

$$k_2{}^* = \frac{k_2 + k_2''(H^+)/K_{1a} + k_2'K_{1b}/(H^+)}{1 + (H^+)/K_{1a} + K_{1b}/(H^+)} \qquad (58a)$$

and similarly

$$k^*_{-2} = \frac{k_{-2} + k''_{-2}(H^+)/K_{2a} + k'_{-2}K_{2b}/(H^+)}{1 + (H^+)/K_{2a} + K_{2b}/(H^+)} \qquad (58b)$$

The other rate constants are defined in a similar manner [cf. Hammes and Schimmel (*61*) and Alberty and Bloomfield (*62*) for further details].

From a careful study of the pH dependence of the rate constants $k_i{}^*$ and k^*_{-i} obtained from measurements of relaxation times, ionization constants may be derived for the free enzyme and enzyme–substrate complexes. This situation is in marked contrast with the information derivable from the pH dependence of steady state kinetic parameters where the apparent ionization constants are often complex functions of rate constants and ionization constants.

7. *Analysis and Interpretation of Relaxation Spectra*

The preceding treatment of relaxation spectra provides a framework within which relaxation spectra may be interpreted in terms of mechanism. The application of these concepts to experimental situations is now briefly considered.

As with any kinetic study probably the most important precept is that data should be obtained over as wide a concentration range as possible. From the examples considered thus far, it is apparent that certain basic types of mechanisms can be readily distinguished from the concentration dependence of the relaxation times: Unimolecular processes have relaxation times that are concentration independent; the reciprocal relaxation time of a bimolecular reaction increases linearly with the sum of the equilibrium reactant concentrations [Eq. (25)]; and the reciprocal relaxation time characteristic of a unimolecular reaction following a rapid bimolecular reaction initially increases linearly with the sum of the equilibrium reactant concentrations and then asymptotes to a limiting value equal to the sum of the rate constants characteristic of the unimolecular reaction [Eq. (38b)].

The behavior of a typical bimolecular reaction is shown in Fig. 7 where a plot of $1/\tau$ vs. $[(\bar{E}) + (\bar{S})]$ is shown for the reaction between lysozyme and di-*N*-acetyl-glucosamine (*63*). In this case an equilibrium constant was assumed, and a plot of $1/\tau$ vs. $[(\bar{E}) + (\bar{S})]$ was constructed. The ratio of the slope to intercept of this plot determined a new equilibrium constant, and the entire process was iterated until a constant value of the equilibrium constant was obtained. In Fig. 8, the concentration dependence of the relaxation time for a conformational change following the binding of ADP by creatine kinase is shown (*58*). An analysis of this curve allows determination of k_2, k_{-2} and k_1/k_{-1}.

When more than one relaxation process is observed, the concentration dependence of each time constant must be investigated. If the relaxation times are well separated on the time axis (usually a factor of 10 is sufficient), then generally each one may be analyzed in a fashion similar to that for a single process: Expressions for the relaxation time may be derived and tested against experimental data without recourse to solving a secular determinant. If two or more of the observed relaxation processes are closely coupled, the corresponding relaxation times must be analyzed by solving a secular determinant. (The method of extracting individual relaxation times from data representing the superposition of two or more relaxation times is not considered here; in general, this will be dependent on the experimental technique employed.) Some invariant properties of

63. D. M. Chipman and P. R. Schimmel, *JBC* **243,** 3771 (1968).

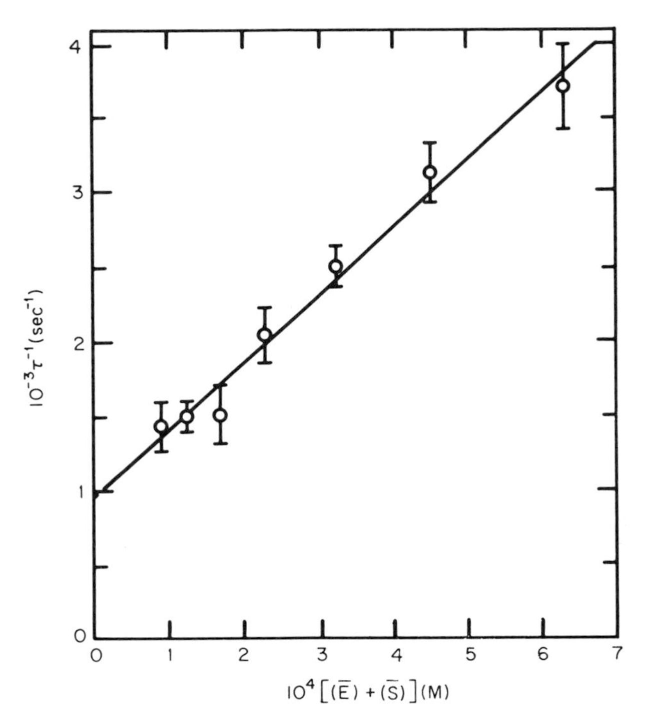

FIG. 7. Concentration dependence of the observed relaxation time in lysozyme-di-*N*-acetyl-glucosamine solutions at pH 6.0, ca. 29°. The data are plotted according to Eq (38a), where $[(\bar{E}) + (\bar{S})]$ was calculated by the reiterative method described in the text. These data give $k_1 = 4.56 \times 10^6\ M^{-1}\ \text{sec}^{-1}$ and $k_{-1} = 950\ \text{sec}^{-1}$ (*63*).

the secular determinant are often useful diagnostic tools for postulating a mechanism. For example, in the case of a two-step mechanism [Eq. (36)],

$$1/\tau_1 + 1/\tau_2 = a_{11} + a_{22} \tag{59a}$$

$$(1/\tau_1)(1/\tau_2) = a_{11}a_{22} - a_{12}a_{21} \tag{59b}$$

For the two-step mechanism of Eq. (33), $1/\tau_1 + 1/\tau_2$ and $(1/\tau_1)(1/\tau_2)$ are linear functions of $[(\bar{E}) + (\bar{S})]$, and from the two slopes and intercepts all four rate constants can be evaluated.

In general the number of invariant properties of a secular determinant is equal to the number of relaxation times (i.e., eigenvalues). These relationships can be most easily derived by writing the determinant in polynomial form:

$$-(1/\tau)^n + A_n(1/\tau)^{n-1} - A_{n-1}(1/\tau)^{n-2} + \ldots + A_1 = 0 \tag{60}$$

The invariants can then be written as

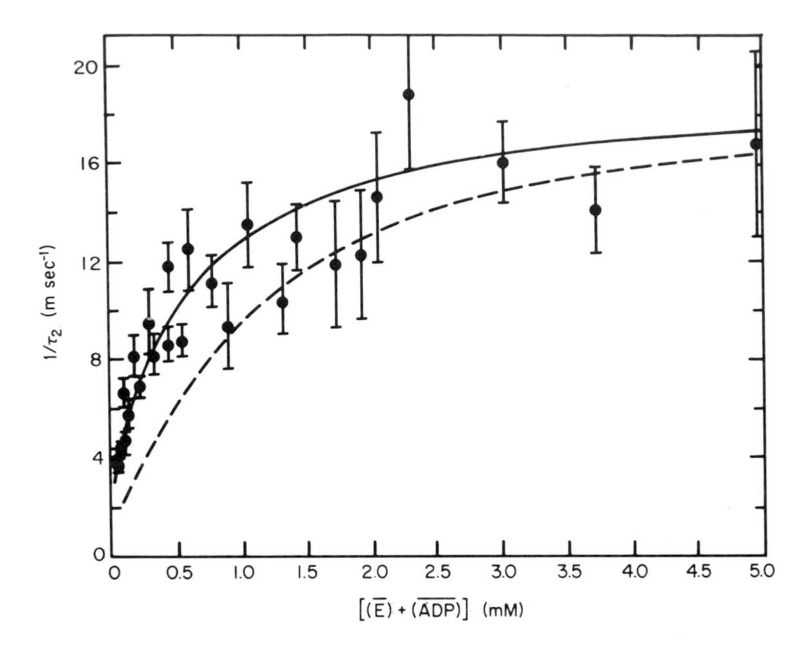

FIG. 8. The reciprocal relaxation time for creatine kinase–ADP interactions, $1/\tau_2$, as a function of the sum of the equilibrium concentration of enzyme and ADP. The solid curve represents the theoretical line for the uncoupled system [Eq. (38b)]. The dashed curve is the corresponding theoretical line for the exact solution of the coupled system. The rate constants used are $k_2 = 1.67 \times 10^4$ sec^{-1}, $k_{-2} = 2.4 \times 10^3$ sec^{-1}, and $k_{-1}/k_1 = 6.0 \times 10^{-4}$ M (*58*).

$$\sum_{i=1}^{n} 1/\tau_i = A_n$$

$$\sum_{j=i+1}^{n} \sum_{i=1}^{n-1} (1/\tau_i)(1/\tau_j) = A_{n-1}$$

$$\sum_{k=j+1}^{n} \sum_{j=i+1}^{n-1} \sum_{i=1}^{n-2} (1/\tau_i)(1/\tau_j)(1/\tau_k) = A_{n-2}$$

$$\vdots$$

$$\prod_{i=1}^{n} 1/\tau_i = A_1 \tag{61}$$

This treatment has been used to analyze the relaxation spectra for the interaction of aspartate aminotransferase with L-α-methylaspartic acid and L-*erythro*-β-hydroxyaspartic acid where three and five closely coupled relaxation times are observed, respectively (*56, 57*). A digital computer was used to calculate the concentration dependence of the individual relaxation times, and the results are in good agreement with experiment.

The actual construction of a plausible mechanism ultimately depends on the ingenuity of the investigator, but the analyses discussed above provide methods for testing postulated mechanisms. The concentration dependence of the relaxation amplitudes can also provide useful information about mechanism. This is considered in Section III,B.

B. Relaxation Amplitudes

Thus far only the time constants associated with the relaxation spectra of chemical reactions near equilibrium have been discussed. Additional information can be obtained from the relaxation amplitudes, which are dependent on the thermodynamic parameters characterizing the chemical reaction. Moreover, an understanding of the factors governing relaxation amplitudes is helpful in optimizing experimental conditions for the observation of relaxation processes. In this section the complete solution to rate equations near equilibrium is discussed in terms of the transformation to normal concentration variables. The thermodynamic formulation of rate processes near equilibrium is then briefly presented, and, finally, application of these concepts to the calculation of relaxation amplitudes is discussed in terms of specific examples.

1. *Transformation to Normal Concentration Variables*

Although solution of the secular determinant provides the characteristic time constants of the integrated rate equations near equilibrium, a more general approach is to consider the transformation of the concentration variables to a new set of *normal concentration variables.* In other words, the coupled rate equations [see Eq. (24) for $n = 1$]

$$\frac{d\Delta(\mathrm{X}_i)}{dt} + \sum_{j=1}^{n} a_{ij}\Delta(\mathrm{X}_j) = \sum_{j=1}^{n} a_{ij}\Delta(\bar{\mathrm{X}}_j) \tag{62}$$

are transformed to a new set of independent rate equations

$$\frac{d\Delta y_i}{dt} + \frac{\Delta y_i}{\tau_i} = \frac{\Delta \bar{y}_i}{\tau_i} \tag{63}$$

Here, Δy_i is a normal concentration which is a linear combination of the $\Delta(\mathrm{X}_i)$'s, and vice versa, i.e.,

$$\Delta(X_i) = \sum_{j=1}^{n} Q_{ij}\Delta y_j, \quad i = 1, 2 \ldots n \tag{64}$$

where the Q_{ij}'s are functions of rate constants and equilibrium concentrations. Equation (63) can now be integrated exactly as previously described for the rate equation of a single reaction (Section III,A,2). This procedure is analogous to that encountered in vibrational spectroscopy where the vibrational characteristics of a molecule are generally discussed in terms of normal modes and normal mode frequencies rather than in terms of the frequency of vibration of individual bonds.

This problem is most easily solved by the use of matrix notation whereby Eq. (64) is written as

$$\mathbf{\Delta X} = \mathbf{Q\Delta y} \tag{65}$$

where $\mathbf{\Delta X}$ and $\mathbf{\Delta y}$ are $n \times 1$ column vectors of the ΔX_i's and $\mathbf{Q}$ is an $n \times n$ square matrix of the Q_{ij}'s. The matrix $\mathbf{Q}$ is defined such that

$$\mathbf{Q}^{-1}\mathbf{aQ} = \mathbf{\Lambda} \tag{66}$$

where $\mathbf{a}$ is the $n \times n$ matrix of a_{ij}'s and $\mathbf{\Lambda}$ is a diagonal matrix whose ith diagonal element is the ith reciprocal relaxation time $1/\tau_i$. ($\mathbf{Q}^{-1}$ is the inverse of $\mathbf{Q}$, i.e., $\mathbf{Q}^{-1}\,\mathbf{Q} = \mathbf{I}$ where $\mathbf{I}$ is the diagonal matrix of ones). The matrix $\mathbf{Q}$ is constructed from the n eigenvectors associated with the n eigenvalues (reciprocal relaxation times) of the matrix $\mathbf{a}$.

As an illustration of the calculation of normal variables, consider a mechanism which can be described by two independent rate equations. In this case

$$\mathbf{a} = \begin{pmatrix} a_{11} & a_{12} \\ a_{21} & a_{22} \end{pmatrix} \tag{67}$$

The eigenvalues of this matrix are $1/\tau_1$ and $1/\tau_2$ and the eigenvectors (or Q_{ij}) can be found by solving the equations

$$(a_{11} - 1/\tau_1)Q_{11} + a_{12}Q_{21} = 0 \tag{68a}$$

$$(a_{11} - 1/\tau_2)Q_{12} + a_{12}Q_{22} = 0 \tag{68b}$$

or equivalently

$$a_{21}Q_{11} + (a_{22} - 1/\tau_1)Q_{21} = 0, \qquad a_{21}Q_{12} + (a_{22} - 1/\tau_2)Q_{22} = 0$$

Only the ratios $Q_{11}:Q_{21}$ and $Q_{12}:Q_{22}$ can be specified so that Q_{11} and Q_{22} can be arbitrarily chosen as unity and

$$\mathbf{Q} = \begin{pmatrix} 1 & \dfrac{a_{12}}{1/\tau_2 - a_{11}} \\ \dfrac{1/\tau_1 - a_{11}}{a_{12}} & 1 \end{pmatrix} \tag{69}$$

The original concentration variables can now be expressed in terms of the normal concentration variables through Eq. (65):

$$\Delta(X_1) = Q_{11}\Delta y_1 + Q_{12}\Delta y_2 \qquad (70a)$$
$$\Delta(X_2) = Q_{21}\Delta y_1 + Q_{22}\Delta y_2 \qquad (70b)$$

The transformation from the $\mathbf{\Delta X}$'s to the $\mathbf{\Delta y}$'s can also be easily found since

$$\mathbf{\Delta y} = \mathbf{Q}^{-1}\mathbf{\Delta X} \qquad (71)$$

These results are considered further in Section III,B,3 where specific calculations of relaxation amplitudes are carried out.

2. *Thermodynamic Effects of Chemical Reactions*

The occurrence of chemical reactions modifies the usual thermodynamic relationships in such a way that additional terms appear. For example, the total differential of the Gibbs free energy G in a closed system is given by

$$dG = V\,dP - S\,dT - A\,d\xi \qquad (72)$$

where V is the volume, P the pressure, S the entropy, and T the temperature. The last term results from the occurrence of a chemical reaction: ξ is the degree of advancement and is defined by the relationship

$$n_i - n_i^0 + \nu_i\xi \qquad (73)$$

where n_i and ν_i are the mole number and stoichiometric coefficient of the ith species, respectively, and n_i^0 is the initial (time-independent) mole number; A is the affinity and is defined by

$$A = -\sum \nu_i\mu_i = -(\partial G/\partial\xi)_{T,P} \qquad (74)$$

where μ_i is the chemical potential of the ith species. At equilibrium $A = 0$ and the usual expression for the Gibbs free energy is obtained. The equations for the total derivatives of the internal energy, enthalpy, etc., can be modified in a similar manner to take into account the occurrence of a chemical reaction.

In order to calculate the amplitude of relaxation effects, the change in ξ when an intensive variable changes must be known in terms of accessible thermodynamic parameters. The desired result is most readily obtained by considering the total differential of the affinity:

$$dA = (\partial A/\partial P)_{T,\xi}\,dP + (\partial A/\partial T)_{P,\xi}\,dT + (\partial A/\partial\xi)_{T,P}\,d\xi \qquad (75)$$

From this equation it is clear that

$$(\partial\xi/\partial Z_1)_{Z_2,A} = [-(\partial A/\partial Z_1)_{Z_2,\xi}]/[(\partial A/\partial \xi)_{Z_1,Z_2}] \tag{76a}$$
$$= \Gamma(\partial A/\partial Z_1)_{Z_2,\xi}/RT \tag{76b}$$

where Z_i is the temperature or pressure and

$$\Gamma = -RT(\partial\xi/\partial A)_{T,P} \tag{77}$$

Here the subscript A implies not only that A equals a constant but also that $A = 0$, i.e., equilibrium is attained. From the usual thermodynamic relationships it can be shown that

$$(\partial A/\partial P)_{T,\xi} = -(\partial V/\partial\xi)_{T,P} = -\Delta V \tag{78}$$
$$(\partial A/\partial T)_{P,\xi} = (\partial S/\partial\xi)_{T,P} = \Delta S = \Delta H/T \tag{79}$$

where ΔV, ΔS and ΔH are the volume, entropy, and enthalpy changes associated with the chemical reaction. Finally, if the chemical potential is written as

$$\mu_i = \mu_i^0 + RT \ln \gamma_i c_i \tag{80}$$

where γ_i is the activity coefficient of the ith species, c_i its concentration, and μ_i its standard chemical potential, then

$$\Gamma = (\partial\xi/\partial \sum_i \nu_i \ln \gamma_i c_i)_{T,P} \tag{81a}$$

Since at equilibrium $\Sigma_i \nu_i \ln \gamma_i c_i = \ln K$, where K is the equilibrium constant, Eq. (81a) may be formally expressed as

$$\Gamma = (\partial\xi/\partial \ln K)_{T,P} \tag{81b}$$

These equations permit explicit evaluation of Γ in terms of known quantities (cf. Section III,B,3).

For small changes Δ in the degree of advancement and temperature and pressure, Eq. (76b) may be written as

$$(\Delta\bar{\xi})_{P,A} = \Gamma\Delta H/(RT^2)\Delta T \tag{82a}$$
$$(\Delta\bar{\xi})_{T,A} = -\Gamma\Delta V/(RT)\Delta P \tag{82b}$$

where the bars indicate this is the shift in the *equilibrium* value of the degree of advancement. Although only temperature and pressure have been considered as intensive variables, a similar treatment can be used for other intensive variables, e.g., electric and magnetic field intensities (cf. *15*).

The time dependence of the shift in the equilibrium value of the degree of advancement is proportional to the time dependence of the perturbation [i.e., ΔP or ΔT in Eqs. (82)]. Although the relationship between the degree of advancement and concentration is dependent on the concentration scale used, for most cases of practical interest a direct proportionality exists *(15)*: for molal (m_i) concentrations $dm_i = \nu_i d\xi/1000$; for mole fractions (γ_i) in sufficiently dilute solutions $d\gamma_i = \nu_i d\xi/n_{\text{solv}}$, where n_{solv} is the number of moles of solvent; for molar (c_i) concentrations at constant volume $dc_i = \nu_i d\xi/V$. In any event when the relationship between ξ and the concentration variable is specified, Eqs. (82) can be inserted directly into the rate equation [Eq. (24)], and the rate equation can be integrated as in Section III,A,2. In this way explicit expressions for the relaxation amplitude can be obtained in terms of thermodynamic quantities.

The extension of this analysis to multistep mechanisms is quite simple. In this case, the normal concentration variables specify a set of ξ's, each of which is described by an independent rate equation such as Eq. (63). The thermodynamic parameters describing the amplitudes are also normal thermodynamic parameters which are linear combinations of those for the individual reactions (see Section III,B,3). A set of Γ's are defined such that

$$\Gamma_i = -RT(\partial\tilde{\xi}_i/\partial\tilde{A}_i)_{T,P,\tilde{\xi}_k(k \neq i)} \tag{83}$$

where the tilde indicates normal quantities. Thus an analysis of the relaxation amplitudes for a complex mechanism can be readily carried out.

In general chemical reactions contribute to any macroscopic thermodynamic properties of the system, for example, heat capacity and compressibility. The chemical contribution to the thermodynamic properties can be easily evaluated by use of the concepts developed above. As an example consider the isobaric thermal expansion coefficient α_P. This quantity can be defined by the equations

$$\begin{aligned} \alpha_P &= \frac{1}{V}\left(\frac{\partial V}{\partial T}\right)_{P,A} \\ &= \frac{1}{V}\left(\frac{\partial V}{\partial T}\right)_{P,\xi} + \frac{1}{V}\left(\frac{\partial V}{\partial \xi}\right)_{T,P}\left(\frac{\partial \xi}{\partial T}\right)_{P,A} \\ &= \alpha_P{}^{\infty} + \bar{\alpha}_P^{\text{ch}} \end{aligned} \tag{84}$$

Here $\alpha_P{}^{\infty}$ represents the thermal expansion coefficient in the absence of a chemical reaction or when the perturbation of the system is applied infinitely fast, i.e., much faster than the rate of the chemical reaction so that the reacting species do not change their concentrations; $\bar{\alpha}_P^{\text{ch}}$ represents the chemical contribution to the thermal expansion coefficient if the reaction is assumed to fully equilibrate while the temperature is changed, i.e., the perturbation is applied very slowly relative to the rate of the chemical reaction. Combination of Eqs. (78) and (82) gives

$$\bar{\alpha}_P^{\text{ch}} = \frac{\Gamma(\Delta V)(\Delta H)}{VRT^2} \tag{85}$$

Analogous expressions can be derived for other thermodynamic quantities (cf. *15, 64*). For example chemical contributions to the constant pressure specific heat $\bar{C}_P^{ch}$, the isothermal compressibility $\bar{\kappa}_T^{ch}$, and the adiabatic compressibility $\bar{\kappa}_S^{ch}$ are

$$\bar{C}_P^{ch} = \frac{\Gamma(\Delta H)^2}{RT^2 \rho V} \tag{86}$$

$$\bar{\kappa}_T^{ch} = \frac{\Gamma(\Delta V)^2}{RTV} \tag{87}$$

$$\bar{\kappa}_S^{ch} = \frac{\Gamma}{RTV}\left(\Delta V - \frac{\Delta H \alpha_P{}^\infty}{\rho C_P{}^\infty}\right)^2 \left(\frac{C_P{}^\infty}{C_P}\right) \tag{88}$$

where ρ is the density. In general the chemical parts of the thermodynamic coefficients are quite small, ca. $<1\%$ of the total, but in some cases they can be measured experimentally to yield thermodynamic data for the reaction (e.g., in ultrasonic experiments).

For multistep reactions, Eqs. (85)–(88) pertain to each normal reaction mode, e.g., $\bar{\alpha}_P^{ch,i} = \Gamma_i(\Delta V_i)(\Delta H_i)/(VRT^2)$. The total contribution of the chemical reactions to thermodynamic quantities is simply the sum over the contributions from each normal variable. In general, of course, the chemical parts of the thermodynamic quantities are frequency dependent (i.e., they depend on the rates of the chemical reactions) so that the chemical part of each normal thermodynamic variable should be multiplied by a frequency-dependent function containing the associated normal mode relaxation time. For example, in the case of ultrasonic absorption, the chemical absorption per wavelength μ^{ch} is given to a good approximation by

$$\mu^{ch} \approx 2\pi \sum_i (\bar{\kappa}_S^{ch,i}/\kappa_S{}^\infty)\omega\tau_i/(1 + \omega^2\tau_i{}^2) \tag{89}$$

The quantity μ^{ch} can be measured experimentally and goes through a series of maxima at $\omega = 1/\tau_i$ with the amplitude being determined by the normal thermodynamic variables of Eq. (88). (If the relaxation times are not sufficiently separated, the maxima may not be resolved.) A more general treatment of dynamic equations of state is given in reference *15*.

The time dependence of chemical reactions near equilibrium can also be formulated in terms of irreversible thermodynamics (cf. *15, 65*). In this case, the relaxation time is proportional to $(\partial A/\partial \xi)$. This derivative is, of course, dependent on which variables are held constant when it is

64. R. O. Davies and J. Lamb, *Quart. Rev. (London)* **11,** 134 (1957).
65. J. Meixner, *Kolloid-Z.* **134,** 3 (1953).

evaluated, for example, $(\partial A/\partial\xi)_{S,P} \neq (\partial A/\partial\xi)_{T,P}$. Thus different relaxation times are obtained depending under what conditions the experiment is done. Moreover, the transformation to normal thermodynamic quantities can become more complex than described above. In practice the difference between the various relaxation times is negligible for dilute liquid solutions so that this matter is not considered further here.

3. *Calculation of Relaxation Amplitudes*

From the preceding discussion it is clear that regardless of the experimental method, Γ must be evaluated in order to determine the amplitude of the relaxation process. Several examples are now considered. In all cases the activity coefficients are assumed to be independent of the degree of advancement and molar concentration units c_i are used with the assumption the volume of the system is constant, i.e., $d\xi$ is replaced by dc_i/ν_i. Then for a single chemical reaction Eq. (81) becomes

$$\Gamma = \left[\partial(c_i/\nu_i)/\partial \sum \nu_i \ln c_i\right]_{T,P} \qquad (90)$$

$$= 1/\sum_i \nu_i^2/\bar{c}_i$$

For a unimolecular reaction such as $X_1 \rightleftarrows X_2$

$$\Gamma = [1/(\bar{X}_1) + 1/(\bar{X}_2)]^{-1}$$

$$= (X_0)\alpha(1 - \alpha) \qquad (91)$$

where $(X)_0 = (X_1) + (X_2)$ and $\alpha = (\bar{X}_2)/(X)_0$. The function $\alpha(1 - \alpha)$ is bell-shaped, goes through a maximum at $\alpha = 0.5$, i.e., when $(\bar{X}_1) = (\bar{X}_2)$, and goes to zero when $\alpha \to 0$ or $\alpha \to 1$. Moreover, Γ is directly proportional to the total concentration. If this relationship is inserted into Eqs. (82), the shift in $\bar{X}_2$, $\Delta\bar{X}_2$, can be calculated. For a temperature change of 10° with $\Delta H = 5$ kcal/mole, $T = 300°$K and $\alpha = 0.5$, $\Delta\bar{X}_2/(X)_0$ is about 0.07. For a ΔV of 5 cm^3/mole, a pressure change of about 1400 atm is required to give the same relative shift in concentration.

For a simple bimolecular reaction such as Eq. (6)

$$\Gamma = [1/(\bar{X}) + 1/(\bar{E}) + 1/(\bar{S})]^{-1} \qquad (92)$$

In general, the maximum relative shifts in concentration are achieved when the concentration of reactants is approximately equal to that of products.

As an example of how the relaxation amplitudes can be calculated for a multistep mechanism, the two-step mechanism of Eq. (33) is again considered. Normal concentration variables have already been obtained for this mechanism, and the shift in concentration of each normal variable is also given by Eqs. (82). In order to calculate Γ_1 and Γ_2, the normal affinities must first be obtained. This can be done by considering a small change in the free energy ΔG from its equilibrium value at constant T and P:

$$\begin{aligned}\Delta G &= \mu_E\Delta(E) + \mu_S\Delta(S) + \mu_{X_1}\Delta(X_1) + \mu_{X_2}\Delta(X_2)\\ &= [\mu_E + \mu_S - \mu_{X_1}]\Delta(E) + [\mu_{X_2} - \mu_{X_1}]\Delta(X_2)\\ &= A_1\Delta(E) - A_2\Delta(X_2)\end{aligned}$$

where the mass conservation relationships $\Delta(S) = \Delta(E) = -\Delta(X_1) - \Delta(X_2)$ have been utilized. If the relationships between $\Delta(E)$ and $\Delta(X_2)$ and the normal variables are now inserted into the above equations [see Eqs. (70) for an analogous transformation except that now the concentration variables are $\Delta(E)$ and $\Delta(X_2)$ rather than $\Delta(X_1)$ and $\Delta(X_2)$] the normal affinities $\tilde{A}_1$ and $\tilde{A}_2$ can be found.

$$\begin{aligned}\Delta G &= [A_1 - A_2(1/\tau_1 - a_{11})/a_{12}]\Delta y_1 + [A_1 a_{12}/(1/\tau_2 - a_{11}) - A_2]\Delta y_2\\ &= -\tilde{A}_1\Delta y_1 - \tilde{A}_2\Delta y_2 \qquad (93)\end{aligned}$$

The quantities Γ_1 and Γ_2 can now be obtained by direct differentiation:

$$\begin{aligned}\Gamma_1^{-1} &= -(\partial\tilde{A}_1/\partial\Delta y_1)_{T,P,\Delta y_2}/RT\\ &= \frac{1}{(\bar{E})} + \frac{1}{(\bar{S})} + \frac{1}{(\bar{X}_1)} + \frac{2(1/\tau_1 - a_{11})/a_{12}}{(\bar{X}_1)}\\ &\qquad + \left(\frac{1/\tau_1 - a_{11}}{a_{12}}\right)^2\left[\frac{1}{(\bar{X}_1)} + \frac{1}{(\bar{X}_2)}\right] \qquad (94)\end{aligned}$$

$$\begin{aligned}\Gamma_2^{-1} &= -(\partial\tilde{A}_2/\partial\Delta y_2)_{T,P,\Delta y_1}/RT\\ &= \left(\frac{a_{12}}{1/\tau_2 - a_{11}}\right)^2\left[\frac{1}{(\bar{E})} + \frac{1}{(\bar{S})} + \frac{1}{(\bar{X}_1)}\right]\\ &\qquad + \left(\frac{2a_{12}}{1/\tau_2 - a_{11}}\right)\frac{1}{(\bar{X}_1)} + \frac{1}{(\bar{X}_1)} + \frac{1}{(\bar{X}_2)} \qquad (95)\end{aligned}$$

where use has been made of the mass conservation equations and the relationships between the concentration variables and the normal concentration variables.

If the first step in the mechanism equilibrates much more rapidly than the second step then $1/\tau_1 \approx a_{11} \gg 1/\tau_2$ and

$$\Gamma_1 = \left[\frac{1}{(\bar{E})} + \frac{1}{(\bar{S})} + \frac{1}{(\bar{X}_1)}\right]^{-1}$$

$$\Gamma_2 = \left\{\left(\frac{a_{12}}{a_{11}}\right)^2\left[\frac{1}{(\bar{E})} + \frac{1}{(\bar{S})} + \frac{1}{(\bar{X}_1)}\right] - \frac{2a_{12}}{a_{11}(\bar{X}_1)} + \frac{1}{(\bar{X}_1)} + \frac{1}{(\bar{X}_2)}\right\}^{-1}$$

The expression for Γ_1 is identical to that for the one-step bimolecular reaction [Eq. (92)], while that for Γ_2 contains the expression for a one-step unimolecular process [Eq. (91)] plus some additional terms resulting from the coupling with the first step.

Relationships between other normal thermodynamic quantities and the thermodynamic quantities of the individual steps may be derived in a manner analogous to that used for deriving the expressions for $\tilde{A}_1$ and $\tilde{A}_2$. An equation similar in form to Eq. (93) is obtained in all cases so that, for example,

$$\Delta\tilde{H}_1 = \Delta H_1 - \Delta H_2(1/\tau_1 - a_{11})/a_{12}$$
$$\Delta\tilde{H}_2 = \Delta H_1 a_{12}/(1/\tau_2 - a_{11}) - \Delta H_2$$

The change in concentrations for any mechanism can now be readily calculated by use of the concepts discussed above. The relationship between the concentration changes and experimentally measured parameters depends on the experimental methods used. For example, if absorption changes are measured following a temperature jump which is rapid compared to the rate of a chemical reaction, the maximum change in absorbency can be calculated in the following manner. If the response of the photomultiplier tube is linear and the changes small, $\Delta V/V = \Delta I/I$ where V is the voltage, ΔV is the experimentally measured amplitude, and I is the light intensity. The relationship between the light intensity and the concentration is given by

$$I - I_0 \exp[-2.303\Sigma\epsilon_i c_i d] \tag{96}$$

where ϵ_i is the decadic extinction coefficient of the ith species present at a concentration of c_i and d is the path length. For small changes in concentration

$$\begin{aligned} -\Delta I/I &= 2.303\ \Sigma\epsilon_i d\Delta c_i \\ &= 2.303\ d\Sigma\epsilon_i\nu_i\Delta\xi \end{aligned} \tag{97}$$

where a single chemical reaction has been assumed. Differentiation of Eq. (97) with respect to temperature gives

$$\begin{aligned} -\Delta I/I &= 2.303\ d\Delta T(\Sigma\nu_i\epsilon_i)(\partial\xi/\partial T)_{P,A} \\ &= 2.303\ d\Delta T(\Sigma\nu_i\epsilon_i)\Gamma\Delta H/(RT^2) \end{aligned} \tag{98}$$

where Eq. (82a) has been utilized. For a bimolecular reaction such as Eq. (6)

$$-\Delta I/I = 2.303\ d\Delta T \frac{\Delta H}{RT^2}[\epsilon_X - \epsilon_S - \epsilon_E]/[1/(\bar{E}) + 1/(\bar{S}) + 1/(\bar{X})]$$

If all the extinction coefficients are known ΔH can be calculated from temperature jump data; conversely, if ΔH is known $\Sigma\nu_i\epsilon_i$ can be determined. Even if both of these quantities are unknown the concentration dependence of the amplitude can be calculated if the equilibrium constant of the reaction is known.

The situation should also be mentioned where coupling occurs such that a temperature jump may cause concentration changes other than those resulting from the temperature dependence of the reaction. For example, in buffered systems where a pH jump is produced by the temperature jump

$$(\partial\xi/\partial T)_{P,A} = (\partial\xi/\partial T)_{C_B,P,A} + (\partial\xi/\partial C_B)_{T,P,A}(\partial C_B/\partial T)_{P,A,\xi} \quad (99)$$

where C_B is the buffer species. The first term on the right-hand side of the equation is the direct effect of the temperature jump on the chemical reaction, while the second term is the effect of the change in the buffer concentration on the chemical reaction. Because of such complexities, relaxation amplitudes must be analyzed with considerable care.

For ultrasonic measurements, the amplitude is proportional to $[\Delta V - \Delta H\alpha_P^\infty/(\rho C_P^\infty)]^2$ [Eqs. (88) and (89)]. In water solutions the volume term usually predominates, while in nonaqueous solutions the enthalpy term is usually larger. Thus, depending on the system, ΔV or ΔH can be obtained from ultrasonic experiments. For complex mechanisms, the experimentally measured parameters can be related to the normal thermodynamic variables and thermodynamic quantities can be extracted from the data (cf. *15, 66, 67*).

IV. Application of Fast Reaction Techniques

Although experimental results are not reviewed extensively in this chapter, a brief summary of some of the most important information obtained about enzymic reactions with fast reaction techniques is now presented [cf. references *16–19* for more comprehensive reviews of experimental results].

A. Enzyme–Substrate Reactions

A class of reactions which has been particularly illuminated by fast reaction methods is the interaction between enzyme and substrate. Table I contains a summary of the second-order rate constants characterizing the formation of the initial enzyme–substrate complex and of the first-

66. G. G. Hammes and W. Knoche, *J. Chem. Phys.* **45**, 4041 (1966).
67. G. G. Hammes and A. C. Park, *JACS* **90**, 4151 (1968).

TABLE I
ASSOCIATION AND DISSOCIATION RATE CONSTANTS OF ENZYME–SUBSTRATE REACTIONS

Enzyme	Substrate	$10^{-8}\,k_1$ (M^{-1} sec^{-1})	k_{-1} (sec^{-1})	Reference
Aspartate aminotransferase	Glutamate, aspartate	>0.1–1	$>10^5$–10^6	*68*
	Oxalacetate, ketoglutarate	>1	$>10^4$	*68*
	α-Methyl aspartate	1.2×10^{-4}	130	*56*
	erythro-β-Hydroxyaspartate	0.031	1.1×10^4	*57*
	NH_2OH	0.037	62	*69*
Chymotrypsin	Proflavin	1.1	2.15×10^3	*70*
	Furylacryloyl-L-tryptophanamide	6.2×10^{-2}	2.7×10^3	*71*
Catalase	H_2O_2	0.05	—	*72*
Catalase-H_2O_2	H_2O_2	0.15	—	*72*
Creatine kinase	ADP	0.22	1.8×10^4	*58*
	MgADP	0.053	5.1×10^3	*58*
	CaADP	0.017	1.2×10^3	*58*
	MnADP	0.074	4.1×10^3	*58*
Glyceraldehyde-3-phosphate dehydrogenase	NAD	0.19, 0.0137	1×10^3, 210	*55*
Lactate dehydrogenase (rabbit muscle)	NADH	~10	$\sim 10^4$	*73*
Lactate dehydrogenase (pig heart)	NADH	0.546	39	*74*
	Oxamate	0.081	17	*74*
	3-Thio-NAD	0.058	410	*74*
Lysozyme	(*N*-Acetyl-D-glucosamine)$_2$[a]	0.0456	950	*63*
	(*N*-Acetyl-D-glucosamine)$_3$	0.0444	28	*63*
Malate dehydrogenase	NADH	5	50	*75*
Old yellow enzyme	FMN	0.015	$\sim 10^{-4}$	*76*
Peroxidase	H_2O_2	0.09	<1.4	*77*
	Methyl H_2O_2	0.015	<2.2	*77*
	Ethyl H_2O_2	0.036	—	*77*
Peroxidase-H_2O_2	Hydroquinone	0.023	—	*78*
	Cytochrome c	1.2	—	*79*
Pyruvate carboxylase-Mn^{2+}	Pyruvate	0.045	2.1×10^4	*80*
Pyruvate kinase-Mn^{2+}	Fluorophosphate	0.13	3.4×10^4	*81*

TABLE I (*Continued*)

Enzyme	Substrate	$10^{-8}\, k_1$ (M^{-1} sec^{-1})	k_{-1} (sec^{-1})	Reference
Ribonuclease	Cytidine 3′-phosphate	0.46	4.2×10^3	*53*
	Uridine 3′-phosphate	0.78	1.1×10^4	*53*
	Cytidine 2′,3′-cyclic phosphate	0.2–0.4	$1–2 \times 10^4$	*51*
	Uridine 2′,3′-cyclic phosphate	0.1	2×10^4	*54*
	Cytidylyl 3′,5′-cytidine	0.14	7×10^3	*52*

[a] β-(1 → 4) linkages.

order rate constants characterizing the reverse process of dissociation. If a pH range has been studied, the maximum measured value of the second-order rate constant is tabulated together with the associated dissociation rate constant.

For physiological substrates the second-order rate constant generally is in the range 10^7–10^8 M^{-1} sec^{-1}, which approaches but is somewhat less than the value expected for a diffusion controlled process. For modified substrates (cf. aspartate aminotransferase entries) this rate constant can be considerably smaller, suggesting that the molecular fit between enzyme and substrate is of crucial importance in the efficiency of this initial step. The dissociation rate constants can be regarded as directly reflecting the strength of the enzyme–substrate interactions.

68. P. Fasella and G. G. Hammes, *Biochemistry* **6,** 1798 (1967).
69. G. G. Hammes and P. Fasella, *JACS* **85,** 3929 (1963).
70. B. H. Havsteen, *JBC* **242,** 769 (1967).
71. G. P. Hess, J. McConn, E. Ku, and G. McConkey, *Phil. Trans. Roy. Soc.* **B256,** 27 (1969).
72. B. Chance, *in* "Curents in Biochemical Research" (D. E. Green, ed.), p. 308. Wiley (Interscience), New York, 1956.
73. G. Czerlinski and G. Schreck, *JBC* **239,** 913 (1964).
74. H. D'A. Heck, *JBC* **244,** 4375 (1969).
75. G. Czerlinski and G. Schreck, *Biochemistry* **3,** 89 (1963).
76. H. Theorell and A. Nygaard, *Acta Chem. Scand.* **8,** 1649 (1954).
77. B. Chance, *Arch. Biochem.* **22,** 224 (1949).
78. B. Chance, *Arch. Biochem.* **24,** 410 (1949).
79. B. Chance, *in* "Enzymes and Enzyme Systems" (J. T. Edsall, ed.), p. 93. Harvard Univ. Press, Cambridge, Massachusetts, 1951.
80. A. S. Mildvan and M. C. Scrutton, *Biochemistry* **6,** 2978 (1967).
81. A. S. Mildvan, J. S. Leigh, and M. Cohn, *Biochemistry* **6,** 1805 (1967).

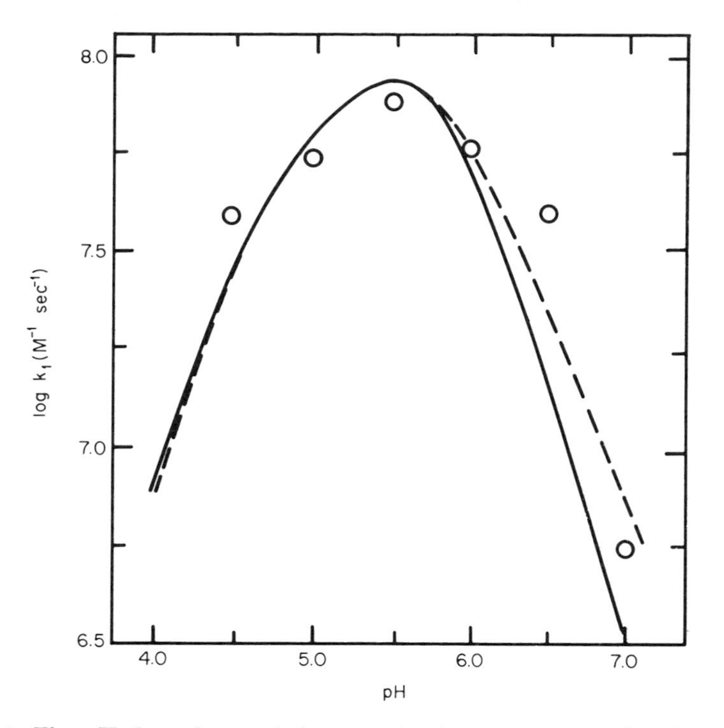

FIG. 9. The pH dependence of the second-order rate constant for the reaction between uridine 3′-phosphate and ribonuclease. The dashed and solid lines represent theoretical curves for two limiting mechanisms (*53*).

The pH dependence of the second-order rate constant is determined by the ionizable groups on the free enzyme and substrate which are of importance in the binding process. The pH dependence of the second-order rate constant characterizing the reaction between uridine 3′-phosphate and ribonuclease A is shown in Fig. 9 (*53*). From these and other data, it has been deduced that ionizable groups on the enzyme with pK values of 5.4 and 6.4 are of importance. The lines in Fig. 9 have been calculated on the basis of binding mechanisms involving these groups (*53*). These pK values can be associated with imidazole residues known to be at the active site.

The pH dependence of the dissociation rate constant is determined by ionizable groups in the enzyme–substrate complex. Because of the presence of substrate, the pK values associated with enzyme–substrate complexes can deviate considerably from those normally encountered; therefore, it is difficult to identify specific ionizable groups on the enzyme from the pH dependence of the dissociation rate constant.

TABLE II
RATE OF CONFORMATIONAL CHANGES OF ENZYMES AND ENZYME–SUBSTRATE COMPLEXES

Enzyme	Substrate	Approximate time constants (sec)	Reference
Alkaline phosphatase	2-Hydroxy-5-nitrobenzyl phosphate	10^{-2}	*82*
Aspartate aminotransferase	α-Methyl aspartate	10^{-2}	*56*
	erythro-β-Hydroxyaspartate	10^{-3} to 10^{-1}	*57*
Chymotrypsin	Proflavin, furylacryloyl-L-tryptophanamide	10^{-4}, 10^{-2}	*70, 71*
Creatine kinase	ADP, MgADP, CaADP, MnADP, and ATP	10^{-4}	*58*
Glyceraldehyde-3-phosphate dehydrogenase	NAD	1	*55*
Lactate dehydrogenase (rabbit muscle)	NADH	10^{-3}	*73*
Liver alcohol dehydrogenase	NADH-imidazole	10^{-3}	*83*
Peroxidase	H_2O_2, methyl and ethyl H_2O_2	10^{-1}	*77*
Pyruvate kinase	None, Mg^{2+}, Mn^{2+}	10^{-4}	*84*
Ribonuclease	None, cytidine 3′-phosphate, uridine 3′-phosphate, cytidine 2′,3′-cyclic phosphate, uridine 2′,3′-cyclic phosphate, cytidylyl 3′,5′-cytidine	10^{-3} to 10^{-4}	*48–54*

B. CONFORMATIONAL CHANGES

In many cases the process of substrate binding involves a two-step mechanism which can be represented as

$$E + S \rightleftarrows X_1 \rightleftarrows X_2 \tag{100}$$

Since covalent changes are not involved, this finding furnishes convincing evidence that the binding of substrate can induce conformational changes in the enzyme and that such processes are a common feature of enzymic reactions. The approximate time constants associated with the second step of Eq. (100) are summarized in Table II. Included in this table are time constants associated with conformational changes of the

82. S. E. Halford, N. G. Bennett, D. R. Trentham, and H. Gutfreund, *BJ* **114,** 243 (1969).
83. G. Czerlinski, *BBA* **64,** 199 (1962).
84. G. G. Hammes and J. Simplicio, *BBA* (1970) (in press).

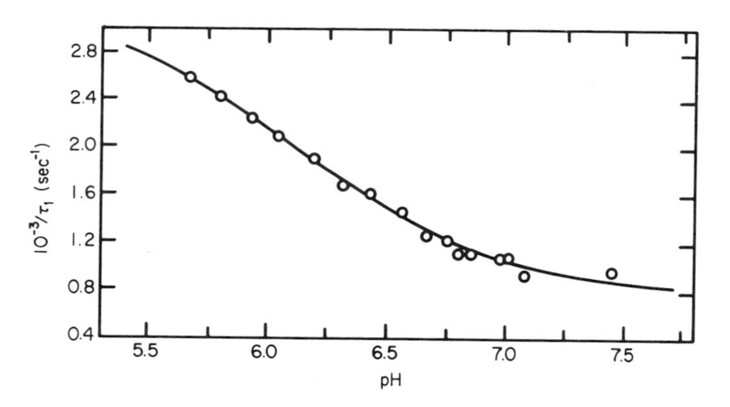

FIG. 10. The pH dependence of the reciprocal relaxation time characterizing the isomerization of free ribonuclease at 25°. The solid line is calculated according to Eq. (102) with the parameters given in the text (*49*).

free enzyme. The rates of almost all the conformational changes are sufficiently rapid so that they could be of significance in the catalytic process.

The pH dependence of kinetic parameters associated with conformational changes can provide information about ionizable groups on the enzyme involved in such processes and may provide some clue to the molecular basis of the change. In Fig. 10, the pH dependence of the reciprocal relaxation time associated with a conformational change in free ribonuclease is shown (*49*). A mechanism consistent with these results is

$$E + H^+ \underset{}{\overset{K_A}{\rightleftharpoons}} EH \underset{k_{-1}}{\overset{k_1}{\rightleftharpoons}} E'H \tag{101}$$

where EH and E′H represent different enzyme conformations. If the protolytic reaction is assumed rapid relative to the rate of the isomerization and hydrogen ion is buffered

$$1/\tau = k_{-1} + k_1/[1 + K_A/(H^+)] \tag{102}$$

The curve in Fig. 10 is calculated from Eq. (102) with $k_1 = 2468$ sec^{-1}, $k_{-1} = 780$ sec^{-1}, and p$K_A = 6.1$ (*49*). These data, together with other considerations, have led to a proposed mechanism which involves the opening and closing of a groove in the structure associated with the active site (*19, 53*).

In virtually all of the enzyme systems which have been extensively investigated with fast reaction techniques, a large number of discrete kinetic processes have been observed. These involve both conformational and covalent changes. For example, the reaction of *erythro*-β-hydroxyaspartate with aspartate aminotransferase to give the pyridoxamine enzyme and dihydroxyfumarate (half of the enzymic transamination reaction) is associated with eight relaxation times (*57*). The occurrence of multiple reaction intermediates may be an important general feature of enzymic catalysis [cf. Hammes (*19*) for a discussion of this point].

C. Concluding Remarks

The most unique information obtained from the study of enzymic reactions with fast reaction techniques can be recapitulated as follows. The binding and specific recognition between enzyme and substrate is very rapid. However, the specific and precise nature of this interaction is illustrated by the finding that small changes in the molecular geometry of the substrate can cause a considerable lowering of the bimolecular rate constant. Stabilization of the enzyme–substrate complex occurs, in part, through conformational changes subsequent to the initial binding reaction.

Multiple, but discrete, enzyme–substrate complexes have been observed, and the lifetimes and rates of interconversion of these states have also been determined. Similarly, the existence of different enzyme conformational states has been demonstrated, and the kinetic parameters associated with interconversion of these states have been obtained. In some cases, the nature of critical ionizable groups can be deduced from the pH dependence of the kinetic parameters.

Delineation of the discrete steps in enzymic catalysis as outlined above has permitted the postulation of detailed reaction mechanisms for enzymic catalysis. When sufficient structural data are available, these mechanisms can be expressed in terms of the molecular structure of enzyme and substrate.

3

Stereospecificity of Enzymic Reactions

G. POPJÁK

I. Introduction

The first stereospecific enzyme catalysis was recorded by Pasteur in 1858 (*1*); he found a form of yeast which fermented dextrorotatory tartaric acid but left the levorotatory acid intact.

The discovery of the dextro- and levorotatory tartaric acids by Pasteur and the stereospecific metabolism of the one enantiomer by microorganisms was followed by the theory of the tetrahedral structure of the carbon atom, proposed independently by Van't Hoff and Le Bell in 1874 [cf. Cohen (*2*)] which was to account for the optical activity observed in many organic substances. In spite of Kolbe's vitriolic attack on Van't Hoff [cf. Krebs (*3*)] the hypothesis of the tetrahedral structure of the valence bonds of carbon triumphed and has been vindicated fully by quantum mechanical calculations and by the establishment of the absolute configuration of the sodium-rubidium salt of (+)-tartaric acid (Fig. 1) by the anomalous X-ray diffraction pattern observed on irradiating the double salt with zirconium $K\alpha$-rays which just excite the rubidium atom (*4*). The absolute configuration of over 100 other reference substances has been determined during the last eighteen years (*5–7*). The latter achievement made possible the representation of the true three-

1. L. Pasteur, *Compt. Rend.* **46,** 615 (1858).
2. E. Cohen, "Jacobus Henricus Van't Hoff. Sein Leben und Wirken." Akad. Verlagsges., Leipzig, 1912.
3. H. A. Krebs, *in* "Current Aspects of Biochemical Energetics" (N. O. Kaplan and E. Kennedy, eds.), p. 93. Academic Press, New York, 1966.
4. J. M. Bijvoet, A. F. Peerdeman, and A. J. von Bommel, *Nature* **108,** 271 (1951).
5. F. H. Allen and D. Rogers, *Chem. Commun.* p. 838 (1966).
6. F. H. Allen, S. Neidle, and D. Rogers, *Chem. Commun.* p. 308 (1968).
7. F. H. Allen, S. Neidle, and D. Rogers, *Chem. Commun.* p. 452 (1969).

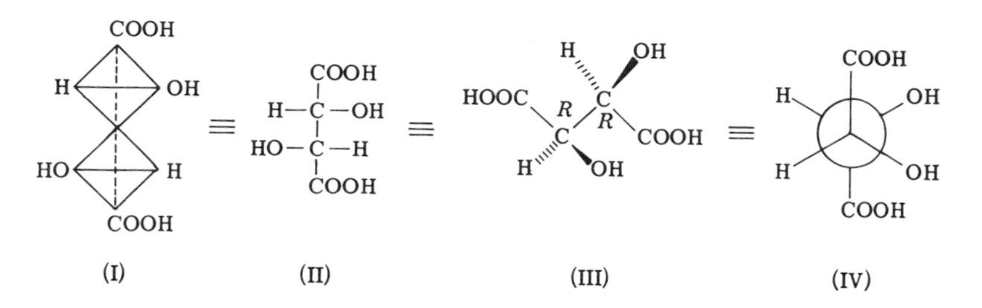

FIG. 1. Stereochemistry of (+)-tartaric acid: (I) and (II) show the configuration chosen arbitrarily by Fischer and represented according to his convention; (III) is the absolute configuration determined by Bijvoet *et al.* (*4*). In (III) the bonds shown by continuous thin lines lie in the plane of the paper, while the bonds represented by the thick arrows and dotted lines are projecting above and below the plane of the paper, respectively. A Newman projection of (III) is shown in (IV) in a conformation when the two carboxyl groups are *antiperiplanar* (*ap*), and hence the OH groups at C-2 and C-2′ are *plus synclinal* (*Psc*) according to the convention of Cahn *et al.* (*34*) which also denotes the absolute configuration of (+)-tartaric acid at *RR* (*31*). The important point to note is that the absolute configuration of (+)-tartaric acid as determined is identical with the one assumed "fortuitously" by Fischer.

dimensional features of organic molecules instead of their configuration relative to an arbitrarily chosen standard as in the Fischer–Rosanoff system (*8*). According to personal inclination, one may attribute it to mere good fortune or to Fischer's intuitive genius that the absolute configuration of (+)-tartaric acid turned out to be identical with the one which followed from a correlation with (+)-saccharic acid and (+)-glyceraldehyde taken by Fischer and Rosanoff (*8*) as standards and to which many other correlations had been made by chemical means. Hence, after the achievements of X-ray crystallography the right hand still remained the right hand and there was no need for an inversion of the thinking about the asymmetrical chemical world.

It is beyond the scope of the present chapter to discuss chemical problems of isomerism, or indeed all the aspects of stereospecificity of enzymic reactions, for each subject has grown to dimensions needing much more space than is available here. A recent treatise in two volumes by Bentley (*9*) presents in detail the available knowledge on both topics.

Parallel with the growth of knowledge of stereochemistry there grew up an increased awareness of the asymmetrical nature of enzyme catalysis. In spite of exceptions [cf. Chapter 6 in Bentley (*9*)], it is generally

8. M. A. Rosanoff, *JACS* **28**, 114 (1906).
9. R. Bentley, "Molecular Asymmetry in Biology," Vols. 1 and 2. Academic Press, New York, 1969, 1970.

true that all forms of life, when faced with the possibility of an asymmetrical carbon atom, elect to synthesize either the D or L series—rarely both—in one and the same cell, or in the same cell compartment. Such observations led to the tacit acceptance of the conclusion, even before concrete knowledge of protein structures existed, that the enzymes were asymmetrical reagents, themselves composed of asymmetrical units. Such notions were much strengthened by the discovery that enzymes can distinguish between two identical ligands in a structure of the $Ca'a''bd$ type. This type of discovery, coupled with refinements of chemical and physicochemical techniques, the use of hydrogen isotopes for stereospecific labeling at prochiral centers (*10*), and sensitive methods for the detection of asymmetry created by such labeling extended the scope of investigating the stereospecificity of enzymic reactions to limits inconceivable twenty years ago. Measuring the optical activity of a substance, resulting entirely from the stereospecific replacement of a hydrogen (protium) atom by deuterium at a prochiral center and giving a specific rotation of (+) or (—) 0.1° at the sodium D line is not at all impossible to achieve on a few milligrams of sample since the measurements in the region of the visible spectrum can be made to an accuracy of ±0.00025° and in the ultraviolet to an accuracy of ±0.001°. Changes in the NMR spectrum of a substance stereospecifically deuterated at a methylene group have also been used to deduce the absolute configuration of the deuterium atom in relation to a neighboring asymmetrical center of known absolute configuration (cf. fumarate hydratase, Section VI,C). A unique achievement, reported recently, has been the stereospecific chemical synthesis of the two enantiomers of CHDT·COOH [(V) and (VI)] (*11, 12*) and the

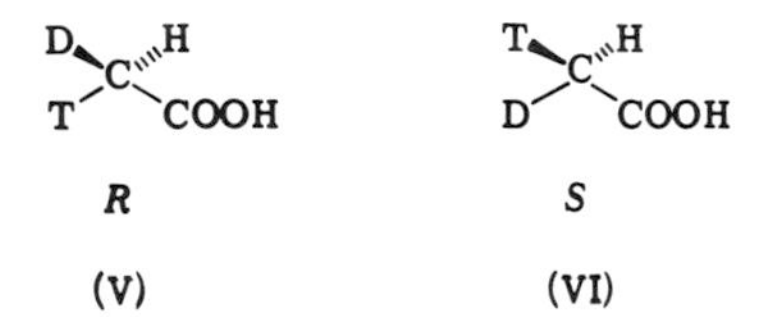

use of acetyl-CoA made from these enantiomers for the study of malate synthetase (EC 4.1.3.2). A kinetic isotope discrimination between protium, deuterium, and tritium has been exploited successfully to demonstrate that the synthesis of malate from acetyl-CoA and glyoxylate occurs with an inversion of configuration at the methyl group of acetyl-

10. For definition of chirality and prochirality, see (*31*).
11. J. W. Cornforth, J. W. Redmond, H. Eggerer, W. Buckel, and C. Gutschow, *Nature* **221**, 1212 (1969).
12. J. Lüthy, J. Rétey, and D. Arigoni, *Nature* **221**, 1213 (1969).

CoA. The use of hydrogen isotopes, coupled with chemical, physical, and physicochemical techniques, has been at the core of many studies of enzyme stereospecificity, and these will be emphasized in this chapter. Because of the very large volume of information accumulated to date it will be possible to review here only selected examples of enzymic reactions; a comprehensive treatment of the subject can be found in Volume II of Bentley's treatise (*9*).

The stereospecificity of some enzymic reactions has been presented in the previous edition of this work (*13*) and was reviewed by Levy *et al.* (*14*) and more recently by Cornforth and Ryback (*15*).

II. Enzyme Specificity and Historical Background to Recent Investigations

The most outstanding characteristics of enzymes are their specificity for substrates; their specificity to catalyze defined chemical reactions, each enzyme being able to carry out only one type of reaction; and their extraordinary catalytic power achieved, still in a mysterious way, by lowering the activation energies of reactions. Whatever might be the ultimate explanation for enzyme catalysis, it cannot be doubted that these characteristic properties are related to the binding of substrates to defined areas of the protein in specific conformations. This is now understood for enzymes whose structure has been fully determined such as lysozyme (*16–18*), ribonuclease (*19–21*), and chymotrypsin (*22*). The ultimate conformation of the enzyme–substrate complex is probably determined by the mutual interaction between protein and substrate, each imposing conformational changes on the other [for discussion cf., e.g.,

13. H. Gutfreund, "The Enzymes," 2nd ed., Vol. 1, p. 233. Academic Press, New York, 1959.
14. H. R. Levy, P. B. Talalay, and B. Vennesland, *Progr. Stereochem.* **3,** 299 (1962).
15. J. W. Cornforth and G. Ryback, *Chem. Soc. (London), Rept.* **62,** 428 (1965).
16. C. C. F. Blake, D. F. Koenig, G. A. Mair, A. C. T. North, D. C. Phillips, and V. R. Sarma, *Nature* **206,** 757 (1965).
17. C. C. F. Blake, L. N. Johnson, G. A. Mair, A. C. T. North, D. C. Phillips, and V. R. Sarma, *Proc. Roy. Soc.* **B167,** 378 (1967).
18. D. C. Phillips, *Sci. Am.* **215,** No. 5, 78 (1966).
19. G. Kartha, J. Bello, and D. Harker, *Nature* **213,** 862 (1967).
20. H. W. Wyckoff, K. D. Hardman, N. M. Allewell, T. Inagami, D. Tsernoglou, L. N. Johnson, and F. M. Richards, *JBC* **242,** 3749 (1967).
21. H. W. Wyckoff, K. D. Hardman, N. M. Allewell, T. Inagami, L. N. Johnson, and F. M. Richards, *JBC* **242,** 3984 (1967).
22. B. W. Matthews, P. B. Sigler, R. Henderson, and D. M. Blow, *Nature* **214,** 652 (1967).

Koshland and Neet (*23*)]: the substrate causing conformational changes in the protein and the protein imposing specific spatial orientation of substrate molecules. The stereospecificity of enzymic reactions also must be the result of such close interactions; if this is the case, an intimate study of enzyme stereospecificity might be expected to furnish information about the space and its geometry around the active and binding regions of enzymes. If such a goal can be achieved, it will furnish vital clues as to enzyme catalysis and structure even before the three-dimensional structure of each enzyme is established. Some efforts in this direction have already been made (*24–28*).

The stereospecificity of enzyme catalysis extends beyond molecules containing asymmetrical centers to differentiation between two identical substituents (a' and a'') on a carbon atom substituted with two further but dissimilar (b and d) groups. Inspection of a model of the $Ca'a''bd$ molecule shows (Fig. 2) that its plane of symmetry, running through groups b, d, and the central carbon atom, bisects it into two halves which are not superimposable. If one looks at the mirror image of this molecule, a superimposable structure is, however, obtained merely by rotating the

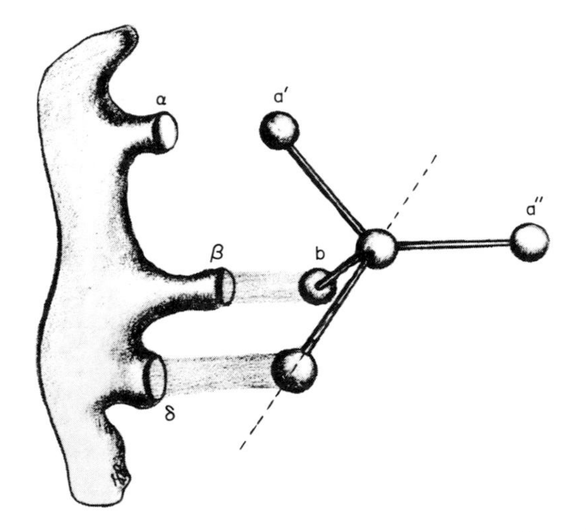

FIG. 2. Model of a $Ca'a''bd$ molecule in conjunction with a hypothetical enzyme area containing binding sites, β and δ, specific for groups b and d, respectively, and a catalytic site α, specific for a' or a''. In the model only group a' can approach the catalytic site.

23. D. E. Koshland, Jr., and K. E. Neet, *Ann. Rev. Biochem.* **37,** 359 (1968).
24. V. Prelog, *Pure Appl. Chem.* **9,** 119 (1964).
25. J. M. H. Graves, A. Clark, and H. J. Ringold, *Biochemistry* **4,** 2655 (1965).
26. H. M. Kagan, L. R. Manning, and A. Meister, *Biochemistry* **4,** 1063 (1965).
27. A. Meister, *Advan. Enzymol.* **31,** 183 (1968).
28. J. Rétey, A. Umani-Ronchi, and D. Arigoni, *Experientia* **22,** 72 (1966).

molecule around an angle of 180°. It is the consequence of the tetrahedral structure of the carbon atom that, although the *Ca′a″bd* molecule is optically inactive, the like groups *a′* and *a″* in it are sterically not equivalent and are distinguished by an asymmetrical reagent such as an enzyme.

Hirschmann (*29*) analyzed in detail the structural requirements for the differentiation of two like groups in a molecule and arrived at the following rule: ". . . a compound permits the differentiation of two structurally identical groups (*a′*) and (*a″*) if and only if it is impossible to superpose a model of the compound upon itself in such a way that the (*a′*) and (*a″*) groups coincide and the two arrangements are otherwise indistinguishable" (*29*, p. 2763, end of second paragraph). It is impossible to find an arrangement for the *Ca′a″bd* molecule in which the *a′* and *a″* groups would coincide and the *b* and *d* groups would occupy at the same time indistinguishable steric positions.

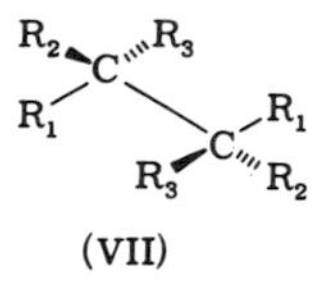

(VII)

The *Ca′a″bd* molecule has some similarity to the *meso* form of a compound (VII) containing two asymmetrical centers of opposite configuration in that it has also a plane of symmetry, but the two halves of the molecule are not superimposable. Because of this similarity, the carbon atom in the *Ca′a″bd* structure became known as the *meso*-carbon atom (*30*). The *meso*-carbon atom is potentially an asymmetrical center, since, if either the *a′* or *a″* group in the *Ca′a″bd* molecule is replaced by a group different from the other and differing also from *b* and *d*, a true optical asymmetry is created. Probably because of the chemists' dislike of the term *meso*-carbon atom, such a potentially asymmetrical center has been renamed a *prochiral center* (*31, 35*).

The first noted example of a substance of the *Ca′a″bd* structure behaving asymmetrically in enzymic reactions was that of citric acid. When

29. H. Hirschmann, *JBC* **235**, 2762 (1960).

30. P. Schwartz and H. E. Carter, *Proc. Natl. Acad. Sci. U. S.* **40**, 499 (1954).

31. For the description of absolute configuration around asymmetrical centers, the *R* and *S* system of Cahn, Ingold, and Prelog (*32–34*) will be used mostly in this chapter. These authors (*34*) revived the term *chiral* (derived from the Greek χειρ, meaning hand, and probably first introduced in the nineteenth century; cf. Webster's Dictionary, 2nd edition) for the description of structures containing asymmetrical carbon atoms; i.e., for such molecules for which neither rotation around the axes of symmetry, nor reflection in planes of symmetry, nor a combination of such maneuvers can bring the molecules into coincidence with themselves. Chirality thus ex-

in 1937 Krebs and Johnson (*36*) demonstrated the synthesis of citric acid in animal tissues from oxalacetate and from a then unknown substance and when the fixation of CO_2 by heterotrophic microorganisms by the carboxylation of pyruvate (Wood–Werkman reaction) was demonstrated (*37, 38*), Wood and Werkman (*38*) suggested that oxalacetate might arise in animal tissues also by the same reaction as in microorganisms. This became highly probable when Krebs and Eggleston (*39*) found that the oxidation of pyruvate and the formation of citrate, α-ketoglutarate, fumarate, and malate were stimulated in pigeon liver mince by the presence of CO_2. Evans and Slotin (*40, 41*) and Wood *et al.* (*42, 43*) proved the foregoing speculation correct by the isolation of isotopically labeled α-ketoglutarate from pigeon liver mince incubated, in the presence of malonate, with pyruvate and $^{11}CO_2$ or $^{13}CO_2$. Because practically all the isotope in the α-ketoglutarate was found in the carboxyl carbon atom adjacent to the carbonyl group, and because of the erroneous assumption that the symmetrical molecule of citric acid reacted randomly around its plane of symmetry in enzymic reactions, it was concluded that citrate could not be an intermediate in the formation of α-ketoglutarate.

It was not until 1948 that Ogston (*44*) questioned the validity of excluding, in experiments with isotopes, symmetrical compounds as intermediates in the formation of asymmetrically labeled substances. Ogston wrote: "These conclusions seem to arise from the fallacy that because symmetrical products arising from the *d*- or *l*-form of an optically active precursor cannot be distinguished, therefore the two identical groups of a

---

presses the necessary condition for the existence of enantiomers of a particular substance. The term *prochiral* was coined from the Greek *pro-*, meaning before in time or place, and *chiral* (*35*); a prochiral center or a prochiral tetrahedral center (= *meso*-carbon atom) could be said to describe a state of the center before acquisition of chirality. Structures of the $C(a)_4$, $C(a)_3b$, and $C(a)_2(b)_2$ type are called *achiral.*

32. R. S. Cahn and C. K. Ingold, *JCS* **612** (1951).

33. R. S. Cahn, C. K. Ingold, and V. Prelog, *Experientia* **12,** 81 (1956).

34. R. S. Cahn, Sir Christopher Ingold, and V. Prelog, *Angew. Chem. Intern. Ed. Engl.* **5,** 385 (1966).

35. K. R. Hanson, *JACS* **88,** 2731 (1966).

36. H. A. Krebs and W. A. Johnson, *BJ* **31,** 645 (1937).

37. H. G. Wood and C. H. Werkman, *BJ* **30,** 48 (1936).

38. H. G. Wood and C. H. Werkman, *BJ* **32,** 1262 (1938).

39. H. A. Krebs and L. V. Eggleston, *BJ* **34,** 1383 (1940).

40. E. A. Evans, Jr. and L. Slotin, *JBC* **136,** 301 (1940).

41. E. A. Evans, Jr. and L. Slotin, *JBC* **141,** 439 (1941).

42. H. G. Wood, C. H. Werkman, A. Hemingway, and A. O. Nier, *JBC* **139,** 483 (1941).

43. H. G. Wood, C. H. Werkman, A. Hemingway, and A. O. Nier, *JBC* **142,** 31 (1942).

44. A. G. Ogston, *Nature* **162,** 963 (1948).

symmetrical product formed from one optical antipode cannot be distinguished. On the contrary it is possible that an asymmetric enzyme which attacks a symmetrical compound can distinguish between its identical groups." Ogston illustrated his argument with the example of aminomalonic acid (VIII) and showed how the two carboxyl groups of this

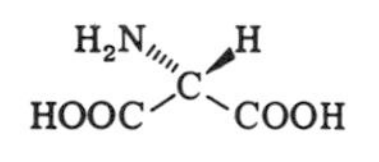

(VIII)

symmetrical substance could be differentiated by its attachment to the enzyme at three points. Ogston's postulate can be best appreciated by the examination of the model of the $Ca'a''bd$ molecule in conjunction with a hypothetical enzyme area in which there are two binding sites, β and δ, specific for groups b and d, and one catalytic site, α, specific for group a' or a'' (Fig. 2). There is only one orientation of this molecule on the enzyme which permits the approach of one, and only one, of the two identical groups to the catalytic site (that of a' in the example of Fig. 2). A differentiation between two identical groups in a substance of this type of structure is clearly possible.

Of course, Ogston's hypothesis is no more than an extension of the polyaffinity theory offered by Bergmann and his colleagues (*45–47*) as an explanation for the stereospecificity of enzyme catalysis involving optically active substrates. Bergmann and his associates postulated that if an enzyme reacted with only one of two antipodes of a substrate, then the enzyme ". . . must contain at least three different atoms or atomic groups which are fixed in space with respect to one another, these groups entering into relation with a similar number of different atoms or atomic groups of the substrate."

Although the polyaffinity theory of enzymic action was formulated originally in relation to antipodes of optically active substrates, it is also clearly applicable to the differentiation by an enzyme of the two like groups in the $Ca'a''bd$ structure. It is not necessary at all that the catalytic site of the enzyme be on the same side of the plane of symmetry of the $Ca'a''bd$ molecule as are the two primary points of attachment β and δ shown in Fig. 2; the catalytic group on the enzyme may be on the opposite side of the plane of symmetry, and in that event the enzyme will attack specifically the a'' group instead of a'.

It seems strange, in retrospect, that neither Van't Hoff's formulation

45. M. Bergmann, L. Zervos, J. S. Fruton, F. Schneider, and H. Schleick, *JBC* **109,** 325 (1935).
46. M. Bergmann, L. Zervos, and J. S. Fruton, *JBC* **115,** 893 (1936).
47. M. Bergmann and J. S. Fruton, *JBC* **117,** 193 (1937).

of the tetrahedral structure of the carbon atom nor Bergmann's hypothesis alerted biochemists sooner than Ogston's call in 1948 to the possibility of an asymmetrical treatment by enzymes of seemingly symmetrical structures.

The three-point attachment of a symmetrical substrate to the enzyme is the simplest way of visualizing differentiation of two like groups around a prochiral carbon atom but is no longer thought to be a prerequisite. Schwartz and Carter (*30*) demonstrated, for example, that even without attachment to a surface, there is a steric preference, although small, in a reaction between an asymmetrical reagent and a symmetrical molecule containing a prochiral carbon atom. They found that the reaction between 3-phenylglutaric anhydride (IX) and (—)-α-phenylethylamine (X) gave an approximately 6:4 mixture of two diastereomeric amides (XI) and (XII). A similar inequality of diastereomeric products between (IX) and (—)-menthol was found by Altschul *et al.* (*48*).

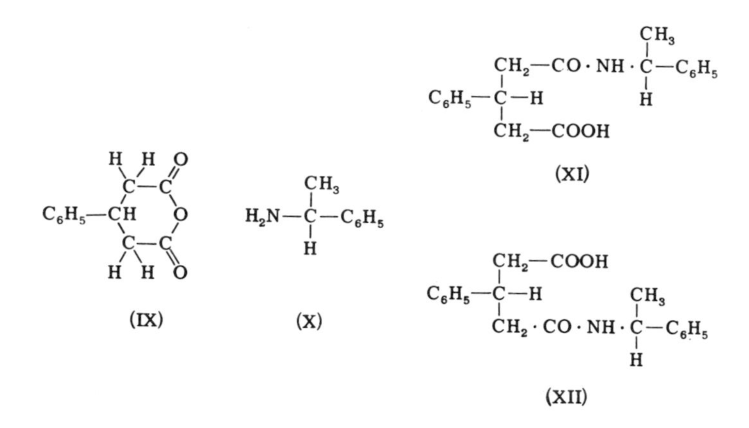

These chemical examples are of very poor stereospecificity in comparison with the very strict stereospecificity of enzymic reactions (*49*). In order to explain such strict stereospecificity it is essential to postulate unique conformation of substrate when bound to the enzyme. However, there is no need to assume that such conformation can be attained only by a three-point attachment of the substrate: Binding by a single group will suffice since a specific conformation can then be imposed on the substrate by the enzyme by steric factors or by interaction with lipophilic or hydrophilic groups. A broader definition of the requirements for the differentiation of like groups in the *Ca′a″bd* molecule in enzymic reac-

48. R. Altschul, P. Bernstein, and S. G. Cohen, *JACS* **78**, 5091 (1956).

49. Within the powers of experimental techniques available the enzymic reactions examined so far, involving prochiral carbon atoms, proved to have absolute stereospecificities.

tions was given by Popják and Cornforth (*50*): ". . . dissymmetric treatment of a substrate by an enzyme can occur whenever the enzyme imposes, whether actively by binding or passively by obstruction, a particular orientation on the substrate at the site of reaction."

Ogston's hypothesis in relation to citrate and α-ketoglutarate was tested by Potter and Heidelberger (*51*) and by Lorber *et al.* (*52*). It was only necessary to repeat the original experiments of Evans and Slotin (*41*) and of Wood *et al.* (*42, 43*) and carry them out in two stages. Citrate was first isolated from rat liver or rat kidney homogenates, incubated with pyruvate, oxalacetate, malonate, and $^{14}CO_2$, and found to be radioactive. This labeled citrate was then added to a second enzyme preparation from which ^{14}C-labeled α-ketoglutarate was isolated as the 2,4-dinitrophenylhydrazone. When this dinitrophenylhydrazone was oxidized with acid potassium permanganate to CO_2 and succinic acid, the entire radioactivity of the specimen was found in the CO_2 derived from the carboxyl carbon adjacent to the carbonyl group (*51*). Another, equally elegant version of this experiment was carried out by Lorber *et al.* (*52*) who used $HO_2{}^{14}C \cdot CH_2 \cdot CO \cdot CO_2H$ as a starting substrate. Radioactive citrate isolated from one enzyme preparation and added to a second one gave again α-ketoglutarate labeled exclusively in the α-carboxyl carbon atom. These experiments proved Ogston's hypothesis correct and, incidentally, reinstated citric acid as a member of the tricarboxylic acid cycle from which it was exiled for about eight years. The stereochemistry of the reactions of the tricarboxylic acid cycle has since been fully elucidated and will be discussed in further detail (Section VI).

The structure of glycerol is analogous to that of citrate: the prochiral central carbon atom of citrate carrying the —OH, —COOH, —CH_2COOH, and —CH_2COOH ligands being equivalent to the one of glycerol substituted by the ligands —H, —OH, —CH_2OH, and —CH_2OH. The first demonstration of asymmetrical behavior of glycerol in enzymic reactions was made, as in the case of citrate, by isotopic labeling. Popják *et al.* (*53*) obtained ^{14}C-glycerol from the milk triglycerides of a lactating goat injected with $CH_3{}^{14}COOH$; they adduced evidence to show that this labeled glycerol was derived in the mammary gland from the metabolism of 3,4-^{14}C-hexose. Degradation of this glycerol by periodate oxidation showed it to contain up to 95% of the isotopic label in C-1 + C-3 and only 5% in C-2. It was not possible to tell by the per-

50. G. Popják and J. W. Cornforth, *BJ* **101,** 553 (1966).
51. V. R. Potter and C. Heidelberger, *Nature* **164,** 180 (1949).
52. V. Lorber, M. F. Utter, H. Rudney, and M. Cook, *JBC* **185,** 689 (1950).
53. G. Popják, R. F. Glascock, and S. J. Folley, *BJ* **52,** 472 (1952).

iodate degradation method whether C-1 or C-3 of the glycerol contained more isotope or whether both positions were equally labeled. However, when this labeled glycerol was fed to rats, starved previously for 48 hr, and the liver glycogen was isolated 1 and 6 hr after the feeding, it was found that most of the radioactivity in the glucose units was confined to carbon atoms 3 and 4 (*54*). Since the specific activity of the glucose was as high as 75–85% of that of the glycerol fed, there could be no doubt that the carbons of glycerol entered the glucose as a single unit and that the ^{14}C-glycerol derived from the previous biological experiments was asymmetrically labeled either at C-1 or C-3.

This experiment, corroborated by the work of Swick and Nakao (*55*), led inevitably to the conclusion that not only was the glycerol used in the experiments asymmetrically labeled but also that it must have been phosphorylated in the rats exclusively to L-α-glycerophosphate (*R*-α-glycerophosphate) (XIII), which after dehydrogenation to dihydroxyace-

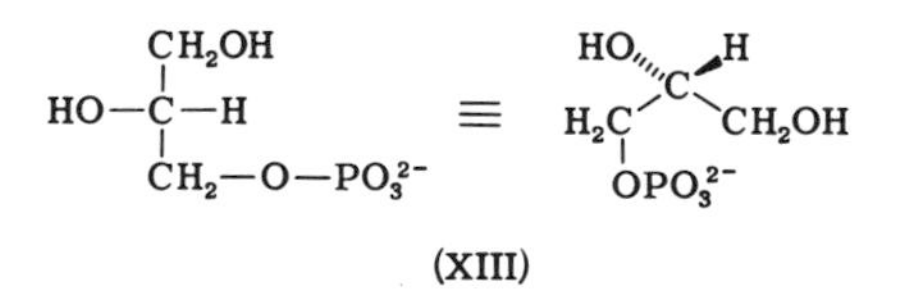

(XIII)

tone phosphate led, through known reactions, to glucose. The results of these experiments were much strengthened by the discovery by Bublitz and Kennedy (*56*) of glycerol kinase which was shown to phosphorylate glycerol with ATP to *R*-α-glycerophosphate. Previous to the work of Bublitz and Kennedy (*56*) *R*-α-glycerophosphate was thought to arise only by the reduction of dihydroxyacetone phosphate; had glycerol kinase been discovered earlier, the asymmetrical treatment by an enzyme of the *Ca′a″bd* molecule might have been appreciated before the experience with citric acid.

Levy *et al.* (*14*), in their excellent review of the steric course of enzymic reactions involving prochiral carbon atoms, pointed out that Van't Hoff's tetrahedral model of the chiral $C(R_1R_2R_3R_4)$ structure (*2*) contained the inherent assumption that there was no spontaneous shift in the relative positions of the four different ligands around the chiral center and that this same principle must apply to identical ligands around achiral or prochiral centers also. The experiments with the ^{14}C-glycerol of biosynthetic origin (*53, 54*) demonstrated this fact thoroughly, for the asymmetrical labeling in the glycerol, and in the substances derived from it, was maintained during synthesis of glycerol in the goat (ultimately

54. P. Schambye, H. G. Wood, and G. Popják, *JBC* **206,** 875 (1954).
55. R. W. Swick and A. Nakao, *JBC* **206,** 883 (1954).
56. C. Bublitz and E. P. Kennedy, *JBC* **211,** 951 (1955).

from $CH_3{}^{14}COOH$), during incorporation of the glycerol into milk triglycerides, during the chemical isolation of the glycerol which included saponifications of triglycerides, benzoylation of glycerol to the tribenzoate, hydrolysis of the tribenzoate, synthesis of glucose from glycerol in the rat, and, finally, in the fermentation of the glucose by *Leuconostoc mesenteroides* according to the reaction of Scheme 1 used for the partial degradation of the glucose isolated from liver glycogen (*57*).

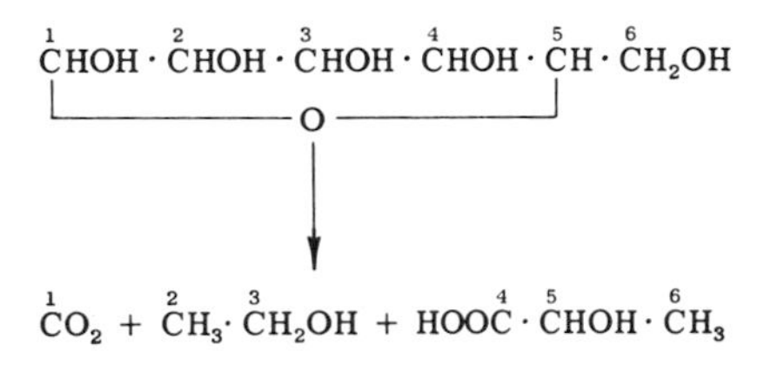

SCHEME 1. Degradation of glucose by *Leuconostoc mesenteroides* (*57*).

The example of glycerol illustrates well that enzymic reactions involving chemically identical ligands at a prochiral center are intrinsically stereospecific and are not artifacts of isotopic labeling, which is often the only means available for the demonstration of differentiation between the identical ligands; the same conclusion was reached from the isotopic labeling of glycerol as from its phosphorylation by glycerol kinase.

The second most important discovery relevant to the stereospecificity of enzymic reactions, next to the discovery of the asymmetrical reactions of citrate, was made by Westheimer and Vennesland and their colleagues. This related to the oxidoreductases (dehydrogenases), which were shown to transfer hydrogen atoms between substrates and diphospho- or triphosphopyridine nucleotide coenzymes in a stereospecific manner, the stereospecificity applying equally to substrate and coenzyme. Thus the dehydrogenases were shown to differentiate between epimeric atoms at a prochiral center of the substrate and to transfer one of these selectively onto the one or other face of the nicotinamide ring at C-4. The discovery was important because it led to a new understanding of the mechanism of action of a large class of enzymes catalyzing closely related reactions, because it gave a renewed interest in reactions around prochiral centers, and because it led (by the use of hydrogen isotopes) to the elucidation of enzymic reactions with great precision. This topic, extensively reviewed previously (*14*, *58–60*), will be one of the main subjects in this chapter.

The stereospecificity of enzymic reactions applies also to unsaturated compounds showing cis- or trans-geometric isomerism. For example,

57. I. C. Gunsalus and M. Gibbs, *JBC* **194**, 871 (1952).

58. B. Vennesland and F. H. Westheimer, *in* "The Mechanism of Enzyme Action"

fumarase (cf. Section VI,C) catalyzes the hydration only of the *trans*-ethylene-1,2-dicarboxylic acid [fumaric acid (XIV)] to L-malic acid [*S*-malic acid (XV)] and is completely unreactive with the cis-isomer [maleic acid (XVI)]. The differentiation by this and other enzymes

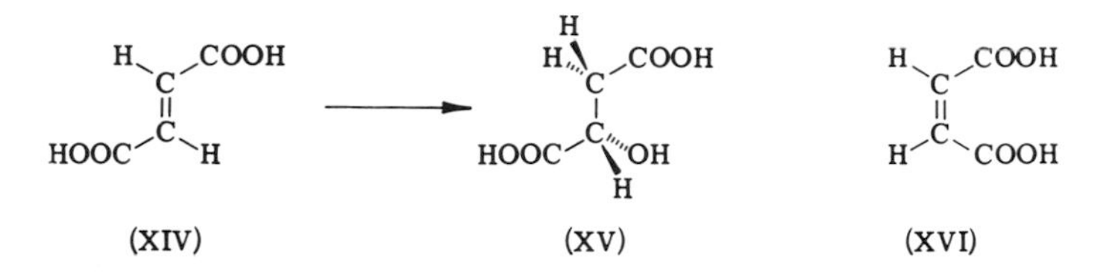

between geometric isomers of substrates is a very strong argument in favor of the postulate that steric factors in the active areas of enzymes are of great importance in the binding of substrates and that these must be bound to their appropriate enzymes in unique conformations (*61*).

There are many analogies of stereospecific additions to double bonds or of elimination reactions on saturated compounds known in organic chemistry, but these are stereospecific only in the sense that the additions or eliminations proceed either in a trans or cis manner without differentiation between two sides of the molecule. Thus addition of Br_2 to a double bond is a trans-addition, but the product of the addition to (R_1R_2) C=C (R_3R_4) is nevertheless racemic (Scheme 2).

The situation is similar in a chemical epoxidation of a double bond:

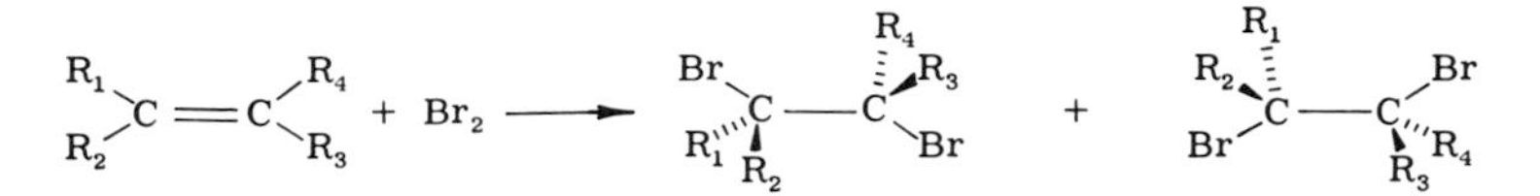

SCHEME 2. Trans-addition of Br_2 to double bond.

although the addition of oxygen to the double bond is cis, the chemical reagent cannot distinguish between the two faces of the double bond and the reaction leads to epimeric 1,2-oxides (Scheme 3; see also preparation of stereospecifically labeled mevalonates, Section VIII,D).

The stereospecificity of enzymic reactions with unsaturated structures

---

(W. D. McElroy and B. Glass, eds.), p. 357. Johns Hopkins Press, Baltimore, Maryland, 1954.

59. B. Vennesland, *J. Cellular Comp. Physiol.* **47,** Suppl. 1, 201 (1956).

60. B. Vennesland, *Federation Proc.* **17,** 1150 (1958).

61. There are only a few known examples of enzymic reactions in which it is necessary to postulate that the substrate might bind to the enzyme in two different conformations, although within the same defined space. One of these is the bacterial propanediol dehydrase which catalyzes the conversion of either *R* or *S*-propane-1,2-diol to propionaldehyde (see Section IX,B).

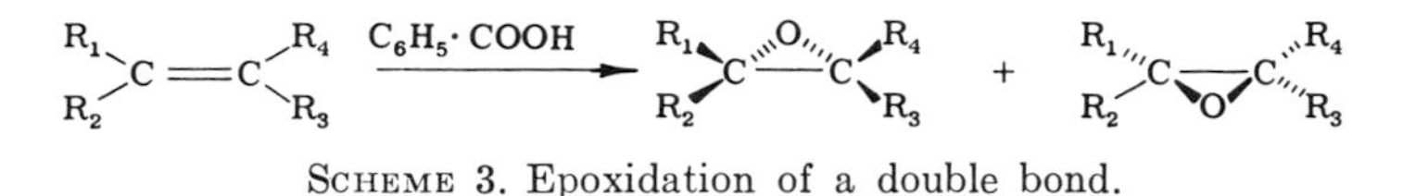

SCHEME 3. Epoxidation of a double bond.

results not only in the selection of cis- or trans-isomers as substrates but also in the selection of the faces of the double bonds onto which the additions are being made. This twofold selectivity of an enzyme further supports the postulate of a unique conformational binding of substrate to the enzyme, the catalytically active residues of the enzyme having one, and only one, spatial relation to the bound substrate. This is the same principle as that required for the differentiation of two chemically identical ligands at a prochiral center.

III. Notations of Molecular Asymmetry

Before the work of Bijvoet and his colleagues (*4*) the configuration of asymmetrical molecules were defined only in relation to (+)-saccharic acid and (+)-glyceraldehyde arbitrarily chosen by Fischer and Rosanoff (*8*) as reference standards.

With the availability of an absolute standard in (+)-tartaric acid (Fig. 1) and with the realization for the need of an unambiguous definition and description of molecular asymmetry, independent of generic relationships prescribed in the Fischer conventions, Cahn and Ingold (*32*) and Cahn *et al.* (*33, 34*) devised a new system for the definition of molecular asymmetry. This system, which became known as the *RS* system and which received wide acceptance, is particularly useful in denoting molecular asymmetry created by isotopic labeling. Since much of the work to be discussed further on in this chapter depended on stereospecific isotopic labeling, the conventions of Cahn *et al.* will be used to describe molecular asymmetries. The definitions of the system will be discussed here only as far as they are pertinent to the material of this chapter. For a full appraisal of the scope of the *RS* convention the reader is referred to the original articles (*32–34*) or to Chapter 2 in Volume I of Bentley's treatise (*9*).

According to the *RS* system the four substituents at a chiral center are arranged in a priority order according to a formula known as the *sequence rule*. The *chirality rule* of the system states that if on viewing the ligands around the chiral center from a direction opposite to the ligand of the lowest priority, the priority order of the three groups seen decreases in the clockwise direction the absolute configuration is said to be *R* (*Rectus*, right handed), and if the priority order decreases in an counterclockwise direction the absolute configuration is S (*Sinister,* left handed).

The steric position of the identical ligands, a' and a'', in the prochiral $Ca'a''bd$ structure can also be described unambiguously by the *prochirality rule* (*35*) which is an extension of the *RS* system: if, e.g., a' is given arbitrarily a priority over a'' and if the decreasing priority order, $d > b > a'$, viewed from a direction opposite to a'', is clockwise, a' is said to be a pro-R and a'' a pro-S ligand [(XVII), (XVIIa), (XVIII),

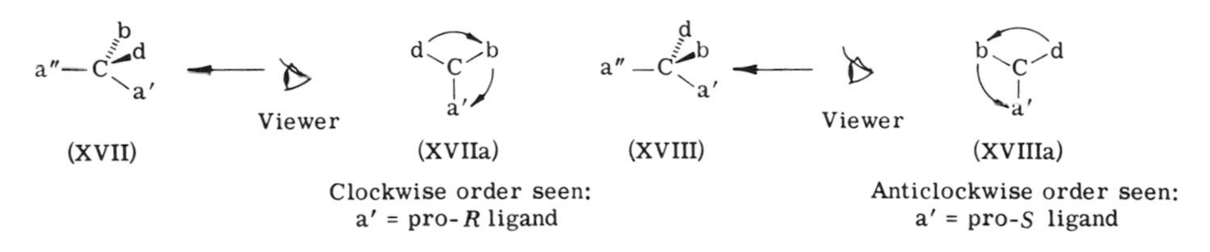

(XVIIIa)]. If a' and a'' were hydrogen atoms they may be described as pro-R and pro-S hydrogen atoms or H_R and H_S. This means also that if H_R were replaced by deuterium or tritium, the prochiral center acquires the R configuration and if H_S is so replaced the center has the S configuration.

For the understanding of the stereochemical ascriptions used in this chapter according to the *RS* system three of the subrules of the sequence rule will suffice. These are as follows:

Subrule (0): Near end (or side) precedes far
Subrule (1): Higher atomic number precedes lower
Subrule (2): Higher mass number (for identical atomic numbers) precedes lower

The only additional information needed here is an explanation of the use of these subrules and the treatment of multiple bonds, or unsaturation, by the sequence rule. The atoms along each bond of the asymmetrical carbon atom are explored in the outward direction from the asymmetrical center, i.e., from the near to the far end, the priorities being assigned according to Subrule (1). Having arrived at an atom, by way of its bond from the asymmetrical carbon atom, all other of its bonds are explored similarly until a decision can be reached about the priorities of the atoms, or of the groups, attached to the asymmetrical carbon atom. When, during such exploration, a double bond is encountered, e.g., at a carbonyl carbon, the double bond is recorded as two single bonds each ending on an oxygen atom, one of these being recorded as a real oxygen atom and another as a duplicate representation. The bonds of the oxygen atom at a carbonyl group are similarly treated: one ending at a real carbon atom and the other at a duplicate carbon atom. The valencies of all duplicate atoms (except hydrogen) are then made up to four by phantom atoms of zero atomic number. All unsaturated structures are treated similarly. Thus, for purposes of priority ordering, the carbonyl group, >C=O, is represented as —C(O)(C)—O, and an ethylenic group, >C=C<, is recorded as —C(C)—C(C)—, the duplicate

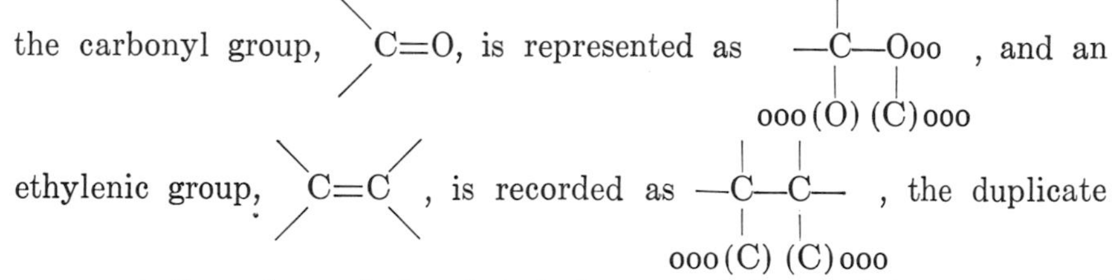

representations shown in brackets being regarded as real for purposes of "accounting." Thus the priority of an aldehyde group, —CHO, precedes that of a carbinol group, —CH_2OH, and similarly the vinyl and methine groups, =CH_2 and =CH—, have priorities over methyl or methylene groups. Bentley [(*9*); Chapter 2, Volume I] has listed very usefully the priority order of groups found most commonly in compounds of biochemical interest: SH > OR > OH > NHCOR > NH_2 > COOR > COOH > CHO > CH_2OH > C_6H_5 > CH_3 > T > D > H.

The use of the *RS* system is illustrated in Scheme 4 with the notations

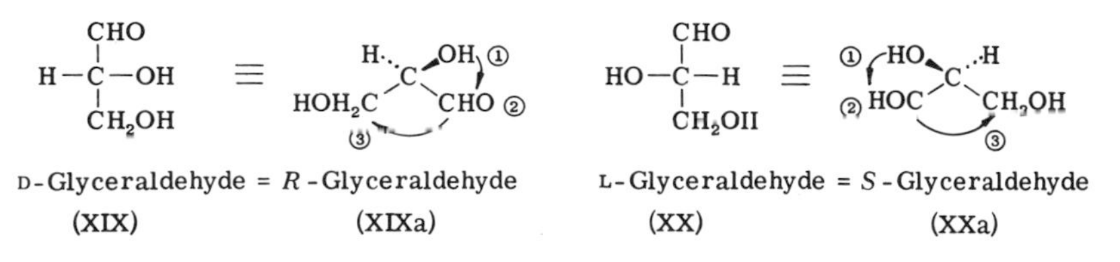

SCHEME 4. Ascriptions of molecular asymmetry for D- and L-glyceraldehyde according to the sequence rule. The priority order of the ligands around the chiral center is OH > CHO > CH_2OH > H.

for D- and L-glyceraldehyde [(XIX) to (XXa)]. The ascriptions of molecular asymmetry by this convention depend critically on the nature of the ligands around the asymmetrical center and, because of this, apparently "identical" absolute configurations in two related substances may have to be assigned antipodal ascriptions. This anomaly may be appreciated by the example of the enantiomeric propane-1,2-diols: The *S*-propane-1,2-diol (XXI) spatially is related to *R*-glyceraldehyde (XIXa) and the *R*-propane-1,2-diol (XXII) to *S*-glyceraldehyde (XXa). In spite of such anomalies, the logical application of the rules gives an unambiguous definition of molecular asymmetry.

Another seeming anomaly of the *RS* system may arise from the strict application of Subrule (2) to chiral substances labeled at a prochiral

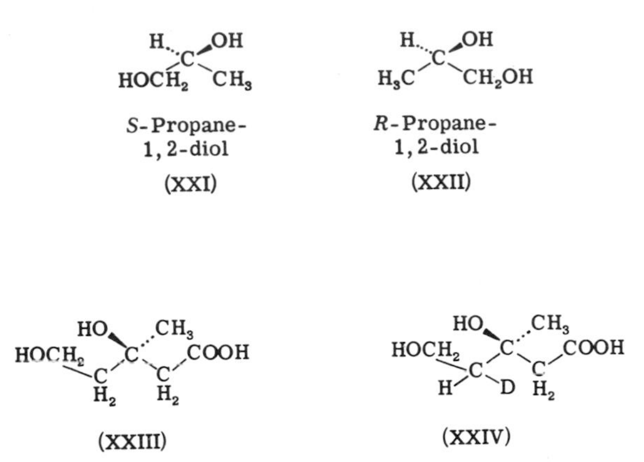

center with an isotope. The best example of this type of anomaly is that of mevalonate labeled with a hydrogen isotope at C-4. The absolute configuration of the enzymically reactive form of mevalonate is 3*R* (XXIII). Introduction of a heavy hydrogen isotope (e.g., D) at C-4 in either the pro-*R* or pro-*S* position would, by the application of Subrule (2) of the sequence rule, change the ascription at C-3 from *R* to *S* because the —CHD group at C-4 acquired, by the isotopic labeling, a priority over the $—CH_2$ group at position 2 (XXIV). However, it was stipulated in the first paper on the *RS* system (*32*) that higher mass numbers are to be taken into account only if a priority order of the ligands cannot be determined otherwise. Isotopic labeling at a ligand of a naturally chiral center does not change, therefore, the ascription of chirality of that center. This may be regarded as another "subrule" and follows logically from the sequence rule which prescribes that on arriving at a particular atom by way of its bond from the chiral center all other of its bonds are to be explored also. Thus, in the case of mevalonate, exploration of the bonds of C-4 leads to —H, —D, and $—CH_2OH$ and of the bonds of C-2 leads to —H, —H, and —COOH. Clearly, the group $—CH_2—COOH$, expanded to

$$\begin{array}{ccc} \mathrm{HooO} & — & \mathrm{(C)ooo} \\ | & & | \\ —\mathrm{C} & — & \mathrm{C}—\mathrm{OH} \\ | & & | \\ \mathrm{H} & & \mathrm{(O)ooo} \end{array}$$

, has precedence over the group

$$\begin{array}{ccc} \mathrm{H} & & \mathrm{H} \\ | & & | \\ —\mathrm{C} & — & \mathrm{C}—\mathrm{OH} \\ | & & | \\ \mathrm{D} & & \mathrm{H} \end{array}$$

and hence the ascription of configuration at C-3 of mevalonate is not affected by the isotopic labeling at C-4.

Stereospecific labeling of a prochiral center with a heavy isotope creates chirality at that center, and this is denoted in accordance with Subrules (1) and (2) (see also Section VIII,C and D).

Hanson (*35*), in his extension of the sequence rule to denoting identical ligands around a prochiral center (the prochirality rule cf. Section III), also devised three rules by which the faces of a trigonal atom (i) Yghi, or of the trigonal atom (ii) Ykki in the system (iii) kkY=Zgh, or the system (iv) kkY=Zgk can be named unambiguously (Fig. 3). These rules should become particularly useful in describing the stereochemistry of enzymic reactions on trigonal carbon atoms.

Rule 1 describes the naming of the faces of the Yghi system: (a) expand all the multiple bonds in the molecular model with replica atoms without altering the configuration around Y or creating new chiral centers; (b) assign priorities to the original three ligands around Y according to Subrule (1) of the sequence rule disregarding replica atoms; and (c) inspect one side of the plane containing Y and the first atoms of ligands ghi. If the priority sequence decreases clockwise the face is said to be *re* (*rectus*), and if the sequence is counterclockwise the *si* (*sinister*) face is seen.

Rule 2 states that the faces of Y in either (iii) or (iv) (Fig. 3) correspond to the faces at Z; the faces at Z are deduced by Rule 1.

Rule 3 specifies the faces of a fixed double bond or of a mesomeric system; this is done by reference to the *re* or *si* faces of particular

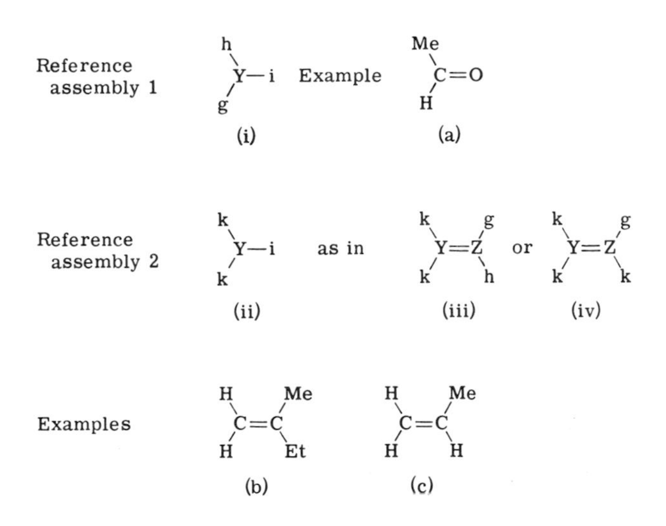

FIG. 3. Reference assemblies of trigonal structures and their examples. The faces seen in the examples are: (a) the *si* face at the carbonyl carbon and the *si–si* faces in both (b) and (c).

trigonal atoms, e.g., the face *re* at C-2 or "the face *re* at the carbonyl carbon."

Subrules (3a) and (3b) specify the naming of faces in structures such as kiY=Zgh. If the face of the double bond is *re* at both trigonal atoms, then this is the *re–re* and the other is the *si–si* face. If the face is *re* at one and *si* at the other trigonal atom then the two sides of the plane are *re–si* and *si–re,* respectively; the face of that trigonal center is called "first" which has ligands of priority around it higher than the other.

Thus the two faces of fumaric acid in the plane of the paper, *si–si* and *re–re,* are represented by formulas (XXV) and (XXVI); the hydration of this acid to *S*-malate (XXVII) can be said to occur by a *si–re* attack.

Another example of the usefulness of this system is illustrated by the reduction of oxalacetate by DPNH to *S*-malate; this can be said to occur by the addition of a hydride ion to the *re* face at the carbonyl carbon of oxalacetate (XXVIII).

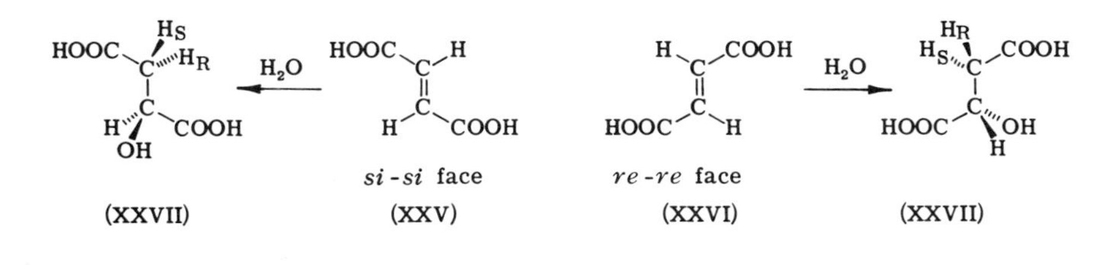

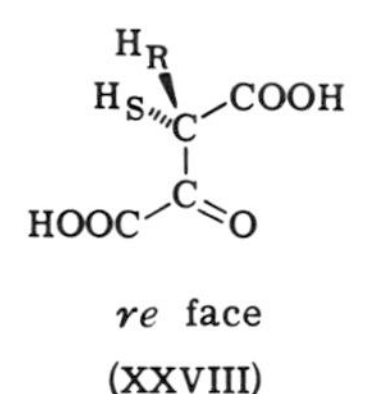

IV. Stereospecificity of Nicotinamide–Adenine Dinucleotide-Dependent Oxidoreductases

A. Historical Background

The subject of the stereospecificity of nicotinamide–adenine dinucleotide-dependent oxidoreductases (dehydrogenases), as well as the much broader topic of the mechanism of the reactions catalyzed by these enzymes, has been reviewed many times in the past (*62, 63*); the most recent accounts of these two topics were those written by Levy *et al.* (*14*) and by Colowick *et al.* (*64*).

Since the main topic to be discussed here is the stereospecificity of the dehydrogenases, only a brief account of the discovery of the nicotinamide–adenine dinucleotide coenzymes and of the historical background to the subject will be given.

The first discovery of a nicotinamide–adenine dinucleotide coenzyme dates back to the early part of this century when Harden and Young (*65*) recognized the existence of a heat-stable cofactor participating in fermentation reactions catalyzed by yeast juice. One component of this heat-stable cofactor was identified as adenosine triphosphate, but the structure of the other, known for many years as *cozymase,* remained unknown until after the coenzyme of glucose-6-phosphate dehydrogenase, the "hydrogen transferring coenzyme," discovered in 1931 by Warburg and Christian (*66*), was identified as a nicotinamide–adenine dinucleotide, containing three moles of phosphate (*67*). Cozymase was then obtained in a highly purified form both in von Euler's and in Warburg's laboratory in 1935–1936; both groups reported that this substance, like the hydrogen-transferring coenzyme, was a nicotinamide–adenine dinucleotide but that it contained only two moles of phosphate per mole (*68, 69*). Warburg and Christian (*69*) proposed the names for the two coenzymes: "diphosphopyridine nucleotide" (DPN) and "triphosphopyridine nucleotide" (TPN). The position of the additional phosphate group in the latter substance at the 2′ position of the adenosine moiety was, however, not established until 1950 (*70*). The terms DPN and TPN, as well as coenzyme I and coenzyme II, have been used for many years to denote the two substances until the recent recommendation of the Commission on Enzymes of the International Union of Biochemistry changed them (*71*) to the chemically more acceptable,

62. N. O. Kaplan, "The Enzymes," 2nd ed., Vol. 3, p. 105, 1960.

63. Nearly all the chapters in Volume 7 of "The Enzymes" (2nd ed., 1963) are devoted to nicotinamide–adenine dinucleotide-dependent oxidoreductases.

64. S. P. Colowick, J. Van Eys, and J. H. Park, *Comp. Biochem.* **14,** 1–98 (1966).

65. A Harden and W. J. Young, *Proc. Roy. Soc.* **B77,** 405 (1906); **78,** 369 (1906).

66. O. Warburg and W. Christian, *Biochem. Z.* **242,** 206 (1931).

67. O. Warburg, W. Christian, and A. Griese, *Biochem. Z.* **282,** 157 (1935).

68. H. von Euler, H. Albers, and F. Schlenk, *Z. Physiol. Chem.* **237,** 180 (1935); **240,** 113 (1936).

69. O. Warburg and W. Christian, *Biochem. Z.* **287,** 291 (1936).

70. A. Kornberg and W. E. Pricer, *JBC* **186,** 557 (1950).

71. "Report of the Commission on Enzymes of the International Union of Biochemistry." Pergamon Press, Oxford, 1961 [reprinted as Vol. 13 of "Comprehensive Biochemistry" (M. Florkin and E. H. Stotz, eds.), 1st ed. Elsevier, Amsterdam, 1964 (2nd ed., 1965)].

though less euphonious, names of "nicotinamide–adenine dinucleotide" (NAD = DPN = coenzyme I) and "nicotinamide–adenine dinucleotide phosphate" (NADP = TPN = coenzyme II). This new nomenclature will be used in this chapter. The oxidized forms of the coenzymes will be abbreviated to NAD^+ and $NADP^+$ and the reduced forms to NADH and NADPH. The structure of NAD^+ and of $NADP^+$, confirmed by synthesis (*72*), is shown in an extended form in formula (XXIX) (NAD:

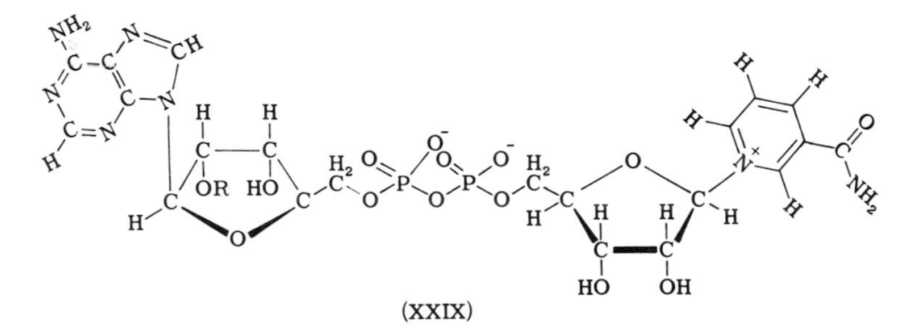

(XXIX)

R = H; NADP: R = $—PO_3^{2-}$); both the nicotinamide and the adenine are joined to the respective D-riboses in β-glycosidic linkages. Dixon and Webb (*73*) pointed out that the structure of these coenzymes has been represented, more often than not, erroneously and incompletely in most textbooks, monographs, and reviews, one or other of the ribose units being shown in the L configuration, or without true regard for the β-glycosidic linkages of both ribose units to the bases. The formula shown here (XXIX) is correct stereochemically for the enzymically reactive β forms of NAD^+ and $NADP^+$. Much evidence has been adduced to show, mostly from fluorescence studies, that the adenine and nicotinamide rings of these coenzymes, when attached to their respective enzymes, are overlapping and that the amino group of the adenine is involved in a hydrogen bonding with the oxygen of the carboxyamide group of the nicotinamide (*74–76*). The folded conformation of reduced NAD, proposed by Velick (*75*), is shown in Fig. 4 (XXX).

Although it was recognized by Warburg and his associates in the mid-1930's that both NAD^+ and $NADP^+$ were hydrogen-transferring coenzymes and that it was their nicotinamide ring which underwent

72. N. A. Hughes, G. W. Kenner, and A. R. Todd, *JCS,* 3733 (1957).
73. M. Dixon and E. C. Webb, "The Enzymes," Fourth Impression 1960; see p. 397. Longmans, Green, New York, 1958.
74. N. O. Kaplan, *in* "Steric Course of Microbiological Reactions" (F. E. W. Wolstenholme and C. M. O'Connor, eds.), p. 37. Churchill, London, 1959.
75. S. F. Velick, *in* "Light and Life" (W. D. McElroy and B. Glass, eds.), p. 108. Johns Hopkins Press, Baltimore, Maryland, 1961.
76. S. Shifrin and N. O. Kaplan, *Nature* **183,** 1529 (1959).

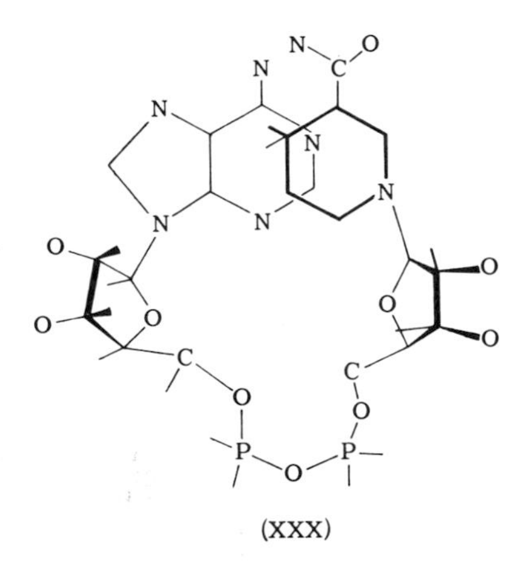

FIG. 4. Folded conformation of NADH according to Velick (*75*). The ring of the adenine and dihydronicotinamide are parallel with the plane of the paper, the dihydronicotinamide being above the adenine. The bonds of other groups are, of course, at various angles to and out of the plane of the paper.

oxidation–reduction reactions (*67, 69, 77*), it was not until twenty years later that the locus of these reactions was established to be C-4 of the nicotinamide ring in either coenzyme (*78–82*). Thus the chemical changes catalyzed by the NAD- and NADP-dependent oxidoreductases (dehydrogenases) are represented by reaction (1) which shows that in the forward direction the nicotinamide ring of the coenzymes is reduced at the expense of an oxidizable substrate AH_2.

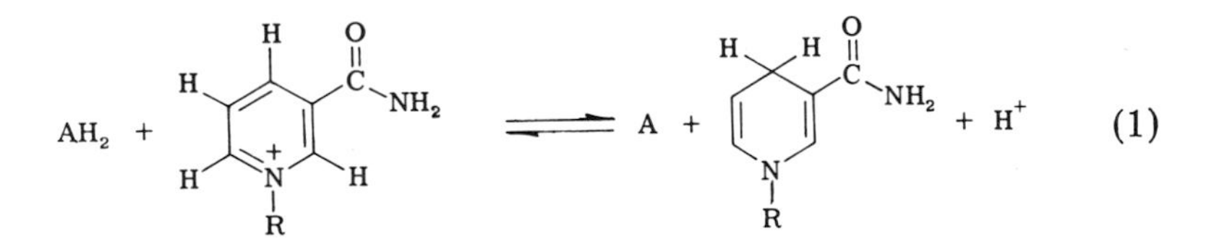

B. Transfer of Hydrogen Atoms between Substrates and Nicotinamide–Adenine Dinucleotide Coenzymes. Stereospecificity of Dehydrogenases with Respect to Coenzymes

It was a major discovery when Westheimer and Vennesland with their colleagues found that, in most of the reactions catalyzed by dehydro-

77. P. Karrer and O. Warburg, *Biochem. Z.* **285**, 297 (1936).
78. M. E. Pullman, A. San Pietro, and S. P. Colowick, *JBC* **206**, 129 (1954).
79. P. Talalay, F. A. Loewus, and B. Vennesland, *JBC* **212**, 801 (1955).
80. F. A. Loewus, B. Vennesland, and D. L. Harris, *JACS* **77**, 3391 (1955).
81. D. Mauzerall and F. H. Westheimer, *JACS* **77**, 2261 (1955).
82. R. F. Hutton and F. H. Westheimer, *Tetrahedron* **3**, 73 (1958).

genases, a direct transfer of hydrogen from the oxidizable substrate AH_2 to NAD^+ and $NADP^+$ or from NADH or NADPH to the oxidized form of substrate A occurred. In the first experiments made with yeast alcohol dehydrogenase (EC 1.1.1.1), it was found that when the reaction (2) catalyzed by this enzyme

$$CH_3CH_2OH + NAD^+ \underset{}{\overset{D_2O}{\rightleftharpoons}} CH_3CHO + NADH + H^+ \tag{2}$$

$$CH_3CD_2OH + NAD^+ \longrightarrow CH_3CDO + NADD + H^+ \tag{2a}$$

was carried out in D_2O the NADH formed did not contain deuterium in a stably bound form. However, when the substrate in the reaction was 1-D_2-ethanol [reaction (2a)], the reduced NAD isolated contained one atom of deuterium per molecule (*83, 84*). The experiments thus demonstrated the direct transfer of a hydrogen atom from substrate to coenzyme in a stably bound form not exchangeable with the protons of water.

Position 4 in the dihydronicotinamide ring is a prochiral center, since if either of the two hydrogen atoms attached to C-4 were replaced with a heavy isotope of hydrogen, an asymmetrical center would be created. Thus, two diastereoisomeric forms of 4-deuterio-NADH (NADD) or 4-deuterio-NADPH (NADPD) can exist: 4*R* (XXXI) and 4*S* (XXXII). On the basis of principles discussed earlier a distinction between the pro-*R* and pro-*S* hydrogen atoms at C-4 of the dihydronicotinamide ring may be expected in enzymic reactions involving these hydrogen atoms.

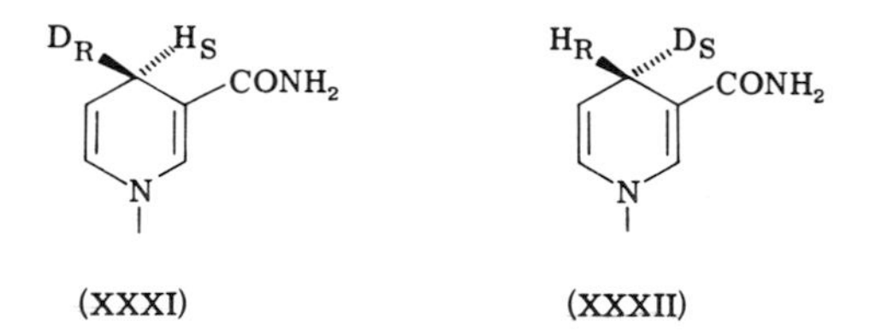

(XXXI) (XXXII)

It was, therefore, an important observation when Fisher *et al.* (*84*) found that when the NADD generated with yeast alcohol dehydrogenase from 1-D_2-ethanol and NAD^+ [reaction (2a)] was used for the reduction of unlabeled acetaldehyde, all the deuterium was transferred from the coenzyme to the substrate to give 1-D_1-ethanol as shown in reaction (3).

$$\underset{\text{[from reaction (2a)]}}{CH_3CHO + NADD + H^+} \rightarrow CH_3CHDOH + NAD^+ \tag{3}$$

83. F. H. Westheimer, H. F. Fisher, E. E. Conn, and B. Vennesland, *JACS* **73,** 2403 (1951).

84. H. F. Fisher, E. E. Conn, B. Vennesland, and F. H. Westheimer, *JBC* **202,** 687 (1953).

This experiment established (a) that the NADD generated with yeast alcohol dehydrogenase from 1-D_2-ethanol and NAD^+ must have been either 4*R*- or 4*S*-4-D_1-NADD and (b) that, in the reversible oxidation–reduction reaction catalyzed by yeast alcohol dehydrogenase, hydrogen was added to or removed from one face only of the nicotinamide ring of the coenzyme. The conclusion that yeast alcohol dehydrogenase reacted in a stereospecific manner at C-4 of the nicotinamide ring of its coenzyme was thus inescapable.

Examination of other dehydrogenases showed that they were all stereospecific irrespective of whether NAD^+ or $NADP^+$ was their preferred coenzyme and that they could be divided into two groups. One group transferred hydrogen from the substrate to C-4 of the nicotinamide ring on the same face of the ring as did yeast alcohol dehydrogenase, but others transferred hydrogen into the epimeric position. Those dehydrogenases which had the same stereospecificity with respect to their coenzyme as yeast alcohol dehydrogenase became known as *A*-side specific and those with the opposite stereospecificity as *B*-side specific enzymes. The hydrogen atoms at C-4 of the reduced nicotinamide ring may be denoted correspondingly as H_A and H_B and the 4-deuterio-NADH obtained by the reduction of NAD^+ with 1-D_2-ethanol as *A*-NADD. *B*-NADD is the diastereoisomer of *A*-NADD. These notations were at first purely arbitrary and were meant to indicate only that H_A and H_B occupied epimeric positions: Their absolute configuration is now known (see Section IV,C).

Vennesland and her colleagues succeeded in establishing a correlation between the *A* and *B* sides of NAD^+ by the use of two enzymes and by the conversion of a stereospecifically labeled NADPH into similarly labeled NADH. Levy and Vennesland (*85*) showed that glutamate dehydrogenase, which catalyzes the oxidation of L-glutamate to α-ketoglutarate and ammonia at similar rates with NAD^+ or $NADP^+$ (*86*), was a *B*-side specific enzyme as determined with NAD^+ as coenzyme. The assumption that the stereospecificity of this enzyme for $NADP^+$ would be the same as for NAD^+ was subsequently proved to be correct (*87*). 4-D-$NADP^+$, prepared by the CN^- exchange reaction of San Pietro (*88*), was first reduced with 2*R*-2*S*-isocitrate and isocitrate dehydrogenase (EC 1.1.1.42) designated earlier as an *A*-side specific enzyme (*89, 89a*).

85. H. R. Levy and B. Vennesland, *JBC* **228**, 85 (1957).
86. J. A. Olson and C. B. Anfinsen, *JBC* **202**, 841 (1953).
87. T. Nakamoto and B. Vennesland, *JBC* **235**, 202 (1960).
88. A. San Pietro, *JBC* **217**, 579 (1955).
89. S. Englard and S. P. Colowick, *JBC* **221**, 1019 (1956).
89a. S. Englard and S. P. Colowick, *JBC* **226**, 1047 (1957).

TABLE I

STEREOSPECIFICITIES OF NICOTINAMIDE–ADENINE DINUCLEOTIDE-DEPENDENT DEHYDROGENASES EFFECTING DIRECT HYDROGEN TRANSFER BETWEEN SUBSTRATES AND COENZYMES

Enzyme and Enzyme Commission No.	Substrate reaction	Source of enzyme	Coenzyme	Reference
	A-Side Specific Enzymes			
Alcohol dehydrogenases (1.1.1.1)	Ethanol $\rightleftharpoons$ acetaldehyde	Yeast	NAD	*(84)*
		Liver (horse)	NAD	*(85)*
		Pseudomonas	NAD	*(90)*
	Propan-2-ol $\rightleftharpoons$ acetone	Yeast	NAD	*(85)*
		Liver (horse)	NAD	*(91)*
	(+)- and (−)-Butan-2-ol $\rightleftharpoons$ butan-2-one	Liver (horse)	NAD	*(91)*
	3-Methylbut-2-enol $\rightleftharpoons$ 3-methylbut-2-enal	Liver (horse)	NAD	*(92, 93)*
	Geraniol $\rightleftharpoons$ geranial	Liver (horse)	NAD	*(94)*
	Farnesol $\rightleftharpoons$ farnesal	Liver (horse)	NAD	*(95)*
Glyoxylate reductase (1.1.1.26)	Glycolate $\rightleftharpoons$ glyoxylate	Spinach leaves	NAD	*(96)*
		Peas	NADP	*(97)*
S-Lactate dehydrogenase (1.1.1.27)	*S*-Lactate $\rightleftharpoons$ pyruvate	Heart muscle	NAD	*(98)*
		Skeletal muscle	NAD	*(99)*
		Potato tuber	NAD	*(99)*
		L. arabinosus	NAD	*(100)*
	S-Glycerate $\rightleftharpoons$ hydroxypyruvate	Skeletal muscle (rabbit) Potato tuber	NAD	*(99)*
R-Lactate dehydrogenase (1.1.1.28)	*R*-Lactate $\rightleftharpoons$ pyruvate	*L. arabinosus*	NAD	*(100)*
	Glycolate $\rightleftharpoons$ glyoxylate	*E. coli*	NAD	*(101)*

R-Glycerate dehydrogenase (1.1.1.29)	*R*-Glycerate ⇌ hydroxypyruvate	Parsley leaf	NAD	*(99)*
Mevaldate reductase (1.1.1.32) (1.1.1.33)	*RS*-Mevaldate → *RS*-mevalonate	Liver (pig)	NAD > NADP[a]	*(95)*
		Liver (rat)	NADP > NAD[a]	*(95)*
Malate dehydrogenase (1.1.1.37)	*S*-Malate ⇌ oxalacetate	Wheat germ	NAD	*(80)*
		Heart muscle	NAD	*(102)*
Malic enzyme (1.1.1.40)	*S*-Malate ⇌ pyruvate + CO_2	Liver	NADP	*(103)*
Isocitrate dehydrogenase (1.1.1.41)	2*S*-3*R*-Isocitrate ⇌ α-ketoglutarate + CO_2	Heart muscle	NADP	*(87, 89a)*
Aldehyde dehydrogenase (1.2.1.3)	Acetaldehyde ⇌ acetate	Liver (bovine)	NAD	*(85)*
Tetrahydrofolate dehydrogenase (1.5.1.3)	5,6,7,8-Tetrahydrofolate ⇌ 7,8-dihydrofolate	Liver (chicken)	NADP	*(104)*
		Leukemia cells (mouse)	NADP	*(104)*
		Streptococcus fecalis	NADP	*(105)*
Methylene tetrahydrofolate dehydrogenase (1.5.1.5)	5,10-Methylene tetrahydrofolate ⇌ 5,10-methenyltetrahydrofolate	Yeast	NADP	*(106)*
Enzymes of fatty acid synthesis (probably β-ketoacyl-*S*-CoA or β ketoacyl-*S*-acyl carrier protein reductase)	β-Ketoacyl-*S*-X ⇌ β-hydroxy-*S*-X (X = CoA or acyl carrier protein)	Lactating rat Mammary gland soluble enzymes (unfractionated)	NADP	*(107)*
NAD and NADP photoreductase	$2NAD^+ + 2H_2O \xrightarrow{h\nu} 2NADH + 2H^+ + O_2$ (Hill reaction)	Spinach chloroplast	NAD	*(108)*
	$2NADP + 2H_2O \xrightarrow{h\nu} 2NADPH + 2H^+ + O_2$	Spinach chloroplast	NADP	*(108)*

TABLE I (*Continued*)

Enzyme and Enzyme Commission No.	Substrate reaction	Source of enzyme	Coenzyme	Reference
	B-Side Specific Enzymes			
α-Glycerophosphate[b] dehydrogenase (1.1.1.8)	*R*-α Glycerophosphate ⇌ dihydroxyacetone phosphate	Skeletal muscle (rabbit)	NAD	*(85)*
Triosephosphate dehydrogenase (1.2.1.12)	*R*-Glyceraldehyde 3-phosphate + P_i ⇌ 1,3-diphospho-*R*-glycerate	Yeast	NAD	*(101)*
		Muscle	NAD	*(101)*
Glutamate dehydrogenase (1.4.1.3)	*S*-Glutamate ⇌ α-ketoglutarate + NH_4^+	Liver	NAD	*(85)*
		Liver	NADP	*(87)*
Glucose dehydrogenase (1.1.1.47)	D-Glucose ⇌ D-gluconolactone	Liver	NAD	*(109)*
Glucose-6-phosphate dehydrogenase (1.1.1.49)	D-Glucose 6-phosphate ⇌ 6-phospho-D-gluconolactone	Yeast	NADP	*(110)*
6-Phosphogluconate dehydrogenase (decarboxylating) (1.1.1.44)	6-Phospho-D-gluconate ⇌ D-ribulose 5-phosphate + CO_2	Yeast	NADP	*(110)*
β-Hydroxybutyryl-CoA dehydrogenase (1.1.1.35)	*S*-β-Hydroxybutyryl pantetheine ⇌ acetoacetyl pantetheine	Heart muscle (pig)	NAD	*(111)*
Transhydrogenase (1.6.1.1)	NADPH + NAD^+ ⇌ $NADP^+$ + NADH	*Pseudomonas*	NAD	*(112)*
17β-Hydroxysteroid dehydrogenase (1.1.1.51)	Testosterone ⇌ androst-4-ene-3,17-dione	*Pseudomonas*	NAD	*(79)*
3α-Hydroxysteroid[b] dehydrogenase (1.1.1.50)	Androsterone ⇌ 5α-androstane-3,17-dione	*P. testosteroni*	NAD	*(113)*
17β-Hydroxysteroid[b] dehydrogenase	Estradiol-17β ⇌ estrone	Placenta (human)	NAD	
			NADP	*(113)*
Squalene synthetase	Farnesyl pyrophosphate → squalene	Liver	NADP > NAD[a]	*(93, 114)*

[a] The notation indicates that the rate of reaction is faster with the first than with the second coenzyme.

[b] With these enzymes the direct transfer of hydrogen from substrate to coenzyme has not been examined although it would be possible to do so.

90. B. Vennesland, *J. Cellular Comp. Physiol.* **47,** Suppl. 1, 201 (1956).
91. F. M. Dickinson and K. Dalziel, *Nature* **214,** 31 (1967).
92. J. W. Cornforth, G. Ryback, G. Popják, C. Donninger, and G. J. Schroepfer, Jr., *BBRC* **9,** 371 (1962).
93. J. W. Cornforth, R. H. Cornforth, C. Donninger, G. Popják, G. Ryback, and G. J. Schroepfer, Jr., *Proc. Roy. Soc.* **B163,** 436 (1966).
94. C. Donninger and G. Ryback, *BJ* **91,** 11P (1964).
95. C. Donninger and G. Popják, *Proc. Roy. Soc.* **B163,** 465 (1966).
96. G. Krakow, J. Ludowieg, J. H. Mather, W. M. Normore, L. Tosi, S. Udaka, and B. Vennesland, *Biochemistry* **2,** 1009 (1963).
97. G. Krakow and B. Vennesland, *Biochem. Z.* **338,** 31 (1963).
98. F. A. Loewus, P. Ofner, H. F. Fisher, F. H. Westheimer, and B. Vennesland, *JBC* **202,** 699 (1953).
99. F. A. Loewus and H. A. Stafford, *JBC* **235,** 3317 (1960).
100. D. Dennis and N. O. Kaplan, *JBC* **235,** 810 (1960).
101. F. A. Loewus, H. R. Levy, and B. Vennesland, *JBC* **223,** 589 (1959).
102. J. L. Graves, B. Vennesland, M. F. Utter, and R. J. Pennington, *JBC* **223,** 551 (1956).
103. W. M. Normore, Thesis, University of Chicago, 1959; quoted by H. R. Levy, P. Talalay, and B. Vennesland, *Progr. Stereochem.* **3,** 299 (1962).
104. E. J. Pastore and M. Friedkin, *JBC* **237,** 3802 (1962).
105. R. L. Blakeley, B. V. Ramasastri, and B. M. McDougall, *JBC* **238,** 3075 (1963).
106. B. V. Ramasastri and R. L. Blakeley, *JBC* **239,** 112 (1964).
107. K. J. Matthes, S. Abraham, and I. L. Chaikoff, *BBA* **70,** 242 (1963).
108. R. N. Ammaraal, G. Krakow, and B. Vennesland, *JBC* **240,** 1824 (1965).
109. H. R. Levy, F. A. Loewus, and B. Vennesland, *JBC* **222,** 685 (1956).
110. B. K. Stern and B. Vennesland, *JBC* **235,** 205 (1960)
111. A. Marcus, B. Vennesland, and J. R. Stern, *JBC* **233,** 722 (1958).
112. A. San Pietro, N. O. Kaplan, and S. P. Colowick, *JBC* **212,** 941 (1955).
113. J. Jarabak and P. Talalay, *JBC* **235,** 2147 (1960).
114. G. Popják, G. L. Schroepfer, Jr., and J. W. Cornforth, *BBRC* **6,** 438 (1962).

H_B in the NADPD produced in this reaction (4)

$$\text{Isocitrate} + [\text{D}]\text{-NAD}^+ \rightarrow \alpha\text{-ketoglutarate} + CO_2 + B\text{-NADPD} \quad (4)$$

must have been deuterium, since by definition the unlabeled hydrogen transferred from substrate to coenzyme occupied the position of H_A. However, it was not known whether or not the designation of H_A and H_B in NADPH meant the same stereochemical configuration as in NADH. The *B*-NADPD obtained from reaction (4) was then oxidized with glutamate dehydrogenase (EC 1.4.1.4) according to reaction (5).

$$B\text{-NADPD} + \alpha\text{-ketoglutarate} + NH_4^+ + H^+ \rightarrow \text{NADP}^+ + \text{glutamate} + H_2O \quad (5)$$

All the deuterium present originally in the coenzyme was found in the glutamate; the nicotinamide in the resulting $NADP^+$ contained no deuterium. The evidence thus showed that isocitrate dehydrogenase and glutamate dehydrogenase had opposite stereospecificities. In a further experiment the *B*-NADPD obtained in reaction (4) was converted into NADD by the action of alkaline phosphatase. This NADD was then oxidized in the glutamate dehydrogenase reaction with α-ketoglutarate and ammonia; once again, all the isotope was transferred from the coenzyme to the glutamate. Since it was unlikely that the alkaline phosphatase affected the steric positions of the hydrogen and deuterium atoms at C-4 of the nicotinamide ring during dephosphorylation of the NADPD, it was concluded that glutamate dehydrogenase had the same stereospecificity for NADP as for NAD. Since this enzyme was shown to be a *B*-side specific enzyme with NAD^+ (*85*), it follows that the *B* side of the dihydronicotinamide ring in NADPH is the same stereochemically as the *B* side in NADH. The experiments of Nakamoto and Vennesland (*87*) also proved that the classification of isocitrate dehydrogenase as an *A*-side specific enzyme was correct.

Table I summarizes the stereospecificities of those dehydrogenases with which a direct transfer of hydrogen between substrates and coenzyme has been demonstrated, or could be demonstrated if tested.

C. Absolute Configuration of Nicotinamide–Adenine Dinucleotide Coenzymes Labeled with Deuterium on the *A* or *B* Side

The meaning of the *A* and *B* sides of nicotinamide–adenine dinucleotide coenzymes in terms of absolute configuration was deduced by Cornforth *et al.* (*115, 116*). These workers devised a method for the

115. J. W. Cornforth, G. Ryback, G. Popják, C. Donninger, and G. L. Schroepfer, Jr., *BBRC* **9**, 371 (1962).
116. J. W. Cornforth, R. H. Cornforth, C. Donninger, G. Popják, G. Ryback, and G. J. Schroepfer, Jr., *Proc. Roy. Soc.* **B163**, 436 (1966).

chemical degradation of NADH by which positions 3, 4, 5, and 6 of the dihydronicotinamide ring could be isolated as succinic acid without affecting the stereochemistry at C-4 of the dihydronicotinamide ring (Scheme 5) and have applied it to two specimens of deuterio-NADH.

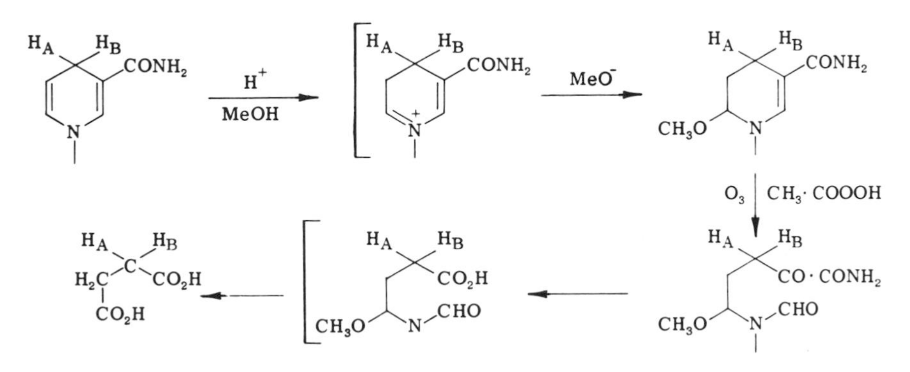

SCHEME 5. Acid catalyzed nucleophilic addition to a dihydronicotinamide and its degradation to succinic acid. The method is equally applicable to 1,4-dihydro-1-methylnicotinamide and to reduced nicotinamide–adenine dinucleotide coenzymes [(*115*, *116*), see also (*62*)]. The intermediary products of degradation shown between the square brackets were not isolated.

One specimen was prepared by the reduction of NAD^+ with 1-D_2-3,3-dimethylallyl alcohol (1-D_2-3-methylbut-2-enol) and crystalline liver alcohol dehydrogenase. Since this enzyme is, by definition, *A*-side specific, the deuterium in the resulting NADH must have occupied the position of H_A. The second specimen was made by reducing 4-deuterio-NAD^+ [prepared by the CN^- exchange method of San Pietro (*88*)] with unlabeled ethanol and liver alcohol dehydrogenase. Since the protium transferred from the ethanol occupies the position of H_A, the deuterium in this second specimen of reduced NAD must occupy the position of H_B. The purified succinic acids obtained from the two specimens were shown, by mass spectrometry, to contain mostly monodeuterated molecules and, by spectropolarimetry, to be optically active, both giving normal optical rotatory dispersion curves but of opposite signs. The deuteriosuccinic acid derived from *A*-deuterio-NADH was levorotatory and the one from the *B*-deuterio coenzyme dextrorotatory (Fig. 5). A correlation was then made with 2-*R*-2-D_1-succinic acid prepared chemically from 2*S*-3*R*-3-D_1-malic acid (cf. Section VI,C and Scheme 16). The reference 2*R*-2-D_1-succinic acid was levorotatory and its optical rotatory dispersion curve indistinguishable from that of the 2-D_1-succinic acid derived from the *A*-deuterio-NADH. The monodeuteriosuccinic acids derived from the *A*-deuterio and *B*-deuterio-NADH specimens have, therefore, the *R* and *S* absolute configurations, respectively

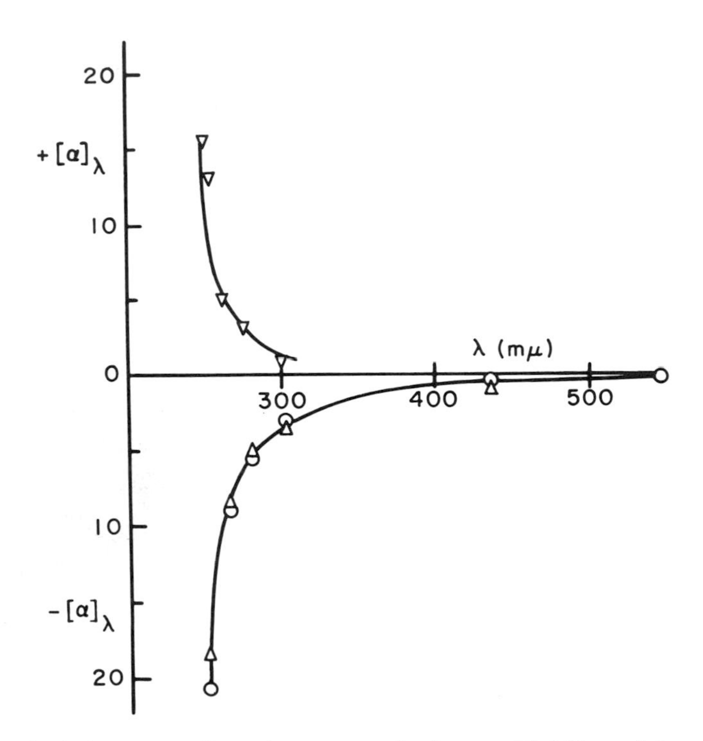

FIG. 5. Optical rotatory dispersion curves of reference 2*R*-2-D_1-succinic acid (○), of D_1-succinic acid derived from *A*-deuterio-NADH (△), and from *B*-deuterio-NADH (▽) (*115, 116*). The measurements were made on the free acids in methanol.

[(XXXIII) and (XXXIV)]. H_A is therefore the pro-*R* and H_B the pro-*S* hydrogen atom at C-4 of the dihydronicotinamide ring of NADH. Since Nakamoto and Vennesland (*87*) have shown that the *A* and *B* sides on the dihydronicotinamide ring of NADH corresponded to the *A* and *B* sides of NADPH, a general rule, applicable to all NAD^+ and $NADP^+$-linked dehydrogenases, was deduced. This rule states that "all *A*-specific dehydrogenases add hydrogen to that side of the nicotinamide ring on which the numbering of the positions from 1 to 6 appears in an anticlockwise order" (*115, 116*). By using the extension of the sequence rule (cf. Section III) for the designation of faces of a plane-trigonal

D, H; H_2C–C–CO_2H; CO_2H	H, D; H_2C–C–CO_2H; CO_2H
(−) *R*-2-D_1-Succinic acid	(+) *S*-2-D_1-Succinic aoid
(XXXIII)	**(XXXIV)**

structure, the above rule may be restated: *A*-specific dehydrogenases add hydrogen to the *re* face and *B*-specific dehydrogenases to the *si* face at C-4 of the nicotinamide ring in either NAD^+ or $NADP^+$ [(XXXV)–(XXXVII)].

D. Stereospecificity of Dehydrogenases with Respect to Substrates

The substrates of many of the enzymes listed in Table I are optically active and each enzyme is specific for only one optical isomer; the stereospecificity of these dehydrogenases with respect to their substrates is thus self-evident. The alcohol dehydrogenases, which catalyze reactions at a prochiral carbon atom, are also stereospecific with respect to the substrates. This was first demonstrated with yeast alcohol dehydrogenase.

Carbon atom 1 of ethanol, like C-4 of dihydronicotinamide, is a prochiral center; hence, two stereospecifically labeled 1-D_1-ethanols can exist: *R* and *S* [(XXXVIII) and (XXXIX)]. The two enantiomeric 1-D_1-ethanols were prepared by Loewus *et al.* (*117*) in two ways. First,

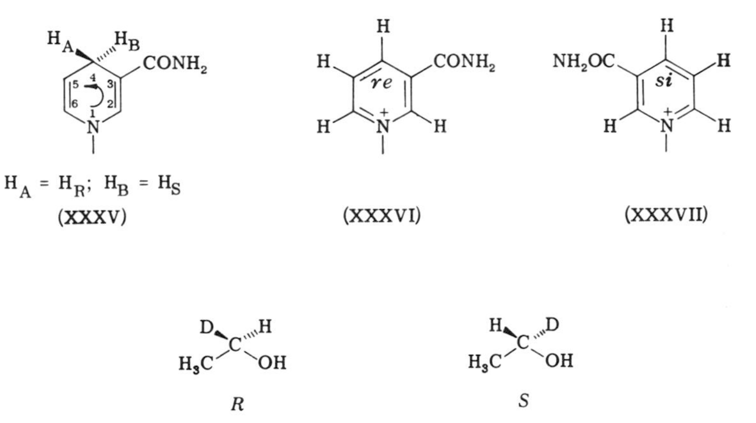

1-D_1-ethanol was made by the reduction of unlabeled acetaldehyde with *A*-NADD (4*R*-4-D_1-NADH) and yeast alcohol dehydrogenase [cf. reaction (6)]; it was isolated from the reaction mixture by distillation. When this 1-D_1-ethanol (specimen 1) was used for the reduction of NAD^+, all the deuterium was transferred from the substrate to the coenzyme [reaction (7)]. In a second experiment another specimen of 1-D_1-ethanol (specimen 2) was prepared by the reduction of 1-D-acet-

117. F. A. Loewus, F. H. Westheimer, and B. Vennesland, *JACS* **75**, 5018 (1953).

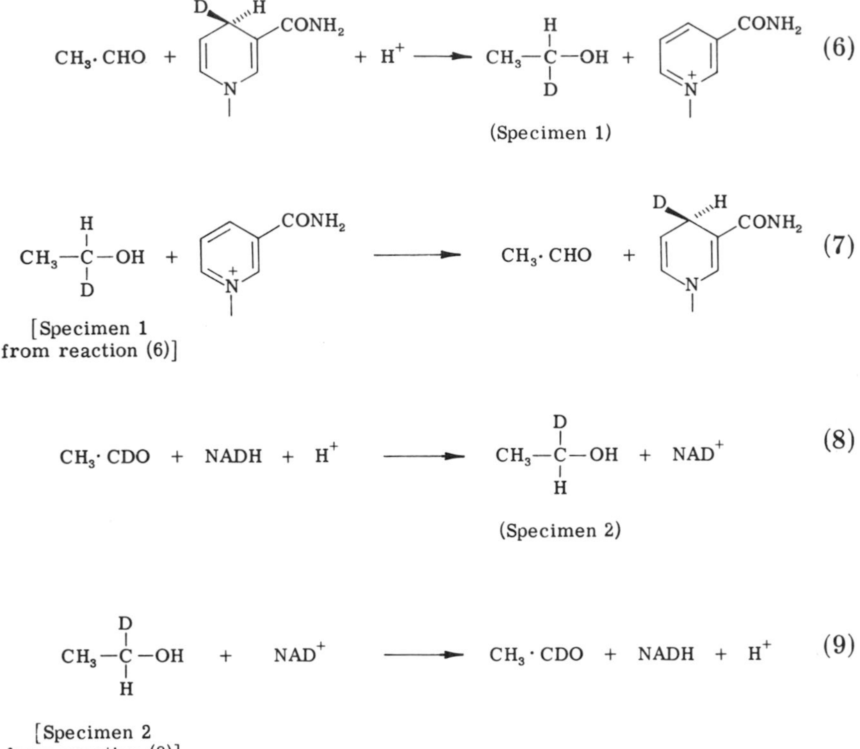

aldehyde with unlabeled NADH and enzyme [reaction (8)]. When this second specimen of 1-D_1-ethanol was in turn oxidized with NAD^+ and alcohol dehydrogenase, it gave 1-deuterioacetaldehyde and unlabeled NADH [reaction (9)]. Thus it became clear that the two 1-D_1-ethanols generated in reactions (6) and (8) were labeled in epimeric positions and that only one specific hydrogen atom at C-1 of ethanol was shuttled between substrate and coenzyme in the alcohol dehydrogenase reaction.

In order to obtain further support for these conclusions, Loewus *et al.* (*117*) inverted the configuration of the 1-D_1-ethanol (specimen 2) obtained in reaction (8) by the solvolysis of its *p*-toluene sulfonate [reaction (10)]. When the 1-D_1-ethanol thus prepared was oxidized with NAD^+ and enzyme, unlabeled acetaldehyde and NADD were formed [reaction (11)], indicating that a stereochemical inversion of the 1-D_1-ethanol had taken place and that the results of reaction (9) could in no way be attributed to an isotope effect.

It remained now only to ascertain the degree of stereospecificity of alcohol dehydrogenase and to deduce the absolute configuration of the

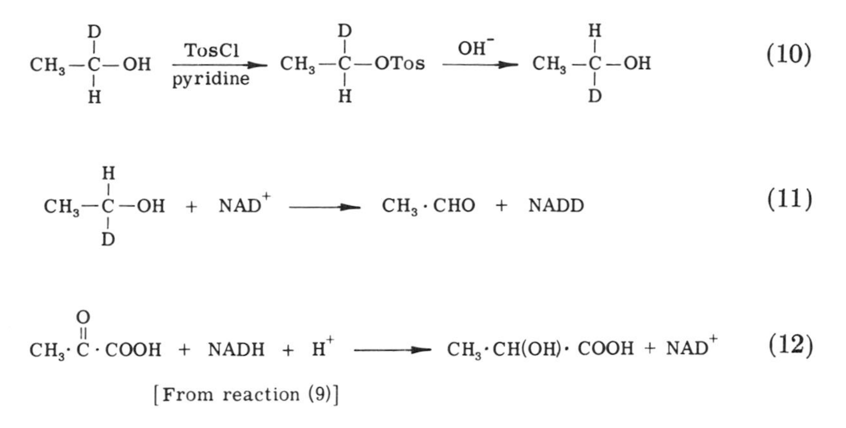

two enantiomeric 1-D_1-ethanols. To this end Levy *et al.* (*118*) prepared 0.9 g of 1-D_1-ethanol by the enzymic reduction of 1-D_1-acetaldehyde [cf. reaction (8)], measured its optical activity, and reoxidized it with NAD^+ and yeast alcohol dehydrogenase [reaction (9)]. If the stereospecificity of alcohol dehydrogenase reaction is absolute, the NADH formed during the reoxidation should contain no isotope. The NADH obtained in reaction (9) was then used to reduce pyruvate to lactate with lactate dehydrogenase [reaction (12] which, like alcohol dehydrogenase, is an *A*-side specific enzyme. The lactate, isolated as the phenacyl derivative, did not contain any measurable amount of excess deuterium, indicating that the NADH obtained in reaction (9) had not acquired any excess of isotope in the position of the pro-*R* hydrogen atom (H_A) at C-4 of the dihydronicotinamide ring. The accuracy of measurements was such that an excess deuterium of one part in a thousand could have been determined. Hence, it appears that the stereospecificity of alcohol dehydrogenase in the transfer of a hydrogen atom between substrate and coenzyme is absolute.

The 1-D_1-ethanol obtained in reaction (8) had a specific rotation of $[\alpha]_D^{28}$ $-0.28 \pm 0.03°$. The absolute configuration of this specimen was at first proposed to be *R* (XL) on the assumption that it must have the same absolute configuration as (—)-1*R*-1-D_1-butanol prepared by Streitwieser (*119*) by an asymmetrical reduction of butyraldehyde.

D, H, C, H_3C, OH	H, D, C, H_3C, OH
R	*S*
(XL)	(XLI)

118. H. R. Levy, F. A. Loewus, and B. Vennesland, *JACS* **79**, 2949 (1957).
119. A. Streitwieser, Jr., *JACS* **75**, 5014 (1953).

However, it soon transpired that this assumption was incorrect. After much argument from several quarters, discussed by Levy *et al.* (*14*), Lemieux and Howard (*120*) prepared, by unambiguous chemical synthesis, a specimen of 1-D_1-ethanol containing a 30% excess of the 1*R* enantiomer; the specimen was dextrorotatory. Similarly, Weber (*121*) made a specimen of 1*S*-1-D_1-ethanol (XLI) by stereoselective synthesis and demonstrated by enzymic reactions that it was identical with the (—)-1-deuterio-ethanol obtained by Levy *et al.* (*118*) by the reduction of 1-D-acetaldehyde with NADH and yeast alcohol dehydrogenase.

Thus with the evidence at hand, the stereospecificity of the alcohol dehydrogenase reaction is fully described by reactions (13) and (14).

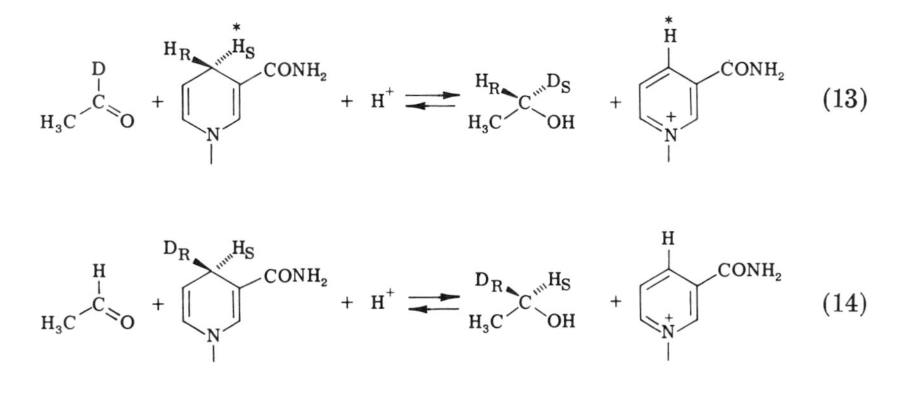

The rule may be formulated: yeast alcohol dehydrogenase transfers a hydrogen atom from the *re* face of the dihydronicotinamide ring of NADH to the *re* face at the carbonyl carbon atom of acetaldehyde, and the same hydrogen atom is shuttled between substrate and coenzyme in the reversal of the reaction. The rule applies equally to liver alcohol dehydrogenase.

A very exceptional determination of substrate stereospecificity of an enzyme was that reported by Johnson *et al.* (*121*) on L-lactate dehydrogenase (EC 1.1.1.27) which catalyzes reaction (15). Pyruvate (XLII) is reduced to *S*-(=L)-lactate (XLIII), clearly, by the transfer of H_R from C-4 of the dihydronicotinamide ring of NADH to the *re* face at C-2 of pyruvate. Lactate dehydrogenase also catalyzes the oxidation–reduction between glyoxylate (XLIV) and glycolate (hydroxyacetate) (XLV) [reaction (16)]. Johnson *et al.* (*122*) tested the hypothesis that the addi-

120. R. U. Lemieux and J. Howard, *Can. J. Chem.* **41**, 308 (1963).

121. H. Weber, Doctoral Thesis (Prom. No. 3591), Eidgenössische Technische Hochschule, Zürich, 1965.

122. C. K. Johnson, E. J. Gabe, M. R. Taylor, and I. A. Rose, *JACS* **87**, 1802 (1965).

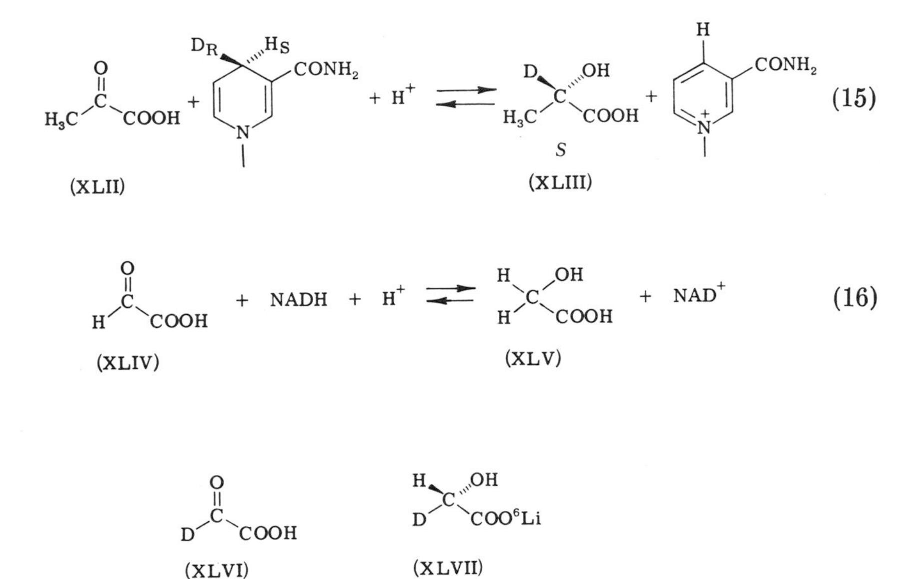

tion of the hydride ion to glyoxylate occurs from the same direction as in the reduction of pyruvate. They prepared a specimen of D_1-glycolate by the reduction of D-glyoxylate (XLVI) with NADH and lactate dehydrogenase from muscle. The crystal structure of the anhydrous ^{6}Li salt of the D_1-glycolate was then determined by X-ray crystallography, and the absolute configuration of the molecule was then deduced from the anomalous neutron-scattering amplitude of ^{6}Li and from the markedly different neutron-scattering amplitudes of H and D. The structure deduced (XLVII) was that of lithium *S*-2-D_1-hydroxyacetate, in complete agreement with the hypothesis. This is so far the only recorded example of the use of X-ray crystallograpy combined with neutron diffraction measurements for the determination of the absolute configuration of a substance the chirality of which was created entirely by the stereospecific labeling at a prochiral center with a heavy isotope of hydrogen.

E. Nicotinamide–Adenine Dinucleotide-Linked Reactions without Demonstrable Hydrogen Transfer

Hydrogen transfer between NAD, NADP, and substrates cannot be demonstrated in several oxidation–reduction reactions. Such is the case in the reactions catalyzed by glutathione reductase, by enzymes containing a flavin prosthetic group, and by the enzymes of the electron transport chain in which the substrate does not acquire hydrogen on reduction. Nevertheless, even these enzymes show distinct stereospecificities with respect to the two hydrogen atoms at C-4 of the dihydronicotinamide ring of NADH or NADPH.

Glutathione reductase (*123–126*) catalyzes the oxidation of NADPH (or NADH) by oxidized glutathione according to reaction (17). Transfer of a hydrogen atom from coenzyme to substrate in this instance cannot be demonstrated because of the rapid exchange of the hydrogen

$$\text{GSSG} + \text{NADPH (or NADH)} + \text{H}^+ \rightarrow 2\text{GSH} + \text{NADP}^+ \text{ (or NAD}^+\text{)} \quad (17)$$

of the SH group of glutathione with the aqueous medium. Stern and Vennesland (*127*) examined the stereospecificities of glutathione reductase from yeast (*126*), reactive with both NADPH and NADH, and from *Escherichia coli* (*128*), which is specific for NADPH. In every instance the oxidation of NADPH and of NADH meant the loss of only the pro-*S* H atom (H_B) from the appropriately labeled coenzymes. Thus, glutathione reductase is a *B*-side specific enzyme.

Demonstration of a direct hydrogen transfer between NADH or NADPH and product is equally difficult in reactions catalyzed by flavoproteins because the hydrogens bound to nitrogen atoms in the reduced form of flavin exchange readily with the hydrogen ions of the medium. A notable feature of several flavoproteins, which oxidize NADH or NADPH, is that they catalyze an exchange reaction between the hydrogen ions of the medium and the reduced nicotinamide–adenine dinucleotide coenzymes (*129–131*) as well as a reduction of analogs of NAD^+ (e.g., of the acetylpyridine analog). In all instances when the exchange can be demonstrated it involves only the pro-*S* hydrogen atom (H_B) of the dihydronicotinamide ring (*132*). Reaction (18) represents the overall

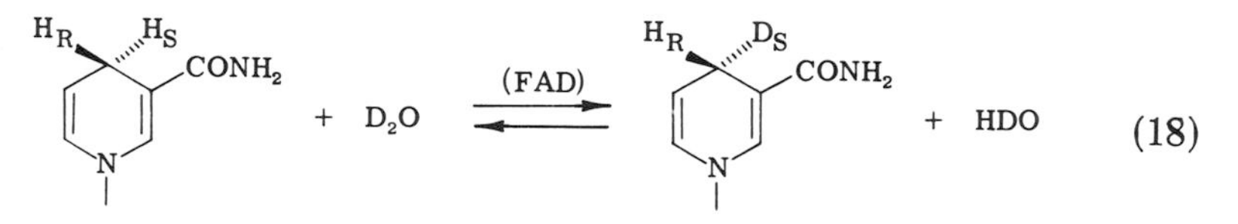

123. L. W. Mapson and D. R. Goddard, *BJ* **49,** 592 (1951).
124. E. E. Conn and B. Vennesland, *JBC* **192,** 17 (1951).
125. T. W. Rall and A. L. Lehninger, *JBC* **194,** 119 (1952).
126. E. Racker, *JBC* **217,** 855 (1955).
127. B. K. Stern and B. Vennesland, *JBC* **235,** 209 (1960).
128. R. E. Asnis, *JBC* **213,** 77 (1955).
129. M. M. Weber, N. O. Kaplan, A. San Pietro, and F. E. Stolzenbach, *JBC* **227,** 27 (1957).
130. G. R. Drysdale, *JBC* **234,** 2399 (1959).
131. M. M. Weber and N. O. Kaplan, *JBC* **225,** 909 (1957).
132. L. Ernster, H. D. Hoberman, R. L. Howard, T. E. King, C. P. Lee, B. Mackler, and G. Sottocasa, *Nature* **207,** 940 (1965).

process of such a *B*-side specific exchange mediated, e.g., by an FAD-containing enzyme.

The glutathione reductase of *E. coli*—in contrast to the reductases of wheat germ, pea seedlings, and mammalian liver—is a flavoprotein with FAD as the prosthetic group (*128*). Stern and Vennesland (*127*) have shown that this *B*-side specific bacterial enzyme catalyzed not only reaction (17) but also reaction (18). However, the rate of the exchange reaction was only about one-thirty-fifth of the rate of the oxidation of NADPH by oxidized glutathione. Stern and Vennesland (*127*) made ingenious use of a coupled reaction to determine the stereospecificity of the exchange reaction: the *E. coli* glutathione reductase was incubated in D_2O with NADPH in the presence of glutamate dehydrogenase, α-ketoglutarate, and ammonia. Presence of deuterium in the glutamate isolated subsequently showed that H_S at C_4 of the dihydronicotinamide ring must have exchanged with the medium, since glutamate dehydrogenase is a *B*-side specific enzyme.

Dihydroorotate dehydrogenase (EC 1.3.3.1) is another flavoprotein containing 1 mole of FMN and 1 of FAD per mole of enzyme (*133*); it catalyzes reaction (19). Although this enzyme used specifically H_A of

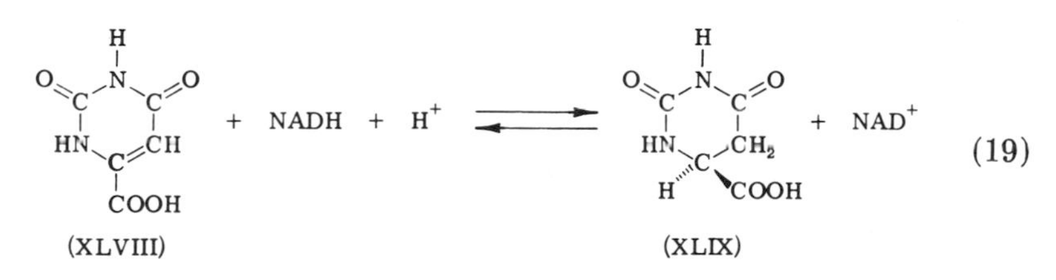

NADH in the reduction of orotate (XLVIII) to *S*-4,5-dihydroorotate (XLIX), transfer of deuterium from deuterio-NADH to the reduced product could not be demonstrated (*134*). Since the flavin prosthetic groups of this enzyme can be reduced by either NADH or by dihydroorotate alone (*135*), and since the reaction catalyzed by it is readily reversible, it is not surprising that a direct hydrogen transfer from NADH to orotate is not observable.

Levy *et al.* (*14*) have pointed out that direct transfer of hydrogen might be demonstrable even in flavoprotein-catalyzed reactions if the substrate and product of the reaction contain stably bound hydrogen and if the rate of the oxidation of NADH or NADPH is faster than the rate of exchange between reduced flavin and the hydrogen ions of the

133. H. C. Friedman and B. Vennesland, *JBC* **233,** 1398 (1958).
134. J. L. Graves and B. Vennesland, *JBC* **226,** 307 (1957).
135. H. C. Friedman and B. Vennesland, *JBC* **235,** 1526 (1960).

medium. Such might be the case if the iso-alloxazine group were in a hydrophobic area of the enzyme. So far the only recorded example of a flavoprotein-catalyzed direct hydrogen transfer is that of microsomal cytochrome b_5 reductase (EC 1.6.2.2), which was shown by Drysdale *et al.* (*136*) to be an *A*-side specific enzyme and to transfer the pro-*R* hydrogen atom (H_A) at C-4 of NADH to the acetyl pyridine analog of NAD into the pro-*R* position at C-4 of the acetyl pyridine moiety. In a nonenzymic reduction of the acetyl pyridine analog of NAD^+ by NADH, hydrogen is also transferred from NADH to the analog (*137*), but the process is not stereospecific (*136*). It may well be that in the microsomal cytochrome b_5 reductase the iso-alloxazine group is in a lipophilic area of the enzyme from which water is excluded, for, unlike other flavoprotein enzymes, it does not catalyze a hydrogen exchange between the hydrogen ions of the medium and NADH. Of course, it is equally possible that a widely different oxidation–reduction potential between NAD/NADH and the enzyme-bound FAD/$FADH_2$ would make the reaction thermodynamically impossible even if the $FADH_2$ exchanged its labile hydrogens with the surrounding medium. This explanation for the absence of a hydrogen exchange seems, however, improbable because Strittmatter (*138*) had shown that reaction (20) between NADH and the microsomal cytochrome b_5 reductase was reversible: the cytochrome

$$\text{Cytochrome } b_5 \text{ reductase}_{ox} + \text{NADH} \rightleftharpoons \text{cytochrome } b_5 \text{ reductase}_{red} + \text{NAD}^+ \quad (20)$$

b_5 reductase, reduced by NADH, could reduce in turn the acetyl pyridine analog of NAD^+. The explanations of the phenomena observed with the microsomal cytochrome b_5 reductase are probably hidden in the as yet unknown structure of this enzyme.

Table II summarizes the properties of some of the enzymes in the category just discussed.

F. Significance of *A*- and *B*-Side Specificity of Dehydrogenases

Bentley [see Chapter 1, Volume 2 of Bentley (*9*)] made a number of generalizations from the properties of the NAD- and NADP-linked dehydrogenases which may be deduced from the data of Tables I and II. These were:

(1) The stereospecificity of a particular reaction is independent of the source of the enzyme which catalyzes it.

136. G. R. Drysdale, M. J. Spiegel, and P. Strittmatter, *JBC* **236,** 2323 (1961).
137. M. J. Spiegel and G. R. Drysdale, *JBC* **235,** 2498 (1960).
138. P. Strittmatter, *JBC* **239,** 3043 (1964).

TABLE II
STEREOSPECIFICITIES OF NADH AND NADPH OXIDATION WITHOUT DEMONSTRABLE HYDROGEN TRANSFER TO SUBSTRATE

Enzyme and Enzyme Commission No.	Source of enzyme	Coenzyme	Steric course of oxidation	Steric course of exchange[a]	Reference
L-4,5-Dihydro-orotate dehydrogenase (1.3.3.1)	*Zymobacterium oroticum*	NADH	A	—	(*134, 135*)
Cytochrome b_5 reductase (1.6.2.2)[b]	Liver microsomes	NADH	A	—	(*136, 138*)
Glutathione reductase (1.6.4.2)	Yeast	NADH	B	—	(*127*)
	Yeast	NADPH	B	—	(*127*)
	E. coli	NADPH	B	B	(*127*)
Lipoyl dehydrogenase (1.6.4.3); known formerly as diaphorase	Pig heart	NADH	B	B	(*129, 132*)
	Spinach	NADH	B	B	(*139*)
Cytochrome c reductase (1.6.99.3)	Pig heart	NADH	B	B	(*130, 140*)
	Rat liver	NADH	B	B	(*130*)

[a] Some of the enzymes listed catalyze a stereospecific exchange of hydrogen between coenzyme and the hydrogen ions of water.
[b] This is the only flavoprotein enzyme known which catalyzes a direct hydrogen transfer between NADH and the acetyl pyridine analog of NAD^+ (see text).

(2) When an enzyme can use either NAD^+ or $NADP^+$ (or their reduced forms), the stereospecificity of the reaction is the same with both coenzymes.

(3) When an enzyme reacts with a range of substrates, the stereospecificity of the hydrogen transfer is the same with each substrate.

The identical stereospecificity of a given reaction catalyzed by enzymes from as divergent sources as *Lactobacillus arabinosus,* potato tuber, and heart muscle may be an expression of evolutionary heritage [cf. Smith (*141*)] as is the structural correlation among proteins fulfilling identical functions in genera and species separated by many millions of years in the evolutionary time scale. Identical stereospecificity with respect to NAD and NADP, when the enzyme can react with either of these coenzymes, and the identical stereospecificity of the hydrogen transfer with a range of substrates (as, e.g., in the case of alcohol dehydrogenases)

139. R. N. Ammeraal, G. Krakow, and B. Vennesland, *JBC* **240,** 1824 (1965).
140. G. R. Drysdale and M. Cohn, *BBA* **21,** 397 (1956).
141. E. L. Smith, Vol. I, Chapter 6, of this treatise.

can only mean that the spatial orientation of the coenzymes in relation to their substrates on the enzyme is the same in every instance. It would be of great interest to know whether reduction of deuterio-glyoxylate to glycolate (hydroxyacetate) by the *R*-lactate (D-lactate) dehydrogenase of *Lactobacillus arabinosus* or of *Escherichia coli* would have the *R* or *S* absolute configuration [see formulas (XLVI) and XLVII)]. The problem of mevaldate reductase acting on the *R* and *S* enantiomers of mevaldate as discussed in detail (see Section VIII,A), and of liver alcohol dehydrogenase exhibiting the same stereospecificity with respect to coenzyme on reacting with the enantiomeric butan-2-ols (cf. Table I), may mean only differences in the binding of the substrates to the enzyme; the spatial orientation of the coenzyme in relation to the reactive atoms (or groups) of the substrates should be the same in each instance.

There is an exception to the generalizations made above. The *S*-lactate dehydrogenase of skeletal muscle reduces not only pyruvate but also glyoxylate [cf. reactions (15) and (16)], the pro-*R* hydrogen (H_A) at C-4 of the dihydronicotinamide ring being added in both reactions to the *re* face of pyruvate or glyoxylate [cf. formulas (XLII), (XLIII) and XLVI), (XLVII)]. The glyoxylate reductase of spinach leaves (EC 1.1.1.26, cf. Table I) is, like *S*-lactate dehydrogenase, an *A*-side specific enzyme; but it adds the hydrogen atom to the *si* face of glyoxylate: the reduction of D_1-glyoxylate by NADH and the spinach enzyme yields the *R*-D_1-glycolate. Thus two enzymes from different sources, catalyzing similar reactions and having identical stereospecificities with respect to their coenzymes, may display opposite stereospecificities with respect to their substrates. This same conclusion may be drawn from a comparison of the *S*-lactate dehydrogenase of muscle and the *R*-lactate dehydrogenase of *L. arabinosus* and *E. coli* (cf. Table I). Perhaps the glyoxylate reductase of spinach leaves is more akin structurally to the bacterial *R*-lactate dehydrogenase than to the mammalian *S*-lactate dehydrogenase.

Levy and Vennesland (*85*) argued that the *A*- and *B*-side specificity of various dehydrogenases may reflect different conformations of the dihydronicotinamide ring of NADH or NADPH on the two sets of enzymes. The folded conformation of NADH or NADPH on the enzymes suggested by Velick (*75*) (cf. Fig. 4) would, of course, permit the direct transfer of H_A to a substrate positioned above the plane of the dihydronicotinamide ring (the plane of the adenine being below the plane of the dihydronicotinamide ring) even if the dihydronicotinamide ring had a planar conformation. It is, however, difficult to envisage the transfer of H_B from such a conformation to a substrate since there would be no space between the dihydronicotinamide and adenine rings to accommodate a substrate even of small dimensions. Levy and Vennesland (*85*) suggested that the dihydronicotinamide ring may have either planar (L) or boat conformations [(LI) and (LII)], H_A and H_B taking up either axial or equatorial positions (*142*). Although this is an attractive hypothesis

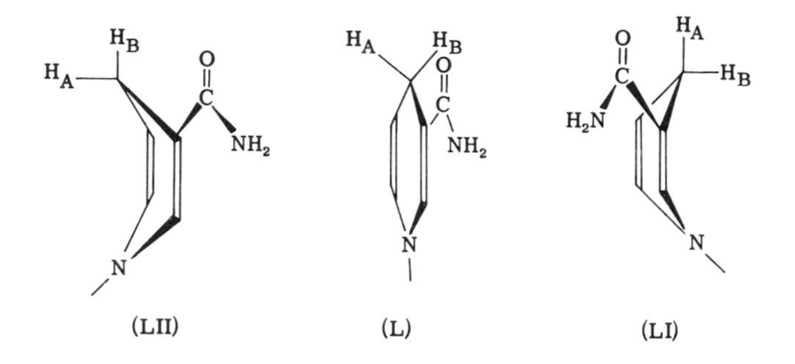

(LII) (L) (LI)

to explain the *A*- and *B*-side specificities of enzymes, it is much simpler to propose that NAD and NADP are, or can be, bound to their enzymes in an extended conformation such that the nicotinamide and adenine rings do not overlap.

V. Asymmetrical Methyl Groups. Stereospecificity of Synthesis of Malate

There are many enzymic reactions in which a methyl group is converted into a methylene group, and conversely a methylene or vinyl group is changed into a methyl group. Typical examples of such reactions are the condensation of acetyl-CoA with oxalacetate to form citrate, the synthesis of malate from glyoxylate and acetyl-CoA, or the carboxylation of acetyl-CoA to malonyl-CoA. Cleavage of a β-ketoacyl-CoA by thiolase transforms the α-methylene group of the β-ketoacyl-CoA into the methyl group of acetyl-CoA, and isopentenylpyrophosphate isomerase transforms the vinyl group of isopentenyl pyrosphosphate into a methyl group on dimethylallyl pyrophosphate. Cornforth *et al.* (*11*) and Lüthy *et al.* (*12*) devised independently two procedures for the preparation of the enantiomeric *R* and *S* protio-deuterio-tritio-acetic acids [(V) and (VI)] for the study of the stereochemistry of these transformations. Cornforth *et al.* (*11*) have argued that when methyl groups, rendered asymmetric by labeling with both deuterium and tritium,

142. The absolute configuration of H_A and H_B was not known when Levy and Vennesland (*85*) made their proposal; hence, the incorrect stereochemistry shown in their drawings for H_A and H_B is understandable.

". . . are transformed by a stereospecific reaction into methylene groups, the stereochemistry can be traced if (a) the chirality of the methyl group is known, (b) the chirality of the derived methylene group can be determined, and (c) there is appreciable discrimination between isotopes in the removal of hydrogen."

These arguments may be best appreciated in conjunction with Scheme 6a and b, which illustrates hypothetical reactions when an incoming X group displaces a hydrogen atom at the methyl group of *R*- (LIII) and *S*-acetate (LV) by inversion of configuration. Isotopic discrimination, if large, will cause the methyl group to assume a conformation favoring the displacement of H and retention of deuterium and tritium at the new methylene group. The products (LIV) and (LVI) formed from the *R*-

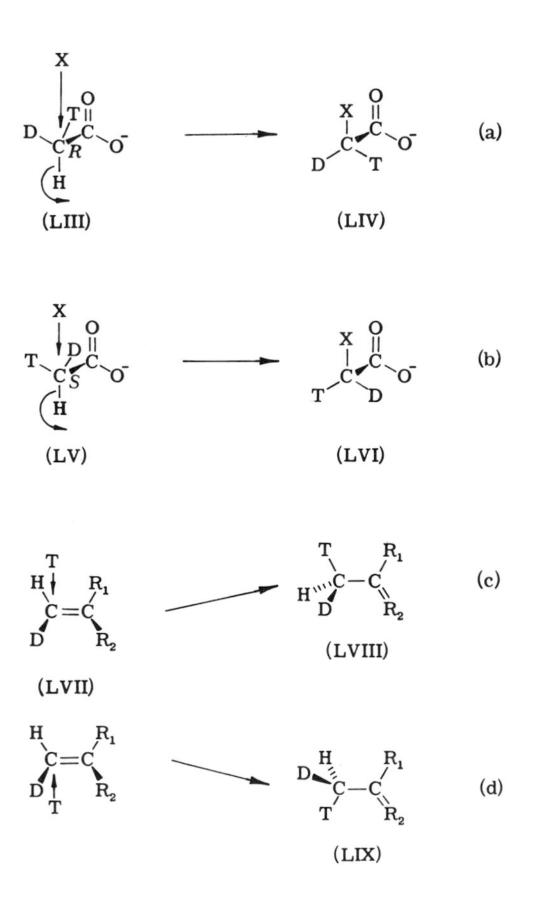

SCHEME 6. Stereospecific conversion of methyl group into methylene group (a) and (b). Stereospecific conversion of vinyl group into methyl group (c) and (d).

and *S*-acetates will be enantiomers, and if the steric positions of deuterium and tritium can be determined in these the stereochemistry of the reaction can be deduced also.

The reverse situation arises when, e.g., a terminal methylene group is changed into a methyl group in a substance of the $CH_2=CR_1R_2$ type, as in isopentenyl pyrophosphate (cf. Section VIII,D). Here the first requirement is the specific replacement of one of the two hydrogen atoms on the vinyl carbon atom with deuterium as shown in (LVII) (*143*). If the addition of a hydrogen atom to a vinyl group thus labeled is stereospecific and if the incoming atom is arranged to be tritium, two antipodally labeled methyl groups may result [(LVIII) and (LIX)], depending on the face of the plane-trigonal structure onto which the tritium atom is added (Scheme 6c and d). It remains now to establish the chirality of the new methyl group: (a) conversion of (LVIII) or (LIX) into acetic acid in such a way that the chiral methyl group becomes, without change of configuration, the methyl group of acetate; and (b) the use of such a specimen in the type of experiments devised by Cornforth *et al.* (*11*) and by Lüthy *et al.* (*12*) would accomplish the task.

The determination of the chirality of a methyl group containing all three hydrogen isotopes in sterically defined positions is difficult, because it is impossible, in common laboratory practice, to use tritium undiluted with normal hydrogen. This restriction does not apply, of course, to deuterium. Thus, in a chemically synthesized specimen, in the preparation of which a maximum inclusion of tritium atoms has been aimed at, very few molecules would contain all three hydrogen isotopes. Optical rotation is therefore ruled out as a measure of chirality because most of the asymmetrical methyl groups in any specimen would consist of $—CH_2D$ groups. The problem has been overcome both by Cornforth *et al.* (*11*) and by Lüthy *et al.* (*12*) by ensuring that in their specimens of *R*- and *S*-acetates practically every tritium-labeled methyl group also contained deuterium and by relating the determination of chirality to measurements of radioactivity.

For the latter purpose both groups of workers resorted to malate synthetase (EC 4.1.3.2) (*144*) which produces *S*-malate (LX) by the condensation of acetyl-CoA with glyoxylate. For the determination of the steric position of the tritium in the *S*-malate, synthesized from glyoxylate and *R*- or *S*-$D_1{}^3H_1$-acetyl-CoA, they have used fumarase (EC 4.2.1.2), which catalyzes the reversible dehydration of *S*-malate [reaction (21)].

143. The *cis*-4-D_1-isopentenyl pyrophosphate shown in formula (LVII) has already been made, cf. Section VIII,D.

144. H. Eggerer and A. Klette, *European J. Biochem.* 1, 447 (1967).

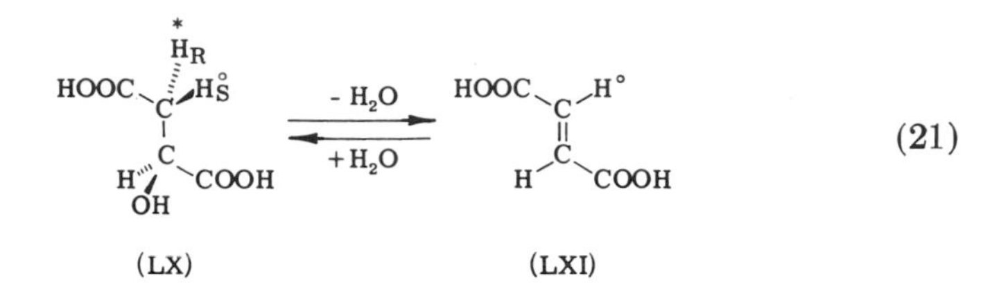

It is known from the work of Anet (*145*) and of Gawron *et al.* (*146*) that the hydroxyl group at C-2 and the pro-*R* hydrogen at C-3 are removed in this dehydration, the process being a trans-elimination. Similarly, the hydration of fumarate to *S*-malate means the addition of the elements of water to fumarate in a trans-manner. Thus *S*-malate, labeled with tritium at C-3, when added to fumarase in normal water, will lose practically all tritium from the pro-*R* position at C-3 [H_R* in (LX)] and retain all tritium that might have been present at the pro-*S* position at that carbon atom [H_S° in (LX)]. It is immaterial that at equilibrium the product of fumarase action will consist of a mixture of malate and fumarate since no carbon bound hydrogen atom of fumarate exchanges with the hydrogen ions of the medium. It follows, therefore, that ". . . the percentage of tritium retained in the organic acids after equilibrium with fumarase is equal to the percentage of 2*S*-3*S*-[3-$^3H_1$] malate in the *S*-malate specimen examined, and the percentage of tritium lost from the organic acids equals the percentage of 2*S*-3*R*-[3-$^3H_1$] malate" (*11*). Because of the very exceptional nature of these experiments, which should open a new area of investigation, the details of the syntheses of the *R*- and *S*-acetates, as far as published, will be recorded here.

Cornforth *et al.* (*11*) prepared the *R*- and *S*-$D_1{}^3H_1$-acetate entirely by chemical methods; Lüthy *et al.* (*12*) engaged also the help of two enzymes. The two procedures are shown in Schemes 7 and 8. Cornforth and his colleagues started the synthesis with deuteriophenylacetylene (LXII) which was reduced with diimide to *cis*-2-D_1-1-phenylethylene (LXIII). The epoxides (LXIVa) + (LXIVb), derived by oxidation of (LXIII) with peroxybenzoic acid, were reduced with lithium borotritide to the mixture of 1*R*-2*R*-[2-$D_1{}^3H_1$]-1-phenylethanol (LXVa) and 1*S*-2*S*-[2-$D_1{}^3H_1$]-1-phenylethanol (LXVb). This racemic mixture was resolved into the (+)-1*R* and (—)-1*S* enantiomers by the crystallization of the brucine phthalates. Oxidation of each enantiomer by chromic acid gave the two enantiomers of acetophenone, (LXVIa) and (LXVIb), which were oxidized with peroxytrifluoroacetic acid to the phenylacetates,

145. F. A. L. Anet, *JACS* **82**, 994 (1960).
146. O. Gawron and T. P. Fondy, *JACS* **81**, 6333 (1959).

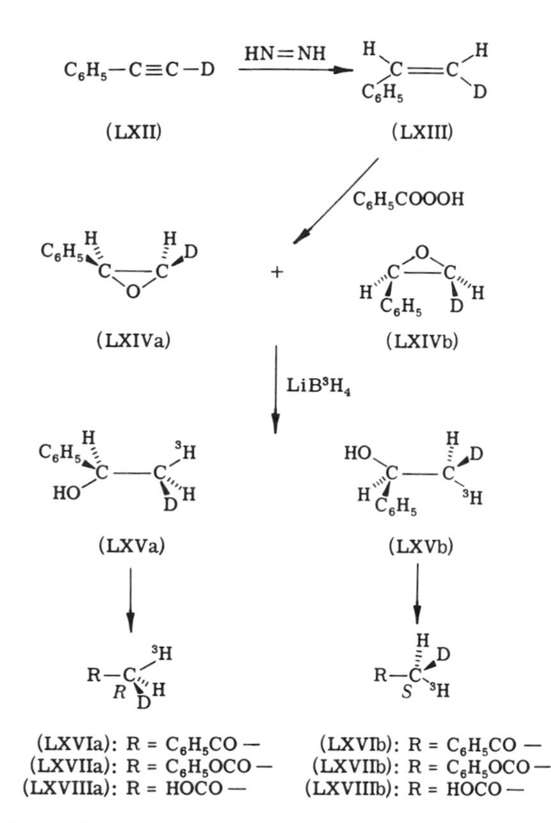

SCHEME 7. Synthesis of *R*- and *S*-$^3H_1D_1$-acetic acids according to Cornforth *et al.* (*11*).

(LXVIIa) and (LXVIIb), from which the *R* and *S*-[$D_1{}^3H_1$] acetic acids [(LXVIIIa) = V and (LXVIIIb) = VI, cf. Section I], respectively, were obtained by mild alkaline hydrolysis. The specific activity of both preparations was 0.07 μCi of $^3H/\mu$mole. It may be calculated that in these specimens one in 450,000 molecules contained tritium, nearly all of which were also labeled with deuterium (*147*).

Lüthy *et al.* (*12*) started their syntheses from the 2*S*- and 2*R*-2-$^3H_1$-glycolates (LXX) and (LXXI) prepared by the reduction of 2-^{3}H-glyoxylate (LXIX) with NADH and muscle lactate dehydrogenase (EC 1.1.1.27) and glyoxylate reductase from spinach leaves (EC 1.1.1.26), respectively. The steps in the preparation of the *R*-acetic acid specimen

147. The number of radioactive atoms in a specimen of one curie is $3.70 \times 10^{10}/\lambda$, where λ is the disintegration constant (sec^{-1}) of the isotope, or $3.70 \times 10^{10} \times 1.44 \times T_{1/2}$, where $T_{1/2}$ is the half-life of the isotope expressed in seconds. The half-life of tritium in the calculations was taken to be 12.26 years $\simeq 3.866 \times 10^8$ sec.

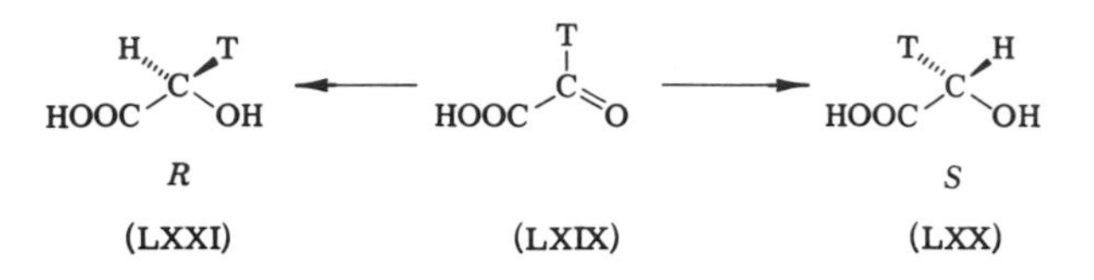

from the 2S-2$^3H_1$-glycolate (LXX) are shown in Scheme 8. Reduction of the methyl ester of (LXX) yielded the 1S-1-$^3H_1$-ethylene glycol (LXXII) which was converted to the mixture of the monobrosylates (LXXIII) and (LXXIV); the reduction of these with $LiAlD_4$ gave the mixture of the ethanols (LXXV) and (LXXVI). The configuration of (LXXV) at C-2 followed from the known inversion of configuration caused by metal hydride reductions (*148*) of tosylates and brosylates.

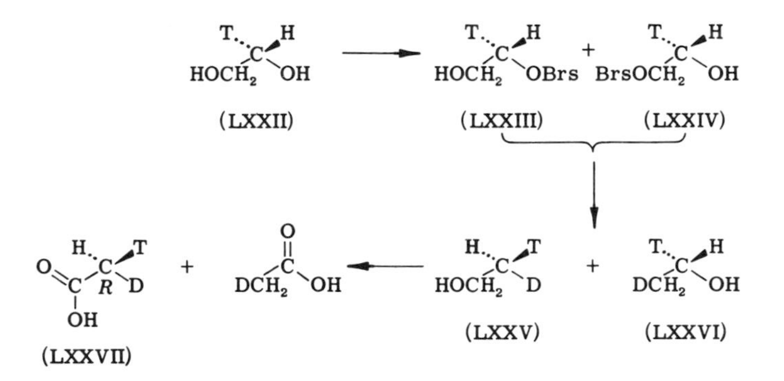

SCHEME 8. Stereospecific synthesis of *R*-$^3H_1D_1$-acetic acid according to Lüthy *et al.* (*12*) (Brs stands for *p*-bromophenylsulfonyl-. Brosylate (see text) stands for *p*-bromophenylsulfonate).

Oxidation of this mixture of the ethanols ". . . in aqueous solution with oxygen in the presence of a platinum catalyst . . ." gave a specimen of acetic acid in which the *R*-$D_1{}^3H_1$-isomer (LXXVII) was the only chiral component. The same sequence of reactions on the 2*R*-2-$^3H_1$-glycolate (LXXI) yielded the *S*-$D_1{}^3H_1$-acetic acid.

Lüthy *et al.* (*12*) did not report the specific activities of their acetate preparations, but they stated that only one in every 10^9 molecules in their specimens contained tritium. If this is a true figure and not a miscalculation, it means that the specific activity of their specimens was only about 71 disintegrations $\times$ min^{-1} $\times$ μmole^{-1} and just at the limit of reasonably

148. E. L. Eliel, *Record Chem. Progr.* (*Kresge-Hooker Sci. Lib.*) **22,** 129 (1961); quoted by Lüthy *et al.* (*12*).

TABLE III

DATA ON THE SYNTHESIS OF MALATE FROM GLYOXYLATE AND R- OR S-$^3H_1D_1$-ACETATE[a]

	Malate from R-$^3H_1D_1$-acetate	Malate from S-$^3H_1D_1$-acetate	Malate from $^3H_1$-acetate
(a) ^3H content of malate before fumarase (cpm/μmole)	724 ± 1.6	726 ± 2.9	706 ± 21
^3H content of malate + fumarate after fumarase (cpm/μmole)	480 ± 8.2	223 ± 1.7	340 ± 6
Retention of ^3H (%)	67.2 ± 1.3	30.7 ± 0.3	48.1 ± 0.9
(b) $^3H/^{14}C$ ratio of malate before fumarase	1.558 ± 0.004	1.560 ± 0.018	1.541 ± 0.024
$^3H/^{14}C$ ratio of malate + fumarate after fumarase	1.071 ± 0.007	0.481 ± 0.003	0.770 ± 0.012
Retention of ^3H (%)	68.7 ± 0.8	30.8 ± 0.5	50.0 ± 1.6

[a] The following are the details of the experiments as described by Cornforth *et al.* (*11*): "Each specimen of acetate was incubated with adenosine triphosphate, acetate kinase, phosphotransacetylase, coenzyme A, glyoxylate and malate synthase. The yield of S-malate, from acetate, was 80–90 per cent. Randomly labelled [$^3H_1$]acetate was also converted into malate. The malic acids were isolated by chromatography on 'Dowex-1' and diluted with unlabelled S-malate. Each sample was divided into two halves for determination, by two methods, of the loss of tritium on incubation with fumarase. A stoichiometrically negligible amount of ^{14}C-malate was added to one half. Aliquots were taken before, and at intervals after, addition of fumarase. The organic acids were separated from inorganic ions and, after complete removal of the aqueous medium, redissolved in water. The ratio of $^3H/^{14}C$ was determined in one set of samples; in the other, counting for ^3H alone was combined with enzyme determination (nicotinamide-adenine dinucleotide + acetyl-coenzyme A + fumarase + malate dehydrogenase + citrate synthase) of total fumarate + malate. In the conditions used, the tritium content of the organic acids became constant after about 10 min."

accurate measurements in ordinary laboratory practice. The unusually high isotope discrimination ratios, $K_H/K_T = 8$–10 and $K_H/K_D = 4$–5, reported by these authors may be attributable to errors in radioactive counting. Although the conclusions of Lüthy *et al.* are qualitatively in agreement with those of Cornforth *et al.* (*11*), the data of the latter workers appear to be more reliable than those of Lüthy *et al.* (*12*). Table III presents the experimental data of Cornforth *et al.* (*11*) which lead, like those of Lüthy *et al.* (*12*), to the inevitable conclusion that the synthesis of malate from acetyl-CoA and glyoxylate occurs with an inversion of configuration at the methyl group and by the addition of this methyl group to the *si* face of glyoxylate (Scheme 9).

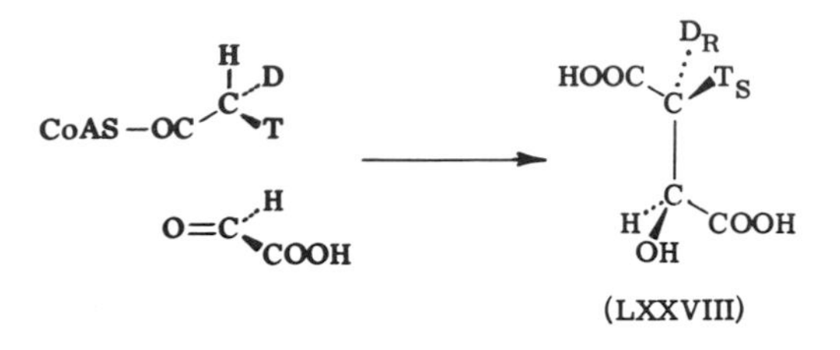

SCHEME 9. Synthesis of 2*S*-3*S*-3-$^3H_1D_1$-malate (LXXVIII) from glyoxylate and *R*-$^3H_1D_1$-acetyl-CoA (*11, 12*). The methyl group is attacked from the *si* face at the carbinol carbon of glyoxylate.

VI. Reactions of the Citric Acid Cycle

A. Synthesis of Citrate and Reactions Catalyzed by Aconitase

The stereochemical aspects of the reactions of the citric acid cycle have been reviewed in detail by Levy *et al.* (*14*) and some of its reactions have been discussed further by Cornforth and Ryback (*15*). The second volume of Bentley's new treatise (*9*) and Volume XIII of "Methods in Enzymology" (*149*), the latter devoted entirely to reactions of the citric acid cycle, contain the most detailed information on this topic. Englard and Hanson (*150*) have summarized in a remarkably succinct manner the preparation of stereospecifically labeled intermediates of the citric acid cycle.

The first experiments on reactions of the citric acid cycle, which drew attention to the ability of enzymes to distinguish between two identical ligands at a prochiral center, have been discussed in the early part of this chapter (cf. Section II). Those early experiments, conducted with oxalacetate labeled with isotopic carbon at the β-carboxyl carbon (LXXIX), established that the carboxyl group of α-ketoglutarate (LXXXI), adjacent to the keto group, originated from the β-carboxyl group (C-4) of oxalacetate, citrate (LXXX) having been "reestablished" as an intermediate in the process (cf. Section II).

Further confirmation of the dissymmetric reactions of citrate was provided by the work of Martius and Schorre (*151, 151a, 151b*) who synthesized the racemic mixture of the two oxalocitramalic acid lactones

149. "Methods in Enzymology," Vol. 13, 1969.

150. S. Englard and K. R. Hanson, "Methods in Enzymology," Vol. 13, p. 567, 1969.

151. C. Martius and G. Schorre, *Ann. Chem.* **570,** 140 (1950).

151a. C. Martius and G. Schorre, *Ann. Chem.* **570,** 143 (1950).

151b. C. Martius and G. Schorre, *Z. Naturforsch.* **5b,** 170 (1950).

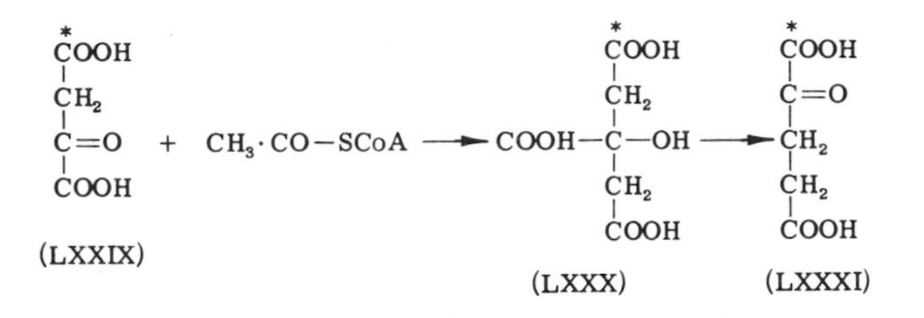

(LXXXII) and (LXXXIII) and resolved the enantiomers by crystallization of their brucine salts. The (—)-4*R*-oxalocitramalic acid lactone (LXXXIII) had a specific rotation of $[\alpha]_{546}^{20}$ —215°, and its (+)-4*S* enantiomer (LXXXII) gave $[\alpha]_{546}^{20}$ +213° (*152*). Equilibration of these specimens with D_2O, and their oxidation with D_2O_2, gave the (—)-3*S*-4-D_2-citric acid (LXXXIV) from the (—)-4*R*-lactone and the (+)-3*R*-2-D_2-citric acid (LXXXV) from the (+)-4*S*-lactone (*153*). Each specimen contained very nearly the theoretical amount of deuterium, 1.922 and

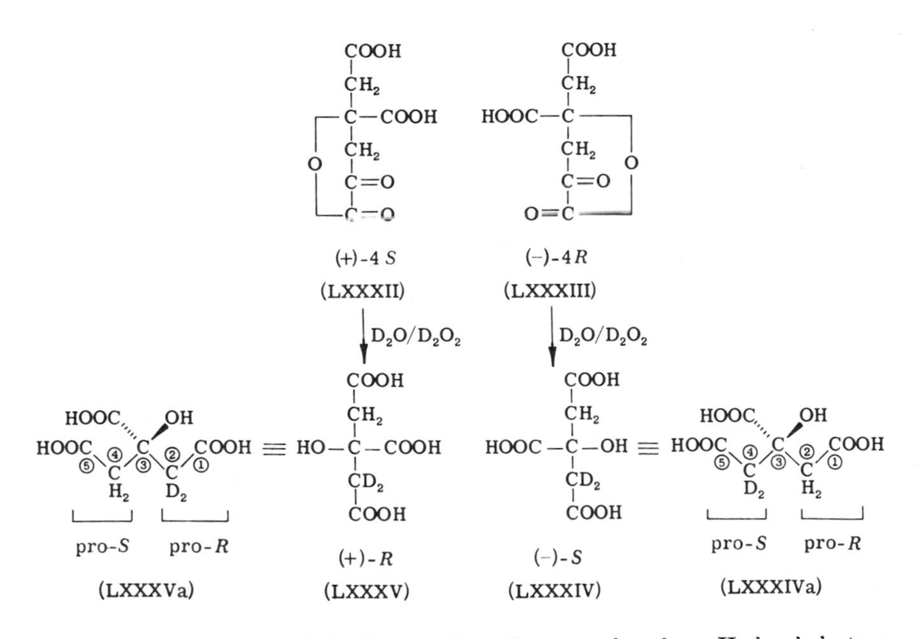

152. The assignments of absolute configuration were based on Hudson's lactone rule (cf. *151a, 151b*); subsequent enzymic work on biosynthesis of citrate fully vindicated these.

153. The numbering of the carbon atoms in citrate shown in formulas (LXXXIVa) and (LXXXVa) follows the proposal of Englard and Hanson (*150*) that the numbering should begin at the carboxyl carbon of the pro-*R* carboxymethylene group of citrate.

1.918 g-atoms per mole. They had a specific rotation, (—) and (+), of about 1° at 546 mμ in water.

In enzymic experiments, the (—)-*S*-4-D_2-citrate gave α-ketoglutarate without any loss of deuterium, whereas the α-ketoglutarate derived from the (+)-*R*-2-D_2-citrate contained no isotope (*151a*). These observations led, of course, to the conclusion that aconitase (EC 4.2.1.3) reacted with the pro-*R* side of citrate (*30*) [cf. formulas (LXXXIVa) and (LXXXVa)]. It could also be concluded from the data, coupled with the information obtained from the experiments made with the labeled oxalacetate referred to earlier, that in the synthesis of citrate from oxalacetate and acetyl-CoA, the acetyl-CoA provided the pro-*S* carboxymethylene ligand at C-3 of citrate. We can go even one step further: The acetyl-CoA must be added to the *si* face at the carbonyl carbon (C-2) of oxalacetate in a manner analogous to the synthesis of malate from glyoxylate and acetyl-CoA (*154*). The mechanism shown in Scheme 9 for glyoxylate synthesis is equally applicable to the synthesis of citrate ex-

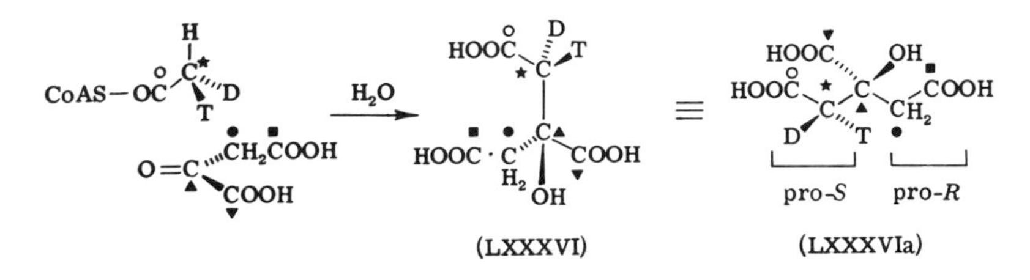

SCHEME 10. Synthesis of 3*S*-citrate (LXXXVI) from oxalacetate and *R*-$D_1{}^3H_1$-acetyl-CoA. The plane of the oxalacetate is meant to be perpendicular to the plane of the paper, the COOH group being in front and the COOH·CH_2 group behind the plane of the paper. The carbon atoms of acetyl-CoA and of oxalacetate are marked by various symbols for identification in the product. The stereochemistry of the synthesis is analogous to that of the synthesis of malate.

cept that the carbon-bound H atom of glyoxylate is to be replaced by the ligand —CH_2COOH (Scheme 10).

Further corroboration of these conclusions was provided by the work of Hanson and Rose (*155*) on aconitase (EC 4.2.1.3) with the aid of stereospecifically labeled citric acid. These workers have prepared first

154. In the reduction of oxalacetate to *S*-malate the hydrogen atom is added to the *re* face of the carbonyl carbon [cf. formulas (XXVII) and (XVIII)]. According to private communication received from Professor J. W. Cornforth, the synthesis of citrate from oxalacetate and *R*- and *S*-$D_1{}^3H_1$-acetyl-CoA proceeds with an inversion of configuration at the methyl group of acetyl-CoA as shown for malate synthetase.

155. K. R. Hanson and I. A. Rose, *Proc. Natl. Acad. Sci. U. S.* **50**, 981 (1963).

6-monotritiated quinic acid (LXXXVIII) by incubating 5-dehydroshikimate (LXXXVII) and NADH with an extract of *Aerobacter aerogenes* in tritiated water. It followed from the known absolute configuration of quinic acid that the citric acid obtained from the tritiated specimen by oxidation must have had the 3*R*-2*R* configuration (LXXXIX).

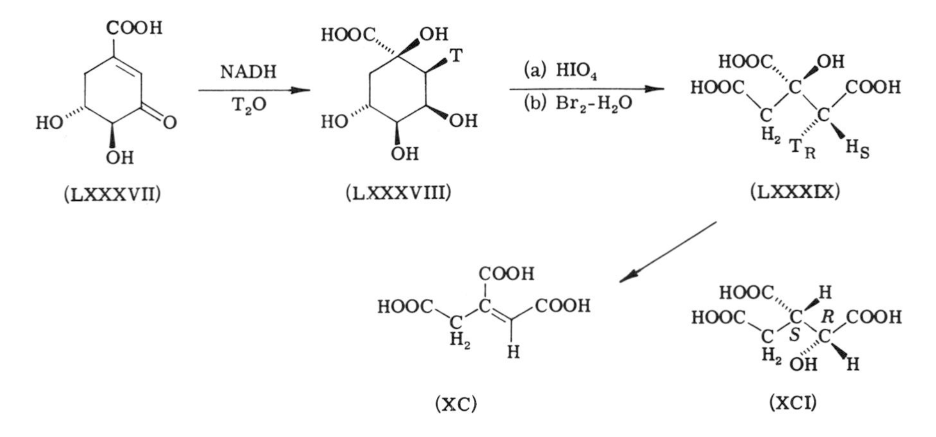

The use of this stereospecifically labeled citrate with aconitase confirmed that this enzyme attacked the pro-*R* end of citrate and that aconitase catalyzed a trans-elimination and trans-addition of water (*156*). This result is in complete harmony with the fact that the enzymically reactive form of aconitate has the cis-geometry (XC), thus the elimination of the hydroxyl group and of the pro-*R* hydrogen atom at C-2 (*153*) leads directly to the cis-configuration. That the hydration of aconitate to (+)-isocitrate is also a trans-addition of water follows, both from the now known absolute configuration, 2*R*-3*S*, of the enzymically reactive form of isocitrate (XCI) (*157–159*) and from the fact that there is no incorporation of deuterium bound to carbon at C-2 of isocitrate even after prolonged equilibration of aconitate with aconitase; deuterium appears only attached to C-3. Equilibration of citrate in D_2O with aconitase leads to the incorporation of only one atom of deuterium at C-2 which must occupy the position of the pro-*R* hydrogen atom since after conversion of

156. The elements of water are added to the ring double bond in 5-dehydroshikimate in a cis manner.

157. C. K. Johnson, A. L. Patterson, D. van der Helm, and J. A. Minkin, *Chem. Eng. News* **39**, 53 (1961); *Program Abstr. Ann. Meeting Am. Cryst. Assoc.* p. 44 (1961).

158. T. Kaneko, H. Katsura, H. Asano, and K. Wakabayashi, *Chem. & Ind.* (*London*) p. 1187 (1960).

159. T. Kaneko and H. Katsura, *Chem. & Ind.* (*London*) p. 1188 (1960).

this 2-D_1-citrate into isocitrate the deuterium is lost. It is the pro-*S* hydrogen atom at C-2 of citrate which becomes the hydrogen atom at C-2 of 2*R*-3*S*-isocitrate and transferable to $NADP^+$ in the isocitrate dehydrogenase reaction. The interconversions of citrate and isocitrate via aconitate are shown in Scheme 11. It follows from these observations that

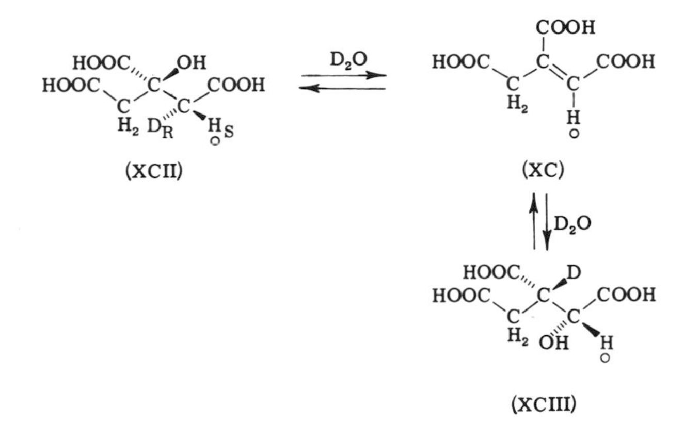

SCHEME 11. Stereochemistry of reactions catalyzed by aconitase. The face of the aconitate molecule seen by the reader is the *re–si* face (*re* face at C-3; *si* face at C-2; for naming the faces of plane-trigonal structures see Section III). Only the changes of carbon-bound hydrogen atoms are shown.

aconitase attacks C-3 of aconitate from the *re–si* face in both the citrate ⇌ aconitate and isocitrate ⇌ aconitate reactions, whereas the attack on C-2 is made from the *si–re* face in both reactions. The interesting consequence of these reactions is that the hydroxyl group of citrate is replaced stereospecifically by a hydrogen atom and the pro-*R* hydrogen atom at C-2 of citrate is replaced by a hydroxyl group in the overall reaction citrate → 2*R*-3*S*-isocitrate.

B. ISOCITRATE DEHYDROGENASE

Isocitrate dehydrogenase (EC 1.1.1.42) catalyzes the conversion of 2*R*-3*S*-isocitrate into α-ketoglutarate and CO_2 with $NADP^+$ as coenzyme. This is an *A*-side specific enzyme (see Table I) transferring the hydrogen atom from C-2 of isocitrate into the pro-*R* position at C-4 of the dihydronicotinamide ring of the coenzyme. As discussed in the preceding section, this hydrogen atom of isocitrate occupied originally the pro-*S* position at C-2 of citrate and is not exchangeable with the hydrogen ions of the medium (*89a*).

When the isocitrate dehydrogenase reaction was conducted in a me-

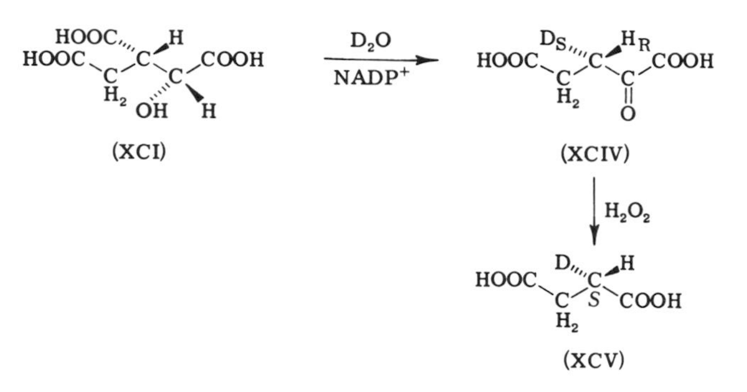

SCHEME 12. Stereochemistry of decarboxylation of 2*R*-3*S*-isocitrate to α-ketoglutarate.

dium of D_2O (*160*), one deuterium atom appeared at C-3 of α-ketoglutarate (XCIV), oxidation of which gave the dextrorotatory 2*S*-2-D_1-succinate (XCV) (*161*) (see Scheme 12). Thus the displaced C-3′ carboxyl carbon of isocitrate was replaced stereospecifically by a hydrogen

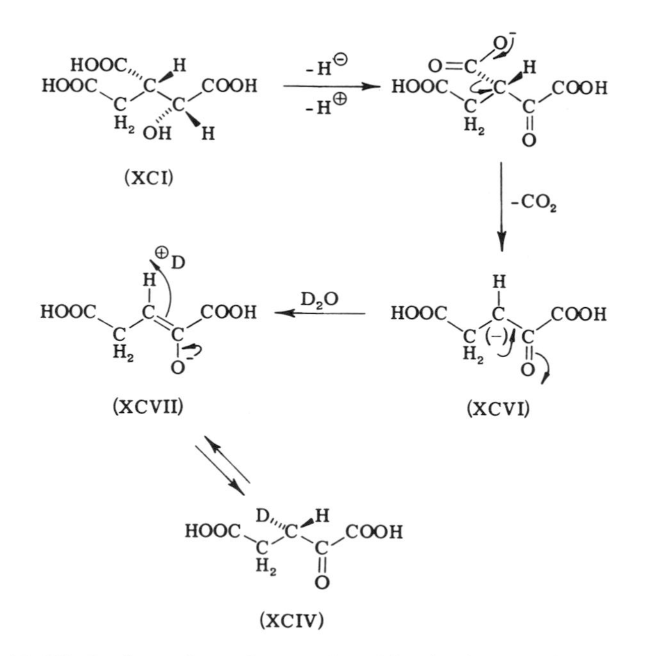

SCHEME 13. Mechanism of reaction catalyzed by isocitrate dehydrogenase according to Lienhard and Rose (*162*).

160. S. Englard and I. Listowsky, *BBRC* **12**, 356 (1963).

161. For the preparation of 2-D_1-succinate of known absolute configuration, see Section VI,C and Cornforth *et al.* (*92*, *93*).

atom. It might be supposed that both a carbanion (XCVI) and the enolate ion (XCVII) of α-ketoglutarate are intermediates in the isocitrate dehydrogenase reaction, the enolate ion being then stereospecifically protonated (*162*) (see Scheme 13). Some supporting evidence to this effect was provided by Lienhard and Rose (*162*) who found that α-ketoglutarate, incubated with NADPH and isocitrate dehydrogenase, exchanged with the medium that hydrogen atom at C-3 which was introduced into it during the oxidative decarboxylation of isocitrate. These workers prepared a specimen of 3-$^3H_2$-α-ketoglutarate (XCIX) from α-ketoglutarate (XCVIII) by the keto–enol exchange reaction in T_2O and

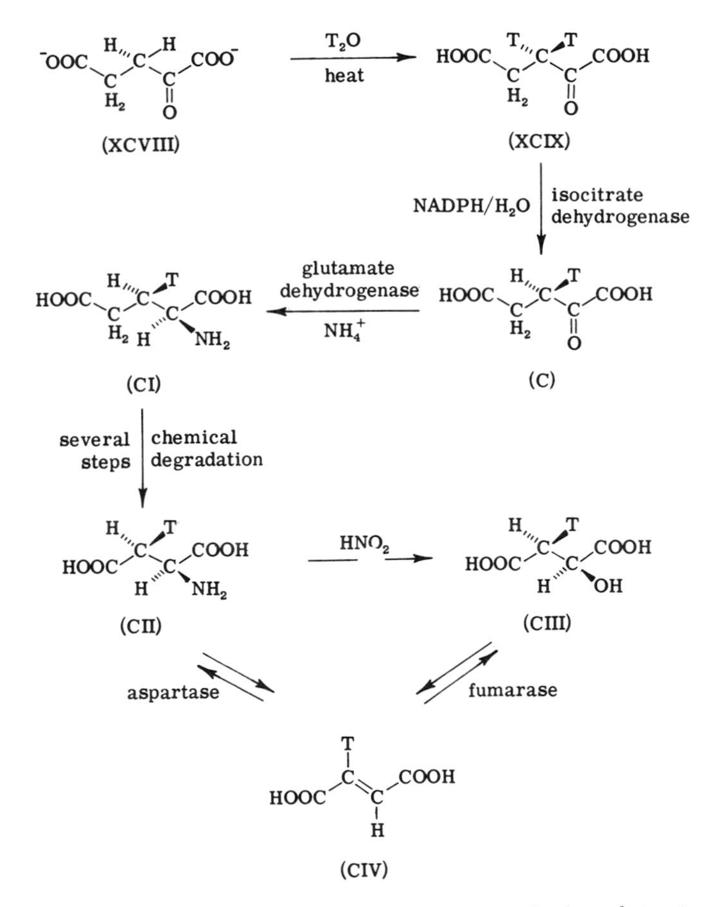

SCHEME 14. Stereospecific hydrogen exchange at C-3 of α-ketoglutarate catalyzed by isocitrate dehydrogenase. Experiment of Lienhard and Rose (*162*).

162. G. E. Lienhard and I. A. Rose, *Biochemistry* 3, 185 (1964).

then reacted it with isocitrate dehydrogenase and NADPH; one labeled hydrogen atom was lost in the process. In order to establish the absolute configuration of the resulting 3-$^3H_1$-α-ketoglutarate (C), the specimen was converted first with glutamate dehydrogenase into *S*-3-$^3H_1$-glutamate (CI) and then by the chemical degradation of the latter into *S*-3-$^3H_1$-aspartate (CII) and 2*S*-3-$^3H_1$-malate (CIII). The $^3H_1$-aspartate (CII) and the $^3H_1$-malate (CIII) were then treated with aspartase and fumarase, respectively. The resulting specimens of fumarate (CIV) from each reaction retained nearly all the tritium contained in the substrates. Since both enzymes are known to catalyze trans-addition eliminations, it followed that the steric position of the tritium remaining in the α-ketoglutarate after equilibration of (XCIX) with isocitrate dehydrogenase in H_2O must have been 3*R* as shown in (C). The mechanism of isocitrate dehydrogenase proposed by Lienhard and Rose (*162*) as well as the hydrogen exchange experiment just discussed are shown in Schemes 13 and 14.

The NAD-specific isocitrate dehydrogenase of bovine heart (EC 1.1.1.41) catalyzes the same stereospecific reactions as the NADP-linked enzyme (EC 1.1.1.42) except the exchange of the pro-*S* hydrogen atom at C-3 of 3-$^3H_2$-α-ketoglutarate (*163*). Thus the general validity of Scheme 13 for the isocitrate dehydrogenase may be in doubt.

C. Fumarase. Reactions of Citric Acid Cycle from Fumarate to Isocitrate. Preparation of D_1-Succinic Acid of Known Absolute Configuration

Several references have been made in previous sections of this chapter to fumarase (= fumarate hydratase = L-malate hydrolyase, EC 4.2.1.2) as an enzyme catalyzing the reversible trans-addition eliminations of water between fumarate and *S*-malate (*164*). This enzyme became, in recent years, a useful "laboratory reagent" after the steric course of the reaction it catalyzes had been established. It was used successfully for deducing the stereochemistry of other enzymic reactions (cf., e.g., Scheme 14) and also for the preparation of deuteriosuccinic acids of known absolute configuration.

The stereospecificity of the reaction catalyzed by fumarase could be inferred from the long established fact that L-(= *S*)-malate was the sole product of the reaction (*165*) and that only one carbon-bound deuterium

163. Z. B. Rose, *JBC* **241,** 2311 (1966).
164. See, e.g., Sections II, III, and V, and Table III.
165. F. B. Straub, *Z. Physiol. Chem.* **275,** 63 (1942).

atom was introduced into malate when the reaction was conducted in D_2O (*89, 166*). Furthermore, this deuterium atom was removed completely from malate on its enzymic reconversion into fumarate.

The conclusion of Farrar *et al.* (*167*), based on erroneous assumptions that fumarase catalyzed a cis-addition of the elements of water to fumarate and reported in the previous edition of this work (*13*), was proved incorrect (*145, 146*). Gawron and Fondy (*146*) and Anet (*145*) prepared the racemic mixture of 2*R*-3*R*-3-*D*₁- and 2*S*-3*S*-D₁-malic acids, (CVIII) and (CIX), by the reduction of 3,4-epoxy-2,5-dimethoxytetrahydrofuran (CV) with $LiAlD_4$, followed by the acid hydrolysis of the racemic product, (CVI) and (CVII), to the malic dialdehydes and oxida-

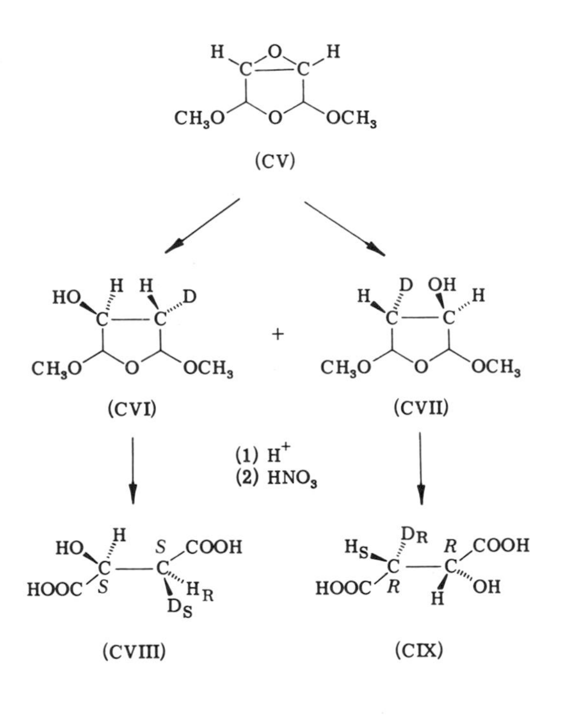

166. H. F. Fisher, G. Frieden, J. S. M. McKee, and R. A. Alberty, *JACS* **77**, 4436 (1955).

167. T. C. Farrar, H. S. Gutowsky, R. A. Alberty, and W. G. Miller, *JACS* **79**, 3978 (1957).

tion of the latter with nitric acid. The protons at C-2 and C-3 in the racemic 3-D_1-malic acids are seen to be *synclinal* (M*sc* and P*sc*) when the carboxyl groups are in an *antiperiplanar* conformation [cf. (CVIII) and (CIX)]. Nuclear magnetic resonance studies showed that the coupling constant for these protons was 4.4 ± 0.2 Hz/sec, a value identical with that found (*168*) for the protons in 2*R*-3-D_1-malic acid obtained by the inversion of configuration at C-2 of 2*S*-3-D_1-malic acid made by the hydration of fumarate with fumarase in D_2O. In contrast, Alberty and Bender (*169*) found a coupling constant of 7.1 Hz/sec for the protons at C-2 and C-3 in 2*S*-3-D_1-malic acid (*170*) prepared by the enzymic hydration of fumarate in D_2O. Hence, the protons at C-2 and C-3 in the latter preparation must be *anticlinal* when the conformation of the carboxyl groups is *antiperiplanar,* and the absolute configuration of such 3-D_1-malic acid should be 2*S*-3*R* (CX). It also follows that

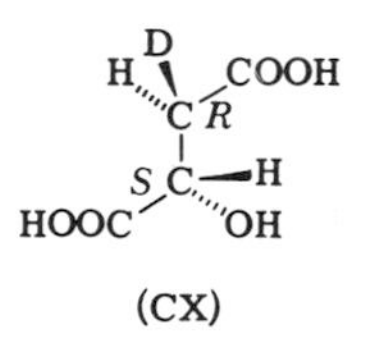

(CX)

fumarase catalyzes not a cis- but a trans-addition of the elements of water to fumarate, the hydroxyl group being added to the *si* face and the proton to the *re* face of the molecule [cf. formulas (XXV)–(XXVII)].

The determination of the correct steric course of the hydration of fumarate, coupled with the knowledge of the stereochemistry of the aconitase reaction, enabled Gawron and Fondy (*146*) to describe the stereochemistry of the reactions of the citric acid cycle from fumarate to isocitrate and to explain why the deuterium contained in 2*S*-3*R*-3-D_1-malate does not appear in 2*R*-3*S*-isocitrate. The reactions are summarized in Scheme 15.

The 2*S*-3*R*-3-D_1-malic acid (CX) has become a valuable starting material for the preparation of a D_1-succinic acid of known absolute configuration. Cornforth *et al.* (*92, 93*) converted the dimethyl ester of 2*S*-3*R*-3-D_1-malic acid (CXI) by treatment with thionyl chloride in pyridine, into dimethyl 2*R*-2-Cl-3*S*-3-D_1-succinate (CXII) (*171*). Reduction

168. A. I. Krasna, *JBC* **233,** 1010 (1958).
169. R. A. Alberty and P. Bender, *JACS* **81,** 542 (1959).
170. The measurements were done on the free acid in D_2O.
171. The reaction with thionyl chloride proceeds with an inversion of configuration at C-2 of malate. The configuration at C-3 is not disturbed, but the notation

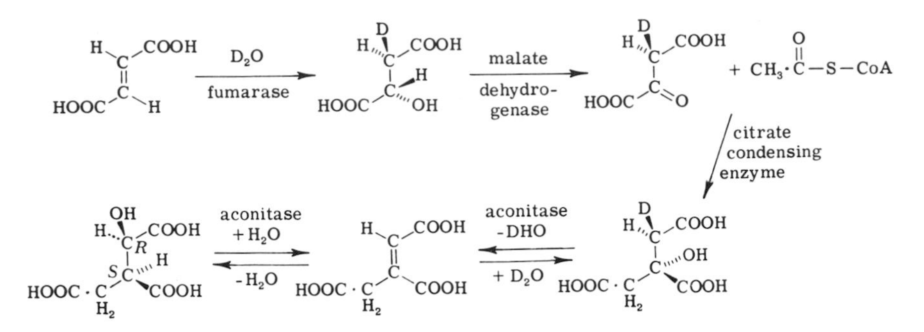

SCHEME 15. Steric course of reactions of citric acid cycle from fumarate to $2R$-$3S$-isocitrate (*146*). See also Schemes 10 and 11.

of the latter with the Zn/Cu couple in acetic acid gave the dimethyl $2R$-2-D_1-succinate (CXIII) from which the $2R$-2-D_1 succinic acid (CXIV) was obtained by acid hydrolysis and repeated crystallization from water (Scheme 16). The preparation obtained contained 93% monodeuteriosuccinic acid and gave a plane optical levorotatory dispersion curve as far in the UV as 248.3 mμ; $[\alpha]^{20°}_{248.3}$ $-21°$ in methanol (see Figs. 5 and 6). This $2R$-2-D_1-succinic acid served as a reference com-

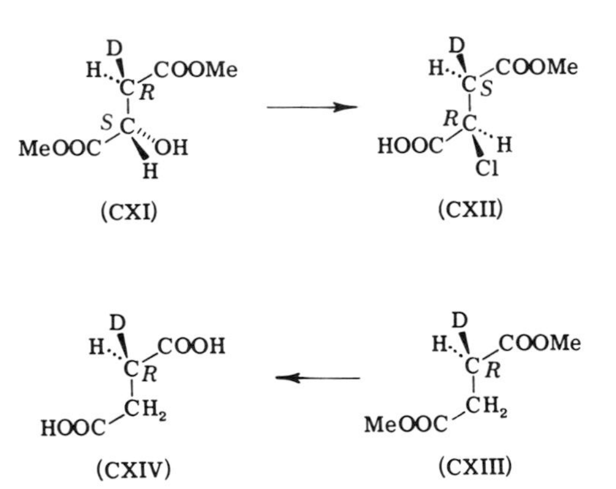

SCHEME 16. Preparation of $2R$-2-D_1-succinic acid from dimethyl $2S$-$3R$-3-D_1-malate (*92, 93*). The changes in stereochemical notations from (CXI) → (CXII) → (CXIII) are explained in footnote (*171*).

---

for this asymmetrical center is reversed according to the *RS* system because the Cl—C—H group has priority over the —COOH (C-4) group, whereas in malate the C-4 carboxyl group has priority over the H—C—OH group.

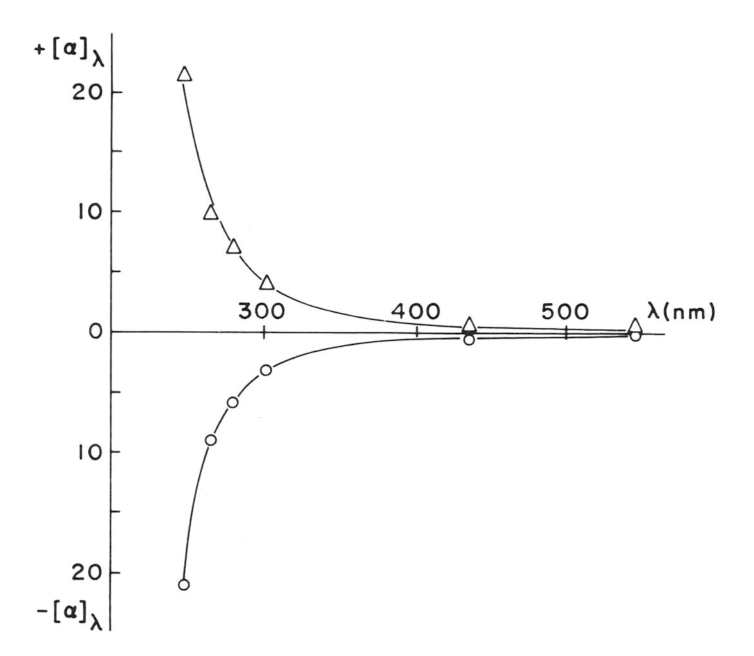

FIG. 6. Optical rotatory dispersion curves of reference $2R$-2-D_1-succinic acid (○), and of D_3-succinic acid (△) derived by ozonolysis from squalene, biosynthesized from 5-D_2-mevalonate (*210, 211*). The measurements were made on the free acids in methanol.

pound in deducing the absolute configuration of H_A and H_B at C-4 of the dihydronicotinamide ring of NADH (cf. Section IV,C), the steric course of isocitrate dehydrogenase reaction (cf. Section VI,B), of the reaction catalyzed by isocitrate lyase (*172*), and of several reactions in the biosynthesis of squalene (cf. Section VIII). With the availability of this reference compound it is safe to predict that a monodeuteriosuccinic acid showing a positive optical rotatory power has the $2S$ configuration; a 2-D_1-3-D_1-succinic acid of the $2R$-$3R$ or $2S$-$3S$ configuration should have a specific rotaton of (—) or (+) 42° at 248.3 mμ and none if its configuration is $2R$-$3S$.

Englard *et al.* (*173*) have also described the preparation of $2R$-2-D_1-succinic acid by a modification of the procedure of Cornforth *et al.* (*92, 93*) from $2R$-$3R$-3-D_1-malic acid obtained by the trans-hydration of maleate in D_2O with maleate hydratase. Syntheses of all the possible stereoisomers of D_1-, 2-D_1-3-D_1-, and 2-D_2-3-D_2-succinic acids, as well as of the optically inactive 2-D_2- and D_4-succinic acids, have been

172. M. Sprecher, R. Berger, and D. B. Sprinson, *JBC* **239**, 4268 (1964).
173. S. Englard, J. S. Britten, and I. Listowsky, *JBC* **242**, 2255 (1967).

achieved; their syntheses have been described by Englard and Hanson (*150*).

D. Succinate Dehydrogenase

The two methylene carbon atoms of succinate are prochiral centers; hence, the pair of hydrogen atoms attached to each are sterically not equivalent: They are H_R,H_S and H_R',H_S' (CXV) and CXVa). The product of succinate dehydrogenase (EC 1.3.99.1) is fumarate, *trans*-ethylenedicarboxylate. Thus the dehydrogenation of succinate to fumarate could proceed either by the trans-elimination of ($H_S + H_R'$) or ($H_R + H_S'$), i.e., of two nonequivalent hydrogen atoms (CXVa) → (CXVIa), or by the cis-elimination of two equivalent hydrogen atoms ($H_R + H_R'$) or ($H_S + H_S'$) to yield (CXVIb) from (CXVa) (Scheme 17).

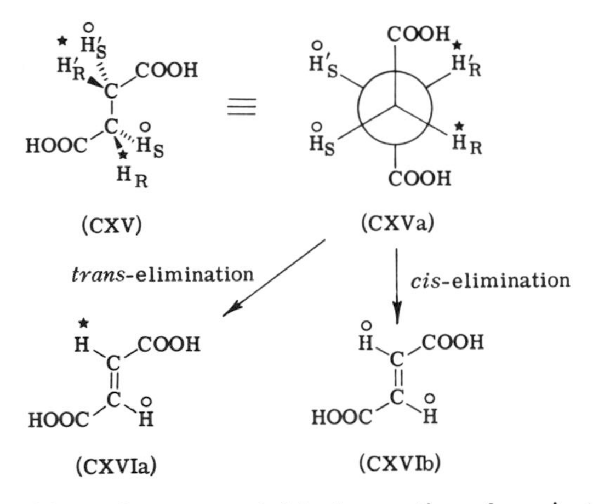

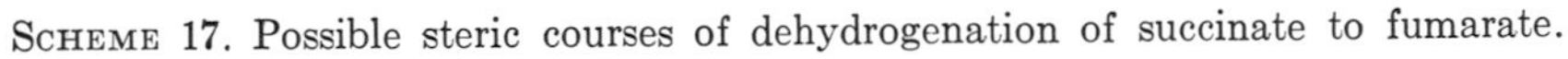
Scheme 17. Possible steric courses of dehydrogenation of succinate to fumarate.

The mechanism of the succinate dehydrogenase reaction engaged the interest of investigators for more than the past twenty years [cf. Levy *et al.* (*14*)], yet the mechanism of this reaction is still not fully understood except that it is now known that fumarate is derived from succinate by a trans-elimination of two nonequivalent hydrogen atoms. The experiments of Tchen and van Milligan (*174*) provided conclusive proof of this. These authors reduced catalytically fumaric and maleic acids in an atmosphere of D_2 gas, a procedure known to result in a cis-addition to double bonds. Thus the reduction of fumarate (CXVII) gave the racemic mixture of 2*R*-3*R*-2-D_1-3-D_1- and 2*S*-3*S*-2-D_1-3-D_1-succinic

174. T. T. Tchen and H. van Milligan, *JACS* **82**, 4115 (1960).

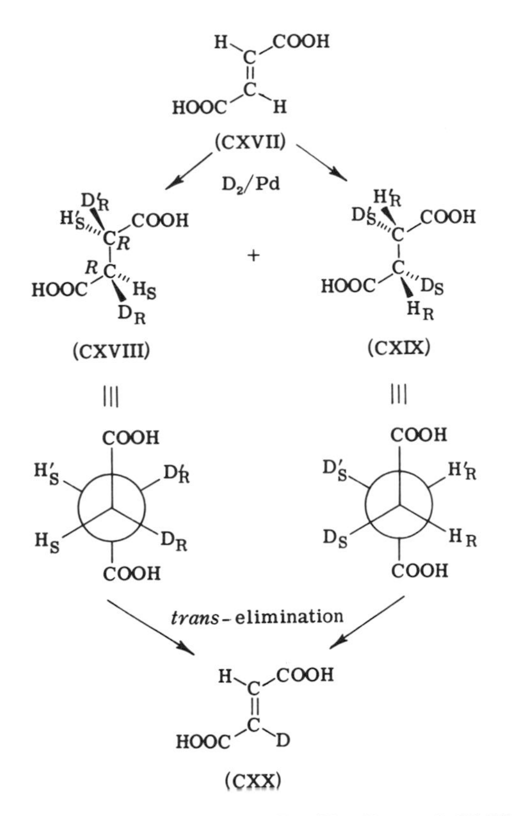

SCHEME 18. Preparation of racemic $2R$-$3R$-2-D_1-3D_1- and $2S$-$3S$-2-D_1-3D_1-succinic acid by catalytic reduction of fumaric acid with D_2. Result of trans-elimination of hydrogen atoms from succinate in the succinate dehydrogenase reaction (*174*) is shown in (CXX).

acids (CXVIII) and (CXIX), respectively (Scheme 18). The similar reduction of maleic acid (CXXI) gave, on the other hand, the *meso*-$2R$-$3S$-2-D_1-3-D_1-succinic acid (CXXII) (Scheme 19). Reference to Scheme 17 will show that trans-elimination of hydrogen atoms from the racemic mixture of $2R$-$3R$- and $2S$-$3S$-2-D_1-3D_1-succinic acids will give monodeuteriofumarate (CXX), whereas a similar procedure from the *meso*-$2R$-$3R$-2-D_1-3-D_1-succinic acid should result in a 1:1 mixture of 2-D_1-3-D_1-succinic acid (CXXIII) and of unlabeled molecules (CXXIV). The experimental findings of Tchen and van Milligan (*174*) supported only the mechanism of trans-elimination. Oxidation of the racemic mixture of dideuteriosuccinate [(CXVIII) + (CXIX), containing 2.4% dideuterio molecules] with heart sarcosomes gave a specimen of fumarate in which only 0.1% of molecules contained two atoms of deuterium. In contrast,

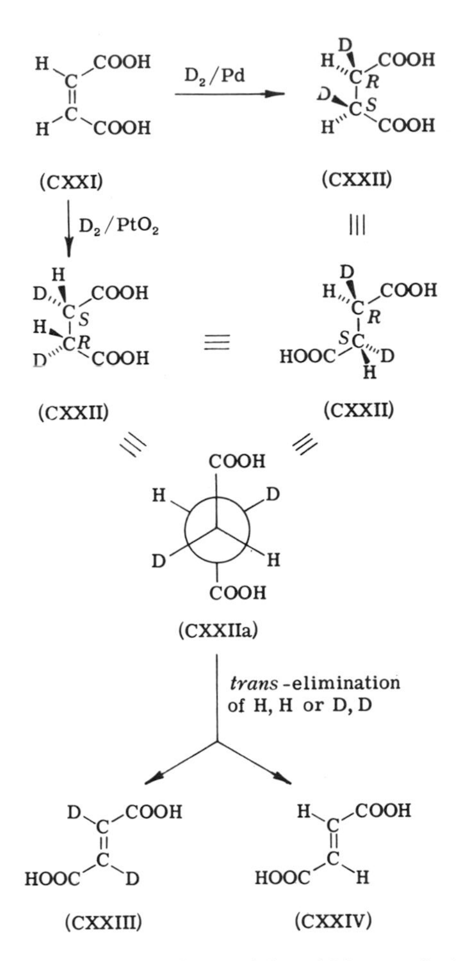

SCHEME 19. Preparation of *meso*-D_2-succinic acid by catalytic reduction of maleic acid with D_2. Results of trans-elimination of hydrogen atoms from succinate in the succinate dehydrogenase reaction are shown in (CXXIII) and (CXXIV) (*174*).

oxidation of the *meso*-dideuteriosuccinate (CXXII), which contained 2.9% dideuterio molecules, resulted in a fumarate preparation in which 1.4% of the molecules were dideuterated.

Further interest in the succinate dehydrogenase reaction was aroused by the observation that under anaerobic conditions succinate dehydrogenase catalyzed an exchange of hydrogen atoms between water and the methylene groups of succinate (*175*) and that fumarate accelerated this process (*176*). The stereochemistry of this hydrogen exchange is subject

175. E. O. Weinman, M. G. Morehouse, and R. J. Winzler, *JBC* **168**, 717 (1947).
176. S. Englard and S. P. Colowick, *JBC* **221**, 1019 (1956).

to conflicting reports. Gawron *et al.* (*177*) claimed that incubation of succinate in D_2O with heart sarcosomes led to a predominantly monodeuteriosuccinate of the *S* configuration. Kahn and Rittenberg (*178*), on the other hand, reported the almost exclusive formation of optically inactive monodeuteriosuccinate under the experimental conditions of Gawron *et al.* (*177*) and inferred that the exchange reaction was a random process. The results of Rétey *et al.* (*179*) differ radically from those of previous workers. Using a purified preparation of succinic dehydrogenase (*180*), Rétey *et al.* (*179*) found that predominantly mono- and dideuteriosuccinate was formed during anaerobic incubations of succinate with the enzyme in D_2O, the percentage of both increasing with increased time of incubation. Significant amounts of trideuterio- and even some tetradeuteriosuccinate were also formed. All preparations showed a remarkably high optical activity and gave plain negative optical rotatory dispersion curves characteristic for (—)-*R*-D_1-, (—)-*R*-2-D_1-3-D_2-, or (—)-*RR*-2-D_1-3-D_1-succinic acids (cf. Section VI,C). Infrared spectra, taken on specimens containing a high percentage of dideuterated species, indicated that the dideuterio molecules were predominantly, though not exclusively, of the optically inactive meso-type, 2*R*-3*S*. The formation of such species is expected if the hydrogen exchange is part of the normal mechanism of the succinate dehydrogenase reaction. However, the exceptionally high optical activity of some of the specimens suggested that a proportion of the dideuterio molecules must have had the (—)-*RR* configuration, the formation of which cannot be related to the normal mechanism of succinate dehydrogenase action. Whatever the mechanism of the succinate dehydrogenase catalyzed hydrogen exchange reaction, the results of Rétey *et al.* (*179*) demonstrated that the H atoms of succinate have different fates, the pro-*R* hydrogen atoms being more prone to exchange than their pro-*S* counterparts.

VII. Reactions at Isolated Double Bonds

A. Conversion of Stearic to Oleic Acid

An interesting case of stereospecificity of an enzymic reaction is the conversion of stearic to oleic acid. Hydrogen atoms are removed in this

177. O. Gawron, A. J. Glaid, III, and J. Francisco, *BBRC* **16,** 156 (1964).
178. J. Kahn and D. Rittenberg, *BBRC* **27,** 484 (1967).
179. J. Rétey, J. Seibl, D. Arigoni, J. W. Cornforth, G. Ryback, W. P. Zeylemaker, and C. Veeger, *Nature* **216,** 1320 (1967).
180. D. V. DerVartanian and C. Veeger, *BBA* **92,** 233 (1964).

instance specifically from C-9 and C-10 of the saturated acid; the hydrogen atoms removed from these two prochiral centers have the same absolute configuration (pro-*R*); and the product, oleic acid, has the cis-configuration (*181*). This example, before any other, emphasizes the unique binding of substrate to enzyme: the selection of positions 9 and 10 for attack could not be explained otherwise.

Schroepfer and Bloch (*181*) have prepared four stereospecifically labeled stearic acids [*S*-9-$^3H_1$: (CXXVIII) $n = 8$, $m = 7$; *R*-9-$^3H_1$: (CXXIX) $n = 8$, $m = 7$; *S*-10-$^3H_1$: (CXXVIII) $n = 7$, $m = 8$; and *R*-10-$^3H_1$: (CXXIX) $n = 8$, $m = 7$] from the enantiomeric pairs of

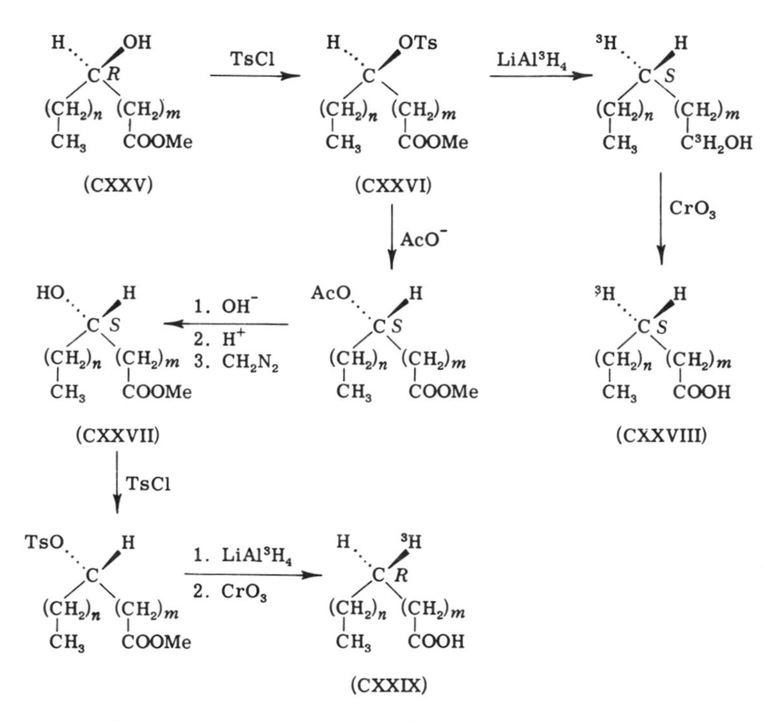

SCHEME 20. Synthesis of stearic acids labeled stereospecifically with ^3H at C-9 and C-10 [(CXXVIII) and (CXXIX)] from methyl (−)-9*R*-9-hydroxyoctadecanoate [(CXXV) $n = 8$, $m = 7$] and from methyl (−)-10*R*-10-hydroxyoctadecanoate [(CXXV) $n = 7$, $m = 8$] according to Schroepfer and Bloch (*181*). Solvolysis and reduction of *p*-toluene sulfonates with $LiAlH_4$ is known to occur with inversion of configuration (*182–184*) (TsCl stands for *p*-toluene sulfonyl chloride; and Ts for *p*-toluene sulfonyl).

181. G. L. Schroepfer, Jr. and K. Bloch, *JBC* **240,** 54 (1965).

182. E. R. Alexander, *JACS* **72,** 3796 (1950).

183. G. K. Helmkamp and R. F. Rickborn, *J. Org. Chem.* **22,** 479 (1957).

184. A. Nickon, J. H. Hammons, J. L. Lambert, and R. O. Williams, *JACS* **85,** 3713 (1963).

methyl 9-hydroxy- and 10-hydroxyoctadecanoates. The starting materials for the syntheses were the methyl esters of (—)-*R*-9-hydroxy- and (—)-*R*-10-hydroxyoctadecanoic acids [(CXXV) $n = 8$, $m = 7$; and (CXXV) $n = 7$, $m = 8$), which by inversion through the *p*-toluene sulfonates (CXXVI) also gave the (+)-*S* enantiomers [(CXXVII) Scheme 20]. The tritium-labeled stearic acids were mixed with 1-^{14}C-stearic acid and were then added to growing cultures of *C. diphtheriae* which convert this saturated acid efficiently into oleic acid (CXXX) (*185*). The results,

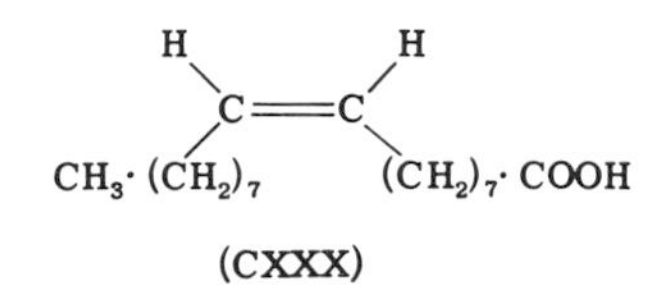

(CXXX)

summarized in Table IV, showed that (a) in the experiments with the *R*-9-$^3H_1$-1-^{14}C-stearic acid there was a significant enrichment of the stearic acid isolated from the bacterial culture in respect of its tritium content but this did not occur with any of the other three stereospecifically labeled stearic acids; and (b) tritium from the 9*R* and 10*R* positions of stearic acid was extensively removed on conversion to oleic acid but it was retained in the oleic acids generated either from the 9*S*- or 10*S*-labeled substrates. The enrichment of the unmetabolized stearic acid with tritium relative to ^{14}C in the experiments with the acid labeled in the 9*R* position, where the hydrogen atom is removed on conversion to oleic acid, indicates a kinetic isotope effect and that the initial rate-limiting step in the desaturation reaction is the eliminaton of the pro-*R* hydrogen atom at C-9 (*186*) and that this is followed by stereospecific removal of the pro-*R* hydrogen atom at C-10. Complete removal or complete retention of tritium on conversion of the labeled stearates into oleate was not observed (cf. Table IV). The differences found between the results obtained and the results that might have been expected for complete removal or retention of the isotopic label were ascribed (*181*) to racemization during the preparation of the substrates. However, unless the degree of racemization during the preparation of the *R*-9-$^3H_1$-stearic acid was much larger than in that of the other stearates, there was a strikingly smaller loss of ^{3}H from the *R*-9-$^3H_1$ substrate than from the *R*-10-^{3}H substrate (*187*). It seems not unlikely that this difference might

185. A. J. Fulco, R. Levy, and K. Bloch, *JBC* **239**, 998 (1964).

186. This conclusion is the logical consequence of the fact that none of the ^{14}C-labeled species of the substrate contained ^{3}H.

187. The maximum racemization is expected to have occurred during the preparation of the *R*-tritio substrates (9-^{3}H and 10-$^3H_1$) because of two inversions involved in the reactions.

TABLE IV

DESATURATION OF STEARIC ACIDS LABELED STEREOSPECIFICALLY WITH TRITIUM AT CARBON ATOMS 9 AND 10 BY GROWING CULTURES OF *C. diphtheriae*[a]

Substrates and products of experiments	$^3H/^{14}C$		
	Expt. 1	Expt. 2	Expt. 3
Experiments with 9*R*-9-^{3}H-1-^{14}C-stearic acid			
Substrate	1.00	1.00	—
Bacterial fatty acids			
Methyl stearate	1.58	1.22	—
Methyl oleate	0.35	0.31	—
Experiments with 9*S*-9-$^3H_1$-1-^{14}C-stearic acid			
Substrate	1.00	1.00	—
Bacterial fatty acids			
Methyl stearate	1.02	0.95	—
Methyl oleate	0.89	0.93	—
Experiments with 10*R*-10-$^3H_1$-1-^{14}C-stearic acid			
Substrate	1.00	1.00	1.00
Bacterial fatty acids			
Methyl stearate	0.99	0.99	1.00
Methyl oleate	0.09	0.10	0.11
Experiments with 10*S*-10-$^3H_1$-1-$_1{}^4$C-stearic acid			
Substrate	1.00	—	—
Bacterial fatty acids			
Methyl stearate	0.85	—	—
Methyl oleate	0.87	—	—

[a] From Schroepfer and Bloch (*181*).

be the result of a "steric" isotope effect as suggested below and in Scheme 21 depicting a hypothetical mechanism of the stearate to oleate conversion. The overall process of the conversion of stearate into oleate according to the quoted observation is equivalent to the cis-elimination of the pro-*R* hydrogen atoms from C-9 and C-10 of stearate. This process is unlikely to be a one-step reaction since cis-eliminations are uncommon. In the absence of known intermediates, it is not unreasonable to propose that all the intermediates are enzyme bound and that the rate-limiting removal of H_R from C-9 of stearate is achieved through the interaction of a nucleophilic $X^{\ominus}$ and an electrophilic $Y^{\oplus}$ group of the desaturating enzyme, followed by a trans-elimination of $X^{\ominus}$ and $H_R{}^{\oplus}$ at C-10 of stearate as shown in Scheme 21. This mechanism is being proposed only as a possible model; the desaturating reaction is likely to be more complex since it requires molecular O_2 and NADPH in addition to FAD (or FMN) and Fe^{2+} (*185*). This mechanism implies a limited mobility of the substituents of C-9 of stearate on the enzyme but not of the sub-

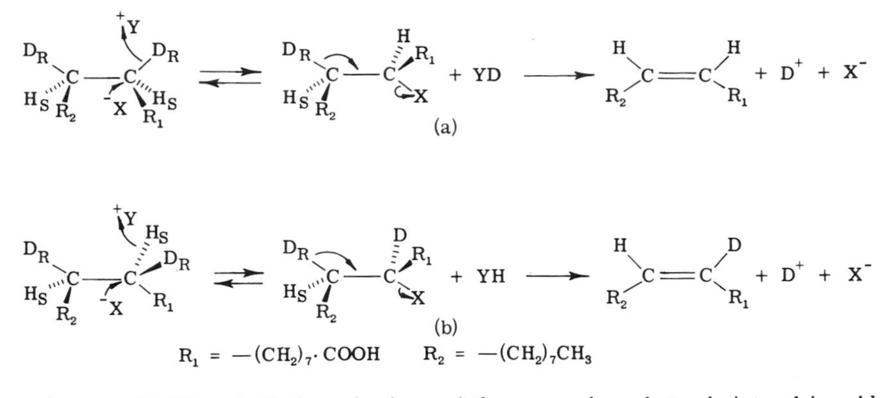

SCHEME 21. Hypothetical mechanisms of the conversion of stearic into oleic acid. Y^+ and X^- are groups of the desaturating enzyme. Mechanism (a) accounts satisfactorily for the preferential elimination of the pro-*R* hydrogen atoms from C-9 and C-10. Mechanism (b) implies that with a small rotation of C-9 around the C-9/C-10 axis the pro-*S* hydrogen at C-9 might also be eliminated when the removal of the pro-*R* hydrogen is energetically unfavorable; it is also necessary to invoke conformational changes in the enzyme molecule.

stituents at C-10; and if 9*R*-10*R*-9,10-$^3H_2$-stearate is offered as substrate to the enzyme, the oleic acid formed could contain, in addition to nonisotopic species, a proportion of monotrito (9-^3H) oleic acid molecules dependent upon the extent of the "steric" isotope effect at C-9 of stearic acid. Schroepfer and Bloch (*181*) have also prepared a specimen of 9*R*-10*R*-9,10-D_2- and 9*S*-10*S*-9,10-D_2-stearic acid by the reduction of oleic acid [illegible] deuterium-labeled diimide (DN=ND). The reduction of olefinic double bonds with diimide is known to be stereospecific in the sense that the addition of hydrogen to either side of the double bond is a cis-addition (*188*). In the specimen produced (diluted with a small amount of 1-^{14}C-stearate) and incubated with a culture of *C. diphtheriae*, 2.9% of the molecules contained no deuterium, 6.3% were monodeuterated, and 90.8% were dideuterated. It may be assumed that 45.4% of the substrate molecules had the 9*R*-10*R*-9,10-D_2 configuration and the same proportion had the 9*S*-10*S*-9,10-D_2 configuration and that the 6.3% monodeuterated molecules were equally distributed among the four possible isomeric species (9*R*-9D_1, 9*S*-9D_1, 10*R*-10-D_1, and 10*S*-10D_1); i.e., 1.575% of each. After incubation the specific radioactivity of the methyl oleate isolated from the bacteria was 2.63×10^4 cpm of ^{14}C per μmole as compared to 4.05×10^4 cpm/μmole of stearate added to the culture. It follows that 65% ($2.63 \times 100/4.05$) of the oleate was derived from the substrate and 35% from endogenous synthesis. It may be cal-

188. E. J. Corey, D. J. Pasto, and W. L. Mock, *JACS* 83, 2957 (1961).

culated from these data that 68.44% ($45.4 \times 0.65 + 35.0 + 2.9 \times 0.65 + 3.15 \times 0.65\%$) of the oleate molecules isolated should contain no deuterium, that 2.05% (3.15×0.65) should be monodeuterated, and that 29.51% should contain two atoms of deuterium (*189*). The found distribution of molecular species labeled, or not, with deuterium were: unlabeled 52%, dideuterated 40%, and monodeuterated 8%. Although the predicted and found values differed much, the greatest divergence (about fourfold) from the predicted value was observed with respect to the monodeuterated oleate. The excess of the monodeuterated oleate molecules found over those predicted is consistent with the "steric" isotope effect postulated above and with the suggestion that the substituents at C-9 of stearate may have a limited mobility in the active area of the desaturating enzyme.

B. Hydration of Oleic to 10*R*-10-Hydroxyoctadecanoic Acid

Another example of a stereospecific reaction at an isolated double bond in an acyclic system is the hydration of oleic to 10-hydroxyoctadecanoic acid effected by a pseudomonad (NRRL-2994). It was shown by Schroepfer and Bloch (*181*) that the absolute configuration of this levorotatory hydroxy acid [methyl ester: $[\alpha]_{546}$ $-0.18° \pm 0.04°$ (S.D.) in methanol was *R* (CXXXI)]. It was found further (*190*) that one atom

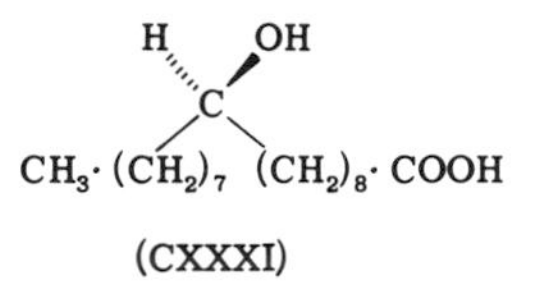

(CXXXI)

of deuterium was incorporated in a stably bound form into the hydroxy acid, when the pseudomonad acted upon the oleic acid in a medium dissolved in 99.8% D_2O. This deuterium atom was located at C-9 of the acid; its absolute stereochemistry was deduced—after conversion of the 10*R*-10-hydroxy-9-D_1-octadecanoic acid (CXXXII) to 9-D_1-stearic acid (CXXXIII) (see Scheme 22)—with the aid of the desaturating enzyme system of *C. diphtheriae,* the stereospecificity of which had been established previously [(*181*), see Scheme 21]. The oleic acid isolated from

189. These calculations are based on the previous conclusion that in the stearate → oleate transformation the pro-*R* hydrogen atoms were eliminated from both C-9 and C-10 of stearate. If the hydrogen atoms eliminated from these two positions of stearate had been the 9-pro-*R* + 10-pro-*S* or 9-pro-*S* + 10-pro-*R* pair, the oleate isolated from the bacteria could not have contained D_2 species.

190. G. L. Schroepfer, Jr., *JBC* **241,** 5441 (1966).

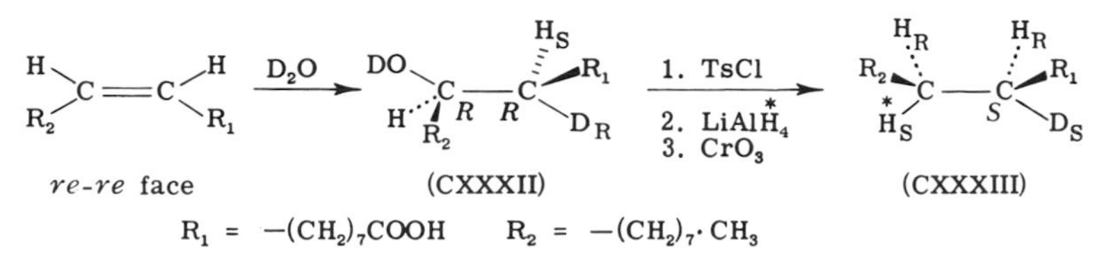

SCHEME 22. Stereochemistry of hydration of oleic acid to 10*R*-10-hydroxyoctadecanoic acid by a pseudomonad (NRRL-2994) in the presence of D_2O and the chemical conversion of the hydroxy acid to stearic acid. The notation for the absolute configuration of the 9-D_1-stearic acid derived from the 10-hydroxy-9*R*-9-D_1-octadecanoic acid becomes 9*S* according to the sequence rule because R_1 in (CXXXIII) acquires priority over (C-10 + R_2), whereas in (CXXXII) C-10 + R_2 have priority over R_1.

the culture of *C. diphtheriae* grown in the presence of the 9-D_1-stearic acid retained all the deuterium contained in the substrate. Since it was known that the bacterial desaturase removed specifically the pro-*R* hydrogen atom at C-9 of stearic acid (cf. Section VII,A), it followed that the absolute configuration of the 9-D_1-stearic acid must have been 9*S* (CXXXIII). This finding establishes that the mechanism of the conversion of oleic into 10*R*-10-hydroxyoctadecanoic acid is equivalent to the trans-addition of the elements of water to the double bond; that the absolute configuration of the acid formed in the presence of D_2O was 10*R*-9*R* (CXXXII); and that the OH (or OD) was added to the *re*, and H (or D) to the *si*, face of the double bond.

Niehaus and Schroepfer (*191*) reported that the hydration of oleic to 10*R*-10-hydroxyoctadecanoic acid was also catalyzed by a soluble extract of the pseudomonad (NRRL-2994) with the incorporation of ^{18}O from $H_2{}^{18}O$ into the hydroxyl group at C-10. Hence, it appears fairly certain that this reaction is truly one of a hydratase and not one of an oxygenase.

The same soluble extract was noted to have hydrated stereospecifically both the *d,l-cis-* and *d,l-trans*-9,10-epoxyoctadecanoic acids, the *cis*-epoxy acid giving rise to the dextrorotatory *threo*-9,10-dihydroxyoctadecanoate (methyl *threo*-9,10-octadecanoate: $[\alpha]_{589}$ $+27.0° \pm 3.80°$ in methanol) and the *trans*-epoxy acid to the *erythro*-9,10-octadecanoate (methyl *erythro*-9,10-dihydroxyoctadecanoate: $[\alpha]_{250}$ $+1.38° \pm 0.11°$ in methanol). These hydration reactions were characterized by both substrate and product stereospecificity: not only were the dihydroxy acids optically active but also the enzyme hydrated only 50% of the racemic substrates. In the case of the *trans*-epoxyoctadecanoate the unreacted

191. W. G. Niehaus, Jr. and G. L. Schroepfer, Jr., *JACS* **89**, 4227 (1967).

substrate recovered from the enzymic reaction mixture was optically active (*trans*-9,10-epoxyoctadecanoic acid: $[\alpha]_{589}$ —14.6° ± 0.40° in methanol). The absolute configuration of the chiral centers in the enzymically reactive substrates and in the products has yet to be established.

VIII. Stereospecificity of Reactions of Polyprenyl Biosynthesis

A. Synthesis of 3-Hydroxy-3-methylglutaryl-CoA. Mevalonate Kinase. Mevaldate Reductase

Every step in the long chain of reactions of polyprenyl and sterol biosynthesis has distinct stereochemical features, chiral and prochiral centers as well as plane-trigonal structures being involved. The reactions of sterol biosynthesis have been discussed by Bloch (*192*) in a 1964 Nobel lecture and have been reviewed by Clayton (*193*). The stereochemical aspects of the same problem have been presented by Cornforth and Ryback (*15*) and by Popják and Cornforth (*194*).

The synthesis of the precursor of mevalonate, 3-hydroxy-3-methylglutaryl-CoA (CXXXIV) itself must be a stereospecific process since the product of its reduction is 3*R*-mevalonate (CXXXV) (*195, 196*). Thus it may be predicted that the absolute configuration of the enzymically synthesized HMG-CoA at C-3 is *S* (*197*), resulting from the addition of acetyl-CoA to the *si* face at C-3 of acetoacetyl-CoA (Scheme 23). Supporting evidence for the stereospecific synthesis of HMG-CoA was provided by the observation that the 3-hydroxy-3-methylglutaryl-CoA cleavage enzyme (EC 4.1.3.4) reacted with only one diastereoisomer of the chemically synthesized substrate, whereas the diastereoisomer of HMG-CoA synthesized enzymically was cleaved completely to aceto-

192. K. Bloch, *Science* **150,** 19 (1965).
193. R. B. Clayton, *Quart. Rev. (London)* **19,** 168 and 201 (1965).
194. G. Popják and J. W. Cornforth, *BJ* **101,** 553 (1966).
195. M. Eberle and D. Arigoni, *Helv. Chim. Acta* **43,** 1508 (1960).
196. R. H. Cornforth, J. W. Cornforth, and G. Popják, *Tetrahedron* **18,** 1351 (1962).
197. According to the sequence rule the reduction of 3*S*-HMG-CoA at the thioester group leads to 3*R*-mevalonate without change of configuration, only of notation, at C-3. This anomaly results from the fact that the —CH_2—CO—S—CoA group in HMG-CoA has a priority over the —CH_2—COOH group, whereas in mevalonate the —$CH_2 \cdot CH_2OH$ group resulting from the reduction of the thioester group of HMG-CoA, has a lower priority than —CH_2COOH.

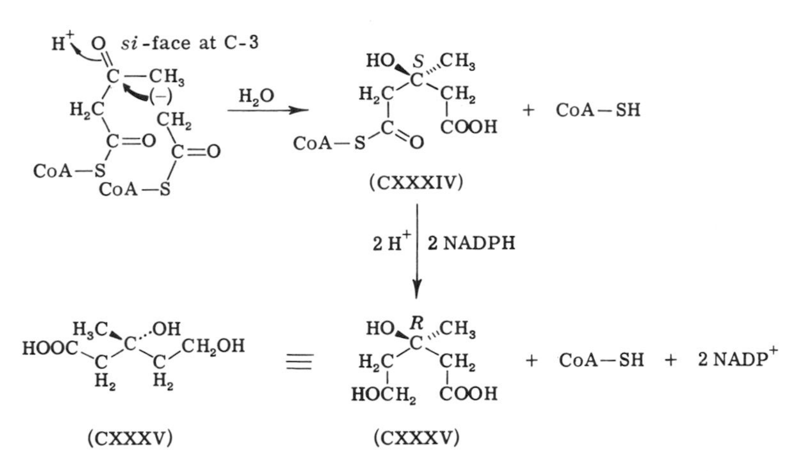

SCHEME 23. Stereospecific synthesis of 3-hydroxy-3-methylglutaryl-CoA and its reduction to mevalonate.

acetate and acetyl-CoA (*198*). It is not known whether HMG-CoA reductase is an *A*- or *B*-side specific enzyme.

The first enzyme acting on mevalonate, the "committed" (*199*) substrate in polyprenyl biosynthesis, has an absolute specificity for the *R* enantiomer (*196*, *200*) and demonstrates very well the principle of enzymic stereospecificity: the unique binding of substrate to the protein. If mevalonate could be bound to mevalonate kinase in more than one way, the absolute specificity of this enzyme for the one enantiomer would be inexplicable. It is safe to assume in this instance that the enzyme has two binding sites for the substrate and that these determine the alignment of the C-5 group of mevalonate to be phosphorylated with the catalytically active enzyme center. The binding of the carboxyl group of mevalonate by ionic and of the 3-hydroxyl group by hydrogen bonding to the enzyme account satisfactority for the stereospecificity of mevalonate kinase. The product obtained from the phosphorylation of *RS*-mevalonate with liver mevalonate kinase is the levorotatory 5-phosphomevalonate ($[\alpha]_{589}^{24}$ —6.1°, measured on the free acid in water) (*201*).

198. L. D. Stegink, Ph.D. Thesis, University of Michigan, 1963.

199. "Committed" is used in the sense that all the known enzymic transformations of mevalonate lead to the synthesis of only prenyl substances: monoprenyls (isopentenyl pyrophosphate and 3,3-dimethylallyl pyrophosphate) and polyprenyls (e.g., farnesyl pyrophosphate, squalene, carotenoids, and rubber). Mevalonate is not known to be used for any other biological purpose.

200. F. Lynen and M. Grassl, *Z. Physiol. Chem.* **313,** 291 (1958).

201. R. H. Cornforth, K. Fletcher, H. Hellig, and G. Popják, *Nature* **185,** 923 (1960).

In contrast one might cite mevaldate reductase (*202*) (EC 1.1.1.32 and 1.1.1.33), which reduces stereospecifically the aldehydo group of mevaldate but reduces the 3*R* and 3*S* enantiomers with equal efficiency: the product of the reduction of 3*RS*-mevaldate by either NADH or NADPH with either pig liver or rat liver mevaldate reductase is 3*RS*-mevalonate (*203*). However, the enzyme from either animal is *A*-side specific with respect to the coenzyme (NADH or NADPH) and the enzymically active component (3*R*) of the 3*RS*-mevalonate formed by the reduction of 3*RS*-mevaldate with either 4*R*-4-D_1 (or 4-$^3H_1$)-NADH or with 4*R*-4-D_1 (or 4-$^3H_1$)-NADPH is the 5*R*-5-D_1 (or 5-$^3H_1$)-mevalonate, irrespective of the source of the enzyme.

Given the fact that mevaldate reductase transfers from either NADH or NADPH only the pro-*R* hydrogen atom from position 4 of the dihydronicotinamide ring of the coenzymes, without discrimination between the two enantiomers of the substrate, one possibility is that the 3-hydroxy group of mevaldate does not partake in the binding to the enzyme; binding of the substrate by the carboxyl and aldehydo groups alone and a unique spatial orientation of the coenzyme on the protein suffice to explain the stereospecificity of the hydrogen transfer from coenzyme to substrate and the disregard by the enzyme of the chirality at C-3 of mevaldate. If these arguments are correct, then the absolute configuration of mevalonate resulting from the reduction of 3*S*-mevaldate, as well as of the 3*R* isomer with 4*R*-4-D-NADH (or NADPH) should be 3*S*-5*R*-5-D_1. *E*xperimental proof on this point is, however, not available; yet information of this kind should give clues as to the spatial orientation of the substrate on the enzyme even in the absence of the knowledge of the structure of the enzyme itself.

An alternative possibility is that mevaldate reductase can bind the two antipodes of mevaldate over two neighboring areas, which are mirror images of one another, one of these being specific for the *R* and the other for the *S* enantiomer. It might be imagined that the binding site for the aldehydo group is one and the same, but the binding sites for the carboxyl group of two enantiomers would be on opposite sides of the group binding the aldehyde; in such a case the 3-hydroxy groups of the two antipodes of mevaldate would also be involved in the binding to the

202. Mevaldate reductase is distinct from the HMG-CoA reductase discussed earlier and is thought not to be an obligatory enzyme in sterol biosynthesis. The substrate specificity of mevaldate reductase is still unexplored, but one of the characteristics of the reaction catalyzed by this enzyme is—just as the reaction catalyzed by HMG-CoA reductase—that it is irreversible.

203. C. Donninger and G. Popják, *Proc. Roy. Soc.* **B163**, 465 (1966).

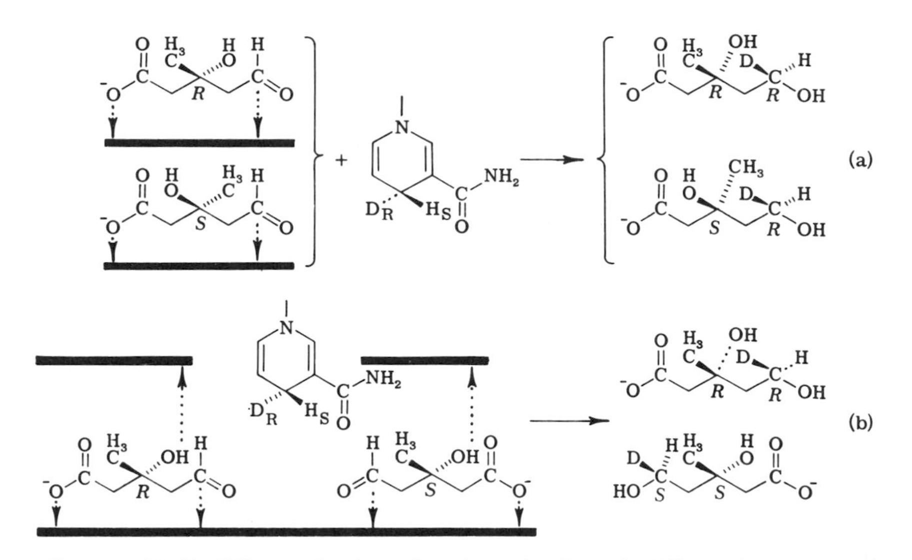

SCHEME 24. Possible mechanisms for the reduction of 3*RS*-mevaldate to 3*RS*-mevalonate coupled with the known stereospecific transfer of the pro-*R* hydrogen atom from NADH or NADPH to substrate. The thick lines below or on two sides of the substrates represent enzyme areas. In mechanism (a) the antipodes of the substrate bind equally to one and the same area of the protein; in mechanism (b) the binding of the carboxylate groups of each enantiomer are on opposite sides of the attachment of the aldehydo groups of the substrates.

protein. Since it is only the pro-*R* hydrogen atom at C-4 of the dihydronicotinamide ring of the coenzymes which is transferred to either enantiomer of mevaldate, it would follow that the reduction of 3*R*-mevaldate with 4*R*-4-D_1-NADH would yield the 3*R*-5*R*-5-D_1-mevalonate and that of the 3*S* enantiomer with the same coenzyme would give the 3*S*-5*S*-5-D_1-mevalonate. The testing of these propositions by experiment would provide an exceptional opportunity for examining Bergmann's polyaffinity theory with relation to substrates possessing a chiral center (cf. Section II). Examples are rare of enzymic reactions which are stereospecific at one center but ignore the chirality of an asymmetrical carbon atom elsewhere in the substrate except among the proteolytic enzymes. These ideas are presented as a theoretical possibility in spite of the fact that it is difficult to imagine that two binding sites of an enzyme specific for two antipodes of a substrate might have identical binding power for the isomers without any kinetic differentiation between the two. The stereochemical consequences of these speculations are summarized in Scheme 24.

B. Asymmetrical Reactions at Prochiral Centers in Squalene Biosynthesis

R-Mevalonate contains three prochiral centers (C-5, C-4, and C-2) all partaking in specific reactions of polyprenyl biosynthesis. The first clue of asymmetrical processes in biosyntheses from mevalonate was provided at a time when the enzymic transformations of this substrate were not known in detail. Thus the presence of isotopic label at defined positions only in the chain—but not in the branched methyl groups of squalene (Fig. 7) biosynthesized from 2-^{14}C-mevalonate (*204*)—may now be interpreted to mean the alignment of the plane-trigonal structure of isopentenyl (CXXXVI) and of 3,3-dimethylallyl pyrophosphate (CXXXVII) in one specific way on the enzyme, isopentenyl pyrophosphate isomerase. If this were not so the isotopic label at C-4 of isopentenyl pyrophosphate (originating from C-2 of mevalonate) might have become redistributed evenly between the *gem*-dimethyl groups of dimethylallyl pyrophosphate, on the one hand, and the methyl and vinyl carbon atoms of isopentenyl pyrophosphate, on the other, and in the appearance of ^{14}C even in the branched methyl groups of squalene.

The first evidence of a stereospecific reaction at a prochiral center in the biosynthesis of squalene from mevalonate was obtained when it was found that squalene biosynthesized from 5-D_2-mevalonate (CXXXVIII) (*205*) or from 1-D_2-farnesyl pyrophosphate (CXXIX) contained only three atoms of deuterium attached to its two central carbon atoms (*206*, *207*). It was shown further that the one atom of deuterium lost from C-1 of farnesyl pyrophosphate during the biosynthesis of squalene was replaced by the pro-*S* hydrogen atom at C-4 of the dihydronicotinamide

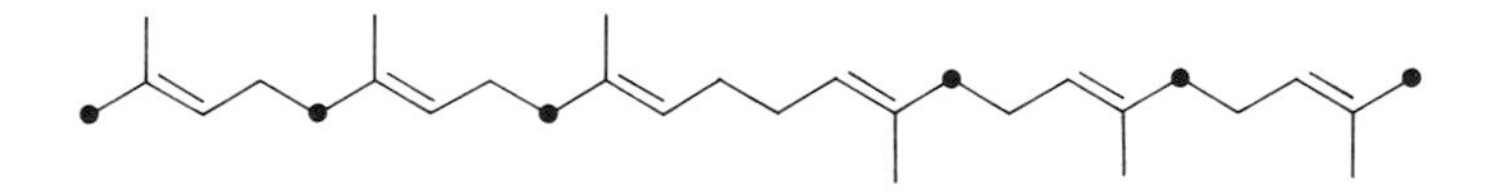

Fig. 7. Labeled positions (●) of squalene biosynthesized from 2-^{14}C-mevalonate.

204. J. W. Cornforth, R. H. Cornforth, G. Popják, and I. Y. Gore, *BJ* **66,** 10P (1957); **69,** 146 (1958).

205. C-5 of mevalonate provides, C-1, C-5, and C-9 of farnesyl pyrophosphate. Farnesyl pyrophosphate biosynthesized from 5-D_2-mevalonate contains six atoms of deuterium (*206*) distributed in three pairs at C-1, C-5, and C-9.

206. G. Popják, J. W. Cornforth, R. H. Cornforth, R. Ryhage, and DeW. S. Goodman, *JBC* **237,** 56 (1962).

207. G. Popják, DeW. S. Goodman, J. W. Cornforth, R. H. Cornforth, and R. Ryhage, *JBC* **236,** 1934 (1961).

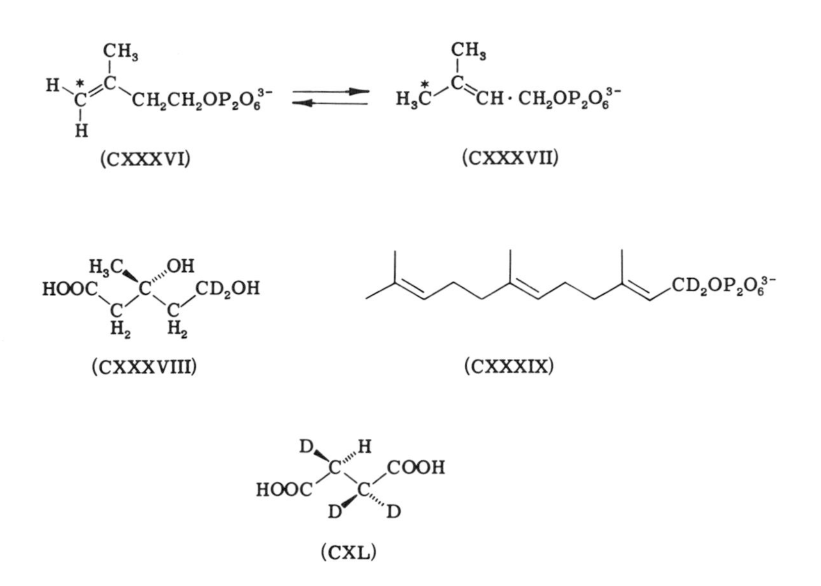

ring of NADPH. The squalene synthetase thus became classified as a *B*-side specific enzyme system (*208*). The stereospecific nature of this "hydrogen exchange" was revealed by the optical activity of the trideuteriosuccinic acid derived by ozonolysis from the four central positions of squalene biosynthesized from the 5-D_2-mevalonate. This succinic acid gave a normal dextrorotatory ORD curve (*209*) ($[\alpha]_{248}$ +21°) which was a mirror image of the levorotatory ORD curve of a reference 2*R*-2-D_1-succinic acid [see Scheme 16 (CXIV)] made by a chemical transformation of 2*S*-3*R*-3-D_1-malic acid (Fig. 6) (*210*). Thus the absolute configuration of the trideuteriosuccinic acid derived from squalene was deduced to be *S* (CXL); it was reasonable to assume that the two atoms of deuterium attached to the second methylene group of succinic acid had no effect on the sign of rotation. The stereochemistry of squalene biosynthesis from farnesyl pyrophosphate will be discussed further in the next section.

The sequence of reactions from mevalonate to squalene (Schemes 25 and 26) shows that all three prochiral centers of mevalonate (C-2, C-4, and C-5) change their bonding at some stage of the biosynthesis. The stereochemical consequences of these changes have been studied with

208. G. Popják, G. J. Schroepfer, Jr., and J. W. Cornforth, *BBRC* **6**, 438 (1961–1962).

209. Here, ORD stands for optical rotatory dispersion.

210. J. W. Cornforth, R. H. Cornforth, C. Donninger, G. Popják, G. Ryback, and G. J. Schroepfer, Jr., *BBRC* **11**, 129 (1963).

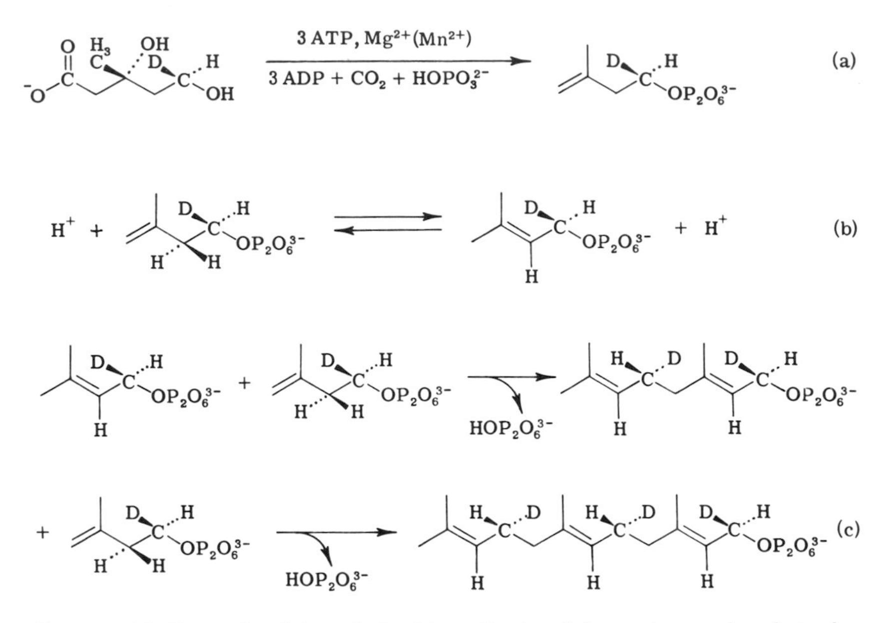

SCHEME 25. Stereochemistry of the biosynthesis of farnesyl pyrophosphate from $5R$-5-D_1-mevalonate. Sequence (a) is catalyzed by three enzymes acting in succession: mevalonate kinase, 5-phosphomevalonate kinase, and 5-pyrophosphomevalonate decarboxylase. Reaction (b) is catalyzed by isopentenylpyrophosphate isomerase, and the two-step sequence (c) is catalyzed by prenyltransferase.

the aid of mevalonates labeled stereospecifically with either deuterium or tritium at the three prochiral carbon atoms (*93, 203, 211, 212*).

C. Preparation and Use of $5R$-5-D_1(5-$^3H_1$)-Mevalonate

The preparation of $5R$-5-D_1 (or 5-T_1)-mevalonate was made by the reduction of RS-mevaldate with $4R$-4-D_1 (or $4R$-4-$^3H_1$)-NADH and mevaldate reductase (cf. Section VIII,A). The absolute configuration of the enzymically reactive $3R$ component of the resulting $3RS$-5-D_1 (or 5-$^3H_1$)-mevalonate at C-5 was deduced enzymically in the following way (*203*). A specimen of $3RS$-5-$^3H_1$-mevalonate, mixed with 4-^{14}C-(RS)-mevalonate, was converted with liver enzymes into farnesyl pyrophosphate, which was hydrolyzed with alkaline phosphatase to the free farnesol and inorganic phosphate. The 1,5,9-$^3H_3$-2,6,10-$^{14}C_3$-farnesol (^3H:

211. J. W. Cornforth, R. H. Cornforth, C. Donninger, and G. Popják, *Proc. Roy. Soc.* **B163**, 492 (1966).

212. J. W. Cornforth, R. H. Cornforth, G. Popják, and L. Yengoyan, *JBC* **241**, 3970 (1966).

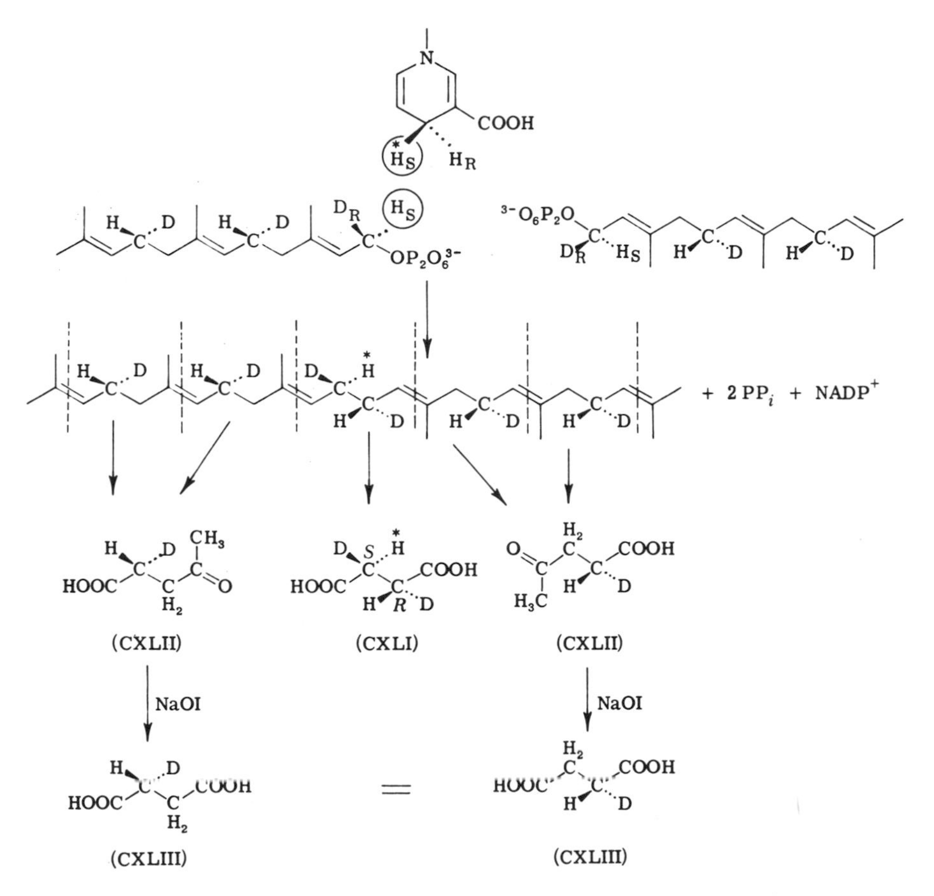

SCHEME 26. Stereochemistry of squalene biosynthesized from $4S$-4-$^3H_1$-NADPH and farnesyl pyrophosphate made enzymically from $5R$-5-D_1-mevalonate. The stereochemistry of the synthesis of farnesyl pyrophosphate is shown in Scheme 25.

^{14}C=3:3) was then oxidized to the aldehyde with NAD$^+$ and liver alcohol dehydrogenase (*213*). The resulting farnesal had a ^3H:^{14}C ratio of 2:3 and the NADH became labeled with ^3H indicating the transfer of a labeled hydrogen atom from C-1 of farnesol to NAD$^+$. Since it was known that liver alcohol dehydrogenase transferred the pro-*R* hydrogen atom from C-1 of ethanol and of geraniol to NAD$^+$ (cf. Table I), it was a reasonable assumption that the enzyme reacted similarly with farnesol and that the absolute configuration at C-1 of the 1,5,9-$^3H_3$-2,6,10-$^{14}C_3$-farnesol biosynthesized from 5-$^3H_1$-mevalonate was *R* and, hence, that the absolute configuration at C-5 of the 3*R* component of the 3*RS*-mev-

213. Yeast alcohol dehydrogenase does not react with farnesol, geraniol, or with 3,3-dimethylallyl alcohol [J. Christophe and G. Popják, *J. Lipid. Res.* **2**, 244 (1961)].

alonate formed by the reduction of *RS*-mevaldate with 4*R*-4-$^3H_1$ (or 4-D_1)-NADH and enzyme was also *R* (*203*) (see also Scheme 24).

The 5*R*-5-$^3H_1$- and 5*R*-5-D_1-mevalonate specimens were used to answer three questions:

(1) Which of the two hydrogen atoms at C-1 of farnesyl pyrophosphate was lost during the synthesis of squalene?

(2) Was the configuration around C-1 of the second farnesyl residue, not involved in the hydrogen exchange, retained or inverted?

(3) What was the nature of the carbon-to-carbon bond formation in the synthesis of farnesyl pyrophosphate?

The ^3H:^{14}C ratios in several specimens of squalene biosynthesized from 1*R*-1,5,9-$^3H_3$-2,6,10-$^{14}C_3$-farnesyl pyrophosphate were identical with the ratio in the farnesol, and squalene biosynthesized from 5*R*-5-D_1-mevalonate was shown by mass spectrometry to contain six atoms of deuterium. Ozonolysis of such hexadeuterio squalene gave of *meso*-(2*S*-3*R*)-2-D_1-3-D_1-succinic acid (CXLI) devoid of optical activity and giving an infrared spectrum identical with the reported (*214*) spectrum of *meso*-2-D_1-3-D_1-succinic acid (*211*). A second product of the ozonolysis of hexadeuteriosqualene was a monodeuteriolevulinic acid (CXLII). The latter, after oxidation with NaOI, gave a levorotatory monodeuteriosuccinic acid, the absolute configuration of which was thus assigned to be *R* (CXLIII) (see Schemes 25 and 26). These observations showed that (a) during synthesis of squalene from two molecules of farnesyl pyrophosphate the hydrogen atom replaced at C-1 of one farnesyl residue by the pro-*S* hydrogen atom at C-4 of the dihydronicotinamide ring was the pro-*S* hydrogen atom; (b) the configuration at C-1 of the second farnesyl residue not involved in the hydrogen exchange was inverted; and (c) during the two-step synthesis of farnesyl pyrophosphate catalyzed by prenyltransferase from 3,3-dimethylallyl pyrophosphate and 2 molecules of isopentenyl pyrophosphate with the intermediary formation of geranyl pyrophosphate, the configuration around the pyrophosphate-bearing carbon atom (C-1) of the intermediary allylic pyrophosphates was inverted, indicating an S_N2 type of reaction mechanism.

D. Preparation and Use of Mevalonates Labeled Stereospecifically with Hydrogen Isotopes at C-4 and C-2

Isopentenyl pyrophosphate loses one hydrogen atom from C-2 (derived from C-4 of mevalonate) both during its isomerization to 3,3-dimethyl-

214. C. R. Childs and K. Bloch, *J. Org. Chem.* **26**, 1630 (1961).

allyl pyrophosphate and during the condensation of the C_5 units [cf. reaction (b) and sequence (c) in Scheme 25] in the synthesis of farnesyl pyrophosphate.

The stereochemistry of these hydrogen eliminations was determined by the use of mevalonate specimens labeled stereospecifically at C-4 with deuterium or tritium achieved through stereospecific chemical syntheses (*211*, *215*).

The starting materials for the syntheses were the geometric isomers of 5-hydroxy-3-methylpent-3-enoic acids [Scheme 27 (CXLIV) and (CL)] readily separable from one another. Each was converted to its diphenylmethylamide (benzhydrylamide) (CXLV) and (CLI); the amides were then epoxidized by peroxybenzoic acid, followed by the reduction of the 3,4-epoxides, (CXLVI, CXLVII) and (CLIIa,b), with either $LiBD_4$ or LiB^3H_4 to mevalonyl benzhydrylamides from which the mevalonates were obtained by alkaline hydrolysis. Each specimen of mevalonate (one derived from the *trans*- and the other from the *cis*-5-hydroxy-3-methylpent-3-enoyl amide) contained, of course, the two enantiomers, $3R$ and $3S$ (CXVIII) (CXLIX) and (CLIII) (CLIV). Because of the absolute specificity of mevalonate kinase for the $3R$ enantiomer, the $3S$ component in each specimen may be ignored. Thus, from the point of view of enzymic reactivity the mevalonate derived from the *trans*-amide is the $4R$-4-H_1^*-mevalonate (CXLVIII) and the one obtained from the *cis*-amide the $4S$-4-H_1^*-mevalonate [(CLIII) Scheme 27]. The enzymic transformations of these two specimens were expected to yield in the first instance $2S$-2-H_1^*- (CLV) and $2R$-2-H_1^*-isopentenyl pyrophosphates (CLVI), respectively (*216*). Depending on whether a hydrogen atom of the same or of epimeric orientation was eliminated from C-2 of isopentenyl pyrophosphate in the isopentenyl pyrophosphate isomerase reaction, as in the synthetic reactions catalyzed by prenyltransferase, one of the 4-D_1-mevalonates was expected to give a trideuterated farnesyl pyrophosphate and the other an unlabeled specimen, or one was expected to give a monodeuterio- and the other a dideuteriofarnesyl pyrophosphate.

The $4R$-4-D_1-mevalonate incubated with liver enzymes and ATP gave, in fact, trideuteriofarnesol with a molecular weight of 225, and the

215. The details of the syntheses of stereospecifically labeled mevalonates, as well as of the preparation of all the known intermediates of squalene biosynthesis, have been described by R. H. Cornforth and G. Popják ("Methods in Enzymology," Vol. 15, p. 359, 1969) and by G. Popják (*ibid.*, p. 393).

216. According to the sequence rule the pro-R hydrogen atom at C-4 of mevalonate becomes the pro-S hydrogen atom at C-2 of isopentenyl pyrophosphate; there is no change in absolute configuration.

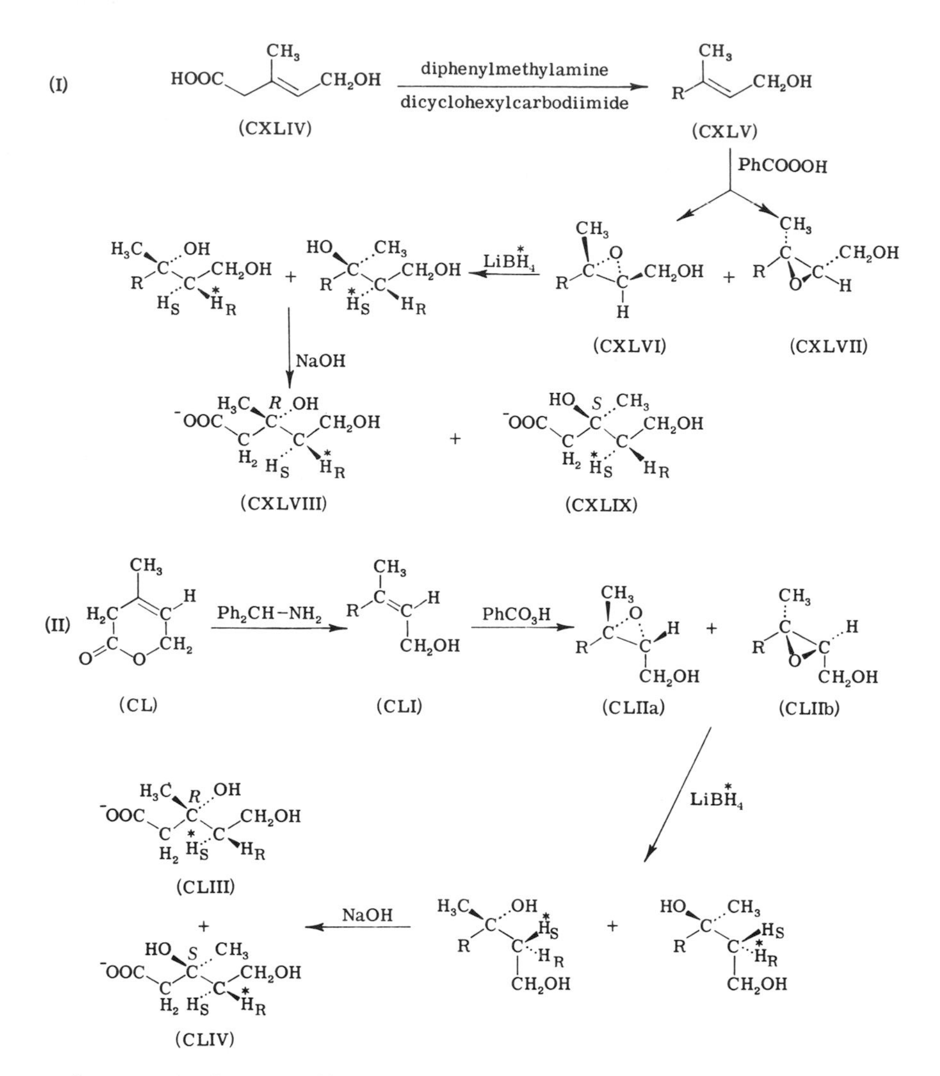

SCHEME 27. Stereospecific synthesis (I) of $4R$-4-H_1*- and (II) of $4S$-4-H_1*-mevalonates. Only the $3R$ enantiomer in each specimen is reactive enzymically. Isotopic hydrogen, deuterium or tritium, is denoted by H*.

$$R = Ph_2{\cdot}CH{\cdot}NH{\cdot}\overset{\overset{\displaystyle O}{\|}}{C}{\cdot}CH_2—.$$

$4S$-4-D_1 specimen gave a farnesol devoid of isotope, molecular weight 222, as determined by mass spectrometry. The observations were also confirmed by the use of mevalonates labeled stereospecifically at C-4 with tritium (*211*). The experiment demonstrated clearly the stereo-

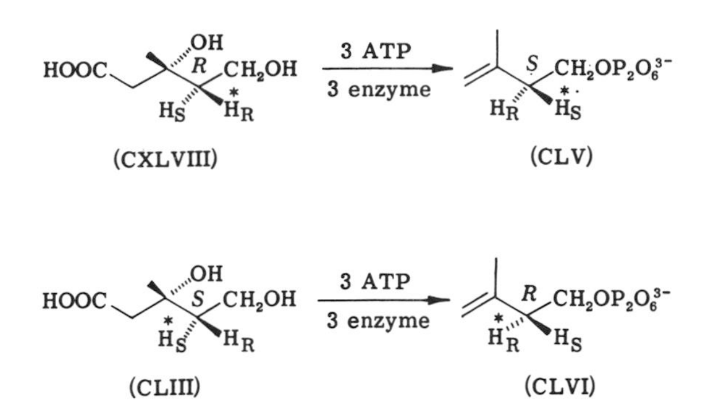

specific displacement of the pro-*R* hydrogen atom from C-2 isopentenyl pyrophosphate in the isomerase as well as in the synthetic reactions (Scheme 28).

The farnesyl pyrophosphate which is intermediate in squalene biosynthesis is the trans-trans-geometric isomer (*217*). The stereochemistry of the hydrogen elimination from C-2 of isopentenyl pyrophosphate just discussed applies only to the biosynthesis of all *trans*-polyprenyl substances, for it was found that during the biosynthesis of cis-rubber, by the enzymes contained in the latex of *Hevea brasiliensis,* the pro-*S* hydrogen atom at C-2 of isopentenyl pyrophosphate (≡ the pro-*R* hydrogen atom at C-4 of mevalonate) was the one eliminated (*218*). The study of *Hevea* latex is of special interest for it contains enzymes which synthesize not only cis-rubber, but also *trans-trans*-farnesyl pyrophosphate. The synthesis of farnesyl pyrophosphate in *Hevea* latex follows the same stereochemical course as its synthesis by liver enzymes. Thus the difference in the stereochemistry of the biosynthesis of rubber and of farnesyl pyrophosphate (or squalene) may be attributed to structural differences between the enzymes designed for the synthesis of *trans*- and *cis*-polyprenyls, respectively. The simplest hypothesis explaining these differences is that the pyrophosphate ester is bound to the enzymes of

SCHEME 28. Stereochemistry of hydrogen elimination from C-2 of isopentenyl pyrophosphate in reactions catalyzed by isopentenylpyrophosphate isomerase (R = H^+) and by prenyltransferase [R = $(CH_3)_2 \cdot C$—$CH \cdot CH_2$— or geranyl].

217. G. Popják, *Tetrahedron Letters* No 19, 19 (1959).

218. B. L. Archer, D. Barnard, E. G. Cockbain, J. W. Cornforth, R. H. Cornforth, and G. Popják, *Proc. Roy. Soc.* **B163**, 519 (1966).

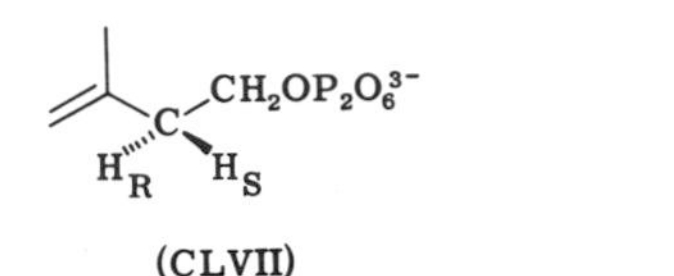

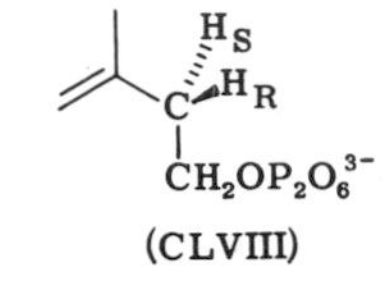

trans-polyprenyl synthesis and of *cis*-polyprenyl synthesis in different conformations, (CLVII) and (CLVIII), respectively. Inspection of the two formulas shows that in (CLVIII) H_S is on the same side of the molecule as the one occupied by H_R in (CLVII). The consequences of the two types of folding are that the elimination of H_R from (CLVII) leads to the trans-configuration and elimination of H_S from (CLVIII) leads directly to the cis-configuration in the products. This question will be discussed further in connection with the stereochemistry of the addition of an allylic residue to the vinylic carbon atom of isopentenyl pyrophosphate in the elongation of polyprenyl chains.

The bonding of C-2 of mevalonate changes in three distinct reactions: (a) in the decarboxylation of 5-pyrophosphomevalonate the C-2 methylene group becomes the vinylic group at C-4 of isopentenyl pyrophosphate; (b) in the isomerization of isopentenyl pyrophosphate to dimethylallyl pyrophosphate the vinylic group of the former is changed to a methyl group on the latter (*219*); and (c) in the elongation of a polyprenyl chain by the condensation of an allylic pyrophosphate (e.g., of geranyl pyrophosphate) with isopentenyl pyrophosphate, the plain trigonal C-4 of the latter changes to a tetrahedral methylene carbon atom in the product. Thus, the two hydrogen atoms at C-2 of mevalonate are expected to change their steric positions according to the mechanism of the reaction catalyzed by the 5-pyrophosphomevalonate decarboxylase and according to the direction from which either a proton or an allylic residue is being added to the vinylic carbon atom of isopentenyl pyrophosphate. Schemes 29 and 30 summarize the stereochemical consequences of alternative reaction mechanisms in the decarboxylase reaction and in the chain elongation of polyprenyls. The possible steric courses of the changing of a vinyl into a methyl group, as in the isopentenyl pyrophosphate isomerase reaction, have been described earlier [cf. Scheme 6, (c) and (d)].

Mevalonates labeled stereospecifically with deuterium (or tritium) at C-2 were made by the inversion of the racemic mevalonates labeled stereospecifically at C-4. The inversion was effected by the conversion of the acids, through their silver salts, into their methyl esters, and by the

219. The stereochemical problems of the transformation of a vinyl into a methyl group are discussed in Section V.

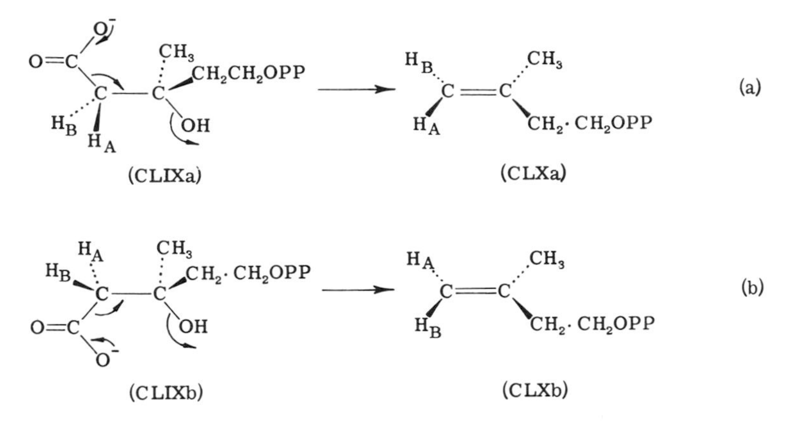

SCHEME 29. Stereochemical consequences of (a) trans- and (b) cis-elimination in the decarboxylation of 5-pyrophosphomevalonate (*212*).

oxidation of the primary alcohol group, with zinc permanganate in acetone, to the carboxyl. The product of this reaction, mono-methyl 3-hydroxy-3-methylglutarate, was reduced exclusively at the ester group with $LiBH_4$. The result of this conversion of mevalonate into mevalonate is that C-5 and C-4 of the starting material became C-1 and C-2 in the

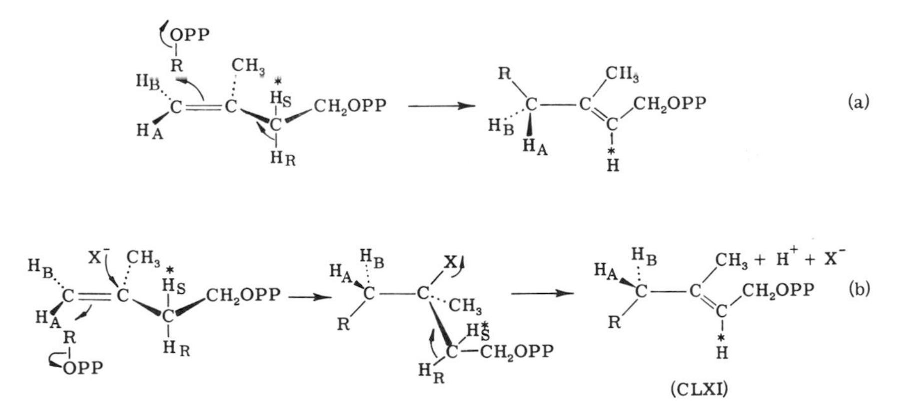

SCHEME 30. Stereochemical consequences of the addition of an R group (e.g., the geranyl of geranyl pyrophosphate) (a) to the *si* face or (b) to the *re* face at the vinylic carbon atom of isopentenyl pyrophosphate. The plane of the molecule is meant to be perpendicular to the plane of the paper, thus the *si* face is above and the *re* face below the plane of the molecule. The scheme takes into account the previously established fact that H_R from C-2 of isopentenyl pyrophosphate is the one eliminated in the synthesis of *trans*-polyprenyls. A simple concerted mechanism for reaction (b), without the intervention of a nucleophilic group X^- of the enzyme, cannot be formulated.

product and that the unnatural 3*S* isomer in the starting mevalonate became the enzymically reactive form 3*R* (Scheme 31). Thus the specimen containing 3*R*-4*R*-4-D_1-mevalonate + 3*S*-4*S*-4-D_1-mevalonate gave, after the inversion, the racemic mixture of 3*S*-2*S*-2-D_1-mevalonate and 3*R*-2*R*-2-D_1-mevalonate (CLXII). The mixture of 3*S*-2*R*-2-D_1-mevalonate + 3*R*-2*S*-2-D_1-mevalonate was derived, by inversion, from the racemic specimen of 3*R*-2*S*-2-D_1-mevalonate + 3*S*-2*R*-2-D_1-mevalonate (*212*).

Cornforth *et al.* (*212*) produced conclusive evidence, by the use of the mevalonates labeled stereospecifically with deuterium at C-2, that in the decarboxylation of 5-pyrophosphomevalonate the carboxyl and the 3-hydroxy group were eliminated in a trans manner since the 3*R*-2-*R*-2-D_1-mevalonate [(CLIXa) $H_A = D$, $H_B = H$] gave *cis*-4-D_1-isopentenyl pyrophosphate [(CLXa), $H_A = D$, $H_B = H$], and the 3*R*-2*S*-2-D_1-mevalonate [(CLIXa), $H_A = H$, $H_B = D$] gave the *trans*-4-D_1-isopentenyl pyrophosphate [(CLXa), $H_A = H$, $H_B = D$]. With this knowledge avail-

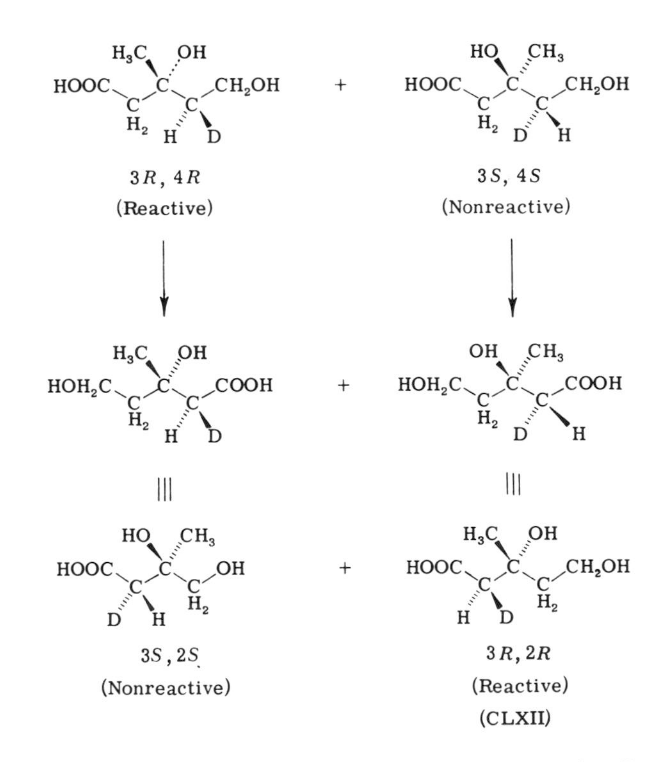

SCHEME 31. Conversion of racemic 4-D_1-mevalonate into racemic 2-D_1-mevalonate. The enzymically unreactive 3*S*-4*S* isomer in the starting specimen becomes the reactive 3*R*-2*R* isomer (*212*).

able, the problem presented in Scheme 30 could be tackled. A specimen of farnesyl pyrophosphate (CLXIII) was synthesized from $3R$-$2R$-2-D_1-mevalonate (CLXII) with a partially purified liver enzyme system.

The steric position of H and D at C-4 and C-8 of the farnesyl pyrophosphate was then deduced by the ozonolysis of the farnesol (liberated with phosphatase from the pyrophosphate) to levulinic acid (CLXIV)

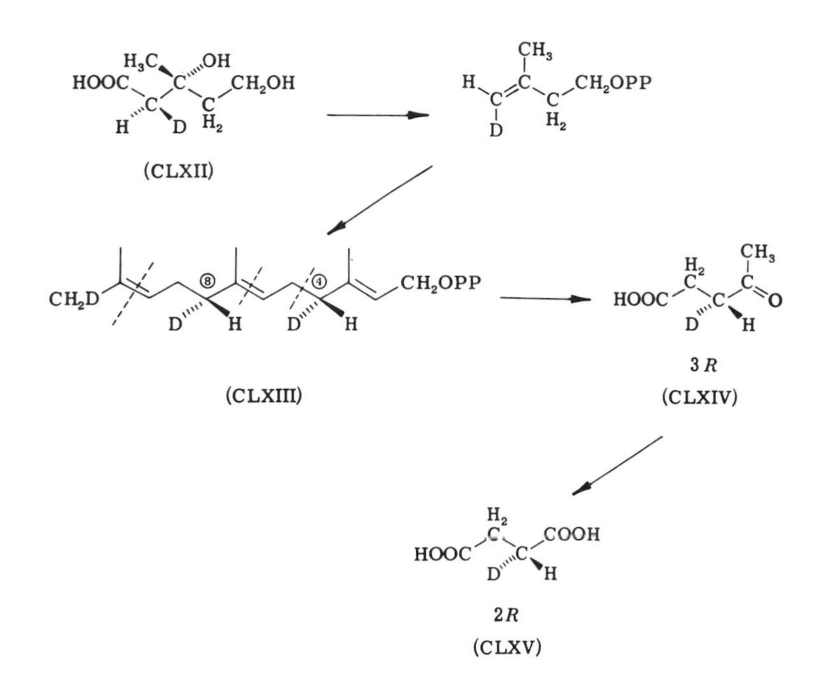

and the conversion of the latter into succinic acid (CLXV). The latter proved to be the levorotatory R-2-D_1-succinic acid (*212*). The result was compatible only with mechanism (b) shown in Scheme 30 [(CLXI), $H_A = D$, $H_B = H$], which means that during the elongation of a polyprenyl chain the new carbon-to-carbon bond between C-1 of an allylic pyrophosphate and C-4 of isopentenyl pyrophosphate is formed on the *re* face of isopentenyl pyrophosphate.

This mechanism can accommodate the synthesis of *cis*-polyprenyls, such as rubber, also with the simple variant that the conformation of isopentenyl pyrophosphate on the rubber-synthesizing enzyme should be that shown in formula CLVIII (see Scheme 32).

Popják and Cornforth (*194*) calculated that there were $2^{14} = 16{,}384$ stereospecific ways by which squalene could be synthesized from six molecules of mevalonate. The ambiguities have been reduced to two

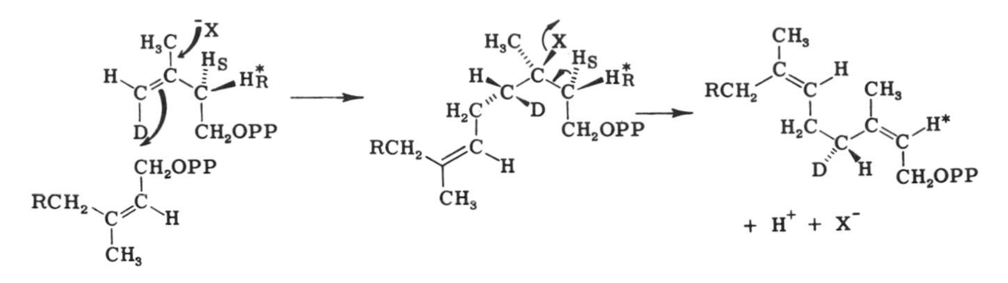

SCHEME 32. Hypothetical mechanism for the biosynthesis of *cis*-polyprenyls, e.g., of rubber. The scheme takes into account that H_S at C-2 of isopentenyl pyrophosphate is lost during biosynthesis of rubber in contrast to the elimination of H_R in the biosynthesis of all-*trans*-polyprenyls. The allylic pyrophosphate is meant to be in a plane behind that of isopentenyl pyrophosphate, which is in the plane of the paper; X^- is in a plane above that of isopentenyl pyrophosphate (*212*).

since it is still not known to which side of the double bond a proton is added during isomerization of isopentenyl pyrophosphate. In spite of this remaining uncertainty one can be confident that there is only one stereochemical path in this biosynthesis. The emphasis here is not so much on any exceptional feature of squalene biosynthesis but rather on a special example illustrating the binding of substrates to the enzyme in unique conformations. The exceptional nature of squalene biosynthesis is that its product has but meager stereochemical features, trans-geometry of its double bonds, and yet nature evolved a precise stereochemical pathway for its synthesis.

Since the introduction by Cornforth, Popják, and their colleagues of mevalonates labeled stereospecifically with hydrogen isotopes at the three prochiral centers of the molecule many new aspects of the biosynthesis of polyprenyl substances (carotenoids, sterols, phytosterols, and cyclic triterpenes) have been investigated with these substrates. A complete review of these topics must be left to specialized articles, but it might be mentioned here briefly that the reaction mechanisms revealed by the work on squalene and rubber biosynthesis probably apply to all syntheses of polyprenyls with a trans- and cis-configuration, respectively. Thus phytoene (CLXVI), biosynthesized from the mevalonates labeled stereospecifically at C-4 with a hydrogen isotope, showed the trans-pattern of labeling (*220*). The same type of labeling was demonstrated also for farnesol synthesized in latex serum (*218*), for squalene and the side chain of ubiquinone biosynthesized by *Aspergillus fumigatus* Fresenius (*221*), and for squalene and phytol biosynthesized by green leaves (*222*).

The work of Hemming and his colleagues on polyprenols of mixed

220. T. W. Goodwin, and R. J. H. Williams, *Proc. Roy. Soc.* **B163,** 515 (1966).
221. K. J. Stone and F. W. Hemming, *BJ* **104,** 43 (1967).
222. A. R. Wellburn, K. J. Stone, and F. W. Hemming, *BJ* **100,** 23c (1966).

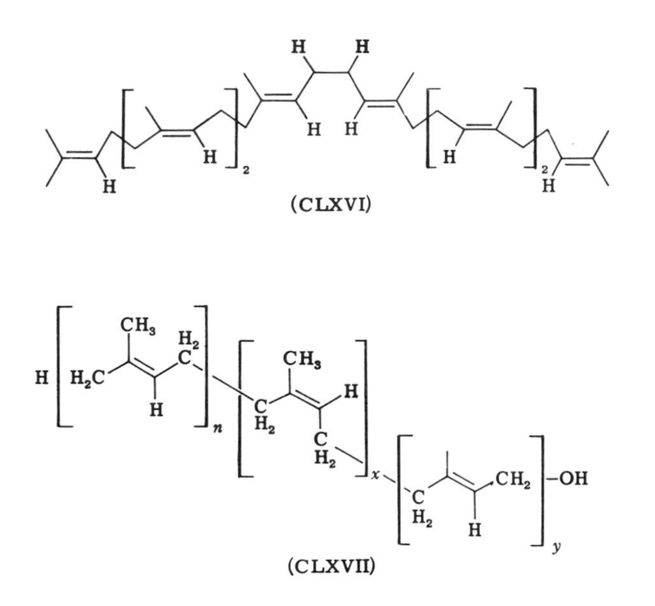

(CLXVI)

(CLXVII)

stereochemistry, the *dolichols,* is of particular interest (*221, 223, 224*). These long-chain alcohols contain prenyl units of the trans-configuration as well as of cis-configuration. The general structure of the dolichols is shown in (CLXVII) in which $n = 3,4$, $x = 11\text{–}16$, and $y = 1\text{–}3$; some of the prenyl units may be in the reduced form. The prenyl units in the dolichols appear to be distinguished also biogenetically: Experiments conducted with 2-^{14}C-4R-4-$^{3}H_1$- and 2-^{14}C-4S-4-$^{3}H_1$-mevalonates gave numerical values of ^{3}H/^{14}C ratios suggesting that those of trans-configuration retained H_S and those of cis-configuration retained H_R (*216*) at C-2 of isopentenylpyrophosphate.

Apart from the many details of carotenoid, sterol, and cyclic triterpene biosynthesis revealed by the use of stereospecifically labeled mevalonates, one observation is particularly worth recording here since it led to the discovery of a hitherto unrecognized intermediate in the conversion of lanosterol into cholesterol. From the pattern of stereospecific labeling of squalene biosynthesized from either 3R-2R-2-$^{3}H_1$- ($H_A = {}^{3}$H) or 3R-2S-2-$^{3}H_1$-mevalonate ($H_B = {}^{3}$H) (*212*), and from the knowledge of the cyclization of squalene-2,3-oxide to lanosterol, it may be deduced that H_A and H_B will occupy in lanosterol the steric positions shown in (CLXVIII). It has been noted both by Canonica *et al.* (*225*) and Gibbons

223. A. R. Wellburn and F. W. Hemming, *Nature* **212,** 1364 (1966).

224. D. P. Gough and F. W. Hemming, *BJ* **105,** 10c (1967).

225. L. Canonica, A. Fiechi, M. Galli Kienle, A. Scala, G. Galli, E. G. Paolotti, and R. Paoletti, *JACS* **90,** 3597 (1968)

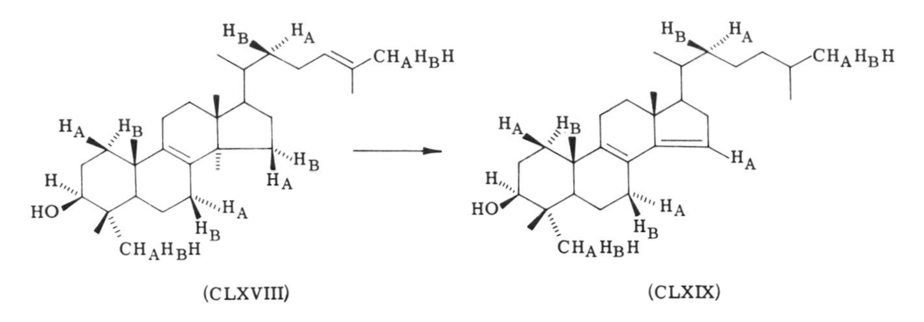

(CLXVIII) (CLXIX)

et al. (*226*), who used the above stereospecifically labeled mevalonates, that during the conversion of lanosterol into cholesterol the 15α-hydrogen atom of lanosterol was lost (*227*). The results suggested that 14-*nor*-lanosta-8(9),14,24-trien-3β-ol, or 14-*nor*-lanosta-8(9),14-dien-3β-ol (CLXIX) might be the first degradation product of lanosterol on the path to cholesterol (*229*). The dienol (CLXIX) has since been identified as a precursor of cholesterol (*228*).

Thus the use of stereospecifically labeled precursors can lead not only to the elucidation of the steric course of enzymic reactions but also to the discovery of previously unchartered pathways.

IX. Coenzyme B_{12}-Dependent Reactions

There are three well-known enzymic reactions dependent on coenzyme B_{12}, (5,6-dimethylbenzimidazolyl)-Co-5′-deoxyadenosylcobamide. These are the reversible isomerizations (a) of *R*-methylmalonyl-CoA (CLXX) to succinyl-CoA (CLXXI) (*230*); (b) of 2*S*-3*S*-methylaspartate

226. G. F. Gibbons, L. J. Goad, and T. W. Goodwin, *Chem. Commun.* No. 20, p. 1212 (1968).

227. Canonica *et al.* misrepresented in their first publication (*225*) the steric positions of H_A and H_B in lanosterol and erroneously concluded that the 15β-hydrogen atom of lanosterol was the one eliminated. The error has been subsequently corrected (*228*).

228. L. Canonica, A. Fiechi, M. Galli Kienle, A. Scala, G. Galli, E. G. Paolotti, and R. Paoletti, *Steroids* **12**, 445 (1968).

229. The reduction of the 24(25)-double bond in the side chain of lanosterol may occur at almost any stage of the lanosterol → cholesterol conversion.

230. R. M. Smith and K. J. Monty, *BBRC* **1**, 105 (1959); E. R. Stadtman, P. Overath, H. Eggerer, and F. Lynen, *ibid.* **2**, 1 (1960); S. Gurnani, S. P. Mistry, and B. C. Johnson, *BBA* **38**, 187 (1960); P. Lengyel, R. Mazumder, and S. Ochoa, *Proc. Natl. Acad. Sci. U. S.* **46**, 1312 (1960).

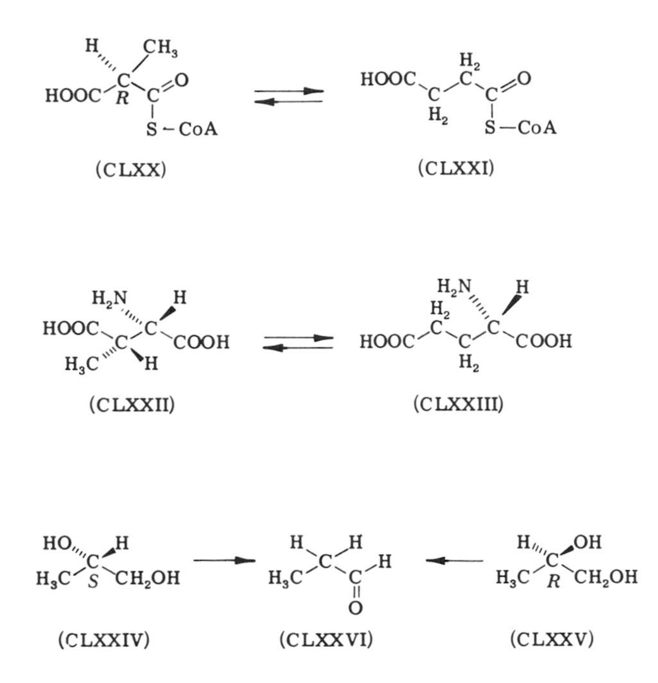

(CLXXII) to *S*-glutamate (CLXXIII) (*231*); and (c) the dehydration of *S*-propane-1,2-diol (CLXXIV) and of *R*-propane-1,2-diol (CLXXV) to propionaldehyde (CLXXVI) (*232*, *233*). The notable feature of all these reactions is a 1,2-hydrogen shift, which, from the evidence of Abeles and of Arigoni and their colleagues (*234*) involves an intermolecular hydrogen transfer between substrate(s) and coenzyme B_{12}.

These three enzymic reactions are of special interest since—contrary to most enzyme-catalyzed reactions—there are no analogies for them in organic chemistry. It may not be inappropriate to recall in this context the almost incredulous reaction, witnessed by the author, of Leopold Ruzicka, one of the greatest chemists of this century, to one of the earliest reports (*235*) on the carboxylation of propionyl-CoA to methylmalonyl-CoA and the conversion of the latter to succinyl-CoA. After the presentation by Ochoa at the Second International Conference on

231. H. A. Barker, H. Weissbach, and R. D. Smyth, *Proc. Natl. Acad. Sci. U. S.* **44**, 1093 (1958).

232. A. M. Brownstein and R. H. Abeles, *JBC* **236**, 1199 (1961).

233. B. Zagalak, P. A. Frey, G. L. Karabatsos, and R. H. Abeles, *JBC* **241**, 3028 (1966).

234. J. Rétey and D. Arigoni, *Experientia* **22**, 783 (1966).

235. M. Flavin, P. J. Ortiz, and S. Ochoa, *in* "Biochemical Problems of Lipids" (G. Popják and E. LeBreton, eds.), p. 208. Butterworths, London, 1956.

Biochemical Problems of Lipids (Ghent, Belgium, July 1955), Ruzicka said (*236*): "Die Isomerisierung von Methylmalonat in Succinat, *die chemisch nicht möglich ist* (my italics), wird biochemisch wohl so verlaufen, dass während der Isomerisierung beide Carboxyle am Enzym gebunden bleiben und in diesem Komplex eine Verschiebung der (+)-Ladung von der α nach der β-Stellung stattfindet."

A. Stereochemical Course of Reactions Catalyzed by Propionyl-Coenzyme A Carboxylase, Methylmalonyl-Coenzyme A Epimerase, Succinyl-Coenzyme A Mutase, and Glutamate Mutase

The enzymic reactions leading from propionyl-CoA via methylmalonyl-CoA to succinyl-CoA have been reviewed by Kaziro and Ochoa (*237*). The methylmalonyl-CoA formed by the carboxylation of propionyl-CoA with crystalline beef heart propionyl-CoA carboxylase (EC 6.4.1.3) (*237, 238*) cannot be converted into succinyl-CoA by purified succinyl-CoA mutase (= methylmalonyl-CoA mutase, EC 5.4.99.2) (*237, 239*). Another enzyme, methylmalonyl-CoA epimerase (= methylmalonyl-CoA racemase, EC 5.1.99.1), intervenes in the sequence and converts the originally formed methylmalonyl-CoA ("form A") into the enzymically reactive "form B" (*237*). The isomerization of form B to succinyl-CoA, catalyzed by succinyl-CoA mutase, is achieved by the intramolecular 1:2 migration of the —CO·S—CoA group (*240, 241*) accompanied by an intramolecular 1:2 hydrogen shift in the opposite direction (*242, 243*) (see Scheme 33; CLXXVI–CLXXX).

The stereochemical course of all three steps has been deduced (*244, 245*) and is shown in Scheme 33. Sprecher *et al.* (*244*), and Rétey and

236. L. Ruzicka, *in* "Biochemical Problems of Lipids" (G. Popják and E. Le Breton, eds.), p. 211. Butterworths, London, 1956.

237. Y. Kaziro and S. Ochoa, *Advan. Enzymol.* **26,** 283 (1964).

238. Y. Kaziro, S. Ochoa, R. C. Warner, and J. Chen, *JBC* **236,** 1917 (1961).

239. R. Mazumder, T. Sasakawa, Y. Kaziro, and S. Ochoa, *JBC* **236,** PC 53 (1961); **237,** 3065 (1962).

240. R. W. Kellermeyer and H. G. Wood, *Biochemistry* **1,** 1124 (1962).

241. E. F. Phares, M. V. Long, and S. F. Carson, *BBRC* **8,** 142 (1962); *Ann. N. Y. Acad. Sci.* **112,** 680 (1964).

242. P. Overath, G. M. Kellerman, F. Lynen, H. P. Fritz, and H. J. Keller, *Biochem. Z.* **335,** 500 (1962).

243. J. D. Erfle, J. M. Clark, Jr., R. F. Nystrom, and B. C. Johnson, *JBC* **239,** 1920 (1964).

244. M. Sprecher, M. J. Clark, and D. B. Sprinson, *BBRC* **15,** 581 (1964); *JBC* **241,** 872 (1966).

245. J. Rétey and F. Lynen, *Biochem. Z.* **342,** 256 (1965).

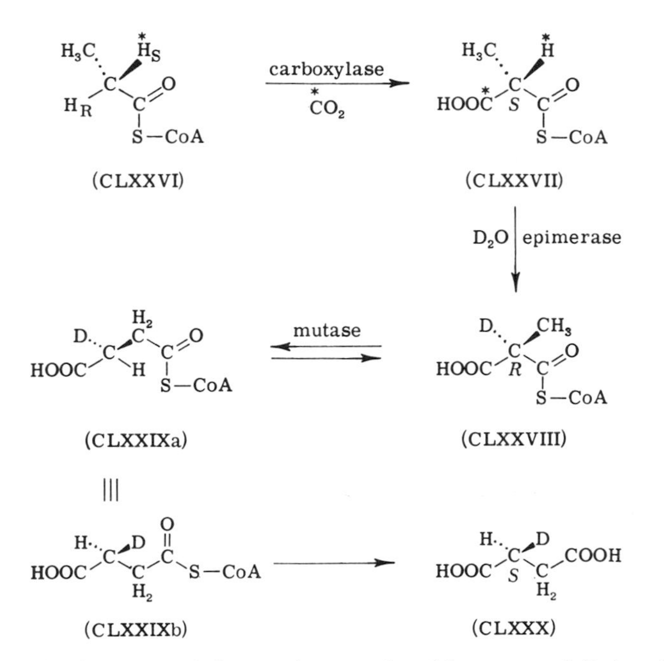

SCHEME 33. Steric course of the reactions catalyzed by proponyl-CoA carboxylase, methylmalonyl-CoA epimerase, and methylmalonyl-CoA mutase.

Lynen (*245*) deduced independently that the absolute configuration of the methylmalonyl-CoA (CLXXVII), synthesized from propionyl-CoA (CLXXVI) by the crystalline carboxylase, was *S*. Sprecher *et al.* (*244*) reduced the methylmalonyl-CoA (CLXXVII) with Raney nickel to a partially racemized, though still optically active, 3-hydroxy-2-methylpropionic acid [(CLXXXI), R = H] and converted this into the levorotatory phenylurethane [(CLXXXI), $R = CO \cdot NH \cdot C_6H_5$] the absolute configuration of which had been established (*246*) to be *R* by synthesis from 2*S*,3*S*-2-amino-3-methylsuccinic acid (= *threo*-3-methyl-L-aspartic acid) (CLXXXII). Rétey and Lynen (*245*) prepared ^{14}C-labeled methylmalonyl-CoA by the enzymic carboxylation of 2-D_2-propionyl-CoA with $H^{14}CO_3^-$. Reduction of the ^{14}C-labeled specimen with Raney nickel gave the doubly labeled hydroxymethylpropionic acid (CLXXXI), R = H), which was converted into the hydrazide [(CLXXXIII), $\overset{*}{H}$ = D]. Portions of this hydrazide were then added to authentic specimens of *R*-, *S*-, and *RS*-hydrazide made from the optical isomers of 3-hydroxy-2-methylpropionic acid. After repeated crystallizations, the ^{14}C, contained in the hydrazide derived from the biological material, remained

246. M. Sprecher and D. B. Sprinson, *JBC* **241**, 868 (1966).

with the *R* and *RS* specimens but was largely excluded from the *S*-hydrazide. It was thus concluded that the ^{14}C-labeled substance must have had the *R* configuration. The conclusion from these experiments, as well as from those of Sprecher *et al.* (*244*), was that form A of methylmalonyl-CoA had the *S* configuration (CLXXVII).

Rétey and Lynen (*245*) have also deduced the stereochemistry of the hydrogen atom displaced from C-2 of propionyl-CoA during carboxylation. They prepared a specimen of 2*S*-3-$^3H_1$-propionic acid (CLXXXVI) by the oxidation with hypobromite of 2*R*-3*S*-3-$^3H_1$-butan-2-ol (CLXXXV), obtained by the reduction of 2*R*-3*R*-2,3-epoxybutane

(CLXXXI) ← 6 stages — (CLXXXII)

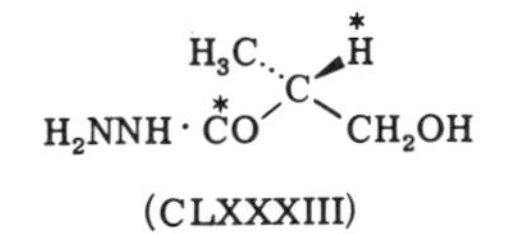

(CLXXXIII)

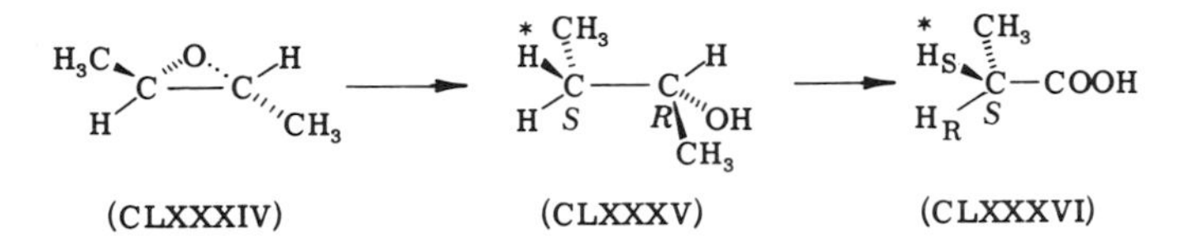

(CLXXXIV) (CLXXXV) (CLXXXVI)

(CLXXXIV) with tritium-labeled $LiAlH_4$. When the coenzyme A derivative (CLXXVI; $\overset{*}{H}_S = {}^3H$) of the 2*S*-2-$^3H_1$-propionic acid (CLXXXVI) was carboxylated, the methylmalonyl-CoA formed retained nearly all the tritium. In contrast, 2*R*-2-$^3H_1$-propionyl-CoA [(CLXXVI), $H_R = {}^3H$, $\overset{*}{H}_S = {}^1H$] lost all its tritium on carboxylation (*247*). Thus the carboxyl group in 2*S*-methylmalonyl-CoA occupies the steric position of the hydrogen atom displaced from C-2 of propionyl-CoA.

Since the absolute configuration of form A of methylmalonyl-CoA was deduced to be *S*, it follows that form B, which is the substrate for the succinyl-CoA mutase, has the *R* configuration [(CLXXVIII), Scheme 33]. It is known (*237*) that the methylmalonyl-CoA epimerase catalyzes

247. D. Arigoni, F. Lynen, and J. Rétey, *Helv. Chim. Acta* **49**, 311 (1966).

the exchange of the tertiary hydrogen atom at C-2 of methylmalonyl-CoA with the hydrogen ions of the medium. Thus if the reaction is carried out in D_2O, deuterium is incorporated into *R*-methylmalonyl-CoA and is retained without loss, and without change of steric position, during conversion into succinyl-CoA. The evidence for this was obtained by Sprecher *et al.* (*244*) who found that the succinic acid obtained from *R*-2-D_1-methylmalonyl-CoA with succinyl-CoA mutase had the (+)-*S* configuration (CLXXX). When *R*-malonyl-CoA is offered as substrate to succinyl-CoA mutase in a medium of D_2O (and in the absence of methylmalonyl-CoA epimerase), deuterium is not incorporated into the succinyl-CoA. Evidently, the hydrogen atom, shifted from the methyl group of methylmalonyl-CoA in the mutase reaction, has in succinyl-CoA the steric position occupied by the —CO—S—CoA group in methylmalonyl-CoA; in other words, the isomerization occurs with a retention of configuration at C-2.

The reaction catalyzed by glutamate mutase (CLXXII)→(CLXXIII) has a formal similarity to the methylmalonyl-CoA mutase reaction (*248*). In this case it is the —CH(NH_2)·COOH group of 2*S*-3*S*-3-methylaspartate (CLXXII) which migrates to the methyl group, and, as in the conversion of *R*-methylmalonyl-CoA to succinyl-CoA, there is no incorporation of deuterium from D_2O either into the product, *S*-glutamate (CLXXIII), or into the substrate, 2*S*-3*S*-3-methylaspartate (CLXXII), during the reaction catalyzed by the mutase (*249, 250*). Again, in analogy with the succinyl-CoA mutase reaction, the 1:2-migration of the —CH(NH_2)·COOH group is accompanied by a 1:2 hydrogen shift, which, apparently, is also intramolecular. The analogy, however, stops here: the isomerization of *R*-methylmalonyl-CoA to succinyl-CoA occurs with a retention of configuration at C-2, whereas the isomerization of 2*S*-3*S*-3-methylaspartate to glutamate involves the inversion of configuration at C-3 of methylaspartate. Sprecher *et al.* (*248*) carried out the conversion of mesaconate (CLXXXVII), in D_2O and in the presence of ammonia, into *S*-4-D_1-glutamate (CLXXXIX) with an extract of *Clostridium tetanomorphum.* The extracts of this microorganism contain also the methylaspartate ammonia-lyase (EC 4.3.1.2), which was shown (*251*) to catalyze the trans-addition and trans-elimination reaction between mesaconate and 2*S*-3*S*-3-methylaspartate. Hence, in the experiment of Sprecher *et al.* (*248*), the 3-D_1-2*S*-3*S*-methylaspartate (CLXXXVIII) must have been the intermediate between mesaconate and glutamate.

248. M. Sprecher, R. L. Switzer, and D. B. Sprinson, *JBC* **241,** 864 (1966).
249. A. A. Iodice and H. A. Barker, *JBC* **238,** 2094 (1963).
250. H. A. Barker, V. Rooze, F. Suzuki, and A. A. Iodice, *JBC* **239,** 3260 (1964).
251. H. J. Bright, R. E. Lundin, and L. L. Ingraham, *Biochemistry* **3,** 1224 (1964).

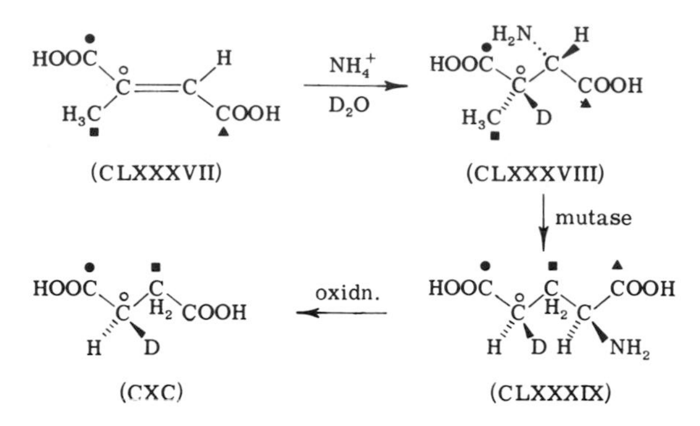

SCHEME 34. Steric course of reactions catalyzed by methylaspartate ammonia-lyase and by glutamate mutase. Various symbols have been used for the identification of carbon atoms. Experiment of Sprecher *et al.* (*248*).

Oxidation of the 4-D_1-glutamate with chloramine-T (*252*) gave a levorotatory monodeuteriosuccinic acid (CXC), the absolute configuration of which was, therefore, *R* (cf. Section VI,C). On the assumption that the 3-D_1-2*S*-3*S*-methylaspartate was the precursor of the glutamate, the result indicates that the hydrogen migrating from the methyl group of methylaspartate attacked C-3 of methylaspartate from a direction opposite to the departing —CH(NH_2)·COOH group; hence, the glutamate mutase reaction occurs with a net inversion of configuration at C-3 of the substrate (Scheme 34).

B. Propane-1,2-diol Dehydrase

It is the unusual property of propane-1,2-diol dehydrase (propanediol hydro-lyase, EC 9.2.1.28) that it converts, with equal efficiency, both the 2*R*- (CXCI) and the 2*S*-propane-1,2-diol (CXCII) into propionaldehyde (CXCIII). This B_{12} coenzyme-requiring dehydrase (*253*) has been purified from *Aerobacter aerogenes* to a state of a homogeneous protein (*254*). Any explanation for the utilization of enantiomeric substrates must be sought, therefore, in the reaction mechanism catalyzed by the enzyme rather than in the participation of a second protein. There is no incorporation of hydrogen from the medium into substrates or products during the dehydration of the propanediols (*253*); therefore, a migration to C-2 of one or other of the two hydrogen atoms from C-1 must accompany the reaction.

In order to examine the hypothesis Frey *et al.* (*255*) and Zagalak

252. Chloramine-T stands for sodium *p*-toluenesulfonchloramide.
253. A. M. Brownstein and R. H. Abeles, *JBC* **236**, 1199 (1961).
254. H. A. Lee and R. H. Abeles, *JBC* **238**, 2367 (1963).

et al. (*256*) prepared the 1*R*-1-D_1-2*R*-propane-1,2-diol (CXCVI) and the 1*R*-1-D_1-2*S*-propane-1,2-diol (CXCVII) by the reduction of *R*- and *S*-lactaldehydes [(CXCIV) and (CXCV), respectively] with liver alcohol dehydrogenase and 4*R*-4-D_1-NADH (*A* side labeled). When the 1-D_1-2*R*-propanediol (CXCVI) was used as substrate with the bacterial dehydrase in the presence of B_{12} coenzyme the resulting propionaldehyde contained deuterium at C-2 (CXCVIII), whereas the 1-D_1-2*S*-propanediol gave the aldehyde containing deuterium at C-1 (CXCIX). The implications of these results were that the dehydration of the pro-

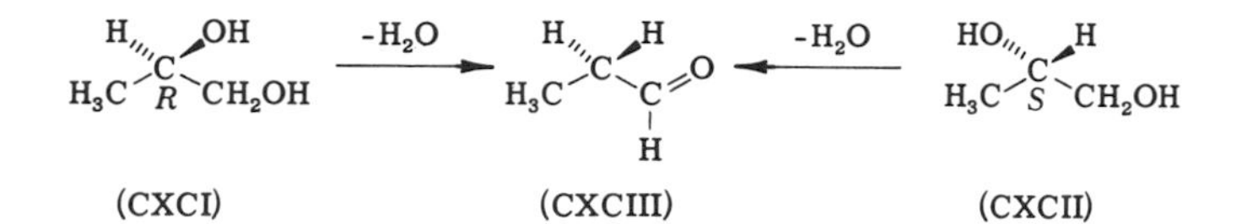

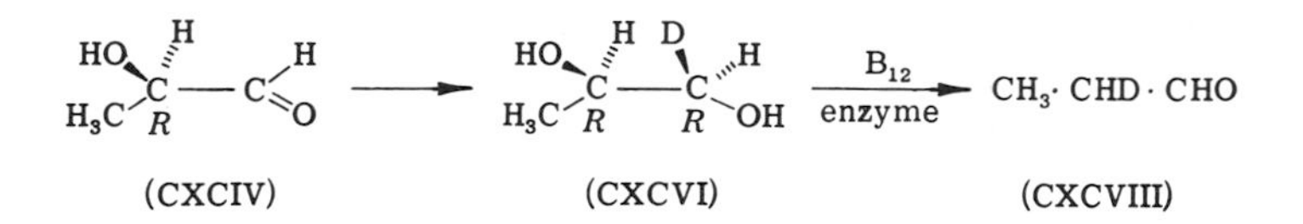

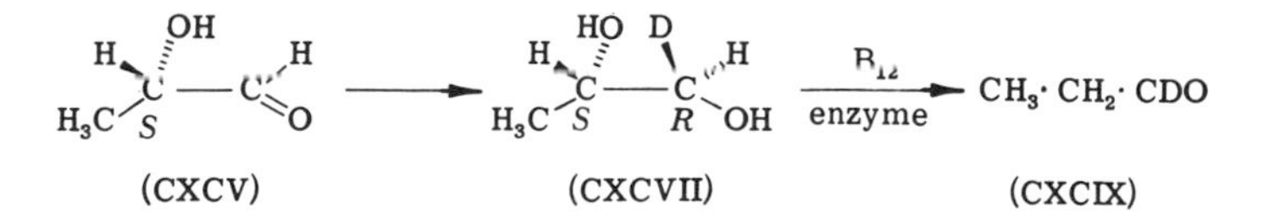

panediols was indeed accompanied by a hydrogen migration and that the steric course of this was determined by the stereochemistry at C-2 of the propanediols since the pro-*R* hydrogen atom from C-1 migrated to C-2 during the dehydration of the 2*R*-propanediol, but during the reaction with the 2*S*-diol it was the pro-*S* hydrogen atom which became shifted (*256a*).

These investigations have been further extended by Arigoni and his colleagues (*257*, *258*). Rétey *et al.* (*257*) reduced pyruvate (CC) with deuterio-NADH and lactate dehydrogenase to the 2-D_1-(+)-*S*-lactate

255. P. A. Frey, G. L. Karabatsos, and R. H. Abeles, *BBRC* **18,** 551 (1965).

256. B. Zagalak, P. A. Frey, G. L. Karabatsos, and R. H. Abeles, *JBC* **241,** 3028 (1966).

256a. G. J. Karabatsos, J. S. Fleming, N. Hsi, and R. H. Abeles, *JACS* **88,** 849 (1966).

257. J. Rétey, A. Umani-Ronchi, and D. Arigoni, *Experientia* **22,** 72 (1966).

258. J. Rétey, A. Umani-Ronchi, J. Seibl, and D. Arigoni, *Experientia* **22,** 502 (1966).

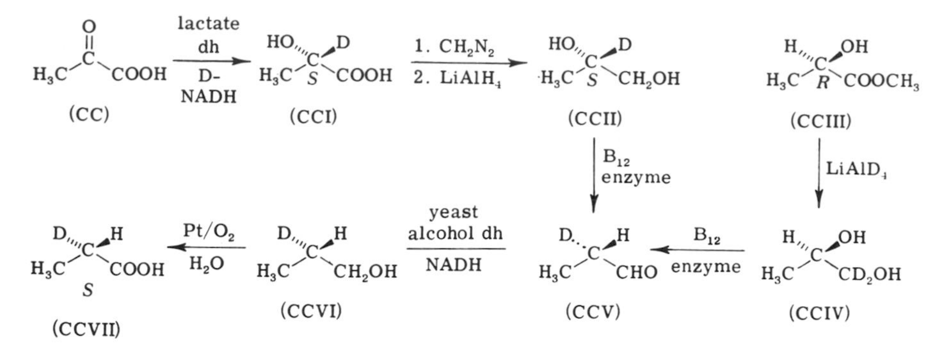

SCHEME 35. Experiment of Rétey *et al.* (*257*) for deducing the steric course of hydrogen migration in the propanediol dehydrase reaction (dh stands for dehydrogenase).

(CCI) and converted the latter to the 2-D_1-(+)-*S*-propane-1,2-diol (CCII) by the reduction of the methyl ester of (CCI) with $LiAlH_4$. Reduction of methyl (—)-*R*-lactate (CCIII) with $LiAlD_4$ gave, on the other hand the 1-D_2-(—)-*R*-propane-1,2-diol (CCIV) (Scheme 35). Incubation of (CCII) and of (CCIV) with a cell-free extract of *Aerobacter aerogenes* (ATCC 8724) in the presence of B_{12} coenzyme gave identically labeled propionaldehydes of the same absolute configuration. The propionaldehyde (CCV) resulting from each experiment was reduced to propan-1-ol (CCVI) with yeast alcohol dehydrogenase and NADH; the alcohol was then oxidized to propionic acid (CCVII) which gave a plane-positive ORD curve established (*245*) for 2*S*-2-D_1-propionic acid. It may thus be inferred that the hydrogen migration from C-1 to C-2 during dehydration of the propanediols induced an inversion of configuration at C-2. Zagalak *et al.* (*256*) arrived at a similar conclusion from experiments made with the 1-D_2-(—)-*R*-propane-1,2-diol (CCIV).

Rétey *et al.* (*258*) have also prepared specimens of *R*- and *S*-propane-1,2-diols labeled with ^{18}O in the primary hydroxyl group by the equilibration of the corresponding lactaldehydes with $H_2{}^{18}O$ followed by the

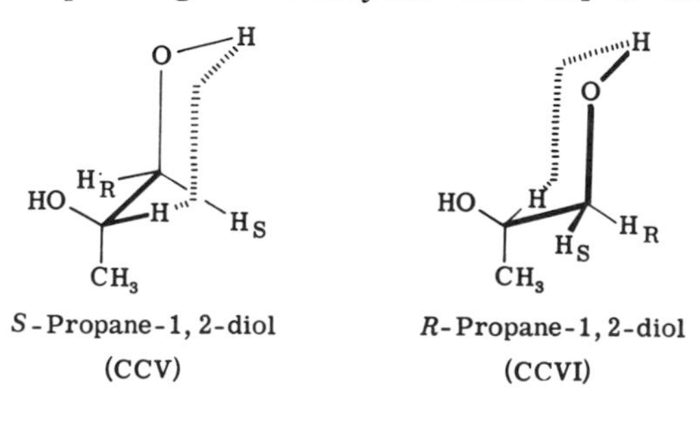

reduction of these with $LiAlH_4$ to the diols. A specimen of the racemic 2-^{18}O-*RS*-propane-1,2-diol was also made by equilibrating acetoxy-acetone with $H_2{}^{18}O$ and reducing it with $LiAlH_4$. Each specimen was incubated with the propanediol dehydrase and B_{12} coenzyme and, in order to avoid the possibility of oxygen exchange between the propionaldehyde and water, yeast alcohol dehydrogenase and NADH were also added to the incubations which reduced the aldehydes to the propan-1-ols. The latter were isolated as the phenylurethanes and their ^{18}O content determined by mass spectrometry. The product obtained from the 1-^{18}O-*R*-propane-1,2-diol retained only 8% of the isotope contained in the substrate, whereas 88% retention was observed when the 1-^{18}O-*S*-propane-1,2-diol was the substrate. The propanol derived from the 2-^{18}O-*RS*-propanediol had an isotope content corresponding to 43% of that in the substrate. Hence, the propanediol dehydrase reaction is characterized not only by a 1:2 hydrogen migration, but also by a 1:2 oxygen shift with the possible intermediary formation of a propane-1,1-diol (hydrated aldehyde) which is dehydrated in a stereospecific manner. The overall mechanism of the propanediol dehydrase catalyzed reactions, as far as discussed, is shown in Scheme 36 from which it may be seen that the two hydroxyl groups in propane-1,1-diol are sterically not equivalent since C-1 is a prochiral center. The retention of one specific oxygen atom on dehydration of the 1,1-diol can be understood only by assuming that this is not a "spontaneous" but an enzyme-catalyzed reaction.

Rétey *et al.* (257) have pointed out that the two propane-1,2-diols may be attached to the same binding area of the enzyme in such a way that the methyl group, the hydroxyl group at C-2, and the migrating hydrogen atom of C-1 have the same steric relationship. On arranging

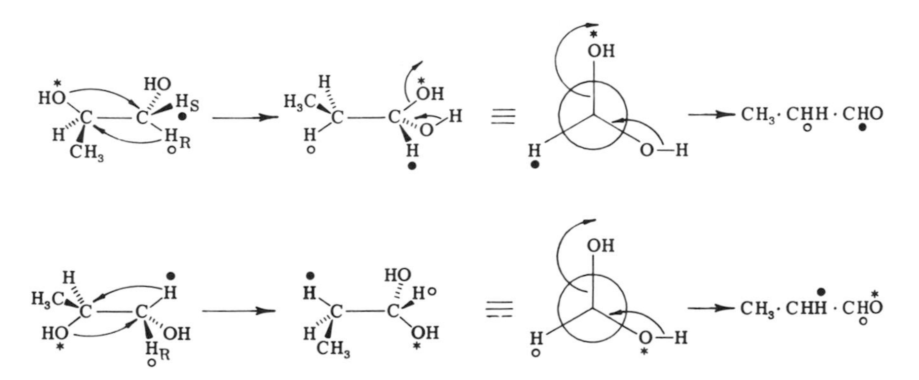

SCHEME 36. Overall mechanism of the propanediol dehydrase catalyzed reactions. In the Newman projections the spatial orientation of ligands around C-1 of propane-1,1-diol are shown.

the *S*-diols (CCV) and the *R*-diols (CCVI) this way, the migrating hydrogen atoms H_S in (CCV) and H_R in (CCVI) are seen to be antiperiplanar to the OH group migrating from C-2.

It will be a most exciting event when the three-dimensional structure of the propanediol dehydrase is established because that structure will have to account not only for the stereospecific binding of epimeric substrates to the enzyme but also for the binding of the B_{12} coenzyme which mediates the hydrogen migrations from C-1 to C-2 in this diol–dehydrase reaction. It has been shown both by Abeles and Arigoni and their colleagues that the hydrogen migrations described in the propanediol dehydrase reaction were not intramolecular as believed at first, but intermolecular, the B_{12} coenzyme acting as a relay between substrate and product, the hydrogen atom migrating from the substrates being accepted by the coenzyme and then returned to the product (*234, 259–262*). The acceptor group of the B_{12} coenzyme has been shown by Abeles and his colleagues to be the C-5′-methylene group of the 5′-deoxyadenosine moiety. The puzzling feature of the reactions is, however, that both the pro-*R* and pro-*S* hydrogen atoms at C-5′ are transferred to propionaldehyde with equal efficiency. It is relevant to mention that the B_{12} coenzyme is likely to mediate the hydrogen transfers, previously thought to be intramolecular shifts, not only in the propanediol dehydrase reactions but also in the reactions catalyzed by succinyl-CoA mutase (methylmalonyl-CoA mutase) and glutamate mutase (*262, 263*).

X. Epilogue

The topics selected for discussion in this chapter have been made with a sense of history as well as to present biochemical reactions of the most diverse nature. We are now at a stage of the development of our science when the stereochemical course of even the most complex of biochemical reaction sequences, as exemplified by the reactions of the propane-1,2-diol dehydrase or of polyprenyl biosynthesis, are, technically and intellectually, well within our reach. We are still far, or so it seems, from being able to predict the forces that will cause a polypeptide chain of

259. P. A. Frey and R. H. Abeles, *JBC* **241,** 2732 (1966).
260. G. T. Cardinale and R. H. Abeles, *BBA* **132,** 517 (1967).
261. P. A. Frey, M. K. Essenberg, and R. H. Abeles, *JBC* **242,** 5369 (1967).
262. P. A. Frey, S. S. Kerwar, and R. H. Abeles, *BBRC* **29,** 873 (1967).
263. H. P. C. Hogenkamp, R. K. Ghambeer, C. Brownson, R. L. Blakely, and E. Vitols, *JBC* **243,** 799 (1968).

specified amino acid sequences to take up one, and only one, particular conformation.

ACKNOWLEDGMENT

The author wishes to thank Professor Ronald Bentley for providing proofs and manuscripts of his two-volume treatise "Molecular Asymmetry in Biology," Academic Press, 1969–1970, before publication.

4

Proximity Effects and Enzyme Catalysis

THOMAS C. BRUICE

I. Introduction

A characteristic of all enzymic reactions involving specific substrates is the initial formation of an enzyme–substrate complex. The rate constant for this process may be diffusion controlled. Following the formation of the initial enzyme–substrate complex (ES_1), conformational changes which involve the functional groups at the active site may occur prior to the bond-breaking and bond-making processes to yield a second enzyme–substrate complex (ES_2) (*1*). This conformational change must be generally anticipated, for it is highly unlikely that the thermodynamically most stable conformations of the enzyme with and without substrate are

1. For a recent review of pertinent literature, see D. E. Koshland, Jr. and K. E. Neet, *Ann. Rev. Biochem.* **37**, 359 (1968).

identical. A rudimentary model for the change of conformation of an enzyme on binding substrate is available in the conformational changes of polyvinylpyrrolidone on complexing with aromatic compounds (*2, 2a*).

The facility of the bond-making and bond-breaking processes of enzymic reactions is to be found in the unique nature of ES_2. It must be supposed that, in general, ES_2 is so constructed that factors of importance in the transition state are provided in the ground state. In this manner $\Delta F^{\ddagger}$ values for bond making and bond breaking are minimized. The formation of ES_2 may be accompanied by (a) juxtaposition in the correct stereochemical fashion of all nucleophilic, general-acid and general-base groups for entrance to the transition state; (b) presentation of appropriately charged or polar species to the substrate at positions where formal or partial charges will occur in the transition state; and (c) conformational alteration of substrate to decrease resonance stabilization of the ground state, to increase resonance stabilization of the transition state, or to provide relief of strain on formation of intermediates or products (*conformational-strain effect*). Factor (a) has been termed the *propinquity effect* (*3*). Factor (b) is a feature possibly available only in a protein tertiary structure wherein a heterogeneous but specific milieu may be presented to the substrate. The result of this feature may be to lower the energy barrier normally required for both desolvation of the ground state and solvation of the transition state. If on the enzyme surface a charged functional group were surrounded by a lyophobic region, then interaction of this charge with developing partial charges in the transition state for an enzymic reaction would be greatly enhanced. This is so since the lower the dielectric constant, the greater are charge–charge interactions. In this manner an essentially lyophobic region could better stabilize a dipolar transition state than the excellent ionizing solvent water. This may be termed the *milieu effect*. The conformational-strain effect may only be important under conditions in which the binding energy of S to E is sufficient such that a portion of this energy may be expended in bond distortion and K_s still remain a reasonably small number. Thus, the small molecule CO_2 does not undergo detectable [by differential infrared spectrometry (*4*)] deformation when bound to carbonic anhydrase. A challenge to the chemist is the evaluation of the relative importance of the propinquity, milieu, and conformational-strain effects to the efficiency of the enzymic process. The magnitude of the contributions of bond distortion and propinquity may be approached through

2. P. Molyneux and H. P. Frank, *JACS* **83,** 3169 and 3175 (1961).
2a. T. Kunitake, F. Shimada, and C. Aso, *JACS* **91,** 2716 (1969).
3. R. Lumry, "The Enzymes," 2nd ed., Vol. 1, p. 157, 1959.
4. M. E. Riepe and J. H. Wang, *JACS* **89,** 4229 (1967).

examination of appropriate chemical models. Evaluation of the milieu effect is much more difficult. Solvent and electrostatic effects may be studied with appropriate models, but the chemist is unable to provide a stereospecific heterogeneous reaction medium. This division into three categories is of course rather arbitrary. The propinquity effect must also influence the nature of the milieu at the active site and, by steric interference, etc., could contribute to the conformational-strain effect.

A most suitable example of the importance of these factors is found in X-ray crystallographic evidence for the mechanism of lysozyme action (*5*). Binding of substrate is proposed to place the bond to be broken in juxtaposition to the carboxyl groups of Asp 52 and Glu 35 (propinquity effect), which are proposed to act as electrostatic and general-acid catalysts, respectively. The difference in polarity on the two walls of the active site is such that the pK_a of Glu 35 is greater than that for Asp 52 (milieu effect). In addition, the *N*-acetylmuramic acid whose anomeric C_1—O bond is to undergo scission, is distorted at the expense of binding energy (*6*) to the half-chair conformation. This distortion has the effect of increasing the resonance energy in the transition state (conformational-strain effect). The importance of propinquity and, most likely, milieu effects is general to all enzymic reactions. The role of the conformational-strain effect, if any, may depend upon the nature of the bond to be made and broken. X-ray crystallographic model building suggests that a twist of the C—N bond of a peptide linkage is important in the mechanism of carboxypeptidase A (*7*). An out-of-plane bending of this type would decrease the resonance stability of the ground state of the substrate and increase its susceptibility to nucleophilic attack. The rapid rates of solvolysis of α-quinuclidones, in which amide resonance is prevented by incorporation of the nitrogen at a bridgehead, serves as a simple kinetic model for the type of destabilization proposed for carboxypeptidase A action (*8*).

In this chapter the extent of rate acceleration brought about by propinquity will be evaluated in terms of rate accelerations found in enzymic reactions. The approach will be primarily through the considera-

5. C. C. F. Blake, D. F. Koenig, G. A. Mair, A. C. T. North, D. C. Phillips, and V. R. Sarma, *Nature* **206,** 757 (1965); D. C. Phillips, *Proc. Natl. Acad. Sci. U. S.* **57,** 484 (1967); L. N. Johnson and D. C. Phillips, *Nature* **206,** 761 (1965).

6. J. A. Rupley, L. Butler, M. Gerring, F. A. Hartdegen, and R. Pecoraro, *Proc. Natl. Acad. Sci. U. S.* **57,** 1088 (1967).

7. W. N. Lipscomb, Awardee Address, *Abstr. 155th ACS Meeting, San Francisco, 1968* M2.

8. J. Bredt, H. Thouet, and J. Schmitz, *Ann. Chem.* **437,** 1 (1924); H. Pracejus, M. Kehler, H. Kedler, and H. Matschiner, *Tetrahedron* **21,** 2257 (1965).

tion of physical organic models. The milieu and conformational-strain effects will also be considered when inseparable from the propinquity effect.

II. Evaluation of Concepts Concerning the Propinquity Effect

A simple mathematical model for evaluation of the propinquity effect has been set forth by Koshland (*1*, *9*). The model is based on the following four assumptions:

(1) Reactant species are about the size of a water molecule.
(2) Packing of both water and reactant molecules is of the twelve nearest neighbors type.
(3) Reactant solutions are sufficiently dilute so that the probability of any more than one reactant molecule being juxtaposed to another is very small.
(4) Reaction occurs only through nearest neighbor pairs.

In what follows, the assumptions of Koshland will be used; however, for the purpose of this review the method of employment will be that of the author's. The alteration of the mode of treatment is necessitated by the desire to compare, in general, rate constants rather than velocities or turnover numbers.

If the reaction of Eq. (1) occurs through formation of nearest pairs

$$\mathrm{A} + \mathrm{B} \xrightarrow{k_2} \mathrm{A{-}B} \tag{1}$$

then it follows that

$$[\mathrm{AB}]_{\mathrm{pairs}} = \frac{[\mathrm{A}][\mathrm{B}]\ 12}{55.5} \tag{2}$$

and

$$\begin{aligned} v &= k_2[\mathrm{A}][\mathrm{B}] = k_2 \frac{55.5}{12} [\mathrm{AB}]_{\mathrm{pairs}} \\ &= k_2 4.6\ [\mathrm{AB}]_{\mathrm{pairs}} \end{aligned} \tag{3}$$

Thus, if A and B are juxtaposed within a single compound, then the rate constant k_1 [Eq. (4)] should only be about 4.6-fold greater than if A and B were

$$\underbrace{\mathrm{A} \quad \mathrm{B}} \xrightarrow{k_1} \underbrace{\mathrm{A{-}B}} \tag{4}$$

9. D. E. Koshland, Jr., *J. Theoret. Biol.* **2**, 75 (1962).

separate entities in solution. The value of 4.6 is then the ratio of the rate constants for a biomolecular reaction compared to its intramolecular counterpart if one assumes that a reaction of A and B occurs regardless of the portions of the surfaces of A and B that come into contact (no orientation factor), and if one ignores the possibility that A and B must be desolvated before they may be brought into contact. Since the velocity of reaction of Eq. (1) is dependent upon the concentrations of both A and B whereas the velocity of Eq. (4) is independent of the concentration of A B, the ratio k_1/k_2 must be dependent upon the standard state employed to express concentration. For the standard state moles per liter:

$$\frac{k_1(\text{time}^{-1})}{k_2(\text{liter mole}^{-1}\ \text{time}^{-1})} = \frac{k_1}{k_2}\ (\text{mole/liter}) \qquad (5)$$

The ratio of Eq. (5) may be considered (*10*) to be the concentration of A required to provide a pseudo-first-order rate of reaction of A with B equal to k_1 of Eq. (4). Thus, if $[A] >> [B]$ then the kinetics for the bimolecular reaction of Eq. (1) become pseudo-first-order:

$$v = k_{obsd}[B] \qquad (6)$$

where

$$k_{obsd} = k_2[A] \qquad (7)$$

For the intramolecular reaction of Eq. (4),

$$k'_{obsd} = k_1 \qquad (8)$$

and the value of k_{obsd} will equal k'_{obsd} at a given concentration of A. It follows from Eq. (3) that this concentration, ignoring solvation and orientation effects, should be 4.6 M. A criticism of this simple mathematical approach becomes immediately evident on rearrangement of Eq. (2) (*viz.*, $12/55.5 = 0.21 = [AB]_{pairs}/[A][B]$). From Eq. (2) one would be led to believe that the equilibrium constant for complexation of reactants in water is universally close to about 0.2 M.

Though Koshland did not consider solvation effects, a means of adjusting for proper orientation for reaction within the AB_{pair} was suggested. If A and B may yield only product from the AB_{pair} when properly aligned, then one may consider that some definable fraction of the surface of A and B must be in contact for reaction (9).

10. M. L. Bender, *Chem. Rev.* **60**, 53 (1960).

(9)

The fraction of the total solid angle of B within which reaction can occur is designated as $1/\theta_B$. A factor f_{AB} is introduced to account for the number of nearest neighbor positions that can react. Since A has twelve nearest neighbors, there are twelve possible constellations of B in contact with A. If A has no preferred orientation it can react in any of the twelve constellations. However, if A has an orientation such that it can react with B in only one of the twelve constellations, the rate of the bimolecular reaction [Eq. (1)] will be only one-twelfth of that where A has no preferred orientation. Employing these orientation factors Eq. (3) becomes

$$v = k_2 \frac{55.5}{12} [AB]_{pairs} \frac{\theta_B}{f_{AB}} \tag{10}$$

Employing this model, the factor f_{AB} can increase the efficiency of an intramolecular reaction [Eq. (4)] over its bimolecular counterpart [Eq. (1)] only by twelvefold. For displacement on ester bonds, because of the location of alkyl and acyl groups in space, the formation of $[AB]_{pairs}$ is restricted to fewer than twelve nearest neighbors. A reasonable assumption is that A and B form only six constellations in which the nucleophilic center of B is in contact with the reactive group of A; two of these constellations allow reaction so that $f_{AB} = 2/6$. Thus, the factor f_{AB} can only account for a 3×4.6 increase in the rate constant for an intramolecular displacement on an ester carbonyl when compared to its bimolecular counterpart. If we now assume that one-fourth of the reactive center of A (i.e., an ester carbonyl group which may be approached by base to provide overlap of the p orbital of carbon) and one-sixth of the reactive center of B (nonbonded sp^3 orbital of an amine) may come into contact, then Eq. (10) becomes Eq. (11):

$$\begin{aligned} v &= k_2 \frac{55.5 \times 4 \times 6}{4} [AB]_{pairs} \\ &= k_2\, 3.33 \times 10^2 [AB]_{pairs} \end{aligned} \tag{11}$$

From Eq. (11) we may conclude that the rate constant for an intramolecular reaction involving an acyl derivative that provided proper orientation of A and B moieties might be about 10^2 greater than the rate constant for the bimolecular counterpart. Thus, under the pseudo-first-order conditions of $[A] \gg [B]$, the concentration of A, in the bimolecular reaction, required to observe a value of k_{obsd} equal to that for the intramolecular reaction would be about 5 to $10^2\,M$, depending upon the extent of proper orientation in the intramolecular reaction. The concentration of $10^2\,M$ is beyond that attainable with any reagent.

An intramolecular reaction bears a close analogy to the conversion of ES_2 to products. In Eq. (12) k_{cat} represents a first-order rate constant

$$E + S \underset{K_s}{\rightleftarrows} ES_1 \rightleftarrows ES_2 \xrightarrow{k_{cat}} E + P \tag{12}$$

for an intracomplex reaction. If the active site possesses one catalytic group (i.e., B) and the substrate is A, then in ES_2 the species A and B exist as AB_{pairs}. It follows that the contribution of the propinquity effect to the rate enhancement for both intramolecular reactions and enzymic reactions (k_{cat}) involving one catalytic species and one substrate should be between 5- and 10^2-fold. The assumptions employed so far are those of Koshland and do not consider the fact that A and B may require desolvation before they may be juxtaposed to form an AB_{pair}. If, in the intramolecular reaction [Eq. (4)] or the hypothetical enzymic reaction involving one catalytic group at the active site, A and B are held together so that solvent is excluded, then the intramolecular or enzymic reaction may be further favored over the bimolecular reaction. In addition, if steric strain is relieved in the intramolecular or enzymic reaction on formation of the transition state these processes will be further favored over their bimolecular counterparts.

The importance of the factors of approximate orientation, desolvation, and steric compression may be tested by examination of physical organic models. *A priori*, based on the previous considerations for the A-B system, the ratio of bimolecular to intramolecular rate constants might be expected to vary as shown in Table I. The difficulty in defining the exact

TABLE I

Factors Effecting the Ratio [Eq. (5)] of Rate Constants for Bimolecular [Eq. (1)] to Intramolecular [Eq. (4)] Reaction of A with B

k_1/k_2 (M)	Conditions
5–10	Little or no orientation in the intramolecular reaction. Solvation of A and B moieties about the same for both processes. No steric compression in intramolecular reaction
10–10^3	Orientation of A and B moieties in intramolecular reaction and/or solvation of A and B less in intramolecular reaction with or without some steric compression
>10^3	Orientation of A and B moieties in intramolecular reaction, less solvation of A and B pairs in intramolecular reaction with or without steric compression

meaning of the propinquity effect becomes obvious. This term must involve solvation since heavily solvated groups cannot be juxtaposed without desolvation. In the enzymic reaction the substrate may be desolvated if the energy gained by specific interactions at the active site equals or exceeds the energy of desolvation (milieu effect) or if the binding of substrate is sufficiently great that a portion of the energy of binding can be expended in forcing the reactive groups together and squeezing out their solvent shells. In intramolecular physical organic models (and of course enzymic processes) the conformational-strain effect cannot always be separated from the propinquity effect. Nonbonded interaction and specific solvation in flexible intramolecular models will favor certain rotamer populations which in turn determine the statistical distance between the A and B moieties.

It is highly unlikely that enzymic reactions have only one catalytic group at the active site. Thus, for lysozyme two carboxyl groups constitute the catalytic groups (*5*), while for the serine esterases an imidazolyl and a hydroxymethyl group are the reactive functions at the active site (*11*). For α-chymotrypsin, X-ray, and sequence studies (*12*) provide evidence that His 57, Ser 195, and Asp 102 constitute the catalytic entities.

To evaluate the contribution of the propinquity effect to the efficiency of enzymic processes, Koshland (*9*) has employed the following rationale: at saturation of enzyme with substrate,

$$v_e = k_E[E_t] \tag{13}$$

where k_E is the turnover number of the enzyme and $[E_t]$ the concentration of total enzyme in mole per liter. For an enzymic reaction involving three catalytic groups (R, S, and T) at the active site and two substrates (A and B), as shown in Fig. 1, the ratio of the rate of the enzymic reaction (v_e) to the nonenzymic reaction (v_o) might be

$$\frac{v_e}{v_o} = \frac{[E_t]\,(55.5)^4\theta_B\theta_R\theta_S\theta_T}{[A][B][R][S][T] \times 12 \times 11 \times 11 \times 10 \times f_{AB} \times f_{BR} \times f_{BS} \times f_{AT}}$$
$$= \frac{[E_t]}{[A]} \times \frac{55.5\,\theta_B'}{[B]} \times \frac{55.5\,\theta_R'}{[R]} \times \frac{55.5\,\theta_S'}{[S]} \times \frac{55.5\,\theta_T'}{[T]} \tag{14}$$

"Assuming substrate and catalysts are all $10^{-3}\,M$, θ' factors are 10, and $[E_T] = 10^{-5}\,M$ gives a v_e/v_o ratio of 9.5×10^{17}. Even if a hydrolytic reaction is involved, the ratio would be 1.7×10^{13}. These are clearly in the range of enzymatic accelerations. In other words, if the reaction in-

11. For a survey of the pertinent literature, see T. C. Bruice and S. J. Benkovic, "Bioorganic Mechanisms," Vol. I, Chapter 2. Benjamin, New York, 1966.

12. D. M. Blow, J. J. Birktoft, and B. S. Hartley, *Nature* **221,** 337 (1969).

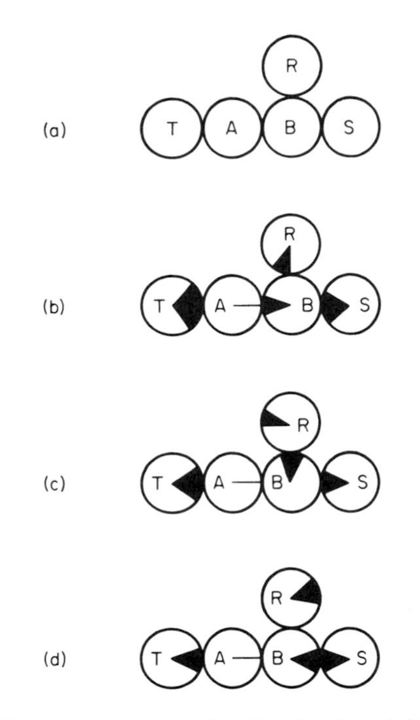

FIG. 1. Some possible arrangements of molecules in solution (all molecules the size of water molecules). (a) Arrangement of molecules A, B, R, S, and T in water lattice without regard to any preferred orientation within the molecules. (b) Arrangement of molecules A, B, R, S, and T. Proper arrangement requires that contact point of A be tangential to black wedge portion ($1/\theta_B$) of B, that appropriate contact point of B be tangential to black wedge portion of S, etc. (c) and (d) Nonreactive arrangements of molecules A, B, R, S, and T; T and S properly oriented, but B and R are not (*9*).

volves a combination of two substrates and three catalytic groups, proximity and orientation alone could account for a large part of the great accelerations caused by enzymes" (*9*). Fault can be found in this conclusion since even a 10^{18} rate acceleration applied to a fifth-order reaction may only provide a small or even an undetectable rate constant. This is so because fifth-order reactions are so improbable that they have never been detected. Now suppose A, B, R, S, and T must be desolvated to provide the correct $ABRST_{quintet}$ for a fifth-order reaction in aqueous solution

$$A + B + R + S + T \rightleftarrows ABRST_{quintet} \rightarrow Prod. + R + S + T \quad (15)$$

If in the process of the eventual formation of ES_2 the necessary desolvation is accomplished then an additional rate factor of perhaps 10^7 would favor the enzymic reaction (assuming about 2 kcal $mole^{-1}$ required to desolvate each species). The rate constant for the enzymic reaction would then be favored over its fifth-order counterpart by perhaps 10^{25}. With the addition of a conformational-strain factor, perhaps a value of 10^{30} would be reasonable. A rate enhancement of this order of magnitude is sufficient to change a first-order reaction with $t_{1/2}$ of $\sim 1.0 \times 10^{14}$ years to a diffusion-controlled process. [It might be noted that the age of the universe has been estimated to be but 5×10^9 years (*13*).] The question remains: Can the summation of the propinquity (if solvation is included), milieu, and conformational-strain effects provide more than sufficient leeway to explain the facility of enzymic processes?

III. Physical Organic Models Establishing the Role of Solvation and Strain

The conjectural aspects of the previous section are, in part, amenable to experimental evaluation. This is perhaps best accomplished via examination of physical organic models. Because of the complexities of synthetic procedures for the preparation of intramolecular models, pertinent work to date centers around the AB_{pair} problem. In this section no attempt will be made to catalog all interesting intramolecular reactions. For a detailed discussion of intramolecular acyl transfer reactions, see Bruice and Benkovic (*11*, Chapters 1 and 3), while for phosphate derivatives, see Bruice and Benkovic (*14*, Chapters 5, 6, and 7).

A. Acyl Transfer to a Carboxyl Anion

Acyl transfer to a carboxyl anion serves admirably for evaluation of propinquity, milieu, and conformational-strain effects for cases involving a carboxyl functional group as the catalytic entity. The reactions to be compared are intermolecular nucleophilic displacement on an acyl derivative by carboxyl anion [Eq. (16)] and its intramolecular

$$\begin{matrix} R{-}C({=}O){-}X \\ + \\ R'COO^{\ominus} \end{matrix} \xrightarrow{k_c} \begin{matrix} R{-}CO \\ \quad \diagdown \\ \quad O \\ \quad \diagup \\ R'{-}CO \end{matrix} + X^{\ominus} \qquad (16)$$

13. W. Benton, "Encyclopedia Britannica," Vol. 6, p. 578. New York, 1966.
14. T. C. Bruice and S. J. Benkovic, "Bioorganic Mechanisms," Vol. II. Benjamin, New York, 1966.

counterpart involving a monoacyl derivative of a dicarboxylic acid [Eq. (17)].

$$\text{(CH}_2\text{)}(\text{C(=O)}{-}X)(\text{COO}^{\ominus}) \xrightarrow{k_a} \text{cyclic anhydride (CO–O–CO)} + X^{\ominus} \qquad (17)$$

For Eqs. (16) and (17), —X must be the same leaving group and the pK_a' values of the participating carboxyl groups must be similar.

The most useful comparison is one for which sufficient experimental evidence exists to obtain data which allows some conclusions to be drawn about the propinquity, conformational-strain, and milieu effects. The

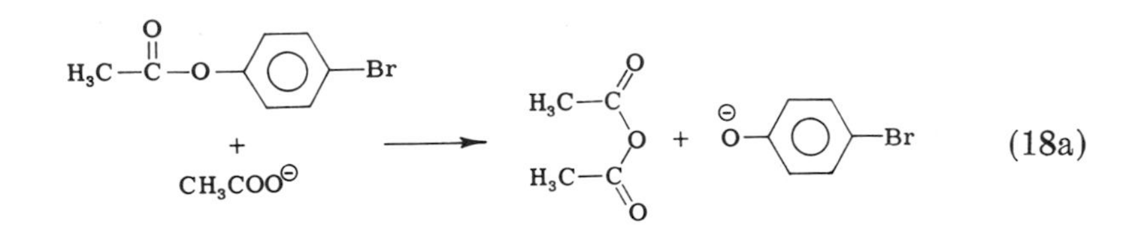

$$\text{exo-3,6-endoxo-}\Delta^4\text{-tetrahydrophthalate mono-}p\text{-bromophenyl ester (COO–C}_6\text{H}_4\text{–Br, COO}^{\ominus}) \longrightarrow \text{anhydride} + {}^{\ominus}\text{O–C}_6\text{H}_4\text{–Br} \qquad (18b)$$

present state of the art suggests the best model is that of Eq. (18). In Eq. (18a) acetate ion plus *p*-bromophenyl acetate yields acetic anhydride and *p*-bromophenolate while in the intramolecular counterpart (18b) the *p*-bromophenyl ester of exo-3,6-endoxo-Δ^4-tetrahydrophthalic acid yields the corresponding anhydride plus *p*-bromophenolate. Very few determinations of k_{intra}/k_{inter} can be made without some recourse to extrapolative procedures employing linear free energy equations. The reason for this is that conversion of a bimolecular reaction of type (18a) into an intramolecular reaction of type (18b) is quite often accompanied by a change of mechanism (*11, 15*). Changes of mechanism may also occur if the basicity of —X is drastically altered (*16*).

For acetate-ion-catalyzed hydrolysis of aryl-substituted phenyl acetates, two mechanisms have been shown to compete (*17*). The two competing mechanisms are direct nucleophilic attack to provide an anhydride

15. T. C. Bruice and W. C. Bradbury, *JACS* **90,** 3808 (1968).
16. J. W. Thanassi and T. C. Bruice, *JACS* **88,** 747 (1966).
17. V. Gold, D. G. Oakenfull, and T. Riley, *JCS*(*B*) **515,** (1968).

intermediate (16) and a general-base assisted attack of water by acetate ion (19).

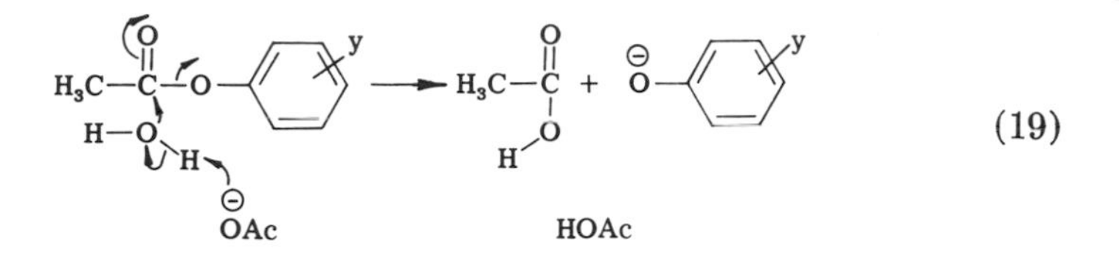

(19)

The fractions of hydrolytic product arising via (16) and (19) depend markedly upon the pK_a' of the conjugate acid of the leaving group (*17*). Acetate ion cannot directly replace a phenolate ion if the conjugate acid of the latter possesses a pK_a' about four units greater than acetate. The free energy barrier for displacement of phenolate ion by acetate increases markedly as the relative basicity of the nucleophile as compared to the attacking group decreases. This allows the electronically less sensitive (*18, 19*) mechanism of general-base catalysis to predominate for substrates with poorer leaving groups. This change from nucleophilic to general-base-catalyzed displacement with decrease in leaving tendency has been noted for other systems as well (*20, 21*).

In order to obtain a comparison of the rate constants for (18a) and (18b) recourse to a linear-free energy extrapolation is necessary since, because of the competing general-base mechanism, the rate constant for (18a) cannot be directly determined. A Hammett $\rho\sigma$ plot for nucleophilic attack of acetate ion upon, 2,6-dinitro-, 2,4-dinitro-, 2,3-dinitro- 3,4-dinitro-, and 4-nitrophenyl acetates provide ρ as $+2.2$. Extrapolation of this plot provides the rate constant for direct attack of acetate upon *p*-bromophenyl acetate as $\approx 5.3 \times 10^{-6}$ liter mole^{-1} min^{-1}. The rate constant for (18b) is too large to measure directly. However, the rate constants of Table II (22) may be employed to extrapolate the rate constant for (18b). A change of the *p* substituent from $-OCH_3$ to —Br increases the rate constant of the intramolecular reaction by a factor of ≈ 23 while going from the mono-*p*-methoxyphenyl esters of succinic to 3,6-endoxo-Δ^4-tetrahydrophthalic acid is accompanied by an increase of rate constant by 2.2×10^4-fold. Thus, the rate constant for (18b) is reasonably $\approx 2.3 \times 10^2$ min^{-1}. The ratio of rate constants for (18a) and (18b) now becomes $\approx 2.3 \times 10^2 / 5 \times 10^{-6} = 5 \times 10^7$ moles/liter. From examination of Table I, one must conclude that the approximation of the

18. T. C. Bruice and M. F. Mayahi, *JACS* **82,** 3067 (1960).
19. T. C. Bruice and S. J. Benkovic, *JACS* **86,** 418 (1964).
20. W. P. Jencks and J. Carriuolo, *JACS* **83,** 1743 (1961).
21. J. F. Kirsch and W. P. Jencks, *JACS* **86,** 833 and 837 (1964).

TABLE II

RATE CONSTANTS FOR INTRAMOLECULAR CATALYSIS OF HYDROLYSIS OF SUCCINATE AND 3,6-ENDOXO-Δ^4-TETRAHYDROPHTHALATE PHENYL ESTERS (*22*)

System	R	k_{rate} (min^{-1})
$CH_2—COOR$ / $CH_2—COO^-$	p-BrC_6H_5—	1.02
	p-$CH_3OC_6H_5$—	4.5×10^{-2}
3,6-endoxo-Δ^4-tetrahydrophthalate (COOR, $COO^\ominus$)	p-$CH_3OC_6H_5$—	1.0×10^1

AB_{pair} as well as orientation in (18b) do not at all suffice to explain this huge rate ratio. Clearly other factors must be considered.

Tables III and IV (*23*, *24*) record experimentally determined relative rate constants for selected monophenyl esters of dicarboxylic acids. Inspection of Table IV reveals that in going from the glutarate ester to the succinate ester the rate of carboxyl anion nucleophilic attack upon the ester bond increases 230-fold. This rate enhancement is accompanied by a decrease from two single bonds allowing free rotation to one single bond allowing free rotation. In going from the succinate ester to the

TABLE III

RELATIVE RATES OF ANHYDRIDE FORMATION FOR THREE ESTERS POSSESSING DIFFERENT ROTATIONAL FREEDOM (*23*)

Ester	Relative rate constants
$H_2C(CH_2—COOR)(CH_2—COO^-)$	1.0
$CH_2—COOR$ / $CH_2—COO^-$	230
3,6-endoxo-Δ^4-tetrahydrophthalate (COOR, $COO^\ominus$)	$230 \times 230 = 53,000$

22. T. C. Bruice and U. K. Pandit, *JACS* **82**, 5858 (1960).
23. T. C. Bruice and U. K. Pandit, *Proc. Natl. Acad. Sci. U. S.* **46**, 402 (1960).
24. T. C. Bruice and W. C. Bradbury, *JACS* **87**, 4846 (1965).

TABLE IV
RELATIVE RATES OF ANHYDRIDE FORMATION FOR MONOPHENYL ESTERS OF 3-R-, 3-R′-SUBSTITUTED GLUTARIC ACIDS[a]

Ester	Relative rate constants
$CH_2(CH_2{-}COOR)(CH_2{-}COO^-)$	1.0
$(CH_3)(H)C(CH_2{-}COOR)(CH_2COO^-)$	3.4
$(C_3H_7)(H)C(CH_2{-}COOR)(CH_2{-}COO^-)$	13
$(i\text{-}C_3H_7)(H)C(CH_2{-}COOR)(CH_2{-}COO^-)$	33
$(CH_3)(CH_3)C(CH_2{-}COOR)(CH_2{-}COO^-)$	23
$(C_2H_5)(C_2H_5)C(CH_2{-}COOR)(CH_2{-}COO^-)$	180
$(C_6H_5)(C_6H_5)C(CH_2{-}COOR)(CH_2{-}COO^-)$	270
$(CH_3)(t\text{-}C_4H_9)C(CH_2{-}COOR)(CH_2{-}COO^-)$	1060
$(i\text{-}C_3H_7)(i\text{-}C_3H_7)C(CH_2{-}COOR)(CH_2{-}COO^-)$	1340

[a] From (*24*).

3,6-endoxo-Δ^4-tetrahydrophthalate ester the rate is increased another 230-fold. The latter rate increase accompanies a decrease from one single bond allowing free rotation to an eclipsed conformation with no rotation of A and B. From these results each single bond that must be frozen to obtain an AB_{pair} increases $\Delta F^{\ddagger}$ by about 3 kcal $mole^{-1}$. Since the results are additive, there is no evidence of steric compression in the ground state for the 3,6-endoxo-Δ^4-tetrahydrophthalate ester. It is most reasonable to conclude that the increase in rate accompanying decrease in freedom of rotation results from a decrease in the population of unprofitable rotomers. It has been concluded, therefore, that the favored ground state for the glutarate and succinate esters is extended and heavily solvated (*25*). For (I) the hydration sphere surrounding the ester

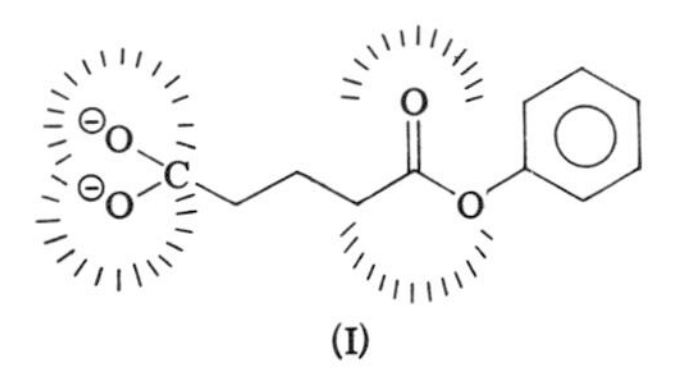

(I)

and particularly the carboxyl anion increases the bulk of these groups with the result that the thermodynamically favored conformation is extended.

Table IV records a representative portion of the rate data for anhydride formation from 3-R, 3-R′-substituted phenyl glutarates (*24*). Examination of the relative rate data of Table IV reveals that as the substitutents in the 3 position increase in bulk, the rate of nucleophilic attack of the carboxyl anion at the ester carbonyl group increases. When $R = R' = C_6H_5$- the rate of reaction is comparable to that of the succinate ester (Table III). It is concluded that the average distance between A and B for the succinate ester and the 3,3′-diphenyl glutarate ester are comparable. When $R = CH_3$- and $R' = t\text{-}C_4H_9$- the rate of reaction exceeds that for the succinate ester so that the A and B reacting groups must be more closely juxtaposed than in the case of the succinate ester. Substitution in the 3 and 3′ positions of the glutarate esters brings about nonbonded interactions of both A and B with the substituent groups. This has the effect of decreasing the space in which A and B may be found and therefore also decreases the distance between A and B. No special effects as bond angle distortion (*26*) occur since the steric effects were found to be additive. Since for the ester of 3,6-endoxo-Δ^4-

25. T. C. Bruice and S. J. Benkovic, *JACS* **85**, 1 (1963).
26. F. G. Bordwell, C. E. Osborne, and R. D. Chapman, *JACS* **81**, 2698 (1959).

tetrahydrophthalic acid water molecules cannot be inserted between the carboxyl anion and ester moieties, it is probable that the water solvent shells (I) are forced to gradually coalesce (II).

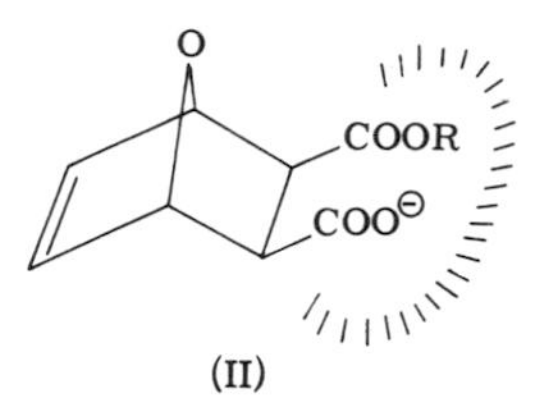

(II)

When A and B represent ester and carboxyl anions, respectively, the process of forming an AB_{pair} in aqueous solution may be represented as in (20).

$$A + B \underset{}{\overset{\text{step 1}}{\rightleftharpoons}} (A\,B)_{pair} \underset{}{\overset{\text{step 2}}{\rightleftharpoons}} AB_{pair} \qquad (20)$$

Step 1 of (20) represents juxtaposition of solvated A and solvated B while step 2 represents the formation of an intimate AB_{pair} which exists within its own solvent cage [much as in the case of solvent separated and intimate ion pairs (*27*)]. The process of (20) is mechanistically akin to the model of Benson (*28*) for ion–dipolar molecule reactions (*29*).

The importance of anion desolvation as a barrier in chemical reactions has been investigated in several laboratories (*30–34*). These workers have examined simple intermolecular reactions involving small anions and have found large rate enhancements on transfer from protic to dipolar aprotic solvents. Free energies of transfer (log γ) from protic to dipolar aprotic solvents, as dimethyl sulfoxide (DMSO), lie in the region of 4–10 for small anions (i.e., they are 10^4 to 10^{10} more solvated

27. For a review, see M. Szwarc, *Accounts Chem. Res.* **2,** 87 (1969).

28. S. W. Benson, "The Foundations of Chemical Kinetics." Chapter XV. McGraw-Hill, New York, 1960.

29. For a general discussion of solvent effects upon ion–dipolar molecule reactions, see E. S. Amis, "Solvent Effects on Reaction Rates and Mechanisms," Chapter II. Academic Press, New York, 1966.

30. J. Miller and A. J. Parker, *JACS* **83,** 117 (1961).

31. A. J. Parker, *Quart. Rev.* (*London*) **16,** 163 (1962).

32. R. Alexander, E. C. F. Ko, A. J. Parker, and T. J. Broxton, *JACS* **90,** 5049 (1968).

33. E. Tommila, *Suomen Kemistilehti* **B37,** 117 (1964).

34. D. D. Roberts, *J. Org. Chem.* **31,** 4037 (1966), and references provided therein.

by protic solvents) (*32*). Since values of log γ are much more positive for anions than for large charge-dispersed transition states, one may conclude that anion desolvation will be more significant in determining reaction rates in dipolar aprotic solvents. It has been established that chloride ion in DMSO–water mixtures exhibits an affinity for water comparable to that of DMSO (*35*). This result establishes that anions are greatly desolvated in aqueous DMSO solutions. A comparison of the rate constants for (18a) and (18b) in DMSO containing one mole of water reveals that the ratio of k_{intra}/k_{inter} is the same as in aqueous solution (*36*). Furthermore, the rate constants for the reaction of acetate ion with phenyl acetates in DMSO solutions containing less than 0.005 M H_2O are comparable to the rate constants determined in DMSO containing 1.0 M H_2O. From these results it can only be concluded that solvation does not explain the rate ratios of (18a) and (18b). Since neither steric compression nor solvation can explain the experimental results, we must conclude that the simple mathematical model described in Section II is quite inexact, as are the predicted rate ratios of Table I.

On the basis that the structural alterations for the compounds of Tables III and IV have an additive effect on the rate constants, it has been reasoned that these structural alterations influence only the proximity of the carboxyl and ester functions (i.e., no relief of steric strain in going to the transition state). Marked accelerations in the rate constants for intramolecular reactions accompanying alteration in rotamer population by substitution have been noted in other systems. Two examples are the 315-fold rate enhancement obtained for the formation of a phthalide upon substitution of methyl groups in the 3 and 5 positions

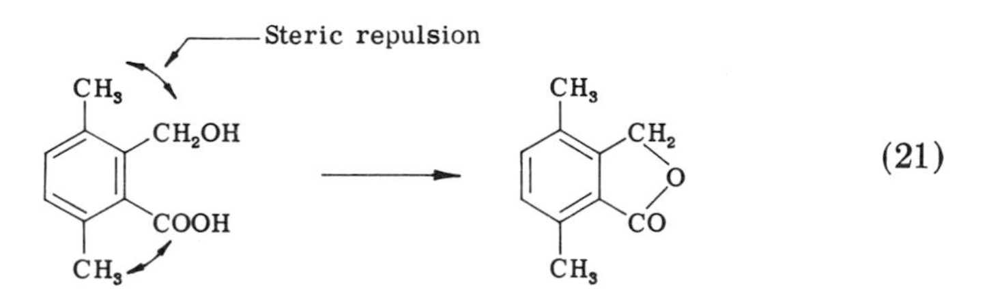

(21)

of 2-hydroxymethyl benzoic acid (*37*) and the about 5×10^6-fold acceleration observed in the Smiles reaction brought about by substitution of a methyl group in the 6 position of 2-hydroxy-2′-nitrophenylsulfones (*38*).

35. C. H. Langford and T. R. Stengle, *JACS* **91,** 4014 (1969).
36. A. Turner and T. C. Bruice, *JACS* **92,** 3422 (1970).
37. J. F. Bunnett and C. F. Hauser, *JACS* **87,** 2214 (1965).
38. T. Okamoto and J. F. Bunnett, *JACS* **78,** 5357 (1956).

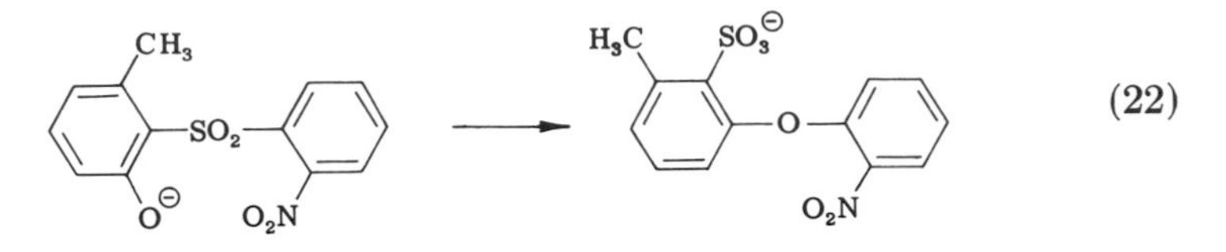

(22)

On increasing the steric requirements of substituent groups which control the statistical proximity of A and B, a point may be reached when further increase in the bulk of substituents will build in a steric strain in the ground state which is relieved on ring formation. For phenyl esters of succinic and glutaric acids some steric compression has most likely been built into the molecules when they are tetra substituted with methyl groups as in (III) and (IV). Thus, the phenyl esters of (III) and (IV)

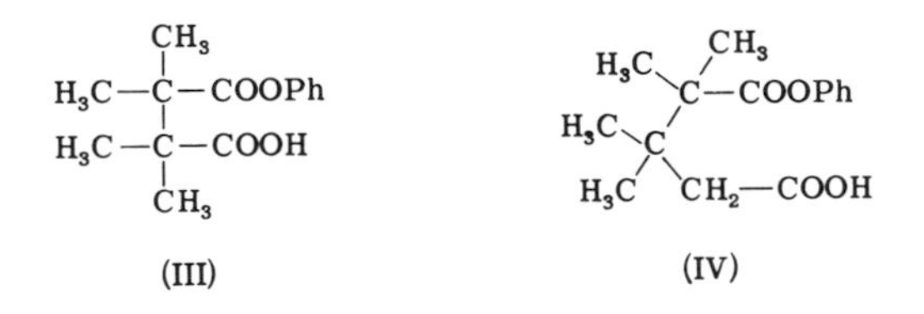

cannot be synthesized (*22*) by the conventional method of treating the corresponding anhydride with phenoxide ion. Presumably the formation of anhydride from ester is too facile to allow the latter's isolation. The hydrolysis of the anilide of tetramethyl succinic acid (V) occurs with

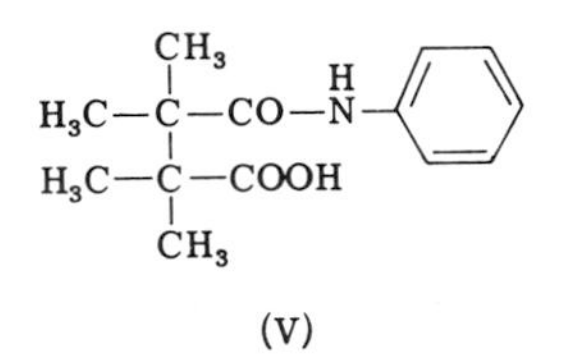

(V)

participation of the undissociated carboxyl group at a rate 1200-fold greater than for the anilide of unsubstituted succinic acid (*39*). By increasing the size of the substituents over that of the methyl group the rate of carboxyl group participation would undoubtedly also increase. In terms of the propinquity effect, once the intimate AB_{pair} is formed steric compression greatly increases the rate of attack of B upon A. Clearly, the ratio of the rate constants for (18a) and (18b) has not been maximized.

Steric acceleration associated with relief of strain in the ground state

39. T. Higuchi, L. Eberson, and A. K. Herd, *JACS* **88**, 3805 (1966).

has been noted in reactions involving carbonium ion formation (*40*). An interesting example of large steric rate accelerations is found in the hydrolysis of the acetals (VI). The rate constants associated with both

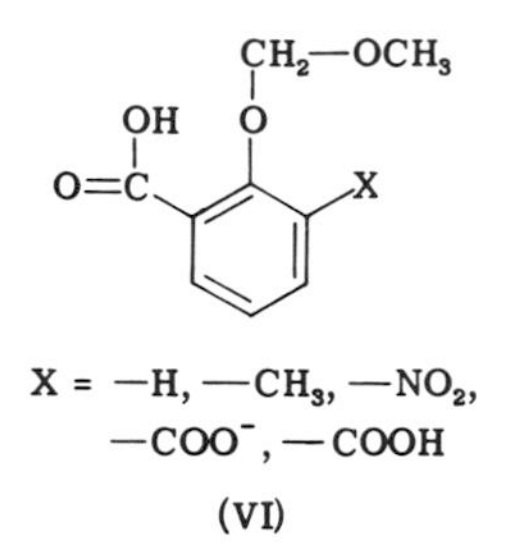

specific-acid and intramolecular general-acid (by carboxyl group) catalyzed hydrolysis increase with increase in the steric requirements of —X (*41*).

B. Acyl Transfer to a Tertiary Amino Group

Acyl group transfer to a tertiary amino group in both bimolecular and intramolecular reactions has been studied (*25*). The rate constants for nucleophilic attack of the tertiary amino group upon the ester carbonyl group can be directly measured for Eqs. (23). The determined rate constants and associated Hammett ρ values are provided in Table V while the determined activation parameters may be found in Table VI. In Table VII the rate ratios for conversion of (23a) to the intra-

TABLE V
Rate Constants (20°) for the Nucleophilic Displacement of Substituted Phenolate Ion by the Trialkylamino Group in Inter- (23a) and Intramolecular (23b and 23c) Models

Substituent	Eq. (23a) (liter mole^{-1} min^{-1})	Eq. (23b) (min^{-1})	Eq. (23c) (min^{-1})
p-NO_2	4.29	21,500	10,000
m-NO_2	0.342	580	255
p-Cl	0.0308	33.1	19.9
H	0.00795	10.0	3.9
p-CH_3	—	3.50	1.28
ρ	2.15	2.54	2.54

40. S. Winstein and N. J. Holmes, *JACS* **77,** 5562 (1955).
41. B. M. Dunn and T. C. Bruice, *JACS* **92,** 2410 (1970).

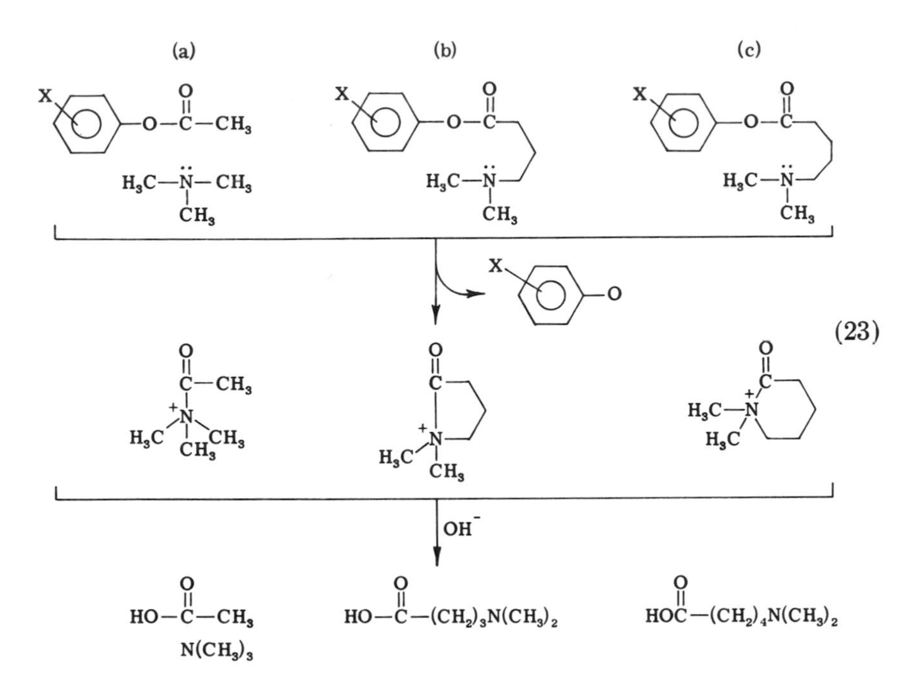

TABLE VI

ACTIVATION PARAMETERS (in kcal/mole) FOR NUCLEOPHILIC DISPLACEMENT BY THE DIMETHYLAMINO GROUP IN Eqs. (23) (25°)

Substituent	Eq. (23a)		Eq. (23b)		Eq. (23c)	
	$\Delta H^{\ddagger}$	$T\Delta S^{\ddagger}$	$\Delta H^{\ddagger}$	$T\Delta S^{\ddagger}$	$\Delta H^{\ddagger}$	$T\Delta S^{\ddagger}$
p-NO_2	12.3	−6.3	11.9	−1.9	11.5	−2.6
m-NO_2	12.1	−8.0	11.5	−4.3	11.8	−4.4
p-Cl	12.5	−9.1	15.9	−2.2	13.8	−4.1
H	12.9	−9.4	12.5	−5.7	12.3	−6.4
p-CH_3	—	—	13.7	−5.1	14.4	−5.5

TABLE VII

RATE RATIOS FOR (23a) AS COMPARED TO (23b) AND (23c)

Substituent	k_b/k_a (M)	k_c/k_a (M)
p-NO_2	5010	2330
m-NO_2	1690	746
p-Cl	1070	646
H	1260	490

molecular counterparts of (23b) and (23c) are recorded. From Table VI it is seen that the decrease in $\Delta F^{\ddagger}$ on converting the bimolecular reaction into an intramolecular reaction primarily results from a decrease in $T\Delta S^{\ddagger}$ (≈ -8 to -4 kcal mole^{-1}) rather than a decrease in $\Delta H^{\ddagger}$. This result taken with the insignificant difference in ρ values for Eqs. (23) support a common mechanism and suggests the rate ratios of Table VII are associated with a gain in translational entropy.

From Table I, the rate ratios of 4.9×10^2 to 5×10^3 moles/liter require that the amino nitrogen and ester group are not only juxtaposed but favorably oriented. From the rate constants of Table V it is obvious that a decrease from three to two single bonds allowing free rotation [(23c) vs. (23b)] is associated with a mere 2.3 ± 0.3-fold rate enhancement. This value is just one hundred times less than the ratio obtained for the glutarate and succinate monoesters (Table III). This difference has been suggested to result from the preferential conformation of the ground states for the dimethyl amines and carboxylic acids. Whereas all evidence indicates that the heavily solvated succinate and glutarate esters prefer an extended conformation (I), the relatively nonpolar tertiary amines have been suggested (*25*) to exist in a coiled conformation (VII) as a result of squeezing of the essentially lyophobic molecules

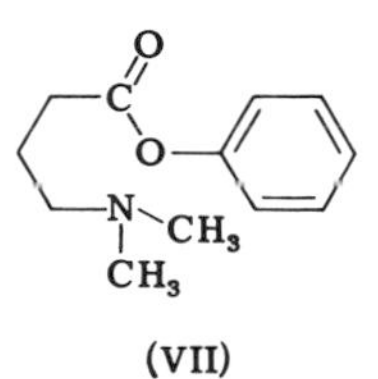

(VII)

by water. The ring closure reaction of (23c) and the glutarate esters both involve formation of a six-membered ring. However, in going from *p*-bromophenyl acetate plus acetate to *p*-bromophenyl succinate the rate ratio is about 1.4×10^5 moles/liter, whereas going from the *p*-chloroester of (23a) to the *p*-chloroester of (23c) the rate ratio is but about 6×10^2 moles/liter.

Numerical computations for the increase in rate constants on conversion of a bimolecular reaction to an intramolecular reaction are quite limited. This is so because (a) either the intramolecular displacement is too facile to record, (b) the bimolecular reaction is so slow that it is not measurable, or (c) the bimolecular reaction is so slow that it does not compete with reactions involving lyate species. Two additional comparisons are provided by Eq. (24) (*42*) and Eq. (25) (*43*).

42. T. C. Bruice and J. M. Sturtevant, *JACS* **81,** 2860 (1959).
43. T. C. Bruice, *JACS* **81,** 5444 (1959).

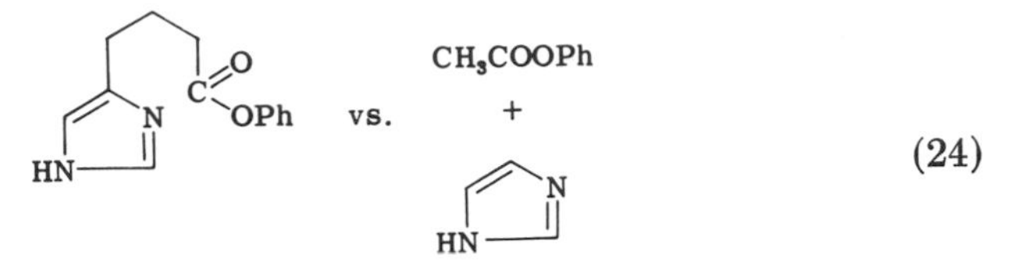

k_{intra}/k_{inter} = 73 moles/liter

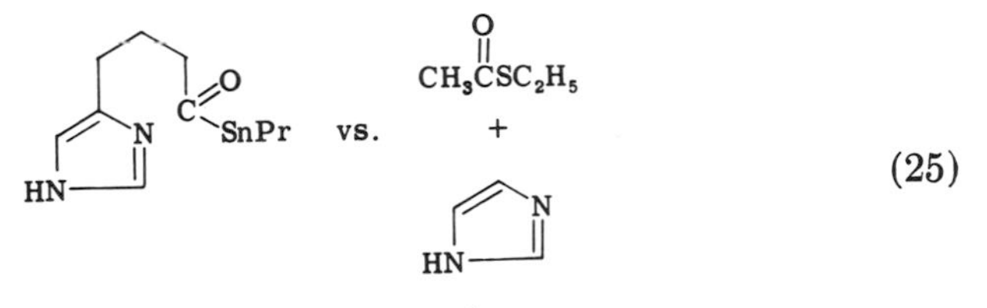

k_{intra}/k_{inter} = 53 moles/liter

IV. Miscellaneous Intramolecular Reactions

In this section a few additional examples of facile intramolecular reactions will be discussed. Unfortunately, no computation of the rate constant increase in going from intermolecular to the corresponding intramolecular reaction can be provided. This is so because the intermolecular reaction rates cannot be generally determined because of competition of lyate species for substrate. For these cases minimal rate ratios may be assigned on the basis of the determined intramolecular rate constant and the knowledge that the intermolecular rate constant must be no more than about 5% of the rate of reaction of substrate with lyate species. As stated previously, general surveys have been provided recently (*11*, *14*) and additional examples are provided in the excellent text of Jencks (*44*). No attempt is made here to duplicate the examples provided in these texts.

The anionic and neutral forms of amide groups are often found to be most effective intramolecular nucleophiles. The rate constant for (26) has been estimated to be minimally 60,000 times larger than the rate constant for the bimolecular reaction of an amide with phenyl acetate under the same conditions (*45*). Since the pK_a of the amide bond of (26)

44. W. P. Jencks, "Catalysis in Chemistry and Enzymology." McGraw-Hill, New York, 1969.
45. M. T. A. Behme and E. H. Cordes, *J. Org. Chem.* **29**, 1255 (1964).

(26)

$k_2 = k_{obsd}/[HO^-] = 1.26 \times 10^6$ liter mole^{-1} min^{-1}

cannot be determined, the rate constant (k_2) can only be expressed as the bimolecular specific base rate constant. The ureido functional group has received attention as a nucleophile because of its presence as imidazolone in biotin (VIII). On the basis that CO_2-biotin enzyme is

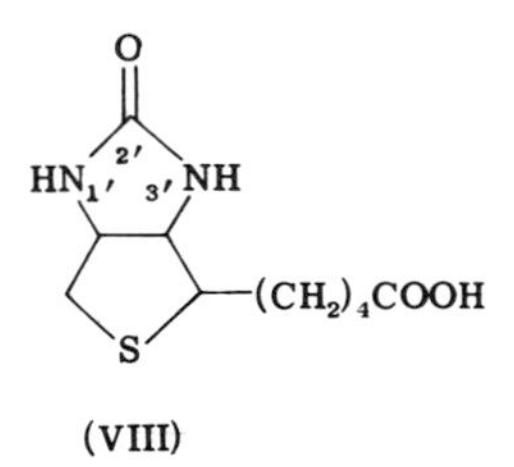

(VIII)

established to contain 1′-*N*-carboxybiotin (*46*), imidazolone has been examined as a base toward *p*-nitrophenyl acetate, acetyl imidazole, 1-acetyl-3-methyl imidazolium chloride, and carbon dioxide (*47*, *48*). No acetylation of imidazolone could be detected with the ester and *N*-acetyl imidazoles and carbon dioxide could not be established to react with imidazolone. In contrast, the ureido anion is a very effective intramolecular nucleophile (*49*). The fact that k_2 for (26) greatly exceeds the corresponding constant for (27), (28), and (29) is attributable to the poorer leaving groups in the latter cases. Reaction (29) is of particular interest since it apparently involves nucleophilic attack on carboxyl anion.

(27)

$k_2 = 5.65 \times 10^4$ liter mole^{-1} min^{-1}

46. J. Knappe, E. Ringelmann, and F. Lynen, *Biochem. Z.* **335**, 168 (1961).
47. M. Caplow, *JACS* **87**, 5774 (1965).
48. M. Caplow, *JACS* **90**, 6795 (1968).
49. A. F. Hegarty and T. C. Bruice, *JACS* **91**, 4924 (1969).

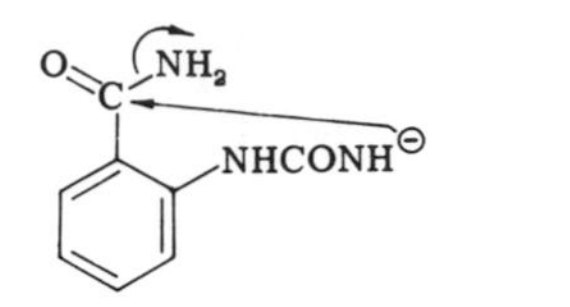

(28)

$k_2 = 7.25 \times 10^1$ liter mole^{-1} min^{-1}

(29)

$k_2 = 3.50 \times 10^{-2}$ liter mole^{-1} min^{-1}

Examples of nucleophilic displacement by neutral amide are found in the hydrolysis of *o*- and *p*-nitrophenyl 2-acetamide-2-deoxy-β-D-glucopyranosides (30) (*50*). The reaction is stereospecific as shown by

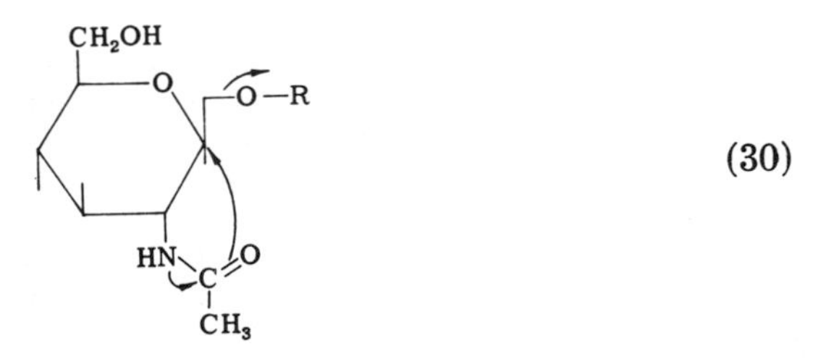

(30)

the fact that the rate of hydrolysis of the β isomer at neutrality is about 10^5 times greater than the rate of hydrolysis of the α isomer. If the leaving group possesses an *o*-carboxyl function the rate of hydrolysis is further increased presumably via (31) (*51*). The *o*-carboxyl group alone

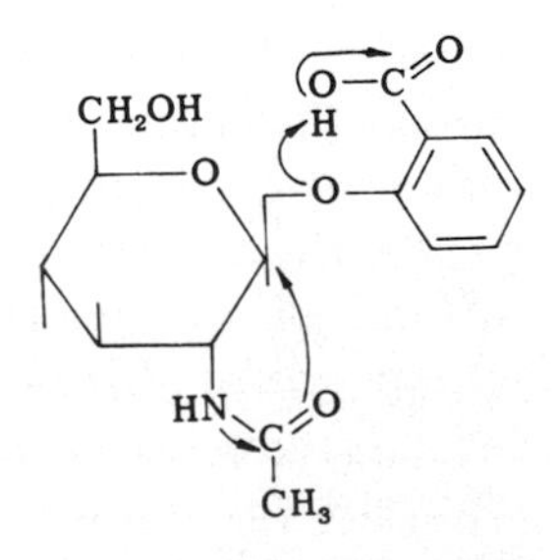

(31)

50. D. Piszkiewicz and T. C. Bruice, *JACS* **89**, 6237 (1967).
51. D. Piszkiewicz and T. C. Bruice, *JACS* **90**, 2156 (1968).

suffices as an effective intramolecular catalyst for the hydrolysis of phenyl glucosides (*51, 52*). These results are of interest from the standpoint of the mechanism of lysozyme action (*5*). When the aglycone possesses little steric requirements specific acid catalysis is assisted by the 2-acetamido group (32) (*53*). The bimolecular counterparts of reac-

(32)

tions (26) to (32) are unknown. In all these cases the propinquity effect on the rate constants must be minimally of the order of 10^4 to 10^6. For reactions (27) to (29) resonance stabilization of the transition state undoubtedly accounts for a portion of the rate enhancements. Reaction (27) closely resembles the intermediate step in a common synthetic procedure (*54*) for the preparation of pyrimidines (33) (*55*).

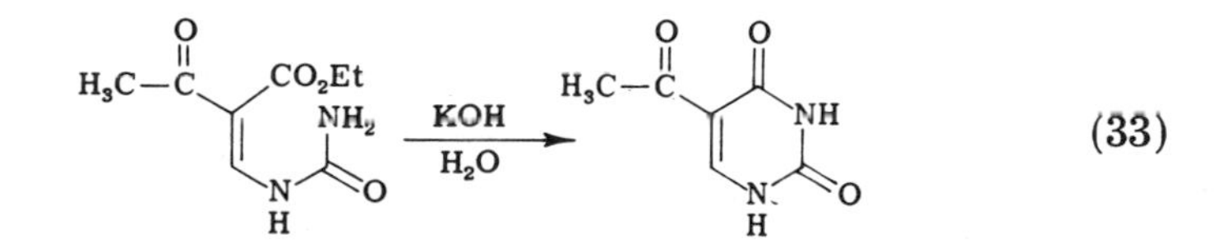

A catalytically inactive molecule may be converted into an efficient intramolecular catalyst by addition to substrate. Examples are found in the benzaldehyde and carbon dioxide catalysis of hydrolysis of nitrophenyl esters of α-amino acids (*56, 57*). These reactions resemble cases in which a normally unreactive neighboring carbonyl group is activated by addition of base (36) (*58–62*).

52. B. Capon, W. G. Overend, and M. Sobell, *JCS* p. 1881 (1961).
53. D. Piszkiewicz and T. C. Bruice, *JACS* **90,** 5844 (1968).
54. D. J. Brown and S. F. Mason, "The Pyrimidines," Chapter II. Wiley (Interscience), New York, 1962.
55. L. Claisen, *Ann. Chem.* **297,** 1 (1897).
56. B. Capon and R. Capon, *Chem. Commun.* p. 502 (1965).
57. R. W. Hay and L. Main, *Australian J. Chem.* **21,** 155 (1968).
58. M. L. Bender, J. A. Reinstein, M. S. Silver, and R. M. Kulak, *JACS* **87,** 4545 (1965).
59. M. S. Newman and S. Hishida, *JACS* **84,** 3582 (1962).

$$C_6H_5CHO + R{-}\underset{NH_2}{CH}{-}COOR' \underset{+H^{\oplus}}{\overset{H^{\oplus}}{\rightleftharpoons}} C_6H_5\underset{O^{\ominus}}{\overset{H}{C}}{-}NH{-}\overset{R}{CH}{-}\underset{OR}{C}{=}O \qquad (34)$$

$$CO_2 + R{-}\underset{NH_2}{CH}{-}COOR' \underset{+H^{\oplus}}{\overset{H^{\oplus}}{\rightleftharpoons}} \underset{O^{\ominus}}{\overset{O}{C}}{-}NH{-}\overset{R}{C}{-}H\,({-}\underset{OR}{C}{=}O) \qquad (35)$$

$$\begin{matrix} C{=}O \\ C{-}X \\ O \end{matrix} + B: \rightleftharpoons \begin{matrix} B \\ O^{\ominus} \\ C{-}X \\ O \end{matrix} \qquad (36)$$

The hydrolyses of monophenyl esters of dicarboxylic acids are known to occur by direct nucleophilic displacement (*15*) as in (17). For these cases the leaving phenoxide ion departs when the intermediate anhydride is formed. In water when the ester is at high dilution the back reaction of phenoxide with anhydride is unimportant (34). For cases in which the leaving group remains attached to the intermediate anhydride the back reaction to reform starting material (37) may be so effective that a

$$XC_6H_5O^{\ominus} + \begin{matrix} {-}C{=}O \\ O \\ {-}C{=}O \end{matrix} \longrightarrow \begin{matrix} {-}\overset{O}{C}{-}OC_6H_5X \\ {-}COO^{\ominus} \end{matrix} \qquad (37)$$

change in the hydrolytic mechanism occurs. A most interesting example is found in the hydrolysis of acetyl salicylates (IX) (*63–66*). On the basis of electronic effects of substituents in the 4 and 5 positions (*63*), lack of ^{18}O incorporation into the carboxyl group of salicylic acid product, values of $\Delta S^{\ddagger}$, and deuterium solvent isotope effects (*64*) the mechanism of hydrolysis was shown to be through Path A of (38) rather than

60. F. Ramirez, B. Hansen, and N. B. De Sai, *JACS* **84,** 4588 (1962).
61. Y. Shaltin and S. A. Bernhard, *JACS* **86,** 2292 (1964).
62. C. N. Lieke, E. G. Miller, Jr., J. J. Zeger, and G. M. Steinberg, *JACS* **88,** 188 (1966).
63. A. R. Fersht and A. J. Kirby, *JACS* **89,** 4853 (1967).
64. A. R. Fersht and A. J. Kirby, *JACS* **89,** 4857 (1967).
65. A. R. Fersht and A. J. Kirby, *JACS* **90,** 5818 (1968).
66. A. R. Fersht and A. J. Kirby, *JACS* **90,** 5826 (1968).

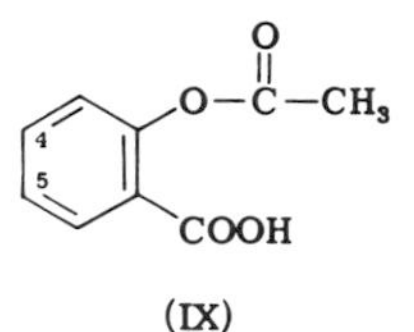

through the competing Path B. Increasing the leaving tendency of the phenolic moiety (as in the case of acetyl 3,5-dinitrosalicylate) was found to favor Path B (*65*). Thus, if the phenolic oxygen is a strong base K_e is far to the left and the mixed anhydride is of too low a concentration for Path B to compete with Path A. Only when the carboxyl and

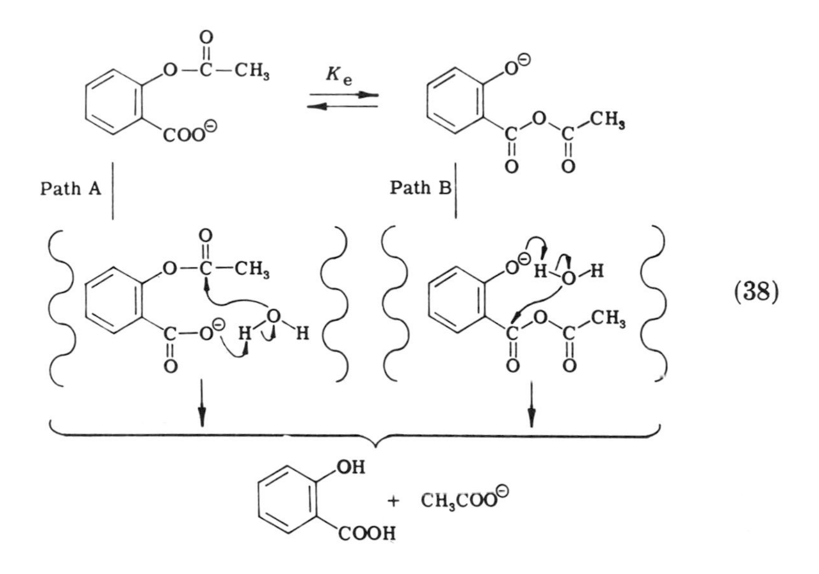

(38)

phenolic groups are of like basicity will Path B become evident. In studies with salicoyl salicylate labeled with ^{13}C it was shown that after partial hydrolysis the label is distributed between the two carboxyl groups (*67*). This establishes the equilibrium of (39). However, with

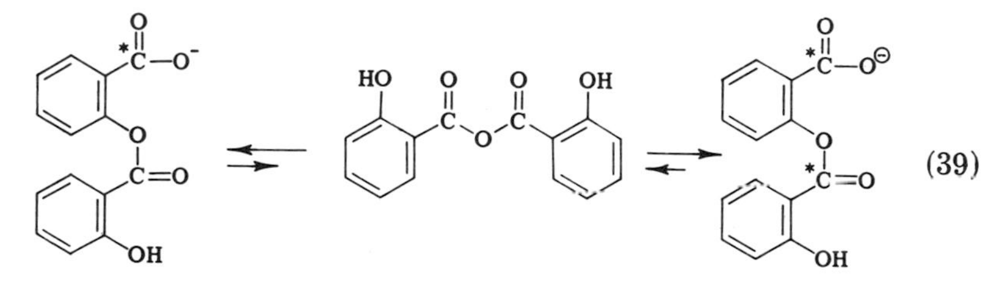

(39)

67. D. S. Kemp and T. D. Thibault, *JACS* **90,** 7154 (1968).

the powerful nucleophile, hydrazine, it could be shown that the mixed anhydride was at too low a concentration to account for acyl hydrazide production.

It is most interesting to note that the neutral unionized species of acetyl 3,5-dinitrosalicylic acid must hydrolyze via an internal catalytic process. This follows from the relative rate constants for structures (X) and (XI) (*66*). Experimental evidence supports the rate-determining step of hydrolysis to involve solvolysis of the mixed anhydride species of (XII). For acetyl salicylate, reaction of weak bases, but not stronger

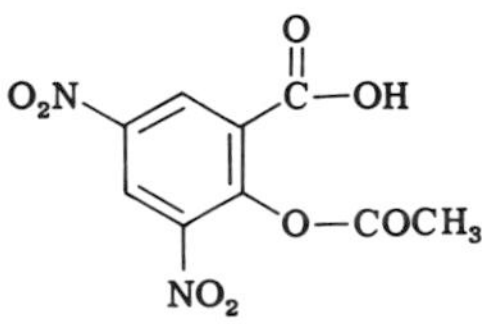

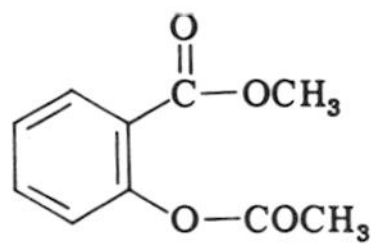

k_{rel} 270 1

(X) (XI)

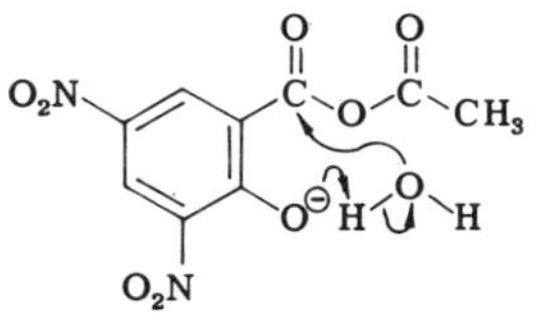

(XII)

bases as normal primary and secondary amines, has been shown to involve intramolecular participation of the dissociated and associated carboxyl group [(40) and (41)] (*68*).

The electron density at the 6 and 8 positions of quinoline are nearly identical (*69*). With this knowledge it would be anticipated that the second-order reaction of nucleophiles with 8- and 6-acetoxyquinoline

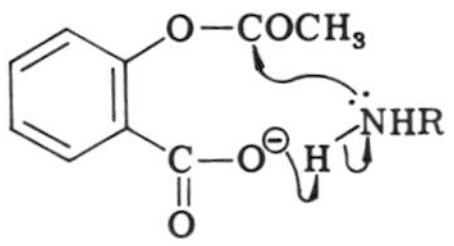

R = NHCONH$_2$ (40)

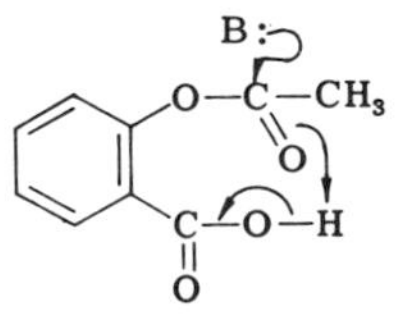

B: = Nicotinamide, $H_2NNHCONH_2$, NH_2OCH_3 (41)

68. T. St. Pierre and W. P. Jencks, *JACS* **90,** 3817 (1968).
69. H. C. Longuet-Higgins and C. A. Coulson, *Trans. Faraday Soc.* **43,** 87 (1947).

should be comparable. This is so when the nucleophiles are hydroxide and tertiary amines but not so if the nucleophiles are water, primary or secondary amines (*70*). The division of the nucleophiles into two separate groups may be explained on the basis that water, primary and secondary amines, which possess acidic hydrogens, react with the 8-isomer via intramolecular general-base catalysis (42) whereas hydroxide

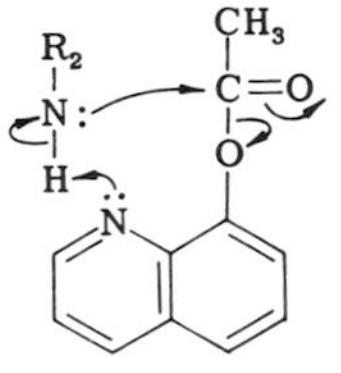

(42)

ion and tertiary amines, which do not possess an acidic hydrogen react with the 8-isomer via unassisted nucleophilic attack. For the 6-isomer, intramolecular catalysis is not possible (43).

(43)

Concerted bifunctional catalysis has always been favored as an explanation of enzymic mechanisms. A nonconcerted bifunctional catalysis is seen in the hydrolysis of 3-acetoxy phthalic acid (44) (*71*). This type of process has been appropriately termed *Series Nucleophilic Catalysis*

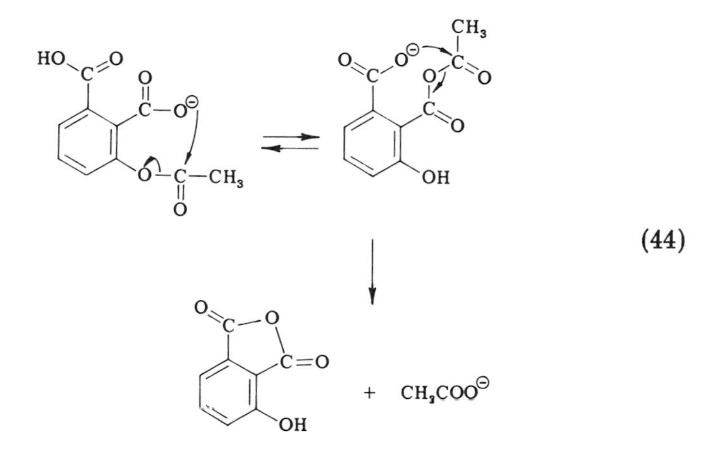

(44)

70. T. T. Bruice and S. M. Felton, *JACS* **91**, 2799 (1969).
71. A. R. Fersht and A. J. Kirby, *JACS* **90**, 5833 (1968).

(*71*). Processes of this type may account for the bell-shaped pH-rate profiles for certain enzymes. The usual interpretation of the bell-shaped profiles involves concerted acid-base catalysis. An appropriate example is the hydrolysis of salicoyl succinate. The pH-rate profile for this hydrolytic reaction was considered to result from a concerted nucleophilic general-acid catalysis (45) (*72*). An alternate mechanism would

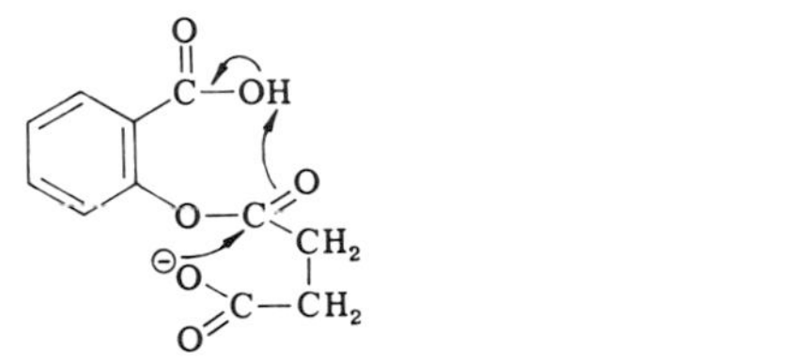

(45)

involve series nucleophilic catalysis (46) (*71*). Other examples of mechanisms that may lead to a bell-shaped pH profile are discussed elsewhere (*11*, *73*).

Until recently the α-pyridone and carboxylic acid catalysis of the mutarotation of tetramethyl glucose (*74*, *75*) has virtually stood alone

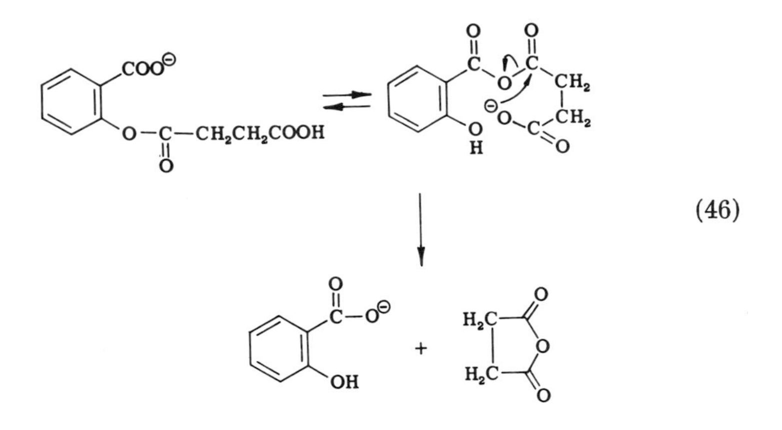

(46)

as an assured example of push–pull catalysis involving a cyclic proton transfer. As a model for an enzymic process this catalysis has received criticism, since its operation is restricted to nonpolar solvents (benzene). Recently, examples of catalysis involving cyclic proton transfer in water have been established.

72. H. Morawetz and I. Oreskes, *JACS* **80,** 2591 (1958).
73. B. Zerner and M. L. Bender, *JACS* **83,** 2267 (1961).
74. C. G. Swain and J. F. Brown, *JACS* **74,** 2534 (1952).
75. P. R. Rony, *JACS* **90,** 2824 (1968).

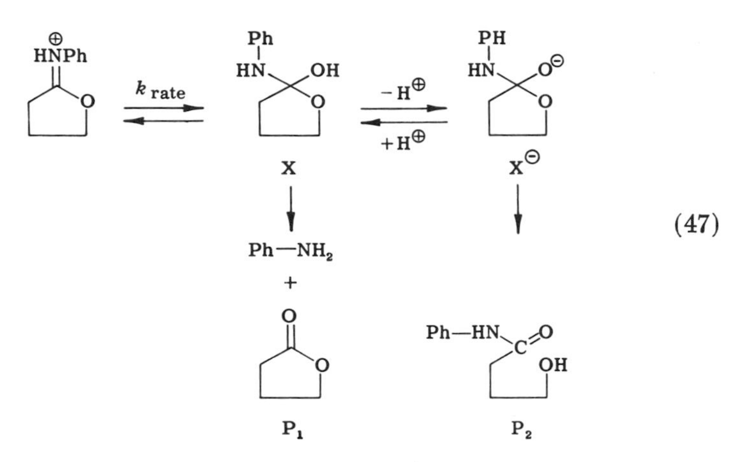

The buffer species $H_2PO_4^-$, HCO_3^-, and CH_3COOH have no effect on the rate of hydrolysis of *N*-phenyliminotetrahydrofuran at low concentrations, but the type of product obtained is dependent upon their concentration (*76*, *77*). The kinetic scheme of (47) was suggested to account for these results. The aforementioned buffer species increase the rate of

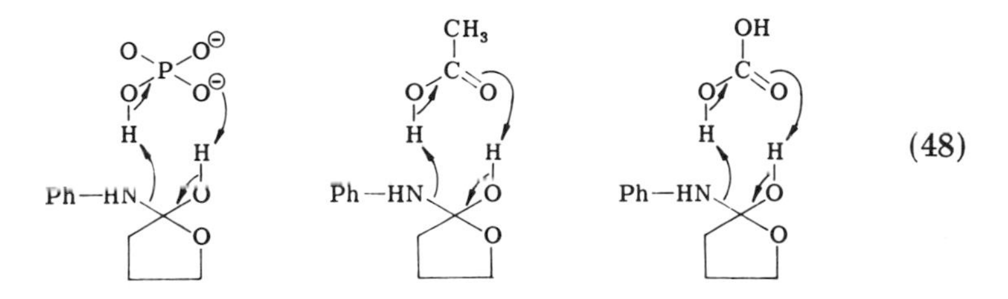

conversion of species X to P_1 by the suggested mechanism of (48). Other reactions which are candidates for cyclic catalytic processes involving proton transfer include: (a) formation of α-pyrrolidone-carboxylic acid from glutamine (49) (*78*) or glutamic acid esters (50) (*79*); (b) amino-

76. B. A. Cunningham and G. L. Schmir, *JACS* **88**, 551 (1966).
77. B. A. Cunningham and G. L. Schmir, *JACS* **89**, 917 (1967).
78. J. B. Gilbert, V. E. Price, and J. P. Greenstein, *JBC* **180**, 209 (1949).
79. A. J. Hubert, R. Buyle, and B. Hargitay, *Helv. Chim. Acta* **46**, 1429 (1963).

(50)

diacylhydrazine rearrangements (which require aprotic solvents but proceed with carboxylic acids, etc., as catalysts) (51) (*80*); and (c)

(51)

hydrolysis of trifluoroacetanilide (52) (*81*). Decomposition of peroxide

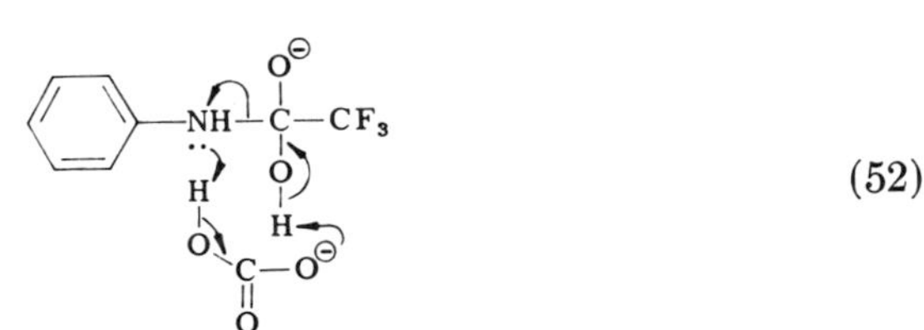

(52)

hemiacetals by carboxylic acids has been suggested to occur through the mechanisms of (53) (*82*). In (53) RCOOH is acting as a specific

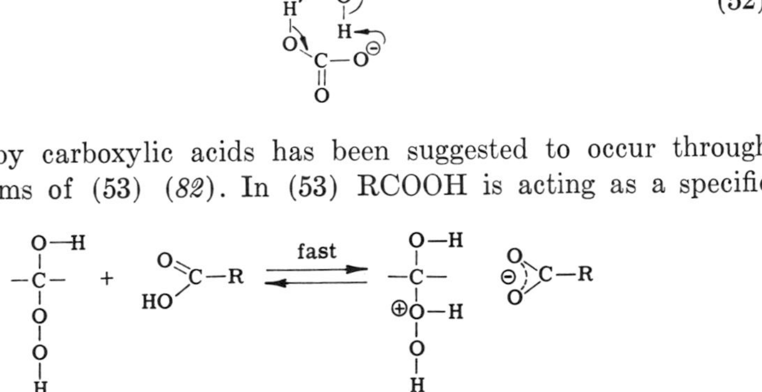

(53)

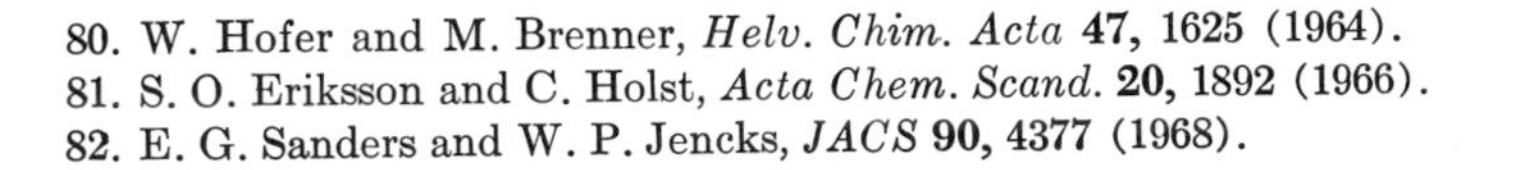

80. W. Hofer and M. Brenner, *Helv. Chim. Acta* **47,** 1625 (1964).
81. S. O. Eriksson and C. Holst, *Acta Chem. Scand.* **20,** 1892 (1966).
82. E. G. Sanders and W. P. Jencks, *JACS* **90,** 4377 (1968).

acid catalyst (Brønsted $\beta = 1.0$). This very novel concept is quite reasonable on the basis of the formation of a complex whose rate of decomposition is very rapid and, in addition, a marked deviation of hydronium

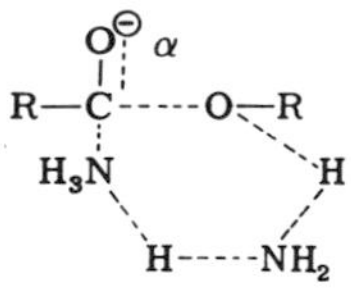

$$\begin{array}{c} O^{\ominus}\ \alpha \\ R\text{—}C\text{-----}O\text{—}R \\ H_3N \qquad\qquad H \\ H\text{-----}NH_2 \end{array} \qquad (54)$$

ion from the Brønsted plot. The cyclic mechanism of (54) has been suggested to account for the sensitivity of the rate constant for amine general-base catalyzed aminolysis to electronic effects on both acyl and phenol portions of phenyl acetates and phenyl benzoates (*18*, *83*). The mechanism of (54) is not unique, however, since alternative transition state structures can also account for the experimental results (*84*).

Cyclic mechanisms for catalysis of concerted bond making and breaking processes which involve proton transfer through hydrogen bonds could well be of importance in enzymic reactions. A large number of enzyme-catalyzed reactions involve either a rearrangement of covalent bonds accompanied by a prototropic shift or general-base assistance to attack of water or a hydroxyl nucleophile. The rate constant for dissociation of an acid is approximately $10^{10}\ K_a$ and the rate constant for protonation of a basic site by water is approximately $10^{10}\ K_w/K_a$, where K_a is the ionization constant of the conjugate acid of the basic site and K_w the autoprotolysis constant of water. It follows that at 25° the optimal rate constant for proton transfer is about $10^3\ \text{sec}^{-1}$ if the pK_a of the basic site is 7. Increase of pK_a above 7 is accompanied by an increase in the protonation rate constant but a decrease in the deprotonation rate constant; decrease in pK_a is accompanied by just the opposite. However, the rate of proton transfer in an enzymic reaction need not be limited to $10^3\ \text{sec}^{-1}$ if transfer occurs along preformed hydrogen bonds in an intramolecular fashion not involving water (*85*, *86*). The jumping frequency of an excess proton along the hydrogen bonds in ice at —10°C (*87*) and the rate constant for intramolecular proton transfer in a hydrogen-bonded network when the proton acceptor is a weaker acid than the donor (*88*) are both about $10^{13}\ \text{sec}^{-1}$. The rate of proton transfer to a basic

83. J. F. Kirsch and A. Kline, *JACS* **91,** 1841 (1969).
84. T. C. Bruice, A. F. Hegarty, S. M. Felton, A. Donzel, and N. Kundu, *JACS* **92,** 1370 (1970).
85. G. G. Hammes, *Accounts Chem. Res.* **1,** 321 (1968).
86. J. H. Wang and L. Parker, *Proc. Natl. Acad. Sci. U. S.* **58,** 2451 (1967).
87. M. Eigen and L. De Maeyer, *Proc. Roy. Soc.* **A247,** 505 (1958).
88. M. Eigen, *Angew. Chem.* **3,** 1 (1964).

site via a direct hydrogen bond may then be of the order of magnitude of 10^4–10^6 sec^{-1}. Facilitation of proton transfer along rigidly held hydrogen bonds as a crucial factor in the determination of efficiency and specificity of several specific enzymes has been considered (*89*).

V. Activation Parameters and Kinetic Order

Conversion of a bimolecular reaction (1) into an intramolecular reaction (4) amounts to a reduction in kinetic order by one. The propinquity effect may then be considered as a reduction of kinetic order without changing the mechanism of reaction. If evaluation of the contribution of the kinetic energy of activation to the free energy of activation could be made, then a means of evaluating the contribution of the propinquity effect would be obtained. In practice this is often not possible. To only a first approximation can kinetic and potential energies of activation be associated solely with experimentally determined $\Delta S^{\ddagger}$ and $\Delta H^{\ddagger}$, respectively (*90*). More importantly in converting (1) to (4), the means of bringing about the approximation must be considered. If approximation is brought about by covalently linking A and B, then nonbonded repulsions in the molecule must be considered. These effects will be evident in both $\Delta S^{\ddagger}$ and $\Delta H^{\ddagger}$ (*90*). Rate enhancement owing to steric compression of A and B species may actually result from a greater decrease in $\Delta H^{\ddagger}$ than increase in $\Delta S^{\ddagger}$. The rate enhancements associated with steric compression in the models of (21) (*37*), (22) (*38*), and (V) (*39*) are primarily because of favorable changes in $\Delta H^{\ddagger}$ rather than $T\Delta S^{\ddagger}$. The position of the hexane–cyclohexane equilibria in the gas phase approaches cyclohexane upon alkyl substitution. In this case ring formation is favored through both entropy and enthalpy effects. This observation has been attributed to a lessening of *gauche* interactions on ring formation (*91*). In contrast, the effect of alkyl substitution on the rate of cyclic anhydride formation from glutarate monoesters (Table IV) (*15*) is associated primarily with a favorable change in $T\Delta S^{\ddagger}$.

Since formation of a ground state intimate AB_{pair} by covalent approximation of A and B requires reorganization of solvent which should be reflected in both $\Delta H^{\ddagger}$ and $T\Delta S^{\ddagger}$, it is likely that no rigorous treatment of propinquity in terms of $T\Delta S^{\ddagger}$ is possible. An empirical approach has,

89. J. H. Wang, *Science* **161**, 328 (1968).
90. R. W. Taft, *in* "Steric Effects in Organic Chemistry" (M. S. Newman, ed.), p. 556. Wiley, New York, 1956.
91. N. L. Allinger and V. Zalkow, *J. Org. Chem.* **25**, 701 (1960).

TABLE VIII

A COMPARISON OF THE VALUES OF $T\Delta S^{\ddagger}$ (kcal $mole^{-1}$)/(4 kcal $mole^{-1}$) TO THE KINETIC ORDER OF DISPLACEMENT REACTIONS ON THE PHENYL AND THIOLESTER BONDS

Reactants	Kinetic order	$\frac{-T\Delta S^{\ddagger a}}{\text{kinetic order}}$	n	Ref.
$RR'C(COO–C_6H_4–Br)(CH_2COO^{\ominus})$	1	4–0.6	10	*15*
$(CH_2)_3$: –COO–C_6H_4–X, –$COO^{\ominus}$	1	4	3	*92*
$(CH_2)_2$: –COO–C_6H_4–X, –$COO^{\ominus}$	1	3	1	*92*
$C_6H_4(COOCH_3)(COOH)$	1	5	1	*16*
–COO–C_6H_4–X, –$N(CH_3)_2$	1	5	5	*25*
–COO–C_6H_4–X, –$N(CH_3)_2$	1	4	5	*25*
8-OAc-quinoline + H_2O	2	4	1	*70*
C_6H_4(OAc)(imidazolyl, N, NH) + H_2O	2	4	1	*70*
$C_6H_4(OAc)(COO^{\ominus})$ + H_2O	2	3	1	*64*
$C_6H_4(OAc)(COO^{\ominus})$ + $AcO^{\ominus}$	2	5	1	*64*

TABLE VIII (*Continued*)

Reactants	Kinetic order	$\frac{-T\Delta S^{\ddagger a}}{\text{kinetic order}}$	n	Ref.
$X-C_6H_4-OAc + (CH_3)_3N$	2	4	4	*25*
$X-C_6H_4-OAc + HO^{\ominus}$	2	3	4	*25*
$X-C_6H_4-OAc + (NH_2)_2$	2	4	4	*19*
$X-C_6H_4-OAc$ + Imidazole	2	6–7	6	*19*
$X-C_6H_4-OAc + 2\,(H_2N)_2$	3	4–5	3	*19*
$X-C_6H_4-OAc + (NH_2)_2 + NH_2NH_3^{\oplus}$	3	6	3	*19*
$X-C_6H_4-OAc$ + 2 Imidazole	3	5	1	*19*
thiolactone (six-membered ring, C=O, S) $+ 2\,(H_2N)_2$	3	5	1	*93*
thiolactone (six-membered ring, C=O, S) $+ (NH_2)_2 + NH_2NH_3^{\oplus}$	3	5	1	*93*
thiolactone (six-membered ring, C=O, S) + 2 morpholine (NH, O ring)	3	5	1	*93*
thiolactone (six-membered ring, C=O, S) + morpholine (NH, O ring) + morpholinium ($^{+}NH_2$, O ring)	3	5	1	*93*

92. F. Gaetjens and H. Morawetz, *JACS* **82,** 5328 (1960).
93. T. C. Bruice and L. R. Fedor, *JACS* **86,** 4117 (1964).

however, been applied with some degree of success. If propinquity is considered to be a change in kinetic order, then comparison of a number of reactions of varying order might be expected to provide some overall trend in the values of $T\Delta S^{\ddagger}$. This is found to be the case when a number of displacement reactions on phenyl and thiol esters are examined (55) (*19*). The equation of (55) is completely empirical and there is no

$$\frac{-(\text{experimentally determined } T\Delta S^{\ddagger})}{4\text{–}5 \text{ kcal mole}^{-1}} = \text{kinetic order} \qquad (55)$$

theoretical rationale for its application, but the approximation holds to a reasonable degree for a series of seventeen different types of displacement reactions on phenyl esters as well as for four on thiol esters. The pertinent data are presented in Table VIII, in which the first column gives the reactants, the second the kinetic order, the third the value of $T\Delta S^{\ddagger}$ (kcal mole^{-1})/(kinetic order), and the fourth (n) the number of substituted esters studied for each reaction in order to obtain the average $T\Delta S^{\ddagger}$ value. Examination of Table VIII reveals that the value of $(-T\Delta S^{\ddagger})$/(kinetic order) is generally between 3 and 7 kcal mole^{-1} with the average value being 4.4 ± 0.8 kcal mole^{-1}. At 25° a value of 4.4 kcal mole^{-1} would amount to a rate decrease of 1.7×10^3 for each additional species incorporated into the transition state. Thus, on the basis of the experimental data of Table VIII, approximation and orientation of A and B should lead to a rate increase of about 10^3 if $T\Delta S^{\ddagger}$ is assumed to be the only index of these factors. This value agrees with the rate enhancement anticipated to accompany approximation and orientation based on other reasoning (Table I). Most surprising is the fact that for the second-order reactions of Table VIII in which the base is a water molecule $(-T\Delta S^{\ddagger})$/(kinetic order) $= 3.7 \pm 0.4$. Thus, though the values of $T\Delta S^{\ddagger}$ of fusion of nonpolar molecules at 25° are typically around 3–4 kcal mole^{-1}, that for water is only $\approx$1.6 kcal mole^{-1} (*94*). This result suggests that perhaps more than one water molecule must be included in the transition states for reactions such as (38), (40), and (42) where water replaces amine.

Since entropy and enthalpy are not completely independent parameters (*90, 95*), a change in one is often reflected in a compensatory change in the other. The ability to determine the rate constants for the third-order reactions of Table VIII (i.e., $k_{gb}[B]^2[E]$ or $k_{ga}[B][BH^+][E]$) which are in competition with second-order reactions involving the same species (i.e., $k_n[B][E]$) results from compensatory changes in the enthalpy and

94. L. L. Schaleger and F. A. Long, *Advan. Phys. Org. Chem.* **1,** 1 (1963).

95. J. E. Leffler and E. Grunwald, "Rates and Equilibria of Organic Reactions," Chapter 9. Wiley, New York, 1963.

entropy of activation. For these systems an increase in kinetic order by one decreases $T\Delta S^{\ddagger}$ by 5 kcal mole^{-1}; but, since the increase in order results from the incorporation of a general-base or general-acid catalytic entity, $\Delta H^{\ddagger}$ is also decreased and by about the same amount (*19*). The values of $\Delta H^{\ddagger}$ for the term $k_{ga}[B][BH^+]$ [thiolactone] and $k_{gb}[B]^2$ [thiolactone], where B and BH^+ represent hydrazine and its conjugate acid, are but 0.6 and —0.4 kcal mole^{-1} while for the case when B and BH^+ represent morpholine and its conjugate acid, the values are 5.3 and 3.7 kcal mole^{-1} (*93*). These values of $\Delta H^{\ddagger}$ approach or are smaller than that for diffusion in water (3–4 kcal mole^{-1}); indeed for the hydrazinolysis reaction $\Delta F^{\ddagger} = T\Delta S^{\ddagger}$ within experimental error.

VI. Approximation through Noncovalent Forces

A. Hydrogen Bonds, London, Debye, and Charge–Transfer, etc., Forces

The formation of complexes of simple organic molecules in water is now a well-established phenomenon (*96*). The free energy of complex formation may be attributed to a summation of factors, the magnitude of the contributions are not, at present, quantitatively definable [a particularly informative discussion of theses factors is provided in Jencks (*44*)]. Hydrogen bonding, in the absence of other attractive forces, may generally be neglected as a factor in complex formation in aqueous solution. This is because of the great tendency of water to act as both a hydrogen acceptor and donor. Indeed, the establishment of stable hydrogen bonds involving but two substituent groups is generally restricted to intramolecular systems in which donor and acceptor are juxtaposed via covalent linkage. A case in point are the dicarboxylic acids. Hydrogen bonding between undissociated carboxylic acids has been reported (*97*); however, the formation constants for such complexes are markedly dependent upon the length of the alkyl chain, suggesting hydrophobic bonding as the chief contributor to complex formation (*98*). As the carboxyl groups in dicarboxylic acids are brought into closer proximity, the value of pK_{a_1} decreases while that of pK_{a_2} increases. This increase in ΔpK_a may be attributed to both electrostatic perturbations (*99, 100*)

96. T. Higuchi and K. A. Connors, *Advan. Anal. Chem. Instr.* **4,** 117 (1965).
97. D. L. Martin and F. J. C. Rossotti, *Proc. Chem. Soc.* p. 60 (1959).
98. E. E. Schrier, M. Pottle, and H. A. Scheraga, *JACS* **86,** 3444 (1964).
99. J. G. Kirkwood and F. H. Westheimer, *J. Chem. Phys.* **6,** 506 and 513 (1938).
100. F. H. Westheimer and M. Shookhoff, *JACS* **61,** 555 (1939).

and intramolecular hydrogen bonding (*101*). It is probable that hydrogen bonding does not become a significant factor in determining ΔpK_a until this value becomes as large as ≈ 4 (*101, 102–107*). The most convincing evidence for intramolecular hydrogen bonding in carboxylic acids arises from measurements of the rates of proton removal by hydroxide ion. The rate constants for reaction of acetic and benzoic acid with hydroxide ion are 4.5×10^{10} and 3.5×10^{10} M^{-1} sec^{-1} and are, therefore, diffusion controlled. In the case of intramolecularly hydrogen-bonded maleate and di-*n*-propylmalonate ions, the corresponding rate constants are 5.3×10^7 and 7.4×10^8 M^{-1} sec^{-1} (*108*). Internal hydrogen bonding is also responsible for a 10^{-3} to 10^{-4} fold effect on the rate of proton removal by hydroxide ion in the case of salicylic acid (*108*) and *rac*-α,α'-di-*t*-butylsuccinic acid monoanion (*109*).

Because of the water solvation of formal positive or negative charges, formal charge–charge electrostatic attraction is unlikely to contribute significantly to the formation of complexes of small molecules in aqueous solution. The measurement of formation constants is not as sensitive a tool as kinetic determinations since the latter can be employed even if the complex is at steady state. There has been considerable interest in the role of formal charge–charge interactions in the determination of the magnitude of bimolecular rate constants. At present no certain conclusions can be reached. Thus, interpretations of experimental results have suggested the importance of formal charge–charge attraction in nucleophilic displacements on the substrates 1-acetoxy-4-methoxypyridinium ion (*110*), *N*-acetylimidazolium ion (*111*), *N*-*trans*-cinnamoyl-*N*-methylimidazolium ion (*112*), *N*-acetyl-*N*-methylimidazolium ion (*113*), and acetyl phenyl phosphate (*114*). However, no charge–charge rate enhancement could be noted for nucleophilic attack upon 1-(*N*,*N*-

101. F. H. Westheimer and O. T. Benfey, *JACS* **78,** 5309 (1956).
102. L. Eberson, *Acta Chem. Scand.* **13,** 211 (1959).
103. M. Shahal, *Acta Cryst.* **5,** 765 (1952).
104. R. E. Dodd, K. E. Miller, and W. F. K. Wynne-Jones, *JCS* p. 2790 (1961).
105. L. Eberson, *Acta Chem. Scand.* **17,** 1552 (1963).
106. D. Chapman, D. R. Lloyd, and R. H. Prince, *JCS* p. 550 (1964).
107. J. L. Haslam, E. M. Eyring, W. W. Epstein, G. A. Christiansen, and M. H. Miles, *JACS* **87,** 1 (1965).
108. M. Eigen and W. Kruse, *Z. Naturforsch.* **18b,** 857 (1963).
109. J. L. Haslam, E. M. Eyring, W. W. Epstein, K. P. Jensen, and C. W. Jaget, *JACS* **87,** 4247 (1965).
110. W. P. Jencks and M. Gilchrist, *JACS* **90,** 2622 (1968).
111. W. P. Jencks and J. Carriuolo, *JACS* **82,** 1778 (1960).
112. S. L. Johnson, *Advan. Phys. Org. Chem.* **5,** 237 (1967).
113. R. Wolfenden and W. P. Jencks, *JACS* **83,** 4390 (1961).
114. G. Di Sabato and W. P. Jencks, *JACS* **83,** 4393 (1961).

dimethylcarbamoyl) pyridinium ion (*115*) and the anion of *o*-nitrophenyl hydrogen oxalate (*116*). For α-substituted *o*-nitrophenyl acetates both charge–charge attraction and repulsion were found to be of importance in the determination of the rate of nucleophilic attack. In this instance, however, it could be shown rather conclusively that the charge–charge effect was a transition state rather than a ground state phenomenon (*117, 117a*). The differentiation between formal charge–charge and ion–dipole or dipole–dipole interaction (*118*) is dependent upon the linear free energy plotting technique employed to analyze the data (*117a*).

The mechanism for lysozyme action, deduced from X-ray data and model building (*5*), provides for general-acid catalysis of glycosidic bond hydrolysis by Glu 35 and electrostatic stabilization of incipient oxocarbonium ion by the anion of Asp 52 (56). It is most interesting to note

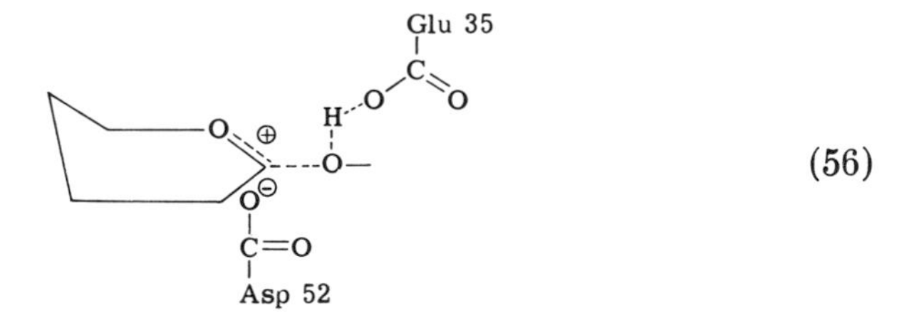

(56)

that in the carboxyl group intramolecular general-acid catalyzed hydrolysis of (VI), electrostatic facilitation is not seen when —X = —COO^- (*41*). The rather unperturbed nature of the pK_{app} of Asp 52, as obtained kinetically with low molecular weight substrates, indicates that its surrounding milieu is not unlike water (*119*). The inability to detect formal charge–charge stabilization of the transition state for the intramolecular general-acid catalysis of (VI) when —X = —COO^- casts doubt upon the role of Asp 52 in the mechanism of lysozyme action.

It is likely that any strong interaction of an ion with a protein or polymer does not arise from simple ion pair formation. Charge–charge interaction is far too weak in aqueous solution to explain, for instance, the free energies of binding of —5 to —7 kcal $mole^{-1}$ for chloride, thiocyanate, and trichloroacetate ions to the high-affinity site of bovine

115. S. L. Johnson and K. Rumon, *JACS* **87,** 4782 (1965).
116. T. C. Bruice and B. Holmquist, *JACS* **89,** 4028 (1967).
117. B. Holmquist and T. C. Bruice, *JACS* **91,** 2982 (1969).
117a. B. Holmquist and T. C. Bruice, *JACS* **91,** 2985 (1969).
118. K. Koehler, R. Skora, and E. H. Cordes, *JACS* **88,** 3577 (1966).
119. K. Meyer, J. W. Palmer, R. Thompson, and D. Khorazo, *JBC* **113,** 479 (1936); J. A. Rupley, *Proc. Roy Soc.* **B167,** 416 (1967); M. R. J. Salton, *BBA* **22,** 495 (1956); J. B. Howard and A. N. Glazer, *JBC* **244,** 1399 (1969).

serum mercaptalbumin (*120, 121*). Strong ion–protein interactions most likely involve sites with multiple charge, additional binding forces such as lyophobic attractions, or charged sites on the protein fixed in a poor ion-solvating environment (*44*). The binding of sulfate to ribonuclease S at low pH involves ionic attraction from the three cationic species of protonated His 12, His 119, and Lys 41 (*122*). The free energy of binding of the quaternary ammonium ion of acetyl choline and various quaternary ammonium ions with the anionic binding site of acetyl choline esterase (*123*) undoubtedly involves lyophobic attraction of the methyl groups of the quaternary amino group. It is known for example that quaternary amines salt in neutral organic molecules whereas cations such as K^+ and Na^+ salt out such species (*124–126*). Also, small inorganic ions are replaced by quaternary ammonium cations in phenolic and polystyrene types of exchange resins (*127, 128*).

The catalysis of hydrolysis of proteins and peptides by sulfonated polystyrene (Dowex-50) has been shown to involve both binding of protonated free amino groups to the polyanion and concentration of hydrogen ion at the polymer surface (*129–132*). Substrate binding to polymer also appears important in the reaction of poly-(4-vinylpyridine) and 3-nitro-4-acetoxybenzene-sulfonate anion (*133*). In contrast to the reference reaction of the sulfonic acid ester with 4-picoline, the reaction with poly-(4-vinylpyridine) exhibits a bell-shaped dependence on the fraction of nitrogen as free base. The bell-shaped curve was rationalized by assuming that protonated nitrogen sites serve to attract or bind the negatively charged substrate to the polymer, where basic nitrogen sites can then serve as catalysts for ester hydrolysis. The behavior of poly-4(5)-vinylimidazole with the same ester is similar (*134*). The reaction

120. G. Scatchard and E. S. Black, *J. Phys. Chem.* **53,** 88 (1949).
121. G. Scatchard, Y. V. Wu, and A. L. Shen, *JACS* **81,** 6104 (1959).
122. H. W. Wyckoff, D. Tsernoglou, A. W. Hanson, J. R. Knox, B. Lee, and F. M. Richards, *JBC* **245,** 305–328 (1970).
123. I. B. Wilson, "The Enzymes," 2nd ed., Vol. 4, p. 501, 1960.
124. F. A. Long and W. F. McDevit, *Chem. Rev.* **51,** 119 (1952).
125. D. R. Robinson and W. P. Jencks, *JACS* **87,** 2470 (1965).
126. H. S. Franck, *Discussions Faraday Soc.* **24,** 133 (1957).
127. T. R. E. Kressman and J. A. Kitchener, *JCS* p. 1190 (1949).
128. H. P. Gregor and J. I. Bregman, *J. Colloid Sci.* **6,** 323 (1951).
129. G. E. Underwood and F. E. Deathrage, *Science* **115,** 95 (1952).
130. J. C. Paulson and F. E. Deathrage, *JBC* **205,** 909 (1953).
131. J. C. Paulson, F. E. Deathrage, and E. F. Almy, *JACS* **75,** 2039 (1953).
132. J. R. Whitaker and F. E. Deathrage, *JACS* **77,** 3360 and 5298 (1955).
133. R. L. Letsinger and T. J. Savereide, *JACS* **84,** 3122 (1962).
134. C. G. Overberger, T. St. Pierre, N. Vorchheimer, and S. Yaroslavsky, *JACS* **85,** 3513 (1963).

of partially protonated poly(*N*-vinylimidazole) with copoly-(acrylic acid-2,4-dinitrophenyl-*p*-vinylbenzoate) follows Michaelis–Menten type of kinetics (*135*). The preequilibrium complex thus formed is presumably held together by the multiple cationic and anionic charges on the substrate and catalytic polymer. The rate of hydrolysis of starch substituted with protonated diethylamino-ethyl groups is over 500 times greater in the presence of polystyrenesulfonic acid than in a hydrochloric acid solution of like acidity (*136*).

Permanent dipole-induced dipole interactions [Debye force (*137*)] are usually very small in the interaction between neutral molecules. Thus, the complexing of purines and pyrimidines with 3,4-benzopyrene in water exhibits no correlation with the dipole moments of the bases (*138*, *139*). This type of force can probably be neglected as an effective contributor to complex formation in aqueous systems.

Dispersion or London forces arise from attraction of induced fluctuating dipoles. This type of interaction is proportional to the polarizability of the interacting molecules and is inversely dependent upon the sixth power of the intermolecular separation (*140*). The extent of attraction via dispersion forces apparently depends on the size of the interacting surfaces. Though π electrons have exalted polarizability, a molecule containing a π system is no more polarizable than a derived molecule obtained by hydrogenation of the π system. This results from a compensating increase in polarizability owing to the added hydrogen atoms (*44*). Thus, the interaction energies of purines with saturated and unsaturated compounds are similar (*141–143*) and the boiling points of benzene and cyclohexane differ by but 1.3°. In a complex of molecules A and B, dispersion forces will add to the binding energy. This stabilizing influence is offset, however, by a loss of dispersion interaction of monomeric A and B with solvent. The resultant of these two compensatory factors determines the contribution of London forces to the stability of the complex. Divergent points of view exist as to whether the resultant of these com-

135. R. L. Letsinger and I. Klaus, *JACS* **86**, 3884 (1964).
136. T. J. Painter, *JCS* p. 3932 (1962).
137. P. Debye, *Physik. Z.* **21**, 178 (1920).
138. H. Weil-Malherbe, *BJ* **40**, 351 (1964).
139. B. Pullman, *in* "Molecular Biophysics" (B. Pullman and M. Weissbluth, eds.), pp. 117–189. Academic Press, New York, 1965.
140. J. L. Webb, "Enzyme and Metabolic Inhibitors," Vol. 1. Academic Press, New York, 1963.
141. A. Munch, J. F. Scott, and L. L. Engel, *BBA* **26**, 397 (1957).
142. P. R. Stoesser and S. J. Gill, *J. Phys. Chem.* **71**, 564 (1967).
143. S. J. Gill, M. Downing, and G. F. Sheats, *Biochemistry* **6**, 272 (1967).

pensatory processes allows a contribution of dispersion forces to the stability of complexes in aqueous solution (*144, 145*).

Charge–transfer complexing (*146*) between electron donors and electron acceptors may be related to the ionization potential of the donor and the reduction potential of the acceptor (*147*). Charge–transfer complexation is characterized by:

(1) The appearance of a new absorption band, accompanied by retention of the characteristic absorption bands of the reactants.
(2) Establishment of an actual complex as shown by the use of double reciprocal plots [to differentiate this phenomenon from contact charge transfer (*148*)].
(3) Variation of the position of the charge–transfer absorption band with the ionization potential of the donor and with the electron affinity of the acceptor [for a linear free energy treatment see Charton (*149*) and Hammond (*150*)].
(4) Sensitivity of the position of the absorption band and the value of the association constant to the polarity of the solvent (*151*).

The stability of charge–transfer complexes as a function of solvent polarity is of particular interest. For the charge–transfer complex formed by trinitrobenzene and *N,N*-dimethylaniline, the stability constant is found to decrease in the following order on change of solvent: cyclohexane > *n*-hexane and *n*-heptane > CCl_4 > $CHCl_3$ > *s*-tetrachloro ethane > 1,4-dioxane (*152*). Formation constants of 1:1 charge–transfer complexes of 1,3,5-trinitrobenzene with hexamethylbenzene, pentamethylbenzene, durene, mesitylene or benzene decrease with change in solvent in the order cyclohexane > *n*-heptane = *n*-hexane > CCl_4 > $CHCl_3$, with a 10- to 20-fold variation in K in passing from cyclohexane to $CHCl_3$ at 20° (*153*). In water it is likely that complex formation derives little stabilization from charge–transfer interaction. Moreover, since the

144. E. Grunwald and E. Price, *JACS* **86,** 4517 (1964).
145. N. C. Deno and H. E. Berkheimer, *J. Org. Chem.* **28,** 2143 (1963).
146. R. S. Mulliken, *JACS* **74,** 811 (1952).
147. J. L. Andrews and R. M. Keefer, "Molecular Complexes in Organic Chemistry." Holden-Day, San Francisco, California, 1964.
148. L. E. Orgel and R. S. Mulliken, *JACS* **79,** 4839 (1957).
149. M. Charton, *J. Org. Chem.* **33,** 3878 (1968).
150. P. R. Hammond, *JCS* p. 471 (1964).
151. E. M. Kosower, "An Introduction to Physical Organic Chemistry." Wiley, New York, 1968.
152. R. Foster and D. L. Hammick, *JCS* p. 2685 (1954).
153. C. C. Thompson, Jr. and P. A. D. deMaine, *J. Phys. Chem.* **69,** 2766 (1965).

charge–transfer band moves to lower wavelength with increasing polarity of the solvent, it can prove very difficult to detect charge transfer within complexes in aqueous solution since the transfer band may be hidden under the strong absorption of reactants (*151*).

Several examples of catalysis by approximation which occur by way of preequilibrium charge–transfer complex formation are known. An example of approximation of nucleophile and substrate by complexing occurs in the methanolysis of tetrachlorophthalic anhydride (*154*). Tetrachlorophthalic anhydride is an excellent π acceptor (*147*) and in methanol forms stable complexes with both *N,N*-dimethylaniline and *O,O*-dimethylhydroquinone. Complexing with the amine is associated with an increase in the rate of methanolysis of anhydride, while complexing with the ether is accompanied by a decrease in the rate of methanolysis. The lessened reactivity of the anhydride-*O,O'*-dimethylhydroquinone complex may be attributed to both steric hindrance of nucleophilic attack at the carbonyl carbon and increased electron density at the carbonyls owing to electron delocalization from the π donor to acceptor. The increased rate of methanolysis in the *N,N*-dimethylaniline-tetrachlorophthalic anhydride complex finds explanation through the mechanism of (57). For

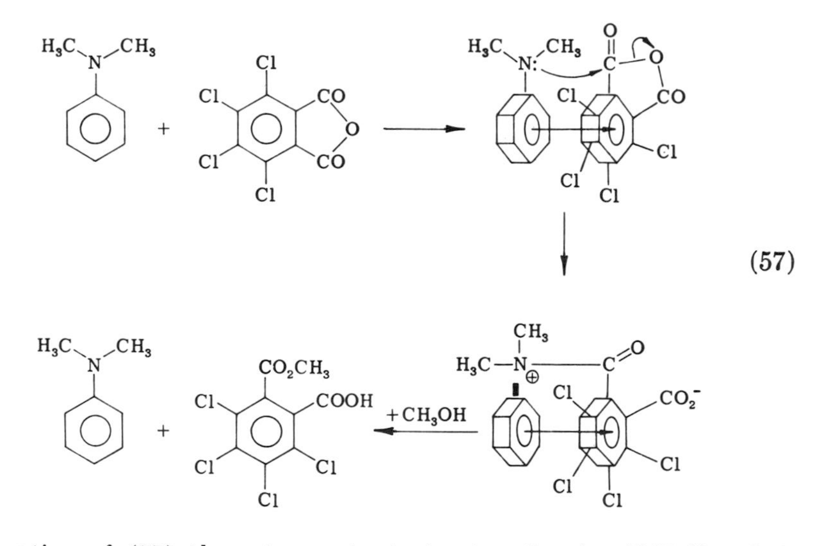

the reaction of (57) the rate constant at saturation by *N,N*-dimethylaniline is approximately twofold greater than in its absence and the equilibrium constant for complex formation is but 2.3 M^{-1}. Alternatively the charge transfer may be unreactive while reaction occurs through un-

154. F. M. Menger, *JACS* **90**, 4387 (1968).

complexed species. Kinetically these two paths may not be differentiated.

The solvolysis of 2,4,7-trinitrofluorenyl *p*-toluenesulfonate (58) in acetic acid proceeds through a transition state with carbonium ion character. Complexing of (58) with π donors such as phenanthrene, hexa-

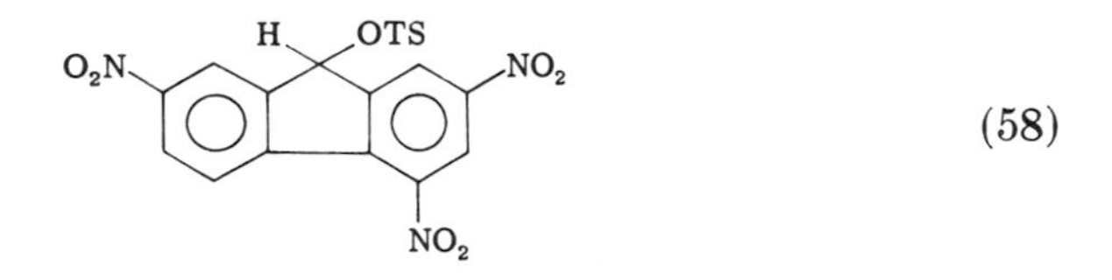

(58)

methylbenzene, naphthalenes, and anthracene increases the electron density in (58) and thus stabilizes the solvolytic transition state (*155–157*). The rate enhancement obtained with 1,5-dimethoxynaphthalene is 1900-fold, and complex formation is associated with an equilibrium constant of 1.0 M^{-1}. In this study the formation of charge–transfer bands was readily discernible, and with phenanthrene the spectrophotometrically determined equilibrium constant agreed well with that determined kinetically.

Complexation and aggregation of lyophobic molecules in water is commonly accepted as resulting from the strong cohesive energy of water molecules per unit volume (squeezing-out effect) and the structuring of water around nonpolar molecules (*145, 158–168*). The cohesive forces of water are readily discernible in the coalescence of oil drops in water and the formation of micelles. Less work is required to make one large cavity in water than to make two or more small cavities. Increased structuring of water upon the addition of substances which do not interact with water as strongly as water (dimethylformamide, *N*-methylformamide, dioxane, and alcohols) is apparent from the initial increase followed by a

155. A. K. Colter and S. S. Wang, *JACS* **85,** 114 (1963).

156. A. K. Colter, S. S. Wang, G. H. Megerle, and P. S. Ossip, *JACS* **86,** 3106 (1964).

157. A. K. Colter and S. H. Hui, *J. Org. Chem.* **33,** 1935 (1968).

158. H. S. Frank and M. W. Evans, *J. Chem. Phys.* **13,** 507 (1945).

159. W. Kauzman, *Advan. Protein Chem.* **14,** 1 (1959).

160. I. M. Klotz, *Science* **128,** 815, (1958).

161. I. M. Klotz, *Brookhaven Symp. Biol.* **13,** 25 (1960).

162. I. M. Klotz and S. W. Luborsky, *JACS* **81,** 5119 (1959).

163. G. Nemethy and H. A. Scheraga, *J. Chem. Phys.* **36,** 1773 (1962).

164. G. Nemethy and H. A. Scheraga, *J. Chem. Phys.* **36,** 3382 (1962).

165. G. Nemethy and H. A. Scheraga, *J. Chem. Phys.* **36,** 3401 (1962).

166. T. Higuchi and J. L. Lach, *J. Am. Pharm. Assoc.* **43,** 465 (1954).

167. J. A. V. Butler, *Trans. Faraday Soc.* **33,** 229 (1937).

168. O. Sinanoglu and S. Abdulnur, *Photochem. Photobiol.* **3,** 333 (1964); *Federation Proc.* **24,** Part II, S-12 (1965).

decrease in viscosity of the solution as solute concentration is increased (*169–171*) accompanied by a near linear increase in the dielectric relaxation time of water (*172*). The squeezing-out effect probably predominates over the water structuring effect in complex formation. The structuring of water around a nonpolar solute molecule is due to the inability of the solvent molecules to hydrogen bond to the solute. The contact water molecules can, however, be fully hydrogen bonded if their specific orientations with respect to their neighbors are restricted to a small number. This restriction amounts to a structuring of the solvent accompanied by a decrease in entropy. Compensation between an increase in enthalpy for cavity formation and a decrease in enthalpy resulting from increased hydrogen bonding around the solute molecule may occur so that the unfavorable free energy of solvation appears as an unfavorable entropy term. If the solute molecule is of sufficient polarity it will have a structure-breaking effect owing to its dipole perturbation of the surrounding water molecules. The unfavorable enthalpy for cavity formation will not be greatly altered and, because of the lessened structuring, a loss of hydrogen bonding will occur so that the unfavorable free energy of solvation will be associated with an unfavorable enthalpy. Aggregation of nonpolar solute molecules will therefore result in a transfer of structured solvent to the bulk solvent and a consequent decrease in solvent structuring. The formation of the "lyophobic bond" between nonpolar molecules is manifested in an increase in entropy with little change in enthalpy because of the compensating changes in enthalpy accompanying less cavity formation and less hydrogen bonding. For polar molecules the favorable free energy for lyophobic bonding is to be found in a decrease in enthalpy owing to a decrease in the number of cavities in the solvent (*96*). An example of the formation of lyophobic bonds between nonpolar molecules is found in the micellation of molecules of nonionic detergents which occur with a ΔH of about +5 kcal mole^{-1} and $T\Delta S$ of +9 kcal mole^{-1} (25°) (*173, 174*). Examples of the aggregation of polar molecules is found in the association of purines ($\Delta H = -2$ to -6 kcal mole^{-1} and $T\Delta S = -1.8$ to -4.8 kcal mole^{-1} at 25°) (*175, 176*) and certain dye

169. A. Fratiello, *J. Mol. Phys.* **7,** 565 (1964).
170. J. A. Geddes, *JACS* **55,** 4832 (1933).
171. F. Franks and D. J. G. Ives, *Quart. Rev.* (*London*) **20,** 1 (1966).
172. G. H. Haggis, J. B. Hasted, and T. J. Buchanan, *J. Chem. Phys.* **20,** 1452 (1952).
173. J. M. Corkill, J. F. Goodman, and J. R. Tate, *Trans. Faraday Soc.* **60,** 996 (1964).
174. M. J. Shick, *J. Phys. Chem.* **67,** 1796 (1963).
175. P. R. Stoesser and S. J. Gill, *J. Phys. Chem.* **71,** 564 (1967).
176. S. J. Gill, M. Downing, and G. F. Sheats, *Biochemistry* **6,** 272 (1967).

molecules (ΔH to -25 kcal mole^{-1} and $T\Delta S$ to -12 to -18 kcal mole^{-1} at 25°) (*177*).

The main driving force for dimerization and aggregation in water is thus to be found in the attractive forces of the solvent rather than the solute. Once complexation has occurred, additional binding energy may be derived from general van der Waal's forces (London and Debye forces), charge–transfer and charge–charge attraction. Within the limits previously discussed, the latter may be of importance in providing specific alignment of interacting molecules.

The formation of complexes in aqueous solution can, in many instances, be detected kinetically as a change in reaction rate of one of the complexing species. Most of the complexes which have been investigated in this manner are composed of a nitrogen heterocycle and an alkaline labile aromatic compound. Generally, complexing of reagent decreases its susceptibility to nucleophilic attack (*178–182*). The ester, 8-acetoxyquinoline, has been found to complex with a number of nitrogen-containing compounds in aqueous solution. The rate of spontaneous hydrolysis of this ester is enhanced when complexed with imidazolium ion, $H_3N^+CH_2CONH_2$, and triglycine. In the case of imidazolium ion a 12.5-fold rate enhancement is obtained, and the equilibrium constant calculated for the formation of a 1:1 complex is 1.2 M^{-1} (*183*). The examples of kinetically detectable complexation provided by Higuchi *et al.* (*178–182*) involve the alteration of reactivity of species A with B when A is complexed with reagent C. In these cases C presumably does not undergo any covalent bond alteration. The same forces which bring about complexation of a substrate with an inert ligand are available to complex substrate with a catalytic or nucleophilic species. In a study of the reaction of a long- and a short-chain amine (*n*-decylamine and ethylamine) with the *p*-nitrophenyl esters of a long- and a short-chain acid (*n*-decanoic and acetic acids), evidence has been provided for rate enhancement resulting from lyophobic bonding of substrate and nucleophile (*184*). Toward the nucleophiles ethylamine and hydroxide ion the acetate ester was found to be more reactive than the decanoate ester. This result is readily accounted for by the expected greater steric hindrance to

177. Y. Tanizaki, T. Hoshi, and N. Ando, *Bull. Chem. Soc. Japan* **38**, 264 (1965).
178. T. Higuchi and L. Lachman, *J. Am. Pharm. Assoc.* **44**, 521 (1955).
179. D. A. Wadke and D. E. Guttman, *J. Pharm. Sci.* **53**, 1073 (1964).
180. F. M. Menger and M. L. Bender, *JACS* **88**, 131 (1966).
181. J. A. Mollica and K. A. Connors, *JACS* **89**, 308 (1967).
182. P. A. Kramer and K. A. Connors, *JACS* **91**, 2600 (1969), and additional references cited herein.
183. S. M. Felton and T. C. Bruice, *JACS,* **91**, 6721 (1969).
184. J. R. Knowles and C. A. Parsons, *Chem. Commun.* p. 755 (1967).

attack on the larger ester. The rate of reaction of the decylamine with the acetate ester was found to be sevenfold greater than the rate of reaction of ethylamine with this ester, and the rate of reaction of the decylamine with the decanoate ester was found to be 700 times greater than the reaction of ethylamine with this ester. These results were interpreted in terms of lyophobic bonding of the ester to decylamine. From competition experiments it was estimated that about 47% of the rate enhancement was owing to aminolysis and the remainder to enhanced hydrolysis. The possibility that either hydrolysis or aminolysis was occurring in a fully formed micelle was dismissed on the grounds that the amine concentration was about three orders of magnitude below the critical micelle concentration and the ester concentration four orders of magnitude below the critical micelle concentration.

B. Micelles

Micelle structure and properties are closely analogous to those of globular proteins. In both cases polar groups are found almost exclusively at the surface while lyophobic groups are contained in the interior away from the aqueous media. Indeed, micelle structure is disrupted by solvents and other agents commonly employed for protein denaturation (*44*). In addition much of current knowledge about lyophobic bonding is derived from studies of micellization.

When the concentration of detergents and other amphiphilic molecules in aqueous solution reaches a certain, individually characteristic concentration, the critical micelle concentration (cmc), the monomers group themselves into aggregates (micelles) of roughly 20–100 molecules (*185–188*). At concentrations above the cmc, further addition of amphiphile results in the formation of additional micelles, the concentration of monomer remaining constant. The cmc of short-chain molecules is relatively broad but grows sharper with increasing chain length (*186*). The ionic or polar "head" groups of the micelle forming agent are at the surface in contact with the aqueous medium; however, the surface is jagged and fluctuating, allowing contact of the aqueous media with the lyophobic portions of the molecule attached to the head groups. It has been shown through ultraviolet and NMR studies that only a small con-

185. J. W. McBain, E. C. V. Cornish, and R. C. Bowden, *JCS* **101,** 2042 (1913).
186. G. S. Hartley, *JCS* **193,** 1968 (1939).
187. G. S. Hartley and D. F. Runnicles, *Proc. Roy. Soc.* **A168,** 420 (1938).
188. L. I. Osipow, "Surface Chemistry, Theory and Industrial Applications." Reinhold, New York, 1962.

centration of water is allowed into the interior of the micelle (*189–192*). Lyophobic molecules are solubilized by micelles, their positions in the micelle apparently being determined by their polarity. Thus, by NMR studies, it has been concluded that cetyltrimethylammonium (CTA^+) micelles absorb cyclohexane and cumene into the inner hydrocarbon portion of the micelle, whereas *N*,*N*-dimethylaniline and nitrobenzene residues are absorbed in the aqueous section of the interior not far from the surface (*193*). The micelle structure is dynamic, undergoing constant and complete exchange with detergent monomers in the solution and possibly with dimers and higher aggregates (*194, 195*). The rate of dissociation of the dodecylpyridinium iodide micelle has been studied by a temperature-jump technique (*196*).

The radii of various micelles have been estimated to be between 10 and 30 Å (*188*), so that for micelles composed of ionic monomers there is a close packing of charged groups around the surface of the sphere. The high charge density at the micelle surface produces such a high electric field that the surrounding solvent molecules are highly polarized and immobilized with a dielectric saturation; that is, there is a large reduction in the dielectric constant of the medium in the immediate vicinity of the surface of the micelle (*197*). The dielectric constant at the surface of dodecylpyridinium iodide micelles has been estimated to be about 36 from the absorption maximum of the pyridinium iodide charge transfer absorption band (*198*). As a result of this decrease in the dielectric constant, both ion–ion and ion–dipole interactions are greatly strengthened. Moreover, counterions from the solution are attracted toward the surface of the micelle owing to the large charge concentration. Even in very dilute solutions, it has been found that 60–85% of the counterions are within a few Ångströms of the micelle surface (*199*). The net result of these effects is the creation of an electrical double layer around the surface of the micelle, in which the dielectric constant varies linearly from that of the hydrocarbon to that of the medium, and in which the electri-

189. N. Sata and H. Sasaki, *Kolloid-Z.* **153,** 41 (1957).
190. J. Clifford, *Trans. Faraday Soc.* **61,** 1276 (1965).
191. N. Muller, R. H. Birkhahn, *J. Phys. Chem.* **71,** 957 (1967).
192. N. Muller and R. H. Birkhahn, *J. Phys. Chem.* **72,** 583 (1968).
193. J. C. Eriksson and G. Gillberg, *Acta Chem. Scand.* **20,** 2019 (1966).
194. P. A. Winsor, *Nature* **173,** 81 (1954).
195. H. B. Klevens, *JACS* **74,** 4624 (1952).
196. G. C. Kresheck, E. Hamori, G. Davenport, and H. A. Scheraga, *JACS* **88,** 246 (1966).
197. P. J. Debye, "Polar Molecules." Dover, New York, 1929.
198. P. Murkerjee and A. Ray, *J. Phys. Chem.* **70,** 2144 (1966).
199. J. N. Phillips and K. J. Mysels, *J. Phys. Chem.* **59,** 325 (1955).

cal zeta potential varies in the same manner. One consequence of these electrical effects is that the pH at the surface of the micelle may vary from that of the bulk solution by an order of magnitude or more (*200*).

There is a steadily growing number of studies of reactions in the presence of micelles. These may be divided into two main classes: those cases in which the micelle itself is inert, and those in which a reactive functionality on the detergent participates in the reaction.

The presence of inert detergents in the reaction mixture may influence the rate through any one or combination of several possible effects. The two most important of these are the micelle-induced changes in concentration and the myriad effects resulting from the presence of the electrical double layer. Because of the different partitioning of various organic and inorganic reactants toward the micelle, their relative concentrations may be drastically altered, with a concomitant effect on the rate. The double layer may affect the rate in many ways: (a) by changing the p*K*'s of the reactants; (b) by providing a medium of different dielectric constant; or (c) by electrostatic destabilization or stabilization of ground and transition states. It is obvious that any effect of rate will depend on the charge or lack of charge on the micelle head groups. A reaction may be catalyzed by a cationic detergent and inhibited by an anionic detergent or vice versa, may be catalyzed or inhibited by one and not affected by the other, or both may have the same effect. Neutral micelles are usually inhibitory. It is rather general that reactions involving one or more negatively charged reactants are catalyzed by micelles with positively charged head groups while reactions involving one or more positively charged reactants are catalyzed by anionic detergents. An abbreviated survey of studies of reactions occurring at or near the surface of micelles follows.

The kinetics of reactions of hydroxide ion and 2,4-dinitrochlorobenzene (*201*) and 2,4-dinitrofluorobenzene (*202*) in micelles of cetyltrimethylammonium bromide, sodium lauryl sulfate (LS^-), and the uncharged polyether Igepal can be explained on the assumption that incorporation of the uncharged substrate into the cationic micelle assists reaction with hydroxide, whereas incorporation into the anionic micelle inhibits it, and incorporation into the uncharged micelle has no effect. The extent of incorporation of 2,4-dinitrochlorobenzene into the micellar phases can be measured directly, and these association constants agree reasonably well with those determined kinetically. On the assumption that only one

200. E. F. J. Duynstee and E. Grunwald, *JACS* **81,** 4540 (1959).
201. C. A. Bunton and L. Robinson, *JACS* **90,** 5972 (1968).
202. C A. Bunton and L. Robinson, *J. Org. Chem.* **34,** 773 (1969).

substrate molecule is taken up in each micelle, the calculated binding constants are $2\text{–}4 \times 10^3$.

Added electrolytes inhibit catalysis by the cationic micelles, largely by making it more difficult for the nucleophilic anion to enter the Stern layer around the micelle. The inhibition depends upon the charge density of the counterion and increases very sharply with decreasing charge density.

Generally similar behavior has been observed for the reactions of hydroxide and fluoride ions with *p*-nitrophenyl diphenylphosphate (*203*, *204*) except that the binding constants are much larger than those found for the halobenzenes, being $10^5\text{–}10^6$. Binding constants of this order of magnitude are quite comparable to those for enzymic reactions. A marked difference between these reactions and those of the halobenzenes is that uncharged micelles of Igepal effectively inhibit the reactions, suggesting that the triaryl phosphate goes into the interior of the uncharged micelle where it is protected from anionic attack.

At high pH, bulky aryl phosphate anions inhibit the reaction of *p*-nitrophenyl diphenylphosphate with HO^- and F^- in the presence of cationic micelles (*204*), and the inhibition decreases in the sequence (59),

$$t\text{-Bu}-C_6H_4-OPO_3^{2\ominus} > C_6H_5-OPO_3^{2\ominus} > H_2PO_4^{2\ominus} \qquad (59)$$

as expected. However, at low pH where the hydroxide ion reaction is relatively unimportant a micellar-catalyzed phosphorylation could be observed (60), with a reactivity sequence the same as the inhibition

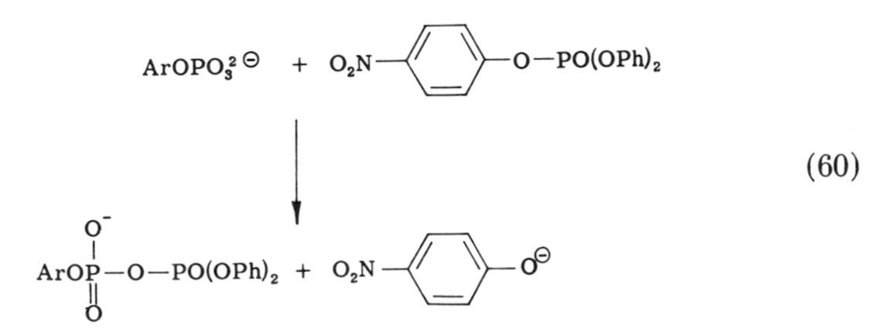

sequence. These results were interpreted as showing that the reactive anion can enter the transition state without losing its interactions with the micelle and that nucleophilicities for micellar-catalyzed reactions depend on the nonbonding interactions between the nucleophile and the

203. C. A. Bunton and L. Robinson, *J. Org. Chem.* **34,** 780 (1969).
204. C. A. Bunton, L. Robinson, and L. Sepulveda, *JACS* **91,** 4813 (1969).

micelle. Spontaneous hydrolyses of the dianions (but not the monoanions) of 2,4- and 2,6-dinitrophenylphosphates are catalyzed strongly by cationic micelles of CTA^+, and again the kinetic form of the catalysis can be explained in terms of incorporation of the substrate into the micellar phase (*205*). In these reactions the binding constants to micelles of CTA^+ are 10^4–10^5, and the inhibition by added anions can be interpreted quantitatively in terms of competition between the inhibiting anions and the substrate dianion for the cationic micelle.

The acid decomposition of α-phenylvinyldiethyl phosphate is effectively catalyzed by the anionic micelle, LS^-, and added cations are good inhibitors (*206*). The inhibition increases with decreasing charge density of the cation.

$$H^+ + CH_2 = CH(Ph)\text{—}OPO(OEt)_2 \xrightarrow{H_2O} CH_3\text{—}COPh + (EtO)_2PO_2H \qquad (61)$$

Catalysis of the reaction of uncharged molecules by micelles occurs in the reaction between aniline and 2,4-dinitrochlorobenzene and 2,4-dinitrofluorobenzene. The reactions are catalyzed by micelles of CTA^+, uncharged micelles of Igepal have slightly smaller effects, and anionic micelles of LS^- are only slightly catalytic (*206*). These catalyses appear to depend upon the ability of the micelles to bring the reacting molecules together but, as expected, the catalyses are smaller than those observed for the corresponding reactions between hydroxide ion and the halobenzenes (*201*, *202*).

Other examples follow: Cationic detergents have been shown to inhibit benzylidene aniline hydrolysis (*207*) and the fading of sulfonthalein dyes (*200*). Anionic detergents have been shown to inhibit hydroxide attack on Schiff bases and substituted benzylidenes (*208*), hydrolysis of the *p*-nitrophenyl esters of acetate, and octanoate, the monoester of 1,10-decanedicarboxylic acid (*209*), and the reaction between 2,4-dinitrofluorobenzene and glycylclycine (*210*). The hydrolyses of benzocaine, homotropine (*211*), and aspirin (*212*) have been shown to be retarded in neutral detergent.

205. C. A. Bunton, E. J. Fendler, L. Sepulveda, and K-U. Yang, *JACS* **90**, 5512 (1968).
206. C. A. Bunton and L. Robinson, *JACS* **92**, 356 (1970).
207. K. G. van Senden and C. Konigsberger, *Tetrahedron Letters* **1**, 7 (1959).
208. M. T. A. Behme and E. H. Cordes, *JACS* **87**, 260 (1965).
209. F. M. Menger and C. E. Portnoy, *JACS* **89**, 4698 (1967).
210. D. G. Herries, W. Bishop, and F. M. Richards, *J. Phys. Chem.* **68**, 1842 (1964).
211. P. B. Sheth and E. L. Parrott, *J. Pharm. Sci.* **56**, 983 (1967).
212. K. S. Murthy and E. G. Pippie, *J. Pharm. Sci.* **56**, 1026 (1967).

On the other hand, cationic detergents catalyze the reaction of hydroxide ion with triphenylmethane dyes (*213*), the attack of hydroxide or leucine on *p*-nitrophenyl acetate and hexanoate (*214*), the attack of cyanide on methyl bromide (*215*), the attack of hydroxide and thiophenoxide on 2,4-dinitrochlorobenzene (*201*), the reaction between *n*-pentyl bromide and thiosulfate (*201*), and the base-catalyzed decomposition of benzaldehyde by the Cannizzaro reaction (*216*). All of these but the reactions of *p*-nitrophenyl acetate are inhibited by LS^-. Anionic detergents have been found to promote the specific acid catalyzed hydrolysis of methyl orthobenzoate, ethyl orthovalerate, and ethyl orthopropionate (*214*), and the incorporation of Cu^{2+} into dimethylprotoporphyrin (*217*). The rate enhancements noted for micelle-catalyzed reactions are generally rather moderate. Thus, for the acid-catalyzed hydrolysis of methyl orthobenzoate in the presence of LS^- at pH 4.76 the rate enhancement is 85-fold (*214*).

Studies of the reactivity of functional head groups of micelles afford a means of investigation of reactions occurring at the Stern layer. A study (*218*) of the hydrolysis of monoalkyl sulfates of varying length has shown that the acid hydrolysis is enhanced by a factor of forty or more and the alkaline hydrolysis is inhibited by a factor of sixty or more when the chain length is sufficient for micelles to form. This result is in accord with attraction of hydronium ion to the negative surface of the micelle and electrostatic restriction of the availability of hydroxide ion. The rates of alkaline hydrolysis of *p*-nitrophenyl laurate and decanoate (*219*) have been shown to decrease with increasing concentration; this effect is not observed in the presence of denaturing agents such as ureas, indicating that it results from an aggregation effect. A very interesting phenomenon is noted with dodecylpyridinium bromide (*220*). Generally, cyanide ion attacks substituted pyridinium ions at the four position to give a diene adduct. With dodecylpyridinium bromide micelles, attack of cyanide gives the dodecylviologen cation radical (XIII), a dimer, in greater than 25% yield.

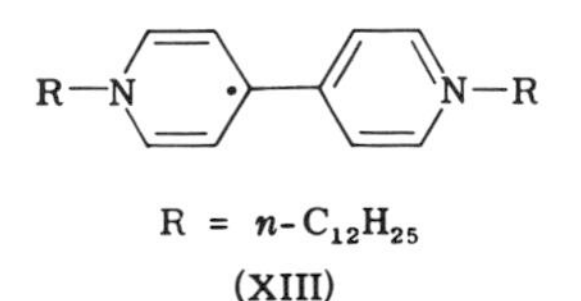

$R = n\text{-}C_{12}H_{25}$

(XIII)

213. E. F. J. Duynstee and E. Grunwald, *JACS* **81**, 4542 (1959).

214. M. T. A. Behme, J. G. Fullington, R. Noel, and E. H. Cordes, *JACS* **87**, 266 (1965).

215. L. J. Winters and E. Grunwald, *JACS* **87**, 4608 (1965).

216. L. R. Cramer and J. C. Berg, *J. Phys. Chem.* **72**, 3686 (1968).

Of particular interest are studies that deal with micelles containing catalytic or nucleophilic groups. These systems bear a rudimentary analogy to enzymes in that the substrate is complexed by the micelle and acted upon by functional groups that are not only part of the micelle itself but exist in a unique microscopic milieu. At pH 7.2, N^{α}-myristoyl-L-histidine (*221*) (MirHis) has a net negative charge owing to the presence of the ionized carboxylic acid. The compound is quite insoluble, and a turbid solution results at a concentration of $6 \times 10^{-5}\,M$; the addition of $CTA^{+}Br^{-}$ results in charge neutralization and an increase in turbidity until complete neutralization occurs at a 1:1 ratio. Further addition of the detergent causes a lowering of the turbidity until at 2:1 ratio the MirHis is completely solubilized and micelles have (presumably) been formed. In the presence of $CTA^{+}Br^{-}$ alone (above the cmc in tris buffer), *p*-nitrophenyl acetate (*p*-NPA), proprionate (*p*-NPP), butyrate (*p*-NPB), valerate (*p*-NPV), and hexanoate (*p*-NPH) are protected from hydrolysis; the inhibition ranges from 2-fold for the *p*-NPA to greater than 50-fold for the longer esters, indicating that the esters partition into the micelles. In mixed micelles of MirHis and $CTA^{+}Br^{-}$, however, the rate of hydrolysis is increased; a maximum rate is obtained at a detergent-to-catalyst ratio of 2:1 and decreases with increasing concentrations of CTA^{+}. The reaction was shown to proceed in two steps, with rapid acylation of the imidazolyl moiety of MirHis followed by rate-determining breakdown of this intermediate. The authors concluded that incorporation of the ester into regions of the mixed micelle where there is no MirHis leads to protection of the ester, whereas catalysis of hydrolysis proceeds through a second-order reaction at a region containing MirHis. This kinetic treatment may be criticized for its assumption of an essentially static micelle; in reality, because of the mobile nature of the micelle, the ester should find itself at one moment surrounded by CTA^{+} and at the next adjacent to MirHis. Nevertheless, catalysis of the ester hydrolysis was observed, and the ratio of the MirHis catalyzed hydrolysis to catalysis by N^{α}-acetyl histidine (NAH) in CTA^{+} was found to increase with increasing chain length of the ester. On the doubtful assumption of a second-order reaction between ester and mixed micelle at the MirHis sites, the ratio of the second-order reaction of *p*-NPH with NAH and mixed micelle is 1:810. If the assumption is made that *p*-

---

217. M. B. Lowe and J. N. Phillips, *Nature* **190,** 262 (1961).
218. J. L. Kurz, *JACS* **66,** 2239 (1962).
219. F. M. Menger and C. E. Portnoy, *JACS* **90,** 1875 (1968).
220. L. J. Winters, A. L. Borror, and N. Smith, *Tetrahedron Letters* p. 2313 (1967).
221. C. Gitler and A. Ochoa-Solano, *JACS* **90,** 5004 (1968).

NPH is incorporated into the mixed micelle and that a resultant intramicellar catalysis ensues, the kinetic propinquity effect thus obtained is about 10-fold greater than noted for (24) and (25), and comparable to that observed for (23c) vs. (23a). A plot of the logarithm of the "calculated" second-order rate constant for the micellar reaction was found to increase linearly with the chain length of the ester allowing calculation of the standard free energy change required per methylene group to account for the increase in the second-order rate constant. The calculated value of 442 cal $mole^{-1}$ is within the range of values expected for lyophobic bonding.

At pH 7.3, N^{α}-stearoylhistidine is reported to be about 2000-fold more effective as a catalyst for the hydrolysis of *p*-nitrophenyl *N*-dodecyl-*N*,*N*-dimethylammonioethyl carbonate bromide (E_{12}) than is N^{α}-acetyl histidine (*222*). The reaction was suggested to occur within a 1:1 complex since surface tension measurements did not reveal mixed micelle formation. From a pH log k_{rate} profile ($[E_{12}] = [N^{\alpha}$-stearoylhistidine$] = 3.3 \times 10^{-5}\ M$) it was established that the histidine imidazolyl group ($pK_{app} = 7.2$) was responsible for the catalysis of the hydrolysis of E_{12}. The reaction exhibited Michaelis–Menten-type kinetics, product inhibition and urea denaturation. Formation of the preequilibrium complex was established to be associated with an enthalpy change of approximately zero and a large positive entropy change. This very interesting system obviously contains many features common to enzymic reactions. The possible presence of mixed micelles of reactants has not been rigorously eliminated.

The effect of charge type and chain length upon the mixed micelle catalyzed acylation of the secondary amines A_4^+, A_{10}^+, and A_4° by a series of esters NE_n^- ($n = 1,5,7,9,15$), E_n^- ($n = 1,9$) E_9^+ and NE_n^+ ($n = 1,7$) has been investigated (*223*). The second-order rate constants for alkaline hydrolysis of the various esters were found to vary as anticipated on the basis of the influence of electronic effects of substituents on the phenyl ring and the steric effect of the alkyl side chains. The absence of change in the rate constants for alkaline hydrolysis with increase in the initial concentration of esters indicated that none of the esters formed complexes or micelles in solution. Complexing or incorporation of the esters into micelles of CTA^+, LS^-, or NDA° drastically reduced their susceptibility to alkaline hydrolysis. Therefore, regardless of the charge on the ester or upon the head group of the micelle, incorporation of ester into the "'nonnucleophilic" micelles decreases the susceptibility of the ester

222. R. G. Shorenstein, C. S. Pratt, C. J. Hsu, and T. E. Wagner, *JACS* **90,** 6199 (1968).

223. T. C. Bruice, J. Katzhendler, and L. R. Fedor, *JACS* **90,** 1333 (1968).

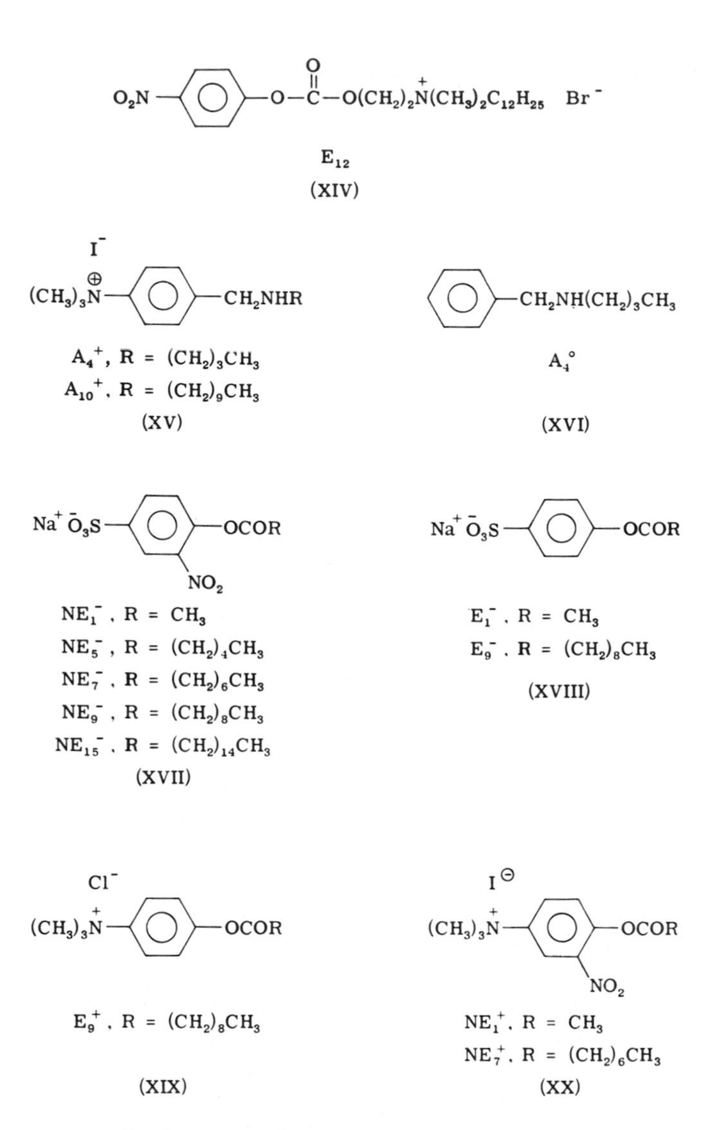

bond to alkaline hydrolysis. This assumption was supported by the finding that incorporation of NE_7^- into micelles of CTA^+ protected the ester from nucleophilic attack by A_4^+. Since the charge of ester and micelle has nothing to do with this decrease in susceptibility to alkaline hydrolysis, the ester bond must be buried below the Stern layer in the

$$HOCH_2CH_2(OCH_2CH_2)_{17}OC_6H_4(CH_2)_{11}CH_3$$

NDA°

(XXI)

micelles. The decrease in the rate of alkaline hydrolysis of esters was found to be dependent upon the second or third power of NDA° or CTA^{+} concentration. This implies that the salt or induced micelle has a ratio of ester:detergent = 1:2 (or 3).

For reaction of the amines with esters NE_1^-, NE_5^-, NE_7^-, NE_9^-, NE_1^+, and NE_7^+ the kinetics were found to be first order in amine and ester when the amines were either A_4^+ or A_4°. For A_4°, the ratio of rate constants for NE_7^+ and NE_7^- was 1.72 and that for NE_1^+ and NE_1^- was 2.1; with A_4^+, the ratios were 1.97 and 1.41, respectively. From these results it was concluded that association of reactants through hydrophobic and electrostatic attraction was not occurring. Unlike A_4^+, however, the amine A_{10}^+ was found to form micelles. For the NE_n^- esters the rate of reaction with A_{10}^+ was found to depend on the fourth power of A_T (A_{10}^+ and its conjugate acid $A_{10}H^{2+}$) at low $[A_T]$ and to be independent of A_T at its higher concentrations. For NE_n^+ and *o*-NPA the increase in rate was found to be dependent upon the second power of $[A_T]$ at lower concentrations and to be independent of $[A_T]$ at the higher concentrations employed. From these results it could be deduced that salts or premicelles of composition $NE_n^-(A_T)_4$ and $NE_n^+(A_T)_2$ are formed which are then incorporated into micelles of A_T and that acylation of A_{10}^+ by the esters occurred within the complexes and micelles. From the pH dependence of the acylation reactions at saturation of esters by A_T it could be established that the rate of disappearance of ester was dependent upon the mole fraction of A_T present as the basic A_{10}^+. The enumerated results are in accord with (62). The importance of charge type and lyophobic bond-

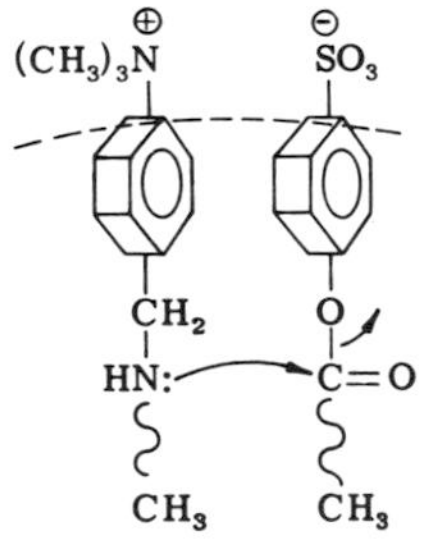

$(CH_3)_3\overset{\oplus}{N}$ $\overset{\ominus}{SO_3}$
CH_2 O
HN: → C=O
CH_3 CH_3

(62)

ing in the intramicellar reaction may be obtained by comparing the concentrations of A_{10}^+ and A_4^+ which provide identical values for the first-order rates of ester disappearance. For the short-chain ester NE_1^+, k_{obsd} for A_{10}^+ never approaches that for A_4^+, even at saturation, whereas for NE_1^-, k_{obsd} for A_{10}^+ is at a saturation level at about $1 \times 10^{-2}\,M$ and the value of k_{obsd} for A_4^+ does not reach this value until $A_4^+ = 4.5 \times 10^{-2}\,M$.

With the acetyl esters charge type is therefore quite important. The values of k_{obsd} for *o*-NPA with A_{10}^{+} and A_4^+ are comparable at about $4.5 \times 10^{-2}\,M$ so that the role of the charge on the ester is most likely one of alignment of ester and amine within the micelle. For the longer chain esters charge type appears to be much less important and alignment of nucleophile and substrate occurs through lyophobic interaction.

A comparison of the importance of electrostatic and lyophobic binding forces in the intramicellar reaction of type (62) to that at a protein surface has been obtained through studies of the catalysis of hydrolysis of esters NE_1^+, NE_1^-, NE_7^+, and NE_7^- as well as *o*-NPA and *p*-NPA by hens' egg white lysozyme (*224*). The esteratic site of lysozyme resides at histidine-15 which is adjacent to positively charged arginine-14 and lysine-96. Several lyophobic regions are also adjacent to histidine-15. For reaction with lysozyme both electrostatic and lyophobic forces could be shown to be involved. Thus, the order of reactivity with lysozyme was found to be $NE_7^- > NE_1^- > NE_7^+ > NE_1^+ >$ *o*-NPA $>$ *p*-NPA as compared to that with N^α-acetyl histidine which is $NE_1^+ \approx NE_7^+ > NE_1^- > NE_7^- >$ *p*-NPA $\approx$ *o*-NPA.

C. Inclusion Compounds

Cyclodextrins or cycloamyloses are cyclic polymers of glucose obtained from starch-containing cultures of *Bacillus marcerans* (*225*). Because of the constraint of the 1-4-linked α-D-glucopyranose units in staggered chair conformation (*226*), the molecules take on the appearance of baskets with open bottoms. Along the top lip of the basket are arranged the 2- and 3-hydroxyl groups and around the bottom lip are arranged the 5-hydroxymethyl groups. Crystalline forms of the cyclodextrins include the so-called α-cyclodextrin (six glucose units, 6 Å diam cavity), β-cyclodextrin (seven glucose units, 8 Å diam cavity), and γ-cyclodextrin (eight glucose units, 10 Å diam cavity). The cyclodextrins are known for their ability to form inclusion compounds of definite stoichiometry with many types of substances including both aliphatic and aromatic carboxylic acids (*227*). In the formation of inclusion compounds, the host cyclodextrin accepts and binds the guest molecule within its basketlike structure. The binding of dyes and many aromatic compounds may be followed spectrophotometrically owing to the shift

224. D. Piszkiewicz and T. C. Bruice, *Biochemistry* **7**, 3037 (1968).
225. D. French, *Carbohydrate Chem.* **12**, 189 (1957).
226. A. Hybl, R. E. Rundle, and D. E. Williams, *JACS* **87**, 2779 (1965).
227. J. F. Brown, Jr. and D. M. White, *JACS* **82**, 5671 and 5678 (1960).

in the spectra of the guest on transfer from water to the less polar cavity of the cyclodextrin molecule. The spectral shifts indicate that the inside of the cavity is very similar to that of a dioxane solution (*228*).

Because the lips of the cyclodextrin basket possess reactive hydroxyl groups, reactions of guest molecules with the cyclodextrin may take place. Studies of cyclodextrins as model enzyme systems have centered around the following: (a) the specificity of binding of the guest substrate; (b) the further reaction of the guest molecule with the reactive hydroxyl groups of the host; or (c) changes in the rate of a reaction owing to either protection of the guest molecule from reaction with agents in the bulk solution or the change of the dielectric of the microscopic media on transfer from water to the cyclodextrin cavity.

The rate of oxidation of furoin to furil is increased in the presence of cyclodextrin (*229*). Spectrophotometrically, it was shown that the cyclo-

$$C_4H_3O-\overset{O}{\overset{\|}{C}}-\underset{H}{\overset{OH}{C}}-C_4H_3O \underset{H_2O}{\overset{HO^-}{\rightleftharpoons}} C_4H_3O-\overset{HO}{C}=\overset{OH}{C}-C_4H_3O \xrightarrow{\frac{1}{2}O_2} C_4H_3O-\overset{O}{\overset{\|}{C}}-\overset{O}{\overset{\|}{C}}-C_4H_3O \quad (63)$$

dextrin preferentially includes the endiol to shift the first equilibrium step to the right and increase the concentration of the oxidizable species. The rate constants for the decarboxylation of a number of α-substituted acetoacetic acids have been determined in the presence of α- and β-cyclodextrins. At a cyclodextrin concentration of $1.65 \times 10^{-2}\,M$ the rate constants for decarboxylation were increased several fold when the β substituent was benzylic, etc. (*230*). The β-cyclodextrin proved most effective. These changes in rate are most likely resulting from a milieu effect since it is known that the rates of decarboxylation also increase on addition of ethanol (*231*).

Symmetrical phenyl esters of pyrophosphate are quite stable in neutral and alkaline solution. In the presence of divalent metal ions they are hydrolyzed to yield monophenyl phosphate as the sole product. In the presence of a cyclodextrin a phosphorylation reaction occurs and only one mole of phenyl phosphate is produced (*232*). Simple sugars have little or no influence on the rate or products of hydrolysis. When unsymmetrical aryl-alkyl-pyrophosphate esters were employed, the phosphorylation reaction was found to exhibit a high degree of specificity. The order

228. R. L. VanEtten, J. F. Sebastian, G. A. Clowes, and M. L. Bender, *JACS* **89**, 3242 (1967).
229. F. Cramer, *Chem. Ber.* **86**, 1579 (1953).
230. F. Cramer and W. Kampe, *JACS* **87**, 1115 (1965).
231. G. A. Hall and F. H. Verhoek, *JACS* **69**, 613 (1947).
232. N. Hennrich and F. Cramer, *JACS* **87**, 1121 (1965).

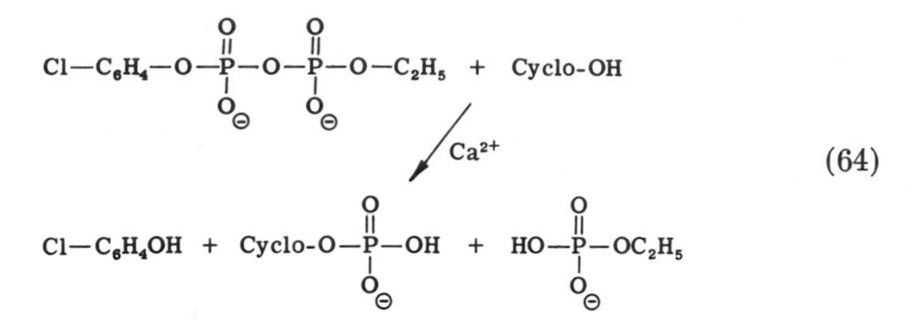

(64)

of effectiveness of the cyclodextrins was determined to be $\beta > \gamma > \alpha$ and the reactions were shown to exhibit saturation phenomena and product inhibition by phenyl phosphates. The specificity of the reaction is undoubtedly derived from the inclusion of the benzenoid moiety of the mixed ester (64) followed by reaction of a 2- or 3-hydroxyl group of the cyclodextrin with the phosphorus closest to the included aromatic ring and thus subsequent hydrolysis of the unsymmetrical cyclodextrin phenyl phosphate ester.

Extensive studies have been made of the hydrolysis of substituted phenyl acetates in the presence of cyclodextrins (*228, 233*). The hydrolysis of a large series of phenyl acetates substituted at various positions of the benzene ring were investigated. Although rate constants for base-catalyzed hydrolysis and hydrolysis in the presence of glucose (basic anomeric hydroxyl group) could be correlated with the σ value of the substituents, no such correlation was obtained in the cyclodextrin-facilitated solvolysis, which exhibited a degree of stereospecificity toward *m* substituents. The order of reactivity and stereospecificity for the cyclodextrins was found to be $\alpha \geq \beta > \gamma$ (Table IX). Since the reaction proceeds via formation of inclusion compounds, the stereospecific effect noted in the rate constants could be accounted for in either the magnitudes of k_{cat} or K_{diss}. Separation of terms revealed that the stereospecific rate acceleration was associated with k_{cat}. As an example, with α-cyclodextrin the ratio of K_{diss} for *p*-*t*-But:*m*-*t*-But is 3:1 whereas the ratio for k_{cat} is 1:240. It should be noted that the equilibrium constants for complex formation of ester and cyclodextrin are considerable, being $5 \times 10^2\ M^{-1}$ for the *m*-*t*-butylphenyl acetate–α-cyclodextrin complex. A comparison of the spectral shifts on inclusion of *p*-*t*-butyl- and *m*-*t*-butylphenol led to the conclusion that the benzene rings of the *p*-substituted phenol was immersed within the cyclodextrin cavity (XXII) but that the benzene ring of the *m* isomer protruded into the aqueous media (XXIII). Steric interference by *m* substitution to the inclusion of an aromatic

233. R. L. VanEtten, G. A. Clowes, J. F. Sebastian, and M. L. Bender, *JACS* **89**, 3253 (1967).

TABLE IX
HYDROLYSIS RATES OF PHENYL ACETATES AT pH 10.6 IN THE ABSENCE (k_{un}) AND PRESENCE (k_{obsd}) OF CYCLODEXTRINS (0.01 M)

Substituent on ester	k_{obsd}/k_{un}		
	α	β	γ
o-CH_3	7.7	6.9	—
m-CH_3	39	16	—
p-CH_3	3.8	6.7	—
m-t-But	221	250	54
p-t-But	1.7	2.2	41
m-Cl	113	18	7.8
m-NO_2	103	54	10
p-NO_2	2.6	6.7	6.2

ring into the cyclodextrin cavity has also been noted with dyes. The placement of substituent groups meta to each other makes it difficult or impossible to thread the benzene ring into the cyclodextrin cavity. Also, inclusion of meta- as compared to para-disubstituted benzenoid compounds requires more solvent reorganization or conformational alter-

(XXII) (XXIII)

ations in the host molecule. Thus, benzene rings substituted in the 1 and 4 positions combine with the cyclodextrin with a rate constant of about 10^8 liter $mole^{-1}$ sec^{-1} which is close to a diffusion-controlled rate (*234*). When the substituent groups are in the 1 and 3 positions it is found that the second-order rate constant for the formation of inclusion compound is more than seven orders of magnitude slower. Both types of compounds exhibit comparable formation constants so that the interfering 3 substituent decreases the rate of the combination and separation steps by similar amounts. These studies provide useful models for diffusion- and nondiffusion-controlled binding of substrate with an enzyme.

Studies (*233*) involving phenyl benzoates were employed to establish that the reaction of phenyl esters with cyclodextrin involved a double

234. F. Cramer, W. Saenger, and H-Ch. Spatz, *JACS* **89,** 14 (1967).

$$m\text{-}NO_2C_6H_4\text{—O—CO—}C_6H_5 + \alpha\text{-cyclo-OH} \xrightarrow[\text{fast}]{\text{pH 10.6}} \underset{390\ m\mu}{m\text{-}NO_2C_6H_4\text{—}O^-}$$

$$+ \underset{245\ m\mu}{\alpha\text{-cyclo-O—CO—}C_6H_5} \xrightarrow[\text{slow}]{} \alpha\text{-cyclo-OH} + C_6H_5\text{—}COO^- \quad (65)$$

displacement mechanism (65). When the *m* substituent on the phenoxy portion of the benzoate esters was altered, it was found that the rate of the acylation step, but not the deacylation step, was altered. The intermediate benzoate ester could be shown to be present on gel filtration and spectrophotometric examination of the reaction mixture of (65) after the acylation step. Methylation of all hydroxyl groups of the cyclodextrin inhibited the rate of ester hydrolysis whereas mesylation of only the 5-hydroxymethyl groups had no effect on rate. The mechanism of (66) is in accord with all experimental observations, and the stereospecificity of the reaction can be appreciated by comparison of (66) with (67).

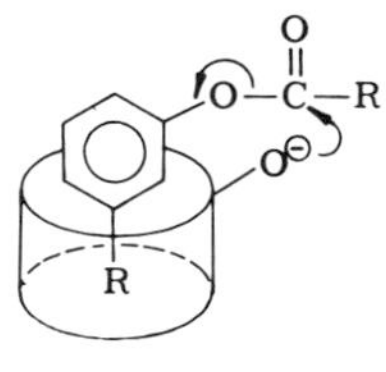

(66)

(67)

The acylation–deacylation process noted in the reaction of cyclodextrins with phenyl esters is akin to the reaction of esteratic enzymes such as α-chymotrypsin. Actually, the cyclodextrin reaction is not a true catalysis because although the disappearance of phenyl ester is facilitated, the acylated cyclodextrin hydrolyzes more slowly than the phenyl ester substrate. In an attempt to overcome this limitation the synthesis of various modified cyclodextrins was carried out (*235, 236*). The modified cyclodextrins contained side chains of various types terminating

235. F. Cramer and G. Mackensen, *Angew. Chem.* **78**, 641 (1966).
236. F. Cramer and G. Mackensen, *Angew. Chem. Intern. Ed. Engl.* **5**, 601 (1966).

in an imidazolyl ring in place of a primary hydroxyl group. These compounds proved, however, to be of little interest as catalysts.

ACKNOWLEDGMENT

Experimental work reported from the author's laboratory was supported by grants from the National Institutes of Health and the National Science Foundation.

5

Enzymology of Proton Abstraction and Transfer Reactions

IRWIN A. ROSE

I. Introduction

The amazing rate accelerations produced in enzyme catalysis are often explained in terms of the acid-base properties of the component amino acids. The concept of acid–base catalysis is a dominant one in the design of model studies of enzyme reaction mechanisms. Direct evidence for base catalysis by an enzyme was first suggested by the pioneering experiments of Talalay and Wang in 1955 with Δ^5-3-ketosteroid isomerase (*1, 1a*). The reaction, as shown, involves the net loss of a proton from C_4 and

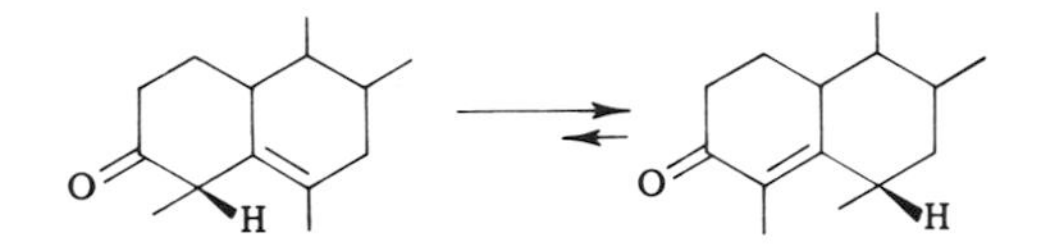

gain of a proton at C_6. The enzyme without the benefit of cofactor or coenzyme completes a conversion in about 10^{-6} sec with little or no incorporation of the protons of the medium into the end product. The implication was drawn that the enzyme supplied the basic group of an amino acid with proper orientation to remove and hold the proton for transfer between the allylic carbons of the resulting carbanion (*1a, 2*).

Further examples of proton transfer within the past decade include the observation of partial exchange of the presumed conjugate acid proton with the medium, found with P-glucose isomerase (*3, 3a*) and pyridoxamine-pyruvate transaminase (*4*), and the observation of intermolecular proton transfer, found with aconitase (*5*). Studies with proline racemase (*6*) and fumarase (*7*) have provided evidence for enzyme acting as a base using kinetic isotope exchange methods. Added to these developments, stereochemical specificity has taken on a new significance when interpreted within the framework of the occurrence of proton transfer.

Certainly proton activation at a tetrahedral carbon is one of the fundamental elementary steps in enzyme catalysis. Its widespread occurrence among the steps of intermediary carbohydrate, lipid, and amino acid metabolism has not, until recently (*8*), made it the focus of attention in the review literature. The present chapter is meant to provide a discussion of developments principally since the 1966 review.

1. P. Talalay and V. S. Wang, *BBA* **18,** 300 (1955).
1a. S. F. Wang, F. S. Kawahara, and P. Talalay, *JBC* **238,** 576 (1963).
2. S. K. Malhotra and H. J. Ringold, *JACS* **87,** 3228 (1965).
3. I. A. Rose and E. J. O'Connell, *JBC* **236,** 3086 (1961).
3a. I. A. Rose, *Brookhaven Symp. Biol.* **15,** 293 (1962).
4. J. E. Ayling, H. C. Dunathan, and E. E. Snell, *Biochemistry* **7,** 4537 (1968).
5. I. A. Rose and E. L. O'Connell, *JBC* **242,** 1870 (1967).
6. G. J. Cardinale and R. H. Abeles, *Biochemistry* **7,** 3970 (1968).
7. J. N. Hansen, E. C. Dinovo, and P. D. Boyer, *JBC* **244,** 6270 (1969).
8. I. A. Rose, *Ann. Rev. Biochem.* **35,** 23 (1966).

II. General Topics

A. Isotope Exchange

The exchange velocity of an isotope such as tritium (T) between a substrate (SH) and water conforms to the rate equation (*9*):

$$v_T = -2.3 \cdot n \cdot t^{-1} \cdot [SH] \log (1 - \text{fraction reacted})$$

Thus, if tritium begins in the substrate one need simply separate the water (from SH by sublimation, ion exchange, extraction, etc.) and count it after times t of reaction and after complete reaction and calculate v_T. If there are three hydrogens in the substrate that the enzyme treats equivalently, $n = 3$, as in the $—CH_3$ group of pyruvate or acetyl-CoA. Since [SH] represents the concentration of substrate, in this case almost exclusively S^1H, v_T calculated from the appearance of tritium in the water will not represent the rate of proton exchange unless corrected for the isotope effect:

$$v_H = v_T \cdot \left(\frac{k_H}{k_T}\right)$$

To determine the isotope effect, one must compare the exchange rate of two hydrogen isotopes under the same conditions. This is easily done by supplying a mixture, SH and ST, to the enzyme in D_2O, or SD and ST in H_2O, and comparing the rates of loss of the two isotopes in the same incubation. From such a determination and $k_T/k_H = (k_D/k_H)^{1.44}$ (*10*), one estimates k_H/k_D by $k_H/k_D = (k_H/k_T)^{0.694}$ and $= (k_D/k_T)^{2.26}$.

Rate equations may be derived for any formal exchange mechanism using the approaches of Boyer *et al.* (*7, 11*), or the more general approaches of Yagil and Hoberman (*12*) and of Cleland (see Chap. 1) for isotope exchange at equilibrium. This equation may be de-

$$E + SH \underset{-1}{\overset{1}{\rightleftharpoons}} ESH \underset{-2}{\overset{2}{\rightleftharpoons}} ES^- + H^+, \quad v_x = \frac{k_{-1}k_2}{k_{-1} + k_2} \cdot \frac{E_T}{1 + [K_1/(SH)] + [K_2/(H^+)]} \quad (1)$$

rived assuming the equilibrium condition since the enzyme, which is

9. H. A. C. McKay, *Nature* **142,** 997 (1938); R. B. Duffield and M. Calvin, *JACS* **68,** 557 (1946).

10. C. G. Swain, E. L. Stivers, J. F. Reuwer, Jr., and L. J. Schaad, *JACS* **80,** 5885 (1958); L. Melander, "Isotope Effects on Reaction Rates," p. 24. Ronald Press, New York, 1960.

11. P. D. Boyer, *ABB* **82,** 387 (1959); P. D. Boyer and E. Silverstein, *Acta Chem. Scand.* **17,** 195S (1963).

12. G. Yagil and H. D. Hoberman, *Biochemistry* **8,** 352 (1969).

present in small amount, will have turned over many times before the reaction is terminated. Since the amount of product, ES^-, is even smaller, equilibrium must exist over most of any period that is practical for doing the experiment. Note, the rate of exchange v_x is governed by a kinetic portion, $k_{-1}k_2/(k_{-1} + k_2)$, made up of constants from each reaction that connects the exchanging species SH and H^+ and a thermodynamic portion, $E_T/1 + [K_1/(SH)] + [K_2/(H^+)]$, which depends on the two dissociation constants of ESH. For —CH bonds, the equilibrium constant for the proton dissociation, even when the substrate is activated in complex with enzyme, is surely going to be small, or $K_2 << (H^+)$. Thus, Eq. (1) becomes (1′):

$$v_x = \frac{k_{-1}k_2}{k_{-1} + k_2} \cdot \frac{E_T}{1 + [K_1/(SH)]} \equiv \frac{V_x}{1 + [K_x/(SH)]} \qquad (1')$$

This rate equation is unchanged in form by a number of changes in the mechanism, such as by complicating the first step by Schiff base formation. This is true also if the enzyme acts as a base:

$$E + SH \underset{-1}{\overset{1}{\rightleftharpoons}} ESH \underset{-2}{\overset{2}{\rightleftharpoons}} \begin{matrix} ES^- \\ \diagdown \\ BH^+ \end{matrix} \underset{-3}{\overset{3}{\rightleftharpoons}} \begin{matrix} ES^- \\ \diagdown \\ B: \end{matrix} + H^+,$$

$$v_x = \frac{k_{-1}k_2k_3}{k_{-1}k_{-2} + k_2k_3 + k_{-1}k_3} \cdot \frac{E_T}{1 + [K_1/(SH)] + K_2 + [K_2K_3/(H^+)]} \qquad (2)$$

Since $K_2 = \left(\begin{matrix} ES^- \\ \diagdown \\ BH^+ \end{matrix}\right) \Big/ (ESH)$ will be $<< 1$ and $K_2K_3 = \left(\begin{matrix} ES^- \\ \diagdown \\ B: \end{matrix}\right) \cdot (H^+)/(ESH)$ will be at $<< H^+$, it is likely that $K_x \cong K_1$, which then has the same meaning as in mechanism (1).

B. Isotope Effects, Primary

The alternatives (1) and (2) cannot be distinguished on the basis of isotope effects or stereochemical studies. Primary deuterium or tritium isotope effects in enzymic proton exchange reactions have not yet been reported. On the basis of model studies (*13*) rate-determining general base abstraction of a proton from a —CH bond is subject to large isotope effects, $k_H/k_D > 4$. Large primary isotope effects have been reported in the aldol condensation of deuterated dihydroxyacetone-P catalyzed by yeast aldolase and carboxypeptidase-treated muscle aldolase (*14*), indicating that the —C—H cleavage step probably controls the rate of isotope exchange. Thus in mechanism (2) $k_2 < k_{-1}$ and $k_{-2} < k_3$ so that

13. R. P. Bell, "The Proton in Chemistry," p. 201. Cornell Univ. Press, Ithaca, New York, 1959.
14. I. A. Rose, E. L. O'Connell, and A. H. Mehler, *JBC* **240,** 1758 (1965).

the result is $V_x = k_2E_T$. Complicating the first step with unreactive conformers of ESH or with intermediates such as those involved in Schiff's base formation will mean that (ESH) may be much less than E_T as substrate becomes saturating and therefore V_x/E_T represents a minimum value for k_2. The fraction of E_T, that is, ESH, the reactant in step 2 of mechanism (2), will be a function of factors that determine the position of this preequilibrium such as pH and other modifiers, which will therefore control V_x without being directly involved in step 2 of mechanism (2) (*15*).

The proton dissociation from the conjugate acid intermediate, $ES^- \rightarrow ES^- + H^+$ (with BH^+ and $B{:}$ attached to E, respectively), step 3 of mechanism (2) cannot be responsible for large deuterium isotope effects (e.g., $k_H/k_D > 4$). Proton transfer to water from acids of the strength of carboxyl, imidazolium, sulfhydryl, or ammonium is more rapid, once diffusive encounter has occurred, than is diffusive separation (*16*, *16a*), which should not be affected by isotope substitutions. Thus, any isotope effect in this process would be limited to

$$E(S^-)(BH^+) + OH_2 \rightleftharpoons \left[\underset{(I)}{E(S^-)(BH^+\cdots OH_2)} \overset{\text{fast}}{\rightleftharpoons} \underset{(II)}{E(S^-)(B{:}\cdots HOH_2)} \right] \xrightarrow{\text{slow}} E(S^-)(B{:}) + H_3O^+$$

that in the equilibrium (II)/(I) and so would be similar to that for dissociative equilibria, i.e., $K_H/K_D \cong 1.4$ (*17*).

The failure to observe an isotope effect in hydrogen exchange implies immediately that step 2 of mechanism (2) is not rate determining. It does not imply that step 3 is slow since step 1 may be complicated by slow isomerizations that are not influenced by substrate concentration.

C. Isotope Effects, Secondary

Secondary hydrogen isotope effect studies are becoming recognized as a useful technique for enzyme mechanism investigation (*18*, *18a–c*). Isotope

15. T. C. Bruice and G. L. Schmir, *JACS* **81**, 4552 (1959).
16. M. Eigen, *Angew. Chem. Intern. Ed. Engl.* **3**, 1 (1964).
16a. W. P. Jencks, "Catalysis in Chemistry and Enzymology," pp. 211 and 265. McGraw-Hill, New York, 1969.
17. The dissociation constants of acids with p*K* 3–7 are about 3-fold lower in D_2O (*13*, p. 188). Therefore $B{:}D + OH_2 \rightleftharpoons B{:} + DOH_2$ should be about $3^{1/3} = 1.4$ less favorable than $B{:}H + OH_2 \rightleftharpoons B{:} + H_3O$.
18. J. F. Biellmann and F. Schuber, *Bull. Soc. Chim. France* p. 3543 (1967).

substitution in a position of the reactants not involved in bond cleavage can influence the rate of a process if, in the transition state of the rate-limiting step of the process being measured, the bond to the isotope differs in energy from that in the ground state. For example, if the state of orbital hybridization of the activated carbon goes from tetrahedral, sp^3, to pure trigonal, sp^2, in the transition state, the out-of-planc vibration of the attached proton may be 500 cm^{-1} lower in the transition state, which corresponds to $k_H/k_D = 1.38$ (*19*). In the reverse direction values of $k_H/k_D = 0.73$ are expected. Since the transition states will differ from pure trigonal or tetrahedral geometry the effects will be smaller. Secondary effects on equilibria can be predicted from infrared spectra.

Two fundamentally distinct approaches are possible in determining a secondary isotope effect, as with the primary effect: The initial rates of the process can be compared in two incubations, one containing deuterium substituted either at the carbon of bond cleavage (α effect) or neighboring to it (β effect), or a single incubation can be used with the isotopic forms, usually tritium and protium in competition and the specific activity of the product at early times then determined. The two approaches do not necessarily test the same portions of the reaction. For example, in a process in which two products are formed from an enzyme saturated with substrate, $\mathrm{ES} \overset{\mathrm{P}}{\underset{2}{\rightleftharpoons}} \mathrm{EQ} \underset{3}{\rightarrow} \mathrm{E} + \mathrm{R}$. The first experiment compares normal and deuterated substrate for their effects on $V_{max} = k_2k_3E_T/(k_2 + k_3)$. An isotope effect at either step 2 or step 3 will alter V_{max} if that step has the smaller rate constant of the two. In a competitive situation an isotope effect in step 3 cannot lead to discrimination since step 2 commits both isotopic species equivalently to the EQ stage which, in the absence of P, can only proceed to R. An isotope discrimination between the two species of EQ at this stage will be quickly compensated for by the change in specific activity of the small pool of EQ and will not be observable as a specific activity difference between R and S. Only if P is present and its reaction with EQ is sufficiently rapid to spread the isotopic discrimination back into the substrate pool could one find a specific activity of R different from that of S. A secondary

---

18a. F. W. Dahlquist, T. Rand-Meir, and M. A. Raftery, *Biochemistry* **8**, 4214 (1969).

18b. D. E. Schmidt, Jr., W. G. Nigh, C. Tanger, and J. H. Richards, *JACS* **91**, 5849 (1969).

18c. J. F. Biellmann, E. L. O'Connell, and I. A. Rose, *JACS* **91**, 6484 (1969).

19. A. Streitweiser, Jr., R. H. Jagow, R. C. Fahey, and S. Suzuki, *JACS* **80**, 2326 (1958); L. Melander, "Isotope Effects on Reaction Rates," p. 89. Roland Press, New York, 1960.

isotope effect in step 2 could be seen by a change in specific activity of initially formed P, even if $k_2 > k_3$, i.e., does not "determine" $-(dS/dt)$ as shown by an isotope comparison study. In such a case one must consider whether the isotope effect relates to a rate or an equilibrium process. Bond hybridization differences will affect stable equilibria in the same way that they affect $K^\ddagger$, and further information will be required to distinguish between these alternatives.

D. Lack of Isotope Exchange

The failure to observe significant isotope exchange in an incomplete system may imply that a second substrate is required for proton activation (see Section V); however, this conclusion is not valid if ES^- is part of a "ping pong" system, that is, if ES^- produces a product without regenerating free E. In the generally accepted mechanism for transaminases, glutamate reacts with enzyme to release a proton, from the α-carbon, and α-ketoglutarate in the sequence:

$$E + H_2NCH \underset{1}{\rightleftharpoons} E{=}NCH \underset{2}{\rightleftharpoons} EN{=}C + H^+ \xrightarrow{H_2O} ENH_2 + \rangle C{=}O,$$

$$v_x = \frac{k_{-1}k_2}{k_{-1} + k_2} \cdot \frac{E_T}{1 + [K_1/(H_2NCH)] + [K_2/(\rangle C{=}O)]}$$

However, proton exchange will be limited by the low level of α-ketoglutarate ($\rangle C{=}O$) that is generated in the reaction. α-Ketoglutarate will appear to be a requirement for the exchange (*20*), although it merely serves as a route from ENH_2 to E, so that the exchange process, steps 1 and 2 can continue.

E. The Role of Electron Delocalization in Enzymic —C—H Fission

Of the factors that have been considered (*21*) to stabilize the developing carbanion structure in C—H bond fission (*s*-orbital character, conjugative electron delocalization, inductive effects of substrate substituents, etc.), the second is probably the most important. It is noteworthy

20. M. A. Hilton, F. W. Barnes, S. S. Henry, and T. Enns, *JBC* **209,** 743 (1954).
21. D. J. Cram, "Fundamentals of Carbanion Chemistry," p. 47. Academic Press, New York, 1965.

that with few exceptions, proton activations occur at carbons alpha, or vinylagous, to a carbonyl group or its equivalent ($-C{=}N-$, $-C(OH){=}O$, $-C({=}O)SR$). Some exceptions are (a) in histidine and phenylalanine deaminase, (b) oleic acid hydratase, and (c) ribulose-5-P-4-epimerase. Furthermore, a number of reactions which might earlier have been included in such a list are now known to proceed through intermediates in which an activating carbonyl or a system vinylagous to it is generated as in the following single enzymic reactions:

(1) The conversion of 3-OH decanoyl-CoA to Δ3 decanoyl-CoA seems to proceed through Δ2 decanoyl-CoA (*22*).

$$-CH_2-CH(OH)-CH_2-C({=}O)-SR \longrightarrow -CH_2-CH{=}CH-C({=}O)-SR \longrightarrow HC{=}CH-CH_2-C({=}O)-SR$$

(2) In a similar vein the activation of the C-5 hydrogen of 2-keto-3-deoxy-L-arabonate (L-KDA) to produce α-ketoglutarate semialdehyde (KGSA) seems to proceed via an allylic system conjugated to the keto group (*23*).

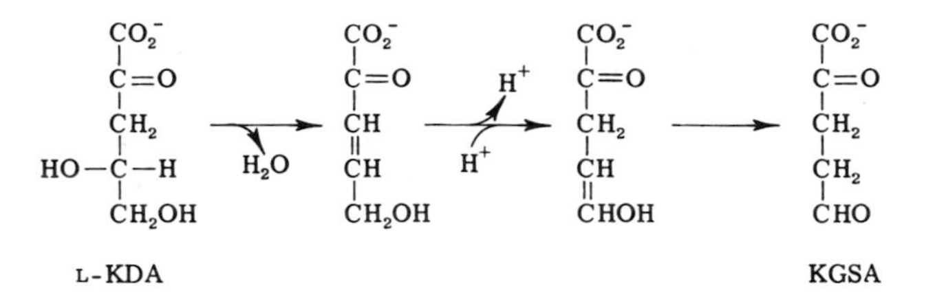

(3) The formation of 6-deoxy sugars in bacteria proceeds via a dehydration mechanism in which a carbonyl is first generated (*24*). The facile dehydration of 4-ketopyranosides has been noted under model conditions (*25*).

The role of the carbonyl in these and the other reactions is to provide forms that delocalize the negative charge left behind by the proton. In

22. R. R. Rando and K. Bloch, *JBC* **243,** 5627 (1968).
23. D. Portsmouth, A. C. Stoolmiller, and R. H. Abeles, *JBC* **242,** 2751 (1967).
24. A. Melo, W. H. Elliott, and L. Glaser, *JBC* **243,** 1467 (1968); O. Gabriel and L. C. Lundquist, *ibid.* p. 1479.
25. O. Gabriel, *Carbohydrate Res.* **6,** 111 (1968).

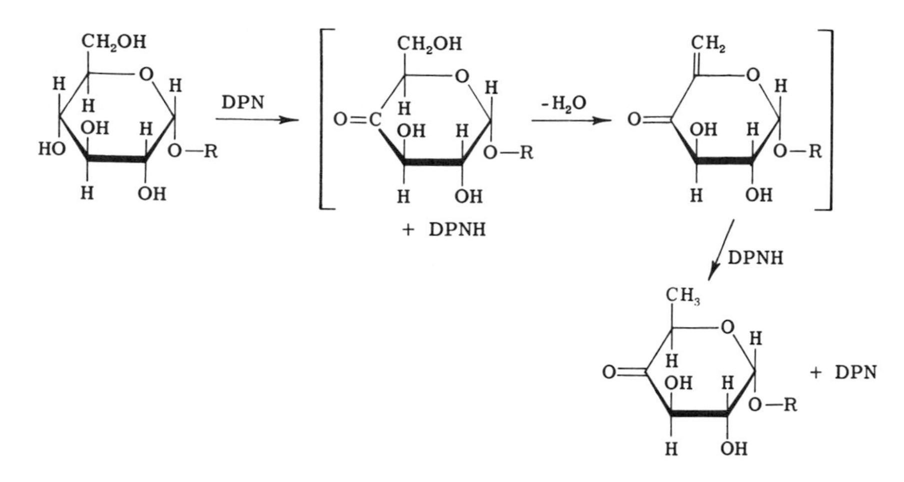

some cases the electrophilic character of the carbonyl is increased by Schiff's base formation (*26*) (see Chapter **7** by Snell and Di Mari) or by association with a metal ion (see Chapter **9** by Mildvan). The role of coenzyme A was first envisaged as a means of strengthening the electrophilicity of the carbonyl of carboxylic acids by thioester formation (*27*). Many of the enzymes that act on carboxylic acid substrates such as isocitratase, β-methylaspartase, and 6-P-gluconate dehydrase are Mg^{2+} or Mn^{2+} dependent and the metal may function to facilitate electron withdrawal from the C—H bond in a carbanion mechanism by complex formation with the carboxyl alpha to the —C—H (*28*). Other of these enzymes, such as fumarase, phenylalanine deaminase, and proline racemase, are not known to require cofactors and must otherwise solve the problem of making the carboxylate anion a good electrophile or else provide the electron delocalization through another group on the substrate (*6*, *29*).

The present chapter attempts to deal primarily with the problem of the role of the enzyme as a base in the direct abstraction of substrate protons. Electron delocalization and proton abstraction are not separable events and are represented together by consecutive curved arrows, $B{:}\ H-C-C=O\ M^{+n} \leftrightarrow BH^+\ C=C-O-M^{+(n-1)}$, where M may be a metal ion, an acidic function of the protein or of the solvent, or may be absent.

Maximum negative charge delocalization in the carbanion occurs when

26. M. L. Bender and A. Williams, *JACS* **88,** 2502 (1966).

27. F. Lynen, *Federation Proc.* **12,** 683 (1953); *J. Cellular Comp. Physiol.* **54,** Suppl. 1, S.33 (1959).

28. H. J. Bright, *JBC* **239,** 2307 (1964).

29. E. A. Havir and K. R. Hanson, *Biochemistry* **7,** 1904 (1968).

the C—H bonding electrons are coplanar with the π orbitals of the carbonyl, which results in the favored geometry of the transition state (*30*):

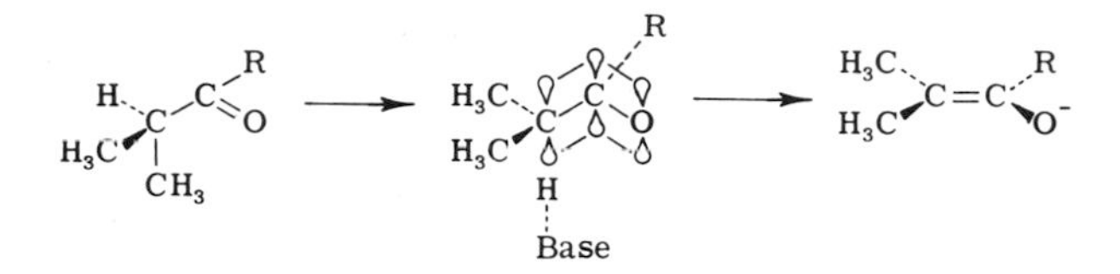

If an additional cation component interacts in a concerted manner at the carbonyl oxygen, it would also be most favored in a position perpendicular to the plane of the molecule at the oxygen and in anticonfiguration with respect to departing proton according to theoretical calculations (*31*). There seems to be no evidence relating to the stereoselectivity of allylic concerted cation displacements, however.

III. Migrations [HS → SH]

A. 1,2-Proton Shifts: Aldo–Keto Isomerases

The best studied example of this class is P-glucose isomerase for which a mechanism of base catalysis is proposed (*3, 3a*). Both transfer of the

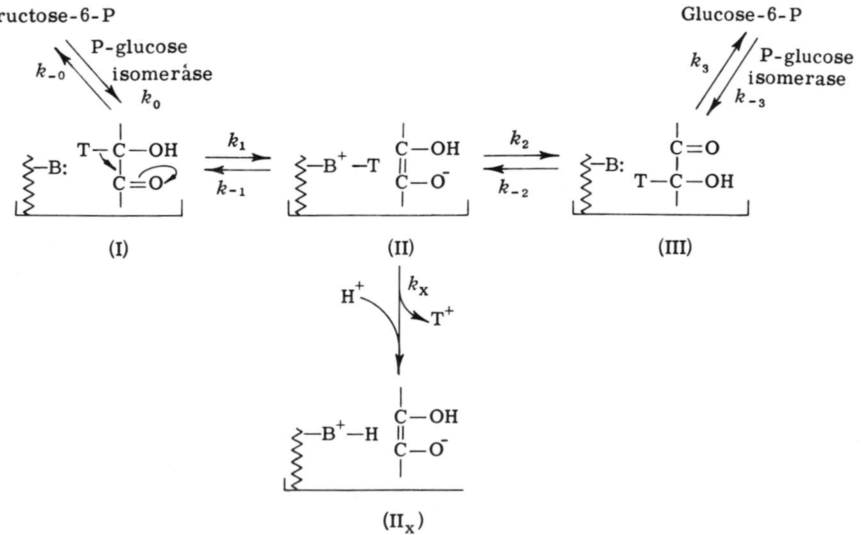

Conjugate acid-enzyme-enediol substrate mechanism

proton and exchange are noted in any statistical number of conversion events. This was rather an unusual observation—an apparent duality in the fate of an atom in an enzymic reaction. The duality was shown to be kinetic in origin since by raising the temperature, complete exchange could be approached and by lowering it, increased retention (80%) was seen. Thus, there was less than a single equivalent of protons diluting the labeled proton in transit through the reaction. This suggested that the basic group on the enzyme was nonprotonic. Since the enzyme was present only in catalytic amount, this conclusion suggested that the transfer was intramolecular, although it did not exclude a mechanism in which there was very limited exchange of protons in the regenerated base (*3*, *3a*, *32*):

$$\left[\begin{array}{l}C{-}T\\ NH_2\end{array}\right] \rightarrow \left[\begin{array}{l}C^-\\ NTH_2^+\end{array}\right] \longrightarrow \left[\begin{array}{l}CH\\ NTH\end{array}\right] \xrightarrow[C'H]{CH} \left[\begin{array}{l}C'H\\ NTH\end{array}\right] \rightarrow \left[\begin{array}{l}C'^-\\ NTH_2\end{array}\right] \rightarrow \left[\begin{array}{l}C'T\\ NH_2\end{array}\right] \rightarrow C'T$$

$$CT + C'H \longrightarrow CH + C'T$$

The transfer was indeed shown to be intramolecular by use of kinetic isotope separation of doubly labeled products (*3*).

The other sugar isomerases show some degree of proton transfer ranging from little for triose-P isomerase (*33*, *33a*) to complete transfer for D-xylose isomerase (*34*). The basis for the large variation in the amount of exchange in the course of the reaction with the different enzymes is not understood. The mechanism may impose the enediol intermediate as a shield between the tritiated group and the medium. The increase in exchange-transfer ratio with temperature (*2*, *3*, *3a*) is difficult to explain on kinetic grounds. A possible explanation might be that the proton dissociation rate is determined by a conformational change with a large activation energy. Attempts to influence the exchange-transfer ratio by specific buffer ions or pH in the P-glucose isomerase case were not successful. Three cases in which very little or no exchange was evident involved nonphosphorylated substrates (*34*). This suggests that the

---

30. E. J. Corey and R. A. Sneen, *JACS* **78**, 6269 (1956); H. E. Zimmerman, *in* "Molecular Rearrangements" (P. de Mayo, ed.), Vol. 1, pp. 345–372. Wiley (Interscience), New York, 1963. H. O. House, "Modern Synthetic Reactions," p. 166. Benjamin, New York, 1965.

31. K. Fukui and H. Fujimoto, *Bull. Chem. Soc. Japan* **39**, 2116 (1966).

32. W. R. Nes, E. Loesser, R. Kirdani, and J. Marsh, *Tetrahedron* **19**, 299 (1963).

33. S. V. Rieder and I. A. Rose, *JBC* **234**, 1007 (1959).

33a. H. Simon, R. Medina, and G. Mullhofer, *Z. Naturforsh.* **23b**, 59 (1968).

34. I. A. Rose, E. L. O'Connell, and R. P. Mortlock, *BBA* **178**, 376 (1969).

phosphate group of the isomerase substrates or the group responsible for binding the phosphate to the enzyme, being in the neighborhood of conjugate acid group, provides a path for proton exchange with water.

Evidence favoring a *cis*-enediol rather than a *trans*-enediol intermediate derives from the relative stereochemistry of the reacting protons in the aldose and ketose pairs for each isomerase (*3a*, *34*, *35*). The single base mechanism with an enediol intermediate requires that proton be added to either C1 or C2 from the same side of the plane of the enediol. As shown in Fig. 1, with the seven enzymes that have been examined, only a *cis*-enediol will satisfy this requirement.

It has been suggested (*3a*, *34*) that the consistency with which the

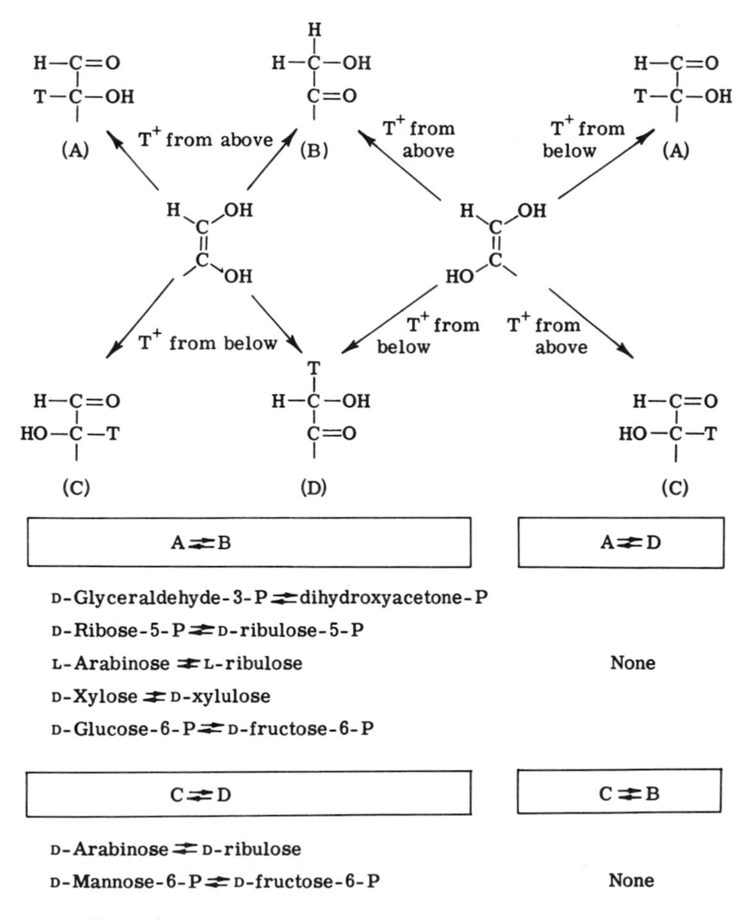

FIG. 1. Stereochemistry of proton addition in the isomerases.

35. I. A. Rose and E. L. O'Connell, *BBA* **42**, 159 (1960).

cis-enediol is found, spanning enzymes of the biological range and including enzymes that do and do not show a metal cofactor requirement, may imply that an acidic group, which may in some cases be water that is associated with a metal ion, operates as an electrophilic center to promote polarization of the carbonyl group at either C1 or C2 as the case may be.

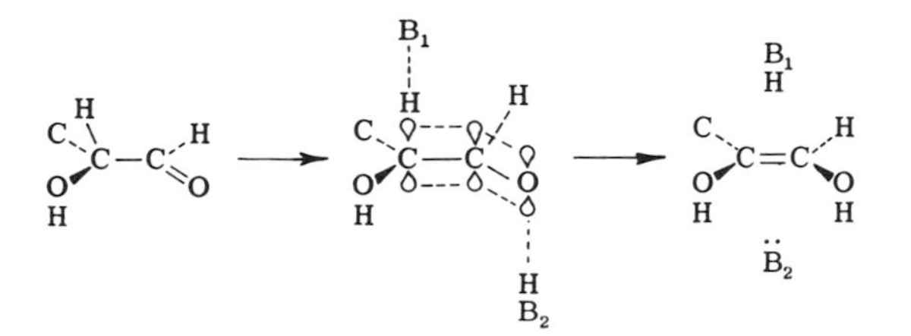

One would expect the basic and acidic groups, B_1 and B_2H of the scheme, to be located at opposite faces of the developing enediol plane in order to avoid their interactions with each other. Alternatively, one might postulate neighboring group assistance by hydrogen bonding as the "reason" for the preservation of the *cis*-enediol in evolution:

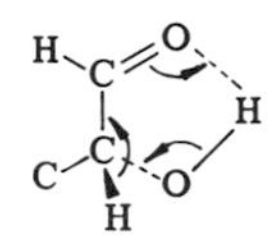

The substrate form that acts in the sugar isomerase reactions has been recently given long overdue attention. The form in which D-glyceraldehyde-3-P reacts with triose-P isomerase is reported to be the carbonyl form (*36*); however, Salas *et al.* (*37*) have presented kinetic evidence that 6-P-α-D-glucopyranose, not the β-anomer, is the true substrate of P-glucose isomerase. Studies with D-xylose isomerase show that the α-anomer (*38*, *38a*), and with L-arabinose isomerase the β-anomer only (*38a*) are reactive substrates. Since the rates of spontaneous mutarotation of the nonphosphorylated sugars are slow, it was possible to examine specificity by a comparison of initial rates with pure and mixed anomers. The conformations of hydroxide at carbons 1 and 2 of the most stable chair (C1) forms of these substrates are as given in the accompanying tabulation. Thus, in the three cases, the C1 and C2 hydroxyls are rela-

36. D. R. Trentham, C. H. McMurray, and C. I. Pogson, *BJ* **114,** 19 (1969).
37. M. Salas, E. Vinuela, and A. Sols, *JBC* **240,** 561 (1965). The report to the contrary by M. S. Feather and M. J. Lybyer [*BBRC* **35,** 538 (1969)] has been retracted (*38*).
38. M. S. Feather, V. Deshpande, and M. J. Lybyer, *BBRC* **38,** 859 (1970).
38a. K. Schray and I. A. Rose, unpublished results, 1970.

	1	2
α-D-Glucose-6-P	a	e
α-D-Xylose	a	e
β-L-Arabinose	a	e

tively cis. It is quite evident that the bond from C-1 to the ring oxygen, as well as the C2 to hydrogen bond, must open in going from aldose to ketose. One may reach the *cis*-enediol intermediate by two basically different processes: (1) direct elimination of the C_2 proton and the C_1 ring oxygen; or (2) ring opening followed by enolization. The elimination,

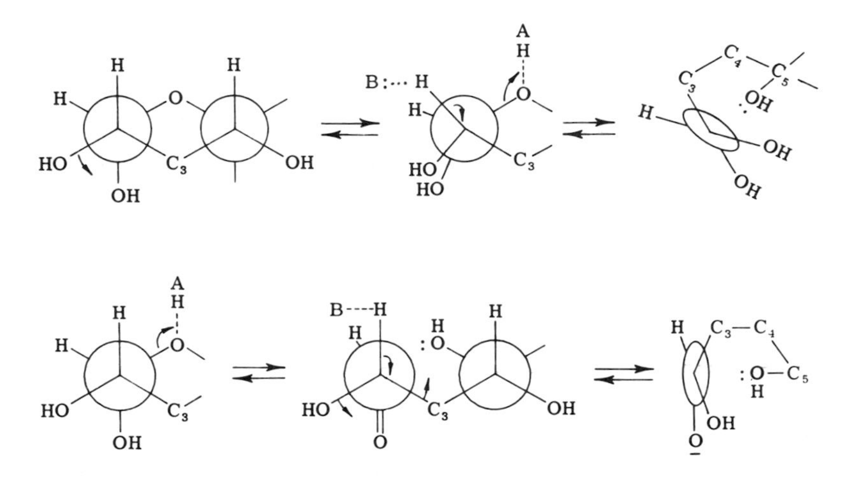

though it seems less promising energetically, is worth considering since the compression of the ring required to obtain the transition to *cis*-enediol geometry is energetically less unfavorable for the cis-relation at C1 and C2 than the increased ring puckering required for the trans-(e,e) relation (*39*). The subsequent elimination is trans. It thus would account for the anomeric specificity.

Mechanism (2), which seems more reasonable, shows that ring closing inevitably leads to a cis-relation between the C_1 and C_2 hydroxyls if the cis-relation of the enediol is maintained in the open chain form of the aldose, that is, if rotation at the C_1-C_2 axis is prevented at the time of ring closing. Thus, ring opening from a *cis*-anomer (C-1 OH and C-2 OH are axial-equitorial or equitorial-axial) leads directly to a cis-orientation of the C-1 and C-2 oxygens in the open form.

39. C. S. Angyal and C. G. MacDonald, *JCS* 686 (1952); L. P. Kuhn, *JACS* **74,** 2492 (1952).

Line broadening of the C-1 proton resonance of xylose by cofactor Mn^{2+} with xylose isomerase has been observed (*40*), indicating a 3–4 bond separation of the two. A role of Mn^{2+} in ring opening and/or enolization has been postulated (see Chapter 9, Section C,3 by Mildvan).

Affinity labeling has been achieved with triose-P isomerase and the agents, 1-iodo-3-hydroxyacetone-P (*41*), 1-bromo-3-hydroxyacetone-P (*42*), and 1,2-epoxipropanol-3-P (*43*). The three laboratories have concluded that inactivation and labeling arise from ester formation with a particular glutamate (*44*), thus suggesting that the γ-carboxyl of glutamate may function as the base in the enolization process. Previous speculations on the basis of iodoacetate- and photoinactivation had favored imidazole catalysis (*45*). Only a carboxyl proton dissociates rapidly enough at neutrality to account for the small amount of tritium transfer observed in the triose-P isomerase reaction. At most, 10% transfer has been reported (*33a*), indicating that $k_x \cong 10\ k_t$. Since the smallest first-order rate constant of the reaction sequence must be greater than the turnover number of an active subunit, $\sim 10^3$ sec^{-1} (*45, 46*), k_x must be greater than 10^4 sec^{-1}. Thus, ammonium, phenol, and imidazolium [$k_{diss} \sim 10$, 10^2, and 10^3 sec^{-1}, respectively (*16*)] are ruled out as proton carriers.

B. 1,1-Proton Shifts: Epimerases

Excluding those enzymes, such as UDP-glucose-4-epimerase, which seem to involve internal oxidation–reduction and intramolecular hydride transfer in their mechanism (*47*), the enzymes of this class may be broadly divided into those that lead to complete proton exchange with the medium and those in which the inversion occurs without exchange. To date there are two examples of the latter: ribulose-5-P-4-epimerase (*48*)

40. A. S. Mildvan and I. A. Rose, *Federation Proc.* **28,** 534 (1969).
41. F. C. Hartman, *BBRC* **33,** 888 (1968).
42. A. F. W. Coulson, J. R. Knowles, and R. E. Offord, *Chem. Commun.* 7 (1970)
43. I. A. Rose and E. L. O'Connell, *JBC* **244,** 6548 (1969).
44. A. F. W. Coulson, J. R. Knowles, J. D. Priddle, and R. E. Offord, *Nature* **227,** 180 (1970); S. G. Waley, J. C. Miller, I. A. Rose, and E. L. O'Connell, *Nature* **227,** 181 (1970); F. C. Hartman, *JACS* **92,** 2170 (1970).
45. P. M. Burton and S. G. Waley, *BJ* **100,** 702 (1966).
46. L. N. Johnson and S. G. Waley, *JMB* **29,** 321 (1967).
47. D. B. Wilson and D. S. Hogness, *JBC* **236,** 1220 (1960); A. U. Bertland and H. M. Kalckar, *Proc. Natl. Acad. Sci. U. S.* **61,** 629 (1968); L. Glaser and L. Ward, personal communication (1969).
48. M. W. McDonough and W. A. Wood, *JBC* **236,** 1220 (1960).

and lactate racemase (*49*, *49a*). In the ribulose-5-P-4-epimerase case there was no evidence for a bound DPN^+ (*48*). In both cases, experiments with $H_2{}^{18}O$ did not show the introduction of isotope into reactants (*48*, *49a*). This evidence can be taken as ruling out OH^- displacement but cannot rule out an oxidation intermediate, the carbonyl of which would be stable or protected from spontaneous exchange with $H_2{}^{18}O$. Thus, UDP-glucose-4-epimerase does not introduce ^{18}O from water into reactants (*50*). However, barring an unusual oxidoreduction pathway (*49*), it seems inevitable that these reactions proceed by a base mechanism with no exchange of the itinerant proton with medium protons. This mechanism requires a rearrangement of the intermediate so that return of the proton occurs from the opposite side of the trigonal carbon formed. Racemization would result if rotation and inversion preceded the return of the proton, mechanism (1):

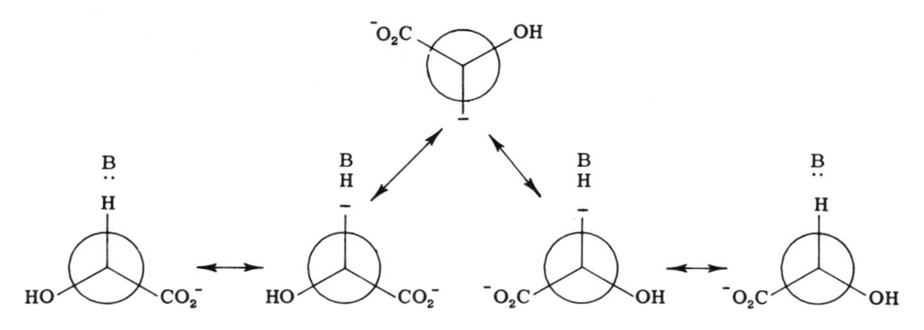

This scheme seems highly unlikely since it requires that the racemic reactants fit interchangeably in the same sites. In the case of ribulose-5-P $\rightleftharpoons$ xylulose-5-P, HO^- and $-CH_2-OPO_3^{2-}$ would have to interchange binding sites on the enzyme. Alternatively, one could have a single conjugate acid change positions with respect to the inverted carbanion, mechanism (2):

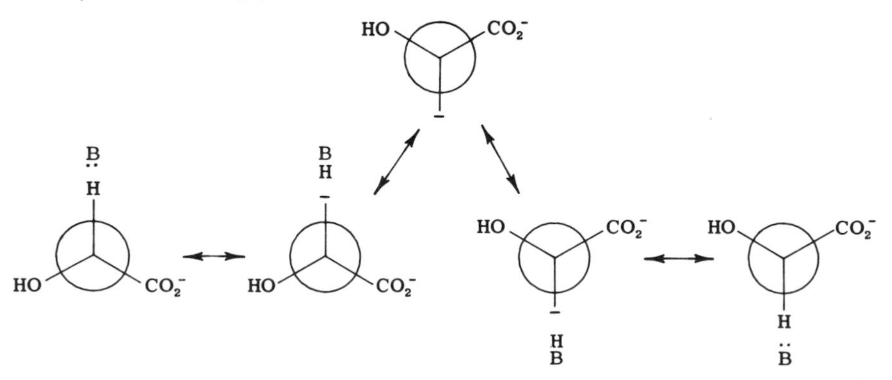

In such a mechanism, B: should be able to move easily, suggesting perhaps the ϵ-NH_2 of lysine or the γ-CO_2^- of glutamate. The presumed conjugate acid should not readily lose its proton to water; hence, —NH_3^+ with $k_{diss} \cong 10\ sec^{-1}$ would be favored over —CO_2H with $k_{diss} \cong 10^7$ sec^{-1} (*16*). NH_3^+ would appear more likely because of its slower dissociation, but it is ruled out since randomization with its hydrogens would limit the extent of transfer to 33%. A two-base mechanism could be postulated with a means for proton migration by way of a helper group (X) that would pass the proton. The isoracemization phenomenon of Cram and Gosser (*51*), in which base-catalyzed racemization greatly ex-

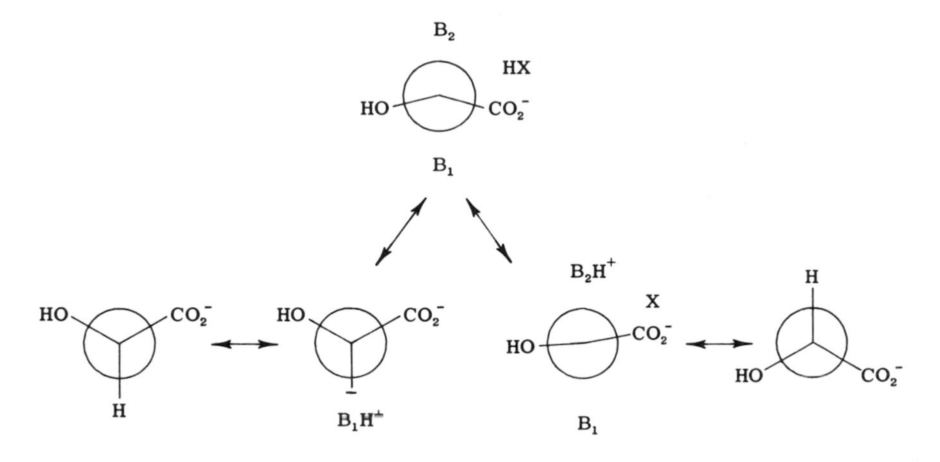

ceeds proton exchange when a proton acceptor group is part of the racemizing molecule, provides a model for this mechanism.

Among the epimerases that show complete proton exchange, ribulose-5-P-3-epimerase (*48*), methylmalonyl-CoA racemase (*52*), and hydroxyproline (*53*) and proline (*6*) racemase, the proline racemase studied by Cardinale and Abeles, may represent a prototype. It has been shown to follow a two-base mechanism:

---

49. S. S. Shapiro and D. Dennis, *Biochemistry* **4**, 2283 (1965).

49a. S. S. Shapiro and D. Dennis, *BBRC* **22**, 635 (1966).

50. L. Anderson, A. M. Landel, and D. F. Diedrich, *BBA* **22**, 573 (1956); A. Kowalsky and D. E. Koshland, *ibid.* p. 575.

51. D. J. Cram and L. Gosser, *JACS* **86**, 2950 (1964).

52. P. Overath, G. M. Kellerman, F. Lynen, H. P. Fritz, and H. J. Keller, *Biochem. Z.* **335**, 500 (1962).

53. E. Adams and I. L. Norton, *JBC* **239**, 1525 (1964).

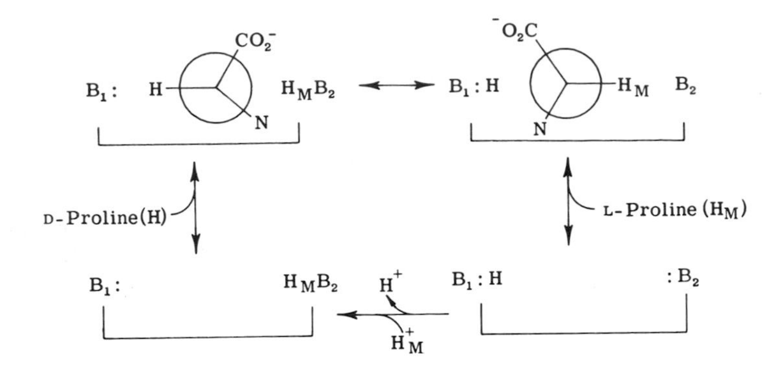

This mechanism is supported by the finding that during racemization in D_2O, deuterium appears in the product immediately and without dilution by substrate proton. Further, at a time when the net flow was largely toward product (small percent of racemization), all of the deuterium was in the product (Table I). This was true in both directions. Thus, the substrate-derived proton does not exchange with the medium until after an "irreversible" step has been taken, the dissociation of product.

TABLE I

COMPARISON OF THE RATE OF ISOMER INTERCONVERSION AND DEUTERIUM INCORPORATION IN PROLINE RACEMASE[a,b]

Substrate	Product in reaction mixture[c] (%)	Deuterium content of D- + L-Proline from reaction mixture	Deuterium content of Proline after D-amino oxidase
L-Proline	5.0	5.2	0.6[d]
L-Proline	9.7	9.6	1.2[e]
D-Proline	6.6–7.2	7.4	—
D-Proline	12.2–13.5	13.4	—

[a] According to Cardinale and Abeles (*6*).

[b] L- or D-Proline was racemized in D_2O to the extent shown. Proline was isolated before and after (in the case of L to D only) treatment with D-amino acid oxidase and the isomer and deuterium content were determined.

[c] Determined directly from the optical rotation of the reaction mixture and by rotation after forming 2,4-DNP-proline.

[d] No D-proline was present in the proline isolated after D-amino acid oxidase.

[e] After D-amino acid oxidase treatment the D-proline was 0.7% of the reisolated proline.

C. 1,3-Proton Shifts: "Allylases"

Two major mechanisms for allylic rearrangement are those in which the proton is first removed, leaving a carbanion which is then protonated at either C1 or C3 with or without transfer:

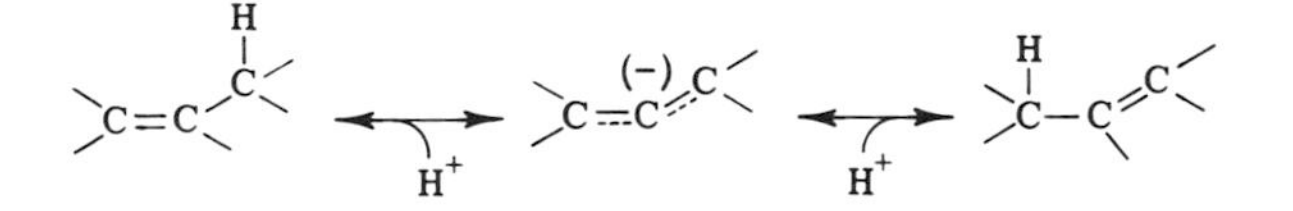

The second possible kind of mechanism is one in which a proton is first added to the vinyl carbon resulting in a carbonium intermediate which loses a proton to form either the product or return to the substrate. No transfer is possible with this mechanism:

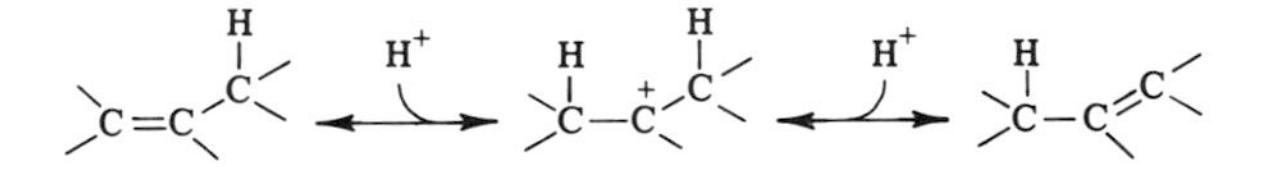

Transfer is clearly shown by two distinctive examples: ketosteroid isomerase (*1, 2*) and pyridoxamine-pyruvate transaminase (*4*). In the former case, proton transfer (see Section I) is complete and stereospecific between axial allylic positions (*2*) of the same molecule (*2*). This stereochemistry as well as the transfer are consistent with a single base mechanism in which the relative motion of the carbanion and the conjugate acid are restricted so that the stereochemistry predicted electronically (*30*) is observed.

The deuterium transfer observed between 2-^{2}H-alanine and pyridoxal is of critical importance to the study of transaminase reactions. It immediately indicates the participation of a basic group on the enzyme that functions in the abstraction of the C-2 proton. The transfer is interpreted as occurring in an allylic rearrangement of the anticipated aldimine intermediate (I) as shown at top of page 300. Ayling, Dunathan, and Snell (*4*) examined by mass spectrometry the pyridoxamine formed in the reaction with α-^{2}H-L-alanine after only a small percent of the equilibrium amount of product had been formed. They found 2–4% of the pyridoxamine to be monodeuterated. The small amount of transfer may be attributed to exchange with the medium at stage (II) or to mixing of D with hydrogens brought in with the base and a subsequent discrimination against D in the formation of (III) (*4*). One could presumably evaluate the latter by examining the isotope con-

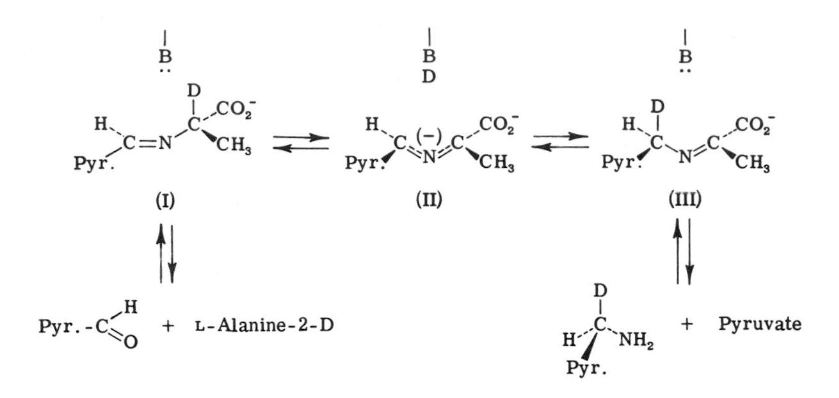

tent of pyridoxamine formed in a mixed H_2O—D_2O medium from normal alanine. A correction for this would allow the calculation of the number of hydrogens present in the base if there were no exchange at stage (II). The statistics of isotope dilution by enzyme protons have been considered in a recent study of coenzyme B_{12} function (*54*).

Dunathan *et al.* (*55*) have found that the transamination from pyridoxamine to α-ketoglutarate by glutamate-oxalacetate transaminase is stereospecific with respect to the methylene protons at C4 of pyridoxamine. The specificity was the same as found for pyridoxamine pyruvate transaminase (*4*). The stereoisomers could be recognized by the 2-fold isotope effect observed with deuterium in the activated position. The isomer of pyridoxamine that is deuterated by exchange in D_2O during transamination by these enzymes was found to be of the *S* configuration (*55, 56*). With this conclusion, and the likely assumption that the aldimine is trans at the C=N, then the proton migration occurs to and from the same face of the allylic anion (II) (*4*). A chemical model showing both a moderate amount of transfer and complete stereospecificity is provided by the 1,3 isomerization studied by Guthrie, Meister, and Cram (*57*):

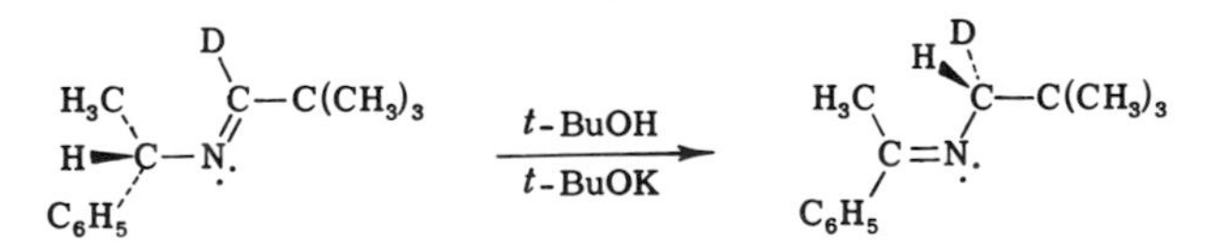

54. W. W. Miller and J. H. Richards, *JACS* **91,** 1498 (1969).
55. H. C. Dunathan, L. Davis, P. G. Kury, and M. Kaplan, *Biochemistry* **7,** 4532 (1968).
56. P. Besmer and D. Arigoni, *Chimia (Aarau)* **23,** 190 (1969).
57. R. D. Guthrie, W. Meister, and D. J. Cram, *JACS* **89,** 5288 (1967).

A carbonium ion mechanism has been proposed for isopentenylpyrophosphate isomerase (*58*):

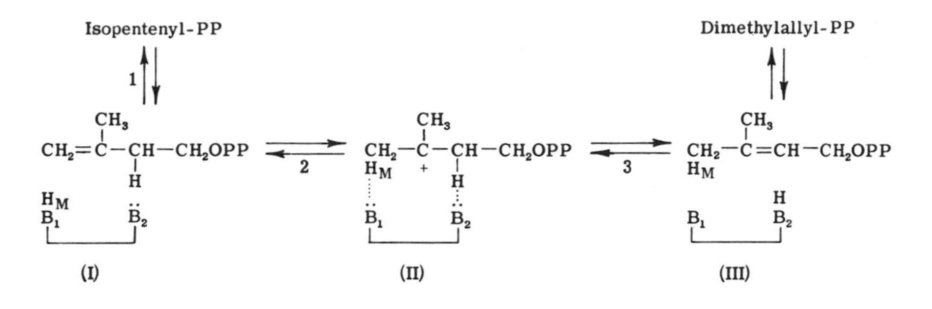

No evidence for proton transfer has been obtained; the data available suggest that the saturating proton is derived from the medium (*58*, *59*). Back-labeling of dimethylallyl-PP was not observed in tritiated water (*59*) and starting with isopentenyl-PP the initial labeling of product was much more rapid than starting material, the labeling of which showed a lag (*58*). A two-base carbanion mechanism in which exchange did not occur with the carbanion intermediate bound to the enzyme would give similar results (*8*). The studies of proline racemase by Cardinale and Abeles (*6*) provide similar kinds of data. A distinction between the carbanion and carbonium pathways may be obtained by detailed kinetic studies. Exchange of the methylene protons of isopentenyl-PP should not show a deuterium isotope effect in the carbonium mechanism with 2-^{2}H-isopentenyl-PP since the path of exchange requires steps 1 and 2 only. In a carbanion mechanism the C_2—H bond must be broken prior to protonation of the methylene group.

An allylic rearrangement must be involved in the formation of β-γ unsaturated decenoyl-CoA from the β-hydroxyl thioester by β-hydroxydecanoyl thioester dehydrase according to Rando and Bloch (*22*). This enzyme also forms the α-β unsaturated ester as a product and this, in the bound form, may be the intermediate between the β-OH ester and the β,γ unsaturated ester. Direct dehydration to the $\Delta\beta,\gamma$ ester does not occur as shown by the loss of one of the (labeled) hydrogens from C2, indicating prior α,β dehydration. On the other hand, dehydration to the $\Delta\alpha,\beta$ ester does not give rise to loss of γ-^{3}H- from the β-hydroxy ester. The stereochemistry of the reactions has not been established, and it will be of special interest to know whether proton transfer (α-^{3}H $\Delta\beta,\gamma$ ester $\rightarrow$ γ-^{3}H

58. D. H. Shah, W. W. Cleland, and J. W. Porter, *JBC* **240,** 1946 (1965).
59. B. W. Agranoff, H. Eggerer, U. Henning, and F. Lynen, *JBC* **235,** 326 (1960).

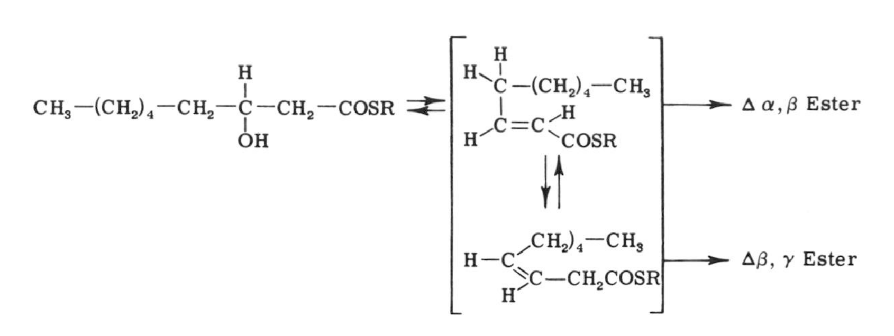

$\Delta\alpha,\beta$ ester) can be detected as an indication of a single base carbanion mechanism.

Active-site labeling and inactivation of this bifunctional enzyme has been achieved by 3-decynoyl thioester, and histidine is reported to be alkylated (*60*).

IV. α,β Eliminations

A. Aconitase

Studies with this enzyme provided the first evidence of basic catalysis by an enzyme of this class (*5*). Aconitase is formally a bifunctional enzyme in that the abstracted proton may either be transferred, isocitrate: citrate isomerase, or if it dissociates, the reaction is an elimination. The reaction scheme, as shown at top of page 303, has the following aspects. The isomerase reaction occurs with no loss of the reacting hydrogen, thus the tritium of citrate formed initially from (2R,3S) [3-^{3}H]isocitrate was undiluted. The transfer is, in effect, intramolecular. In this reaction the oxygen of (2-^{18}O)isocitrate was not transferred. The dissociation of *cis*-aconitate occurs prior to that of tritium since otherwise incomplete tritium transfer would be seen. Furthermore, one can demonstrate intermolecular tritium transfer from [2-^{3}H]2-methylcitrate to *cis*-aconitate to produce [3-^{3}H]isocitrate.

An unusual feature of the isomerase reaction mechanism comes from the stereochemistries of citrate (*61, 61a*) and isocitrate (*62*). Both hydra-

60. K. Bloch, *Accounts Chem. Res.* **2,** 193 (1969).
61. S. Englard, *JBC* **235,** 1510 (1960).
61a. K. R. Hanson and I. A. Rose, *Proc. Natl. Acad. Sci. U. S.* **50,** 981 (1963).
62. T. Kaneko and H. Katsura, *Chem. & Ind.* (*London*) p. 1188 (1960); A. L. Patterson, C. K. Johnson, D. van der Helm, and J. A. Minkin, *JACS* **84,** 309 (1962); O. Gawron, A. J. Glaid, III, A. LoMonte, and S. Gary, *ibid.* **80,** 5856 (1958).

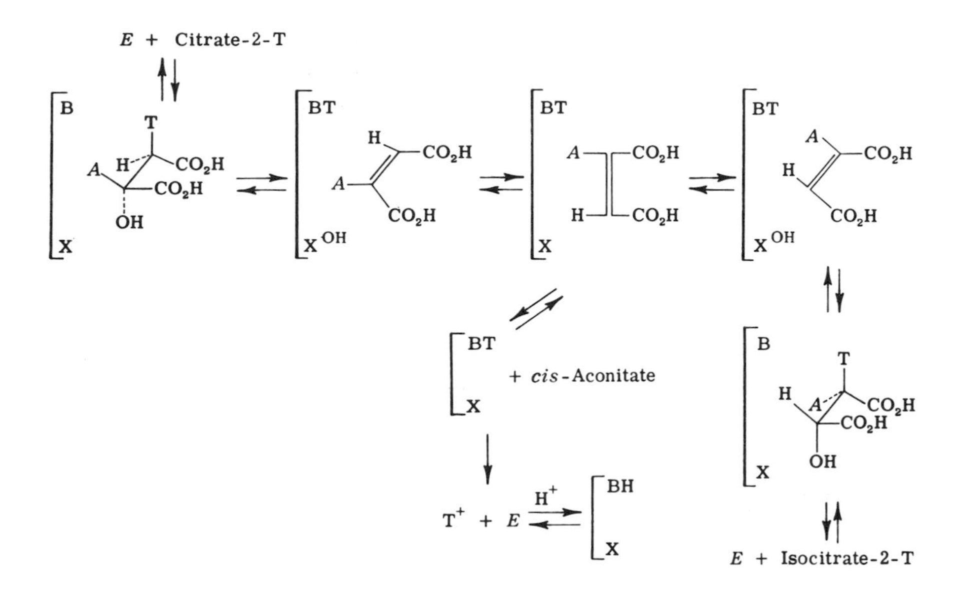

tions of *cis*-aconitate are anti-additions, but they are such that the proton, as well as the hydroxyl, approach alternative faces of *cis*-aconitate in making either citrate or isocitrate as shown in the scheme. This requires that the *cis*-aconitate assumes different geometries, with respect to the proton-donating site at least. Three mechanisms have been proposed for achieving the rearrangement of the intermediate *cis*-aconitate formed in the interconversion of the hydroxyacids which occurs without dissociation of the *cis*-aconitate from the medium: (1) the *cis*-aconitate flips over while remaining attached by the $—CH_2—CO_2^-$ group (*63*); (2) (shown in the scheme above) the *cis*-aconitate flips over in a space within the protein from which it either returns without mixing with external medium or diffuses out to form free *cis*-aconitate (*5*); and (3) the *cis*-aconitate bound to the enzyme at the 1 and 4 carboxyls and the Fe^{2+} at the central carboxyl undergoes a 90° rotation to expose opposite faces to EH and to a H_2O ligand of the Fe^{2+} (*64*). This mechanism, pictured in Fig. 2 and elsewhere in this book [see Chapter 9 (Fig. 16) by Mildvan], is consistent with known conformations of citrate and isocitrate and the metal chelates that have been studied in the crystalline state and provides both a catalytic and structural role for the Fe. Its great ingenuity lies in providing an explanation for the necessity of

63. O. Gawron, A. J. Glaid, III, and T. P. Fondy, *JACS* **83**, 3634 (1961).
64. J. P. Glusker, *JMB* **38**, 149 (1968).

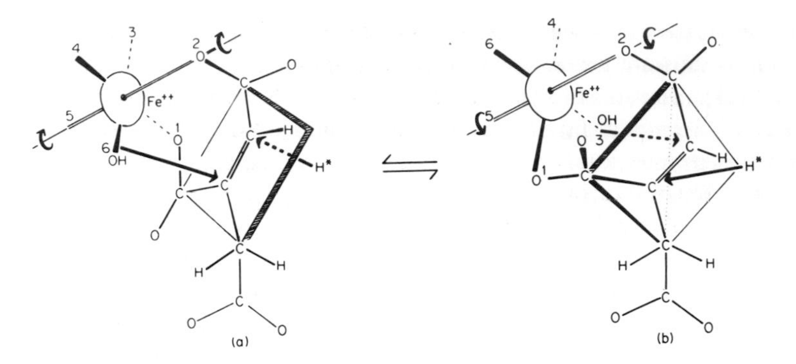

FIG. 2. Fe^{2+}-*cis*-Aconitate complexes for aconitase in the "citrate-like" (a) and "isocitrate-like" (b) conformations, proposed by Glusker (*64*).

alternate face attacks by the —OH groups and by the single acid group donating the proton.

B. FUMARASE

In a definitive isotope exchange study Hansen, Dinovo, and Boyer (*7*) have established the following sequence of events in the fumarase reaction:

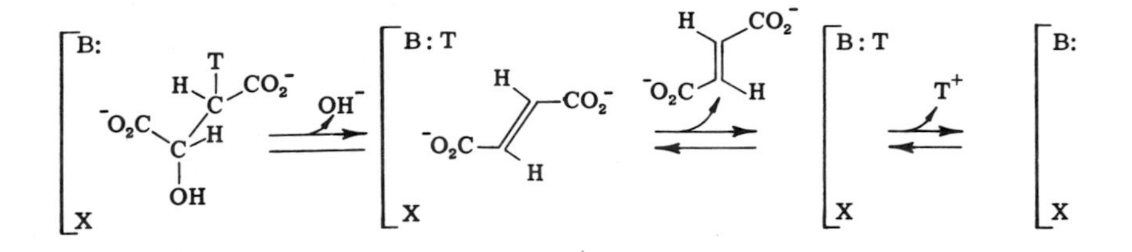

Studying the exchange of malate α-hydroxyl in $H_2{}^{18}O$, it was found that ^{18}O entered malate much more rapidly than would have been expected if the only route had been the return of newly formed free fumarate. The fact that no lag was observed in the exchange indicated that the steady state for exchange was established immediately, that it depended only on the concentration of an intermediate. Exchanges were measured at equilibrium in a medium of equivolumes of glycerol and water, which provided the greatest range of differences in rates. The results (Table II) show that tritium exchange between 3-(R)^{3}H-malate and water is less than half the rate of ^{14}C-fumarate–malate exchange at equilibrium containing 0.05 M fumarate. There is no discrimination between tritiated and normal malate in the fumarase reaction (*7*, *65*) and the preferential fumarate exchange depended on the presence of high fumarate. This must mean that in the reaction of malate, fumarate may be liberated to the medium for exchange prior to liberation of the tritium. Thus the hydrogen resides on the enzyme long enough to be captured for return to malate by free fumarate. It has been noted (*66*) that the turnover rate of fumarase, about 2×10^3/sec, is similar to the rate of imidazolium dissociation and the protonation of imidazole as measured by temperature jump (*16*), suggesting that dissociation of the conjugate acid of imidazole limits V_{max}. The lack of a deuterium rate effect is not pertinent since, as indicated in Section II,B, one does not expect isotope effects in prototropic reactions involving such acids. On the other hand, it seems unlikely that proton dissociation could be rate limiting by the following reasoning. According to the data of Hansen *et al.* (*7*), it requires about $10^{-2}\,M$ fumarate to halve the (3-^{3}H)malate:H_2O exchange rate relative to the ^{14}C-fumarate–malate exchange rate. The fumarate dissociation constant of enzyme fumarate is reported to be $5 \times 10^{-6}\,M$ (*67*), hence the effect of $10^{-2}\,M$ fumarate would be to lower enzyme-^{3}H by 2000-fold ($10^{-2}/5 \times 10^{-6}$). If, as seems indicated (below), the presence of fumarate in the complex prevents dissociation of the tritium, it is clear that this step must not be rate limiting since it should require only the K_s concentration of fumarate to halve enzyme-^{3}H. As seen in Table II, the stability of the tritiated conjugate acid group seems to be dependent on the conditions of the medium since 0.3 M $(NH_4)_2SO_4$ added to the medium increased the ^{3}H exchange rate substantially. This may indicate that conformational reactivity of the enzyme limiting the accessibility of water may limit the proton exchange rate.

When similar equilibria isotope exchange studies were conducted at low concentrations of the substrates the ^{3}H exchange was reported to be significantly greater than the ^{14}C exchange (*7*). This result suggests that proton can leave the enzyme before the fumarate dissociates. In other words, the $E^{H+}_{fumarate}$ complex may dissociate randomly. This would require that the conjugate acid can be exposed to the medium even with fumarate in position on the enzyme. This is important in view of the known anti-relation of —OH and proton additions to the plane of fumarate. One would have to say that both faces of fumarate are exposed

65. R. A. Alberty, W. G. Miller, and H. F. Fisher, *JACS* **79,** 3973 (1957).

66. E. M. Kosower, "Molecular Biochemistry," p. 118. McGraw-Hill, New York, 1962.

67. R. A. Alberty and W. H. Pierce, *JACS* **79,** 1526 (1957); R. A. Alberty, "The Enzymes," 2nd ed., Vol. 5, p. 531, 1961.

TABLE II

EQUILIBRIUM EXCHANGE RATES OF ^{18}O, ^{14}C, AND ^{3}H IN THE FUMARASE REACTION[a]

$(NH_4)_2SO_4$ (M)	Exchange	Relative rate
0	^{3}II(malate) ⇄ ^{3}H(water)	1.0
0	^{14}C(malate) ⇄ ^{14}C(fumarate)	2.3
0	^{18}O(malate) ⇄ ^{18}O(water)	9.2
0.3	^{3}H(malate) ⇄ ^{3}H(water)	2.3
0.3	^{14}C(malate) ⇄ ^{14}C(fumarate)	2.7
0.3	^{18}O(malate) ⇄ ^{18}O(water)	11.6

[a] A reaction mixture contained 0.104 M L-malate containing tritium at the 3-R position, 0.050 M ^{14}C-fumarate (1.5 μCi), 0.03 M tris acetate, and 110 μg/ml fumarase in 7.2 ml of 1 v:1 v glycerol:water at pH 7.3 and 25°. Rates were calculated from five measurements for each exchange at appropriate time intervals ranging from 6 to 300 min. Glycerol was found to increase the ^{18}O-exchange relative to fumarate formation in initial rate studies (*7*); data from Hansen *et al.* (*7*).

for diffusive exchange with the medium. A critical experiment concerning the prior release of proton is whether indeed any proton exchange into malate is observed when malate is forming fumarate. The work of Alberty's group suggests an upper limit of 10% for this exchange rate relative to the rate of fumarate production (*65*). More sensitive tests using $^{3}H_2O$ and earlier times would improve the sensitivity of these experiments. Such an experiment at very high malate concentration for a short interval to minimize the back reaction of formed fumarate showed that proton exchange, if any, was not greater than 1% of the forward rate (*68*).

It is worth emphasizing that there are no data that focus on the fundamental question of whether the abstraction of proton from malate precedes, follows, or is concerted with breaking the C—OH bond. Previous optimism in the use of isotope exchange to examine this question (*65*) must be tempered with the likelihood that a lack of proton exchange may

68. I. A. Rose, unpublished (1969): In 0.2 ml at pH 7.2 were incubated Na malate (0.4 M, freed of traces of fumarate), fumarase (0.015 unit), and $^{3}H_2O$ (125,000 cpm/μatom). After 0.104 μmole of fumarate was formed, the incubation was diluted with H_2O (100-fold) and passed through a Dowex-1-Cl column for the isolation of malate. The malate was treated with fumarase and the total distillable radioactivity of 164 cpm was obtained, corresponding to 0.0013 μmole of ^{3}H incorporated. A determination of back reaction that would be expected under these conditions was obtained in a similar incubation to which the time average amount of ^{14}C-fumarate was added initially. An exchange of 0.0003 μmole was found. Thus, one may say that any proton exchange, if it occurs, cannot be greater than 1% of the forward rate: 0.001 μmole, compared with 0.1 μmole of fumarate produced.

result from a slow dissociation of the proton from the enzyme, and similar restrictions on the interpretation of —OH exchange rates may be required (*68a*).

In the fumarase case it is clear that the C—OH bond breaking step must precede the rate-limiting step since the exchange of $H_2{}^{18}O$ with malate occurs prior to completion of the reaction sequence. However, in spite of the fact that barely detectable proton exchange can be observed under these conditions, it is possible that —C—H bond cleavage precedes the slow step and that proton exchange is simply limited by release of the proton from the enzyme. The lack of a deuterium isotope effect taken with the occurrence of ^{18}O exchange clearly indicates that neither bond breaking step can be rate limiting for the net reaction.

Schmidt *et al.* (*18b*) have measured secondary isotope effects in the dehydration of malate by isotope competition using 2-^{3}H-malate and malate prepared from 2,3-^{3}H-fumarate, both mixed with ^{14}C,^{1}H-malate. The $^3H/^{14}C$ ratio of fumarate formed initially was compared with that of the malate and found to be about 11% lower for both mixtures. This isotope effect, corresponding to $k_H/k_T = 1.12$, relates to the portion of the mechanism, ordered as above according to Hansen *et al.*, between malate and free fumarate and does not include the final portion in which proton is released from the enzyme which follows the irreversible release of fumarate,

$$E + \text{malate} \underset{1}{\rightleftharpoons} E_{\text{malate}} \overset{{}^-OH}{\underset{2}{\rightleftharpoons}} E^H_{\text{fumarate}} \rightarrow E^H + \text{fumarate}$$

Since the return of E^H_{fumarate} to E + malate was found by Hanson *et al.* (*7*) to be at least 2.3 times the rate of release of fumarate from E^H_{fumarate}, the sequence made up of steps 1 and 2 must be at least 2.3/3.3 = 70% toward equilibrium. Thus a large portion of the equilibrium isotope effect for the reaction (*18b*)

$$\text{Malate-2,3-}^3\text{H} \rightleftharpoons \text{fumarate-2,3-}^3\text{H} + H_2O,\ K_H/K_T = 1.23$$

should be observed in the rate studies due to the quasi-equilibrium,

68a. As explanation of incorporation of oxygen but not hydrogen from water into unreacted malate, P. D. Boyer (personal communication) suggests consideration of formation of a discrete E·X⁺, where X⁺ is a carbonium ion, as follows:

$$E_{\text{malate}} \rightleftharpoons E \cdot X^+ + OH^-$$
$$EX^+ \rightleftharpoons E^H_{\text{fumarate}}$$

The additional $E \cdot X^+$ step allows incorporation of oxygen without the otherwise anticipated incorporation of H^+ and fumarate from interchanges of E^H_{fumarate} with medium components.

$$E + \text{malate} \rightleftharpoons E^{H}_{\text{fumarate}} + OH^-$$

This may be the explanation of the equivalent discrimination seen with tritium at either C-2 or C-3 of malate (*68b*).

Since the α-OH–water exchange is the fastest measured rate, one could try to get at the chemistry by looking for primary and secondary deuterium isotope effects in the malate-$H_2{}^{18}O$ exchange either at equilibrium or initial rate conditions. Other approaches might be to alter the enzyme or the reaction kinetics in such a way as to make one or both of the bond breaking steps rate determining.

The aconitase and fumarate reactions, despite the Fe^{2+} requirement for aconitase, have several features in common. For both, the exchange rates at equilibrium are $^{18}OH > {}^{14}C > {}^{3}H$. For both, there are no primary isotope rate effects in the dehydration of deuterated substrates (*65, 69*). One may ask if there is with the fumarase reaction anything analogous to the apparent "flip over" of *cis*-aconitate seen in the aconitase. This would not be observed as an isomerization, as in the case of aconitase because of the symmetry of fumarate, but rather as a randomization of a ^{14}C-carboxyl label in malate without dissociation of fumarate:

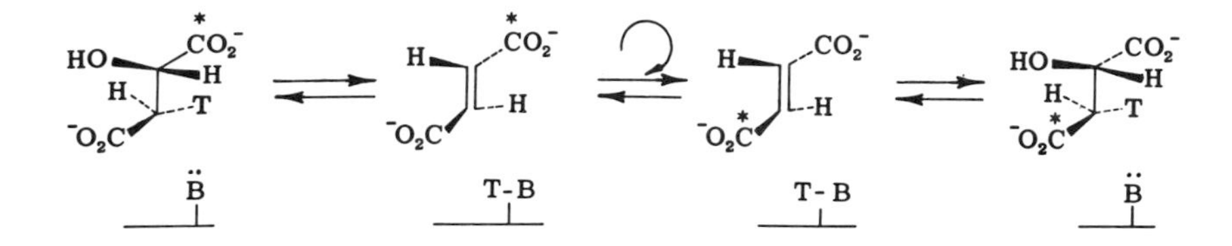

The results of such an experiment (*69a*) show no evidence for rotationally free fumarate that is not completely off the enzyme. Similar results were obtained with aspartase of *Pseudomonas* (*69a*).

68b. A further discussion of the fumarase reaction mechanism as revealed by these isotope studies is given by J. H. Richards, Chapter 6 of this volume.

69. J. F. Thomson, S. L. Nance, K. J. Bush, and P. A. Szczepanik, *ABB* **117**, 65 (1966).

69a. I. A. Rose, unpublished (1966): A typical experiment with fumarase is the following: two incubations at 40° contained: P_i buffer (40 m*M*, pH 7.3), malate (0.6 m*M*), fumarate (0.2 m*M*), and 0.013 unit of pig heart fumarase/ml with either 1-^{14}C-malate or 1,4-^{14}C-fumarate. Samples were taken, malate and fumarate were separated and counted, and the ^{14}C at C-4 of malate was determined as CO_2 by converting malate to oxalacetate and specific β-decarboxylation with analine citrate. At 8 min, for example, 6.7% of the counts initially in malate were in fumarate, and 20% of the counts initially in fumarate were in malate. Only 0.25% of the initial malate counts were in C-4. This is somewhat *fewer* than can be explained by the route of ^{14}C-malate to free fumarate and back again randomly. Thus, *all* of the randomly returned malate came from the mixed pool of free fumarate.

C. Stereochemistry of Elimination Reactions

Most of the reactions in which substrates are dehydrated or deaminated lead to anti, "trans-"eliminations. The list includes fumarase (*63, 70*), aspartase (*71*), argininosuccinase (*72*), adenylosuccinase (*73*), malease (*74*), aconitase (*61, 61a, 62*), histidine deaminase (*75*), enolase (*76*), and oleic acid hydratase of *Pseudomonas* (*77*). In addition the elimination of CO_2 and OH^- from 5-pyrophosphomevalonate is anti (*78*). There are three arguments that may be considered to explain the preponderance of anti-eliminations:

(1) Concerted syn-processes are forbidden by orbital symmetry rules (*31, 79*); however, as indicated already, there is no evidence available to define these enzymic reactions as concerted.

(2) The attack at a double bond by a nucleophile or electrophile results in a carbanion or carbonium ion intermediate in which the stereochemistry at the β carbon is anti. Orbital symmetry factors forbid the syn-relation (*31, 79*). Thus, ^-OH attack on fumarate leads to the carbanion (II):

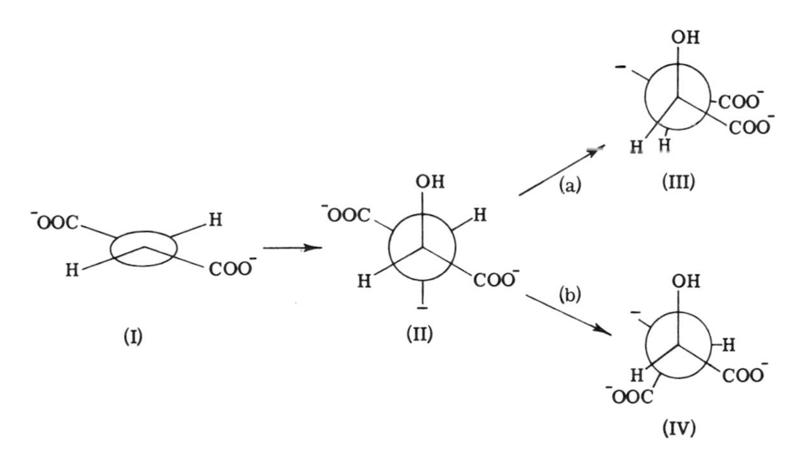

70. F. A. L. Anet, *JACS* **82,** 994 (1960).

71. S. Englard, *JBC* **233,** 1003 (1958). A. I. Krasna, *ibid.* p. 1010.

72. H. D. Hoberman, E. A. Havir, O. Rachovansky, and S. Ratner, *JBC* **239,** 3818 (1964).

73. R. W. Miller and J. M. Buchanan, *JBC* **237,** 491 (1962).

74. S. Englard, J. S. Britten, and I. Listowsky, *JBC* **242,** 2255 (1967).

75. I. L. Givot, T. A. Smith, and R. H. Abeles, *JBC* **244,** 6341 (1969).

76. M. Cohn, J. Pearson, E. L. O'Connell, and I. A. Rose, *JACS* **92,** 4095 (1970).

77. G. L. Schroepfer Jr., *JBC* **241,** 5441 (1966).

78. J. W. Cornforth, R. H. Cornforth, G. Popjak, and L. Yengoyan, *JBC* **241,** 3970 (1966).

The specificity of fumarase for fumarate and analogs as inhibitors indicates that binding sites for the carboxyls are fixed relative to each other. Thus, free rotation, reaction (a), at the C_2—C_3 bond of the hypothetical carbanion is limited. In any case, free rotation followed by proton addition to the sp^3 carbanion would not be evaluated as syn-addition. Inversion to (IV) for an uncomplexed carbanion involves a 6–10 kcal/mole energy barrier (*80*) and must be greater when an amino acid side chain, whose interaction with the carboxyl contributes to the substrate specificity, must be moved as well. In the case of reactions in free solution the interconversion of (II) and (IV) may be rapid compared with the second addition, the stereoselectivity of which will depend on steric and electrical effects. However, in the enzymic process, the inversion may be considerably slower relative to the second addition, so that the orbital restrictions dominate the stereochemistry.

(3) A third factor relates not to the "binding sites" but to the stereochemistry of the catalytic sites. The chance of a neutralization between the acid and base groups of the catalysis should be excellent if the two groups are disposed at the same face of the plane of the double-bonded conjugate system.

It is of interest that these two factors, orbital symmetry rules and the separation of catalytic centers, come into conflict in the case of 1,4 additions across a four carbon chain containing two double bonds. Here, concerted additions in the anti-relation are forbidden (*79*). It has recently been shown (*81, 81a*) that chorismate synthase promotes the anti-elimination of H^+ and OPO_3^-:

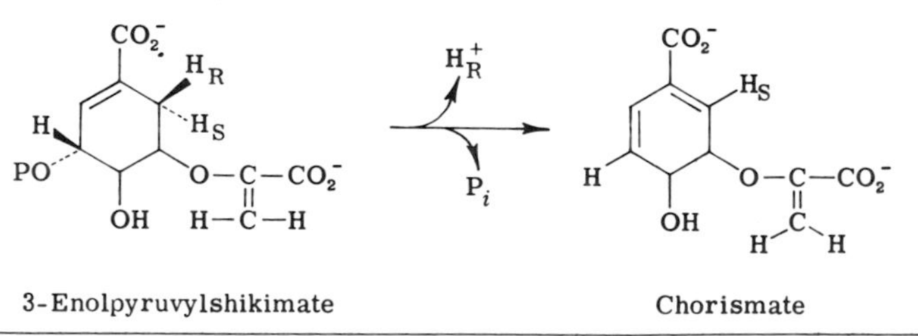

79. S. I. Miller, *Advan. Phys. Org. Chem.* **6**, 185 (1968).
80. G. W. Koepple, D. S. Sagatys, G. S. Krishnamurthy, and S. I. Miller, *JACS* **89**, 3396 (1967).
81. R. K. Hill and G. R. Newkome, *JACS* **91**, 2893 (1969).

Thus, concerted elimination is thereby ruled out. One would conclude that the preferred spacial separation of catalytic acid and base controls the stereochemistry of the reaction by allowing reaction with the carbanion or carbonium ion only after steric inversion. An alternate scheme has been proposed (*81a*) involving SN2′ replacement by a nucleophile, expected to be syn (*82*) followed by anti-elimination:

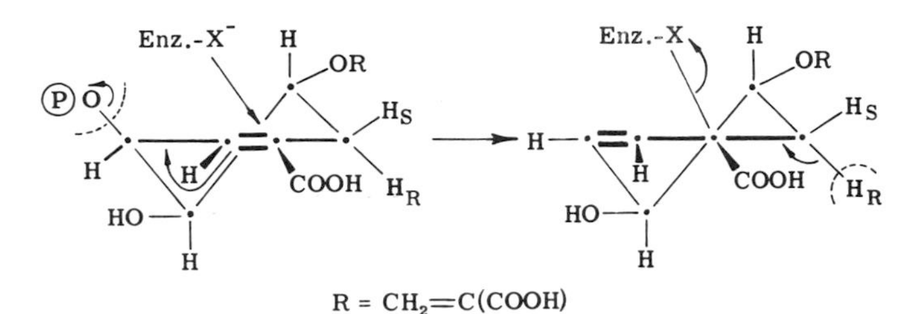

[After Onderka and Floss (*81a*).]

Having said this, the known exceptions to anti-addition to a single double bond must be considered. The first case reported was β-methylaspartase where both L-*threo*- and L-*erythro*-β-methylaspartate are produced from mesaconate in a K^+ and Mg^{2+} requiring reaction (*83*). The V_{max} of syn-elimination with the L-*erythro* isomer was only 1% of the deamination of L-*threo*-β-methylaspartate. Subsequent work on this enzyme by Bright and co-workers (*28, 84*) has presented evidence for a carbanion mechanism based primarily on the exchange of the β proton of the L-*threo* isomer with water that was independent of the net formation of either product. Notwithstanding the limitations of such evidence, explicitly stated in these papers, the carbanion mechanism offers a reasonable explanation of the apparent syn-elimination pathway. In this mechanism the enzyme is both a deaminase and an epimerase (*85*).

Two further examples of apparent syn-elimination may also be written as carbanion mechanisms, but there are no experiments related to this question. These are dehydroshikimate hydratase (*61a*) and *cis-cis*-muconate cycloisomerase (*86*).

81a. D. K. Onderka and H. G. Floss, *JACS* **91**, 2894 (1969).

82. G. Stork and F. H. Clarke, *JACS* **78**, 4619 (1956); G. Stork and W. N. White, *ibid.* p. 4609.

83. H. A. Barker, R. D. Smyth, R. M. Wilson, and H. Weissbach, *JBC* **234**, 320 (1959).

84. H. J. Bright, L. L. Ingraham, and R. E. Lundin, *BBA* **81**, 476 (1964).

85. K. R. Hanson and E. A. Havir, *Recent Advan. Phytochem.* **3**, (1970) (in press).

86. G. Avigad and S. Englard, *Federation Proc.* **28**, 345 (1969).

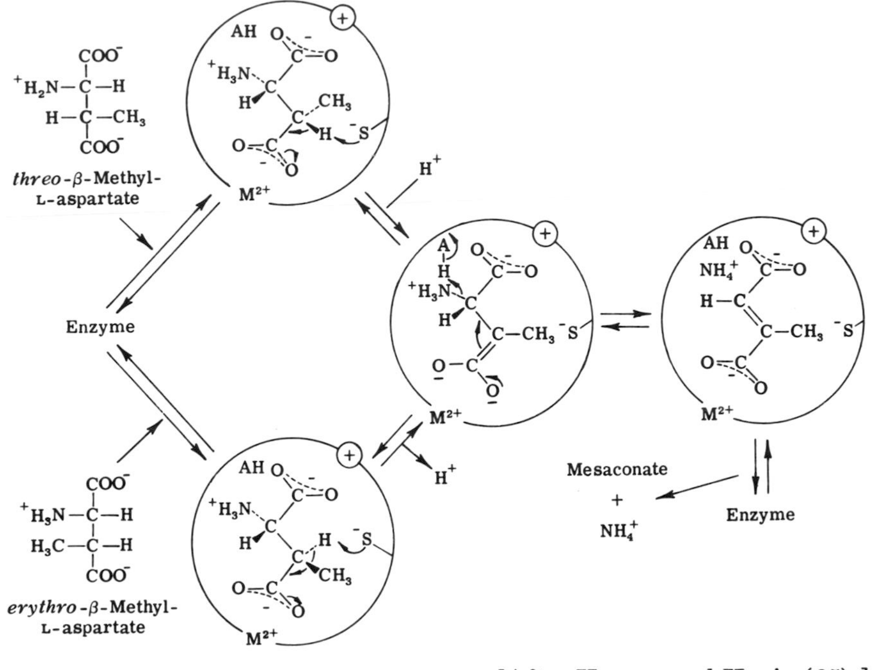

[After Hanson and Havir (*85*).]

V. Proton Dissociation–Replacement Reactions

$$\left[\;\rangle\text{C—H} + \rangle\text{C=O} \rightarrow \rangle\text{C—C}\langle\text{OH}\;\right]$$

The reactions of this class have in common the absence of a position in the product to which the activated proton of the substrate can be stably fixed. Therefore, one cannot demonstrate the enzyme to be acting as a base by the method of isotope transfer. The mechanisms may be formulated in several sequence patterns that might occur in parallel or not:

(1) Cosubstrate R is not required for H exchange (H^x).

$$\text{(a)}\quad E \xrightarrow{SH} E^{SH} \xrightarrow{-H^+} E^{S} \xrightarrow{R} E^{SR} \xrightarrow{-SR} E$$

At equilibrium: $H^x > S^x$, $R^x > S^x$, $S^x = H^x \cdot R^x / H^x + R^x$
high $(R + SR)$ leads to $H^x \rightarrow 0$ and $S^x \rightarrow 0$.

(2) Cosubstrate R is required for H exchange.

$$\text{(a)}\quad E \xrightarrow{SH} E^{SH} \xrightarrow{R} E^{SH}_{R} \xrightarrow{-H^+} E^{SR} \xrightarrow{-SR} E$$

At equilibrium: $H^x > S^x$, $R^x > S^x$, $v_1 = R^x \cdot S^x/R^x - S^x$
high (R + SR) leads to $H^x \to 0$ and $S^x \to 0$.

$$\text{(b)}\quad E \xrightarrow{SH} E^{SH} \xrightarrow{R} E_{H}^{SR} \xrightarrow{SR} E_H \xrightarrow[(4)]{H^+} E$$

At equilibrium: $R^x > S^x > H^x$, $v_4 = H^x \cdot S^x/S^x - H^x$
high (R + SR) leads to $H^x \to 0$ and $S^x \to 0$.

$$\text{(c)}\quad E \xrightarrow[(1)]{R} E_R \xrightarrow{SH} E_{R}^{SH} \xrightarrow{H^+} E_{R}^{S} \xrightarrow{SR} E$$

At equilibrium: $H^x > S^x > R^x$, $v_1 = R^x \cdot S^x/S^x - R^x$
high (R + SR) does not suppress H^x or S^x.

$$\text{(d)}\quad E \xrightarrow{R} E_R \xrightarrow{SH} E_{R}^{SH} \xrightarrow{SR} E_H \xrightarrow[(4)]{H^+} E$$

At equilibrium: $S^x > R^x$, $S^x > H^x$, $v_4 = S^x \cdot H^x/S^x - H^x$
high (R + SR) suppresses H^x but not S^x.

Only in the case that sequences (2b) and (2d) pertain would it be possible to demonstrate intermolecular proton transfer:

$$SH^* + S'R \to S'H^* + SR$$

This demonstration could be made by (a) using a proton acceptor S′ (as S′R) that differed from the donor SH chemically [analogous to the use of 2-methyl-*cis*-aconitate as an acceptor of tritium from 2-^{3}H-citrate in the aconitase studies (*5*)], (b) using an acceptor S*R that differed in S by its content of an isotope, such as deuterium that would be identifiable by mass spectrum or kinetic isotope separation (*3*), or (c) by determining the effect of increased SR and SH together on the exchange rates at equilibrium of SH* with water and R* with SR, etc. [analogous to the use of increased fumarate to suppress malate-(H*)–water exchange relative to malate–fumarate exchange in the fumarase studies (*7*)]. No examples of intermolecular proton transfer have been reported for this class of reactions.

A. Stereochemistry of Proton Replacement Reactions

Table III gives the stereochemical course of proton replacement reactions (*87–96a*). It will be noted that both retention and inversion are seen and, in fact, the two examples of β-oxidative decarboxylases show opposite stereoselectivities suggesting that their stereochemical course is not an obligatory facet of this mechanism. The fact that enolization (proton

TABLE III
STEREOCHEMISTRY OR PROTON REPLACEMENT REACTIONS

Reaction ($SH + R \rightarrow SR + H^+$)	Requirements for $H^+ + SH^* \rightleftarrows SH + H^*$	Stereochemistry	Ref.
α-Ketoglutarate + $CO_2 \xrightarrow{TPNH}$ isocitrate (isocitrate dehydrogenase)	TPNH, Mg (not CO_2)	Retention	*87*
Ribulose-P + $CO_2 \xrightarrow{TPNH}$ 6-P gluconate (6-P gluconate dehydrogenase)	TPNH, Mg (not CO_2)	Inversion	*88*
Dihydroxyacetone-P + glyceraldehyde-3-P → FDP (yeast and muscle aldolase)	Not glyceraldehyde-3-P	Retention	*89*
Dihydroxyacetone-P + D-lactaldehyde → rhamnose-1-P (rhamnose-1-P aldolase)	Not D-lactaldehyde	Retention	*90*
Glycine + $H_2C{=}O$ → serine (serine aldolase)	Not formaldehyde	Retention	*91*
Propionyl-CoA + CO_2 → methylmalonyl-CoA (propionyl-CoA carboxylase)	Carboxy biotin	Retention	*92*
Succinate + glyoxalate → isocitrate (isocitratase)	Requires glyoxalate	Inversion	*93, 93a*
Acetyl-CoA + glyoxalate → malate (malate synthase)	Requires Mg^{2+} and glyoxalate	Inversion	*94, 95*
Acetyl-CoA + oxalacetate → citrate (citrate synthase)	Requires analog of oxalacetate	Inverson	*96, 96a*

87. Z. B. Rose, *JBC* **235,** 928 (1960); S. Englard and I. Listowsky, *BBRC* **12,** 356 (1963); G. E. Lienhard and I. A. Rose, *Biochemistry* **3,** 185 (1964).
88. G. E. Lienhard and I. A. Rose, *Biochemistry* **3,** 190 (1964).
89. I. A. Rose, *JACS* **80,** 5835 (1958); I. A. Rose and S. V. Rieder, *JBC* **231,** 315 (1958).
90. T. H. Chiu and D. S. Feingold, *Federation Proc.* **26,** 835 (1967); *Biochemistry* **8,** 98 (1969).
91. M. Akhtar and P. M. Jordan, *Tetrahedron Letters* p. 875 (1969); *Chem. Commun.* p. 1691 (1968).
92. J. Retey and F. Lynen, *Biochem. Z.* **342,** 256 (1965); D. J. Prescott and J. L. Rabinowitz, *JBC* **243,** 1551 (1968).
93. M. Sprecher, R. Berger, and D. B. Sprinson, *JBC* **239,** 4268 (1964).
93a. H. H. Daron, W. J. Rutter, and I. C. Gunsalus, *Biochemistry* **5,** 895 (1966).
94. H. Eggerer and A. Klette, *European J. Biochem.* **1,** 447 (1967).
95. J. W. Cornforth, J. W. Redmond, H. Eggerer, W. Buckel, and C. Gustow, *Nature* **221,** 1212 (1969); J. Luthy, J. Retey, and D. Arigoni, *ibid.* p. 1213.
96. H. Eggerer, *Biochem. Z.* **343,** 111 (1965).
96a. J. P. Klinman and I. A. Rose, unpublished (1970); H. Eggerer, W. Buckel, H. Lenz, P. Wunderwald, J. W. Cornforth, C. Donninger, R. Mallaby, J. W. Redmond, and G. Gottschalk, *Nature* **226,** 517 (1970); J. Rétey, J. Lüthy, and D. Arigoni, *Nature* **226,** 519 (1970).

exchange) is independent of $R(CO_2)$ in these cases is consistent with variable stereochemistry.

The occurrence of retention, seen in the dihydroxyacetone-P aldolases suggests the possibility that a single base may be involved in the two

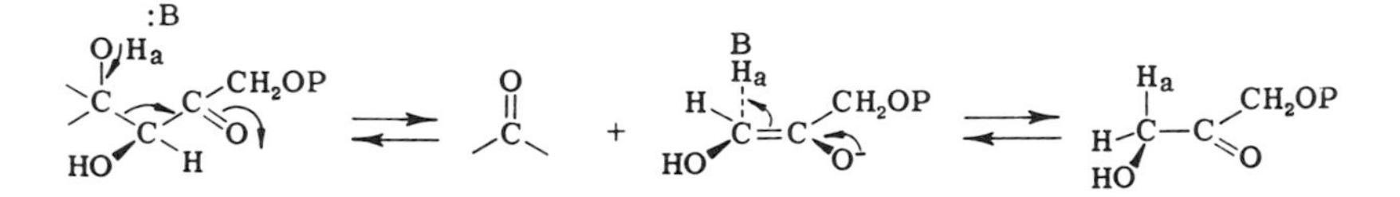

prototropic steps. It is of interest in this respect that the retention mode is found in the three dihydroxyacetone-P aldolases that have been examined—the muscle and yeast fructose-diP aldolases and rhamnulose-P aldolase—although the last two differ in stereospecificity with respect to the C-3 protons of dihydroxyacetone-P (*90*). It will be relevant to examine the other enzymes of this class as well as the acetaldehyde aldolases such as deoxyribose-P aldolase and the pyruvate aldolases such as 2-keto-3-deoxy-6-P gluconate aldolase now that the problem of the absolute stereochemistry of asymmetrically labeled —CH_3 has been solved by use of the kinetic isotope effect of malate synthase (*95*). Since all of these aldolases belong to class (1), in which proton abstraction and replacement are independent events, one would expect to see some examples of stereo-inversion in the absence of some mechanistic determinant such as the single base possibility.

The citrate and malate synthase reactions and isocitratase all require the presence of acceptor for proton activation. However, they also result in inversion of stereochemistry at the α-carbon making a simple single base mechanism unlikely and suggesting a two base sequence:

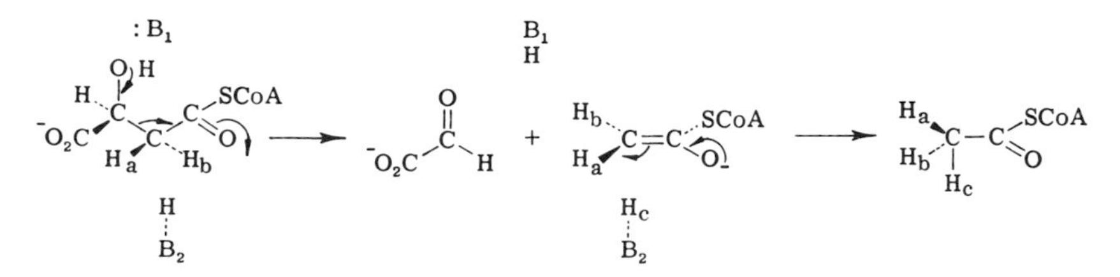

With the isocitratase reaction, Daron *et al.* (*93a*) found that during isocitrate formation in tritiated water succinate became labeled more rapidly than could be attributed to reversal from formed isocitrate, indicating that proton exchange at B_2H in stage II occurs prior to release of isocitrate. This was not the case, however, in the malate synthase studies of Eggerer and Klette (*94*) where remaining acetyl-CoA failed to become labeled in the presence of glyoxalate.

B. Citrate Synthase and Malate Synthase

These reactions, studied by Eggerer and his co-workers (*94, 96*) and Srere (*97*), show a large or complete dependence on cosubstrate for "enolization" of acetyl-CoA. In seeking to establish the roles of the oxalacetate and glyoxalate in these proton activations a number of substrate analogs were tested. For citrate synthase only L-malate (not D-malate, maleate, succinate, itaconate, citrate, or pyruvate) was effective in place of oxalacetate (*96, 97*). For malate synthase the exchange of tritium from water into acetyl-CoA was 0.1% of the rate of synthesis obtained with glyoxalate present. This exchange was clearly enzymic and required Mg^{2+}. Pyruvate in the presence of Mg^{2+} stimulated the exchange 1000-fold but was not a condensation partner (*96*). A mechanism was proposed to explain the role of magnesium and cosubstrates (*96, 98*):

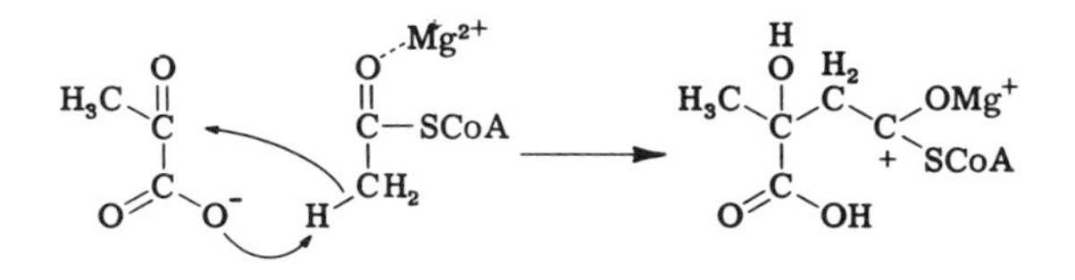

The main features of the mechanism are the role given to Mg^{2+} to act as an electrophile or Lewis acid, and the postulate that cosubstrate or substrate analog acts as a general base in the enolization. Subsequent work has established the stereochemistry of proton substitution to be that of inversion in the case of malate synthase (*95*) and citrate synthase as well (*96a*). This would tend to contraindicate the proposed mechanism where attack on the methyl carbon of acetyl-CoA by the carbonyl is concerted with proton abstraction by the carboxyl of glyoxalate (or oxalacetate) if both processes are required to occur antipodally. The possibility that the effects of added cosubstrates in stimulating exchange results from conformational effects is supported, in the case of citrate synthase, by results of Srere (*97*) showing that L-malate confers a degree of stabilization on the enzyme toward denaturation in urea. That the thio ester linkage is hydrolyzed subsequent to the aldol condensation is suggested by the finding that citryl-CoA is both a substrate for aldol cleavage and for hydrolysis (*99*). Attempts to demonstrate the

97. P. A. Srere, *BBRC* **26,** 609 (1967).

98. J. Bove, R. O. Martin, L. L. Ingraham, and P. K. Stumpf, *JBC* **234,** 999 (1959).

99. H. Eggerer and U. Remberger, *Biochem. Z.* **337,** 202 (1963); P. A. Srere, *BBA* **77,** 693 (1963).

formation of an anhydride have been negative; oxygen exchange between water and citrate, in the absence of CoA, implied by the last step was not observed (*100*). In the forward reaction in $H_2{}^{18}O$ the carboxyl of citrate derived from acetyl-CoA contains the ^{18}O, indicating that if the anhydride is formed its hydrolysis occurs specifically at the acetyl-CoA–derived end (*100*).

C. Hydrogenase

Undoubtedly the first enzyme to be studied by isotope exchange (*101*) was hydrogenase of bacteria. The simplicity of its substrate and the availability of interesting catalytic model systems makes this an important system for investigation. Most studies formulate the first step of the reaction as a heterolytic split:

$$H{:}H + E \rightleftharpoons E{:}H + H^+$$

although the results in support of this are suggestive rather than definitive.

(1) The reaction, $H_2 \xrightarrow[D_2O]{} HD$, can be demonstrated with purified enzyme and cell suspensions indicating the reversal of the bond rupture step and the availability of one of the hydrogens for exchange with protons of the medium (*101, 102*).

(2) In these experiments one also observes the formation of D_2 without a lag and at an initial rate that is somewhat less than the formation of HD. This indicates that one hydrogen is more stable toward exchange with the medium (*37*) but that both positions exchange, i.e., the formation of D_2 does not proceed through free HD. It is not clear whether the more stable hydrogen is the proton or hydride in the expanded mechanism:

$$E + H{:}H \rightleftharpoons E\genfrac{}{}{0pt}{}{:H}{H^+} \begin{cases} \xrightarrow{H_m} E\genfrac{}{}{0pt}{}{:H_m}{H^+} \begin{cases} \longrightarrow E + H_m{:}H \\ \xrightarrow{H_m} E + H_m{:}H_m \end{cases} \\ \xrightarrow{H_m} E\genfrac{}{}{0pt}{}{:H}{H_m^+} \begin{cases} \xrightarrow{H_m} E + H_m{:}H_m \\ \longrightarrow E + H{:}H_m \end{cases} \end{cases}$$

(3) The enzyme catalyzes the interconversion of the nuclear spin

100. C. H. Suelter and S. Arrington, *BBA* **141,** 423 (1967).
101. A. Farkas, L. Farkas, and L. Yudkin, *Proc. Roy. Soc.* **B115,** 375 (1934).
102. H. D. Hoberman and D. Rittenberg, *JBC* **147,** 211 (1943).

isomers of H_2 (ortho-para conversion), indicating the reversible separation and return of the nuclei as also demonstrated by the isotope exchange studies. However, when para-hydrogen conversion was examined in a D_2O medium with *E. coli* (*103*) or *Proteus vulgaris* (*104*) cells, there was little or no isomerized H_2, rather the converted hydrogen was HD and D_2. This indicates that the proton either goes directly to the medium or if it is abstracted by a basic group on the enzyme its subsequent dissociation is much more rapid than the return step.

(4) Recent experiments (*105*) with thoroughly dried hydrogenase from *Desulfovibrio desulfuricans* provide evidence in support of basic catalysis, i.e., $E + H_2 \rightleftharpoons E\genfrac{}{}{0pt}{}{:H}{\diagdown H^+}$ The enzyme was reduced with $Na_2S_2O_4$ and dried at 10^{-6} mm of pressure. Significant catalysis of the para → ortho reaction was observed. It was also observed that the conversion rate was much greater than the reaction: $H_2 + D_2 \rightarrow HD$. This result rules out the presence at the active site of a pool of water with which the abstracted proton could mix. Such a pool would soon equilibrate with the two isotopic species and lead to hybrid molecules. Likewise, this result requires that the base involved be nonprotonic (*105*).

IV. C—H Fission in Other Catalyses

We have summarized the evidence for a basic catalytic role of enzyme in —C—H bond fission in the following classes of reactions: enolization, racemization, elimination, and allylic rearrangement. There is not yet sufficient evidence to establish this point for the variety of aldolases or for dehydrogenases such as succinate dehydrogenase. A few final interesting examples that have not been considered in this chapter because the data are either incomplete or preliminary could be mentioned briefly:

(1) Guggenheim and Flavin (*106*) have shown that the conversion of *O*-succinyl homoserine to succinate and α-ketobutyrate by bacterial cytathionine synthease in 100% D_2O results in deuteration of the α-ketobutyrate at both C-3 and C-4. However, there was only 0.2 atom of deuterium in the C-4 position, indicating that a proton had been con-

103. A. Farkas, *Trans. Faraday Soc.* **32,** 992 (1936).
104. A. I. Krasna and D. Rittenberg, *JACS* **76,** 3015 (1954).
105. T. Yagi, M. Tsuda, Y. Mori, and H. Inokuchi, *JACS* **91,** 2801 (1969).
106. S. Guggenheim and M. Flavin, *BBA* **151,** 664 (1968).

served from either the C-2 or C-3 position of the homoserine part of the substrate:

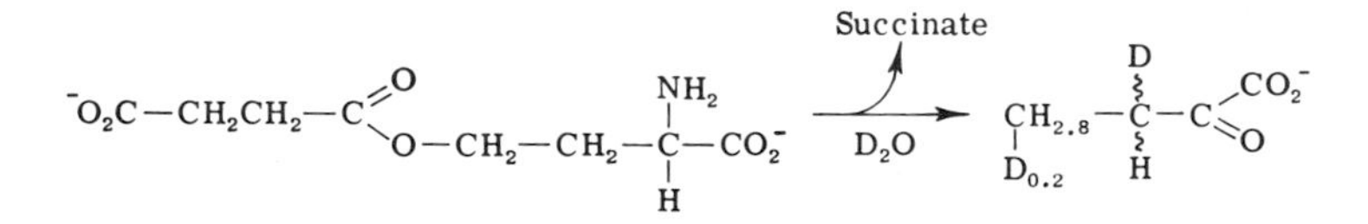

(2) Portsmouth, Stoolmiller, and Abeles (*23*) have looked for the conservation of tritium in the conversion of 2-keto-3-deoxy-L-arabonate-5-^3H to α-ketoglutarate semialdehyde in the reaction previously discussed (see Section II,E). They report that the product had a specific activity of 6.6×10^3 dpm/μmole, whereas the substrate was 1.9×10^6 dpm/μmole. The average specific activity of the water present in this incubation would have been about 2.5×10^2 dpm/μmole of protons. Therefore, the product was about 25$\times$ more active than could have been expected if there were no transfer, but it was only 0.3% as active if there had been no exchange. As in many cases, one has the choice of which extreme he wishes to emphasize. However, in view of the expected purity of the labeled substrate, one finds it difficult to explain the appearance of tritium in the product except by a transfer-exchange competition.

(3) Hermann and Lehman (*107*) have reported the conversion of D-TDP-glucose-5-^3H to the C-6 methyl group of D-TDP-4-keto-6-deoxyglucose with an impure preparation of D-TDP-glucose oxidoreductase, the mechanism of which is very briefly mentioned earlier (Section II,E). Although the extent of transfer reported was only a few per cent and the transfer of tritium from C-4 by a dehydrogenase-type reaction seems well established (*24*), this observation can be mentioned in this category.

(4) In a different kind of study, Brand, Tsolas, and Horecker (*108*) have shown that yeast transaldolase is photoinactivated at pH 7.6 in the presence of Rose Bengal with the parallel loss of one of the two histidine residues. Enzyme that has been incubated with fructose-6-P in ^3H_2O and then isolated on Sephadex is found to contain about one atom of tritium per enzyme·dihydroxyacetone-P·Schiff base complex. The tritium is found to reside at the carbinol carbon of β-glyceryl lysine that is derived by acid hydrolysis of the borohydride-reduced complex. The native complex was immeasurably more stable to photoinactivation than the free enzyme. It was postulated that the sensitive histidine is protected by close association with the Schiff base and may, in fact, be largely proto-

107. K. Hermann and J. Lehman, *European J. Biochem.* **3,** 369 (1968).
108. K. Brand, O. Tsolas, and B. L. Horecker, *ABB* **130,** 521 (1969).

nated by the same proton that appears in dihydroxyacetone-P after reduction and hydrolysis. It could be shown that this histidine in the free enzyme is protected from photooxidation at pH's that would lead to its acidification.

ACKNOWLEDGMENTS

The author is indebted to Dr. K. R. Hanson and Dr. J. Glusker for permission to use figures from papers prior to publication. In addition, he has profited from discussions with Drs. A. Krasner, S. England, O. Gabriel, L. Glaser, and especially Dr. Judith Klinman.

Experimental results from the author's laboratory and preparation of this chapter were supported by Public Health Service Research Grant No. CA-07818 from the National Cancer Institute, and also by grants to this Institute: NIH grants CA-06927 and FR-05539, and an appropriation from the Commonwealth of Pennsylvania.

6

Kinetic Isotope Effects in Enzymic Reactions

J. H. RICHARDS

I. Introduction

During the conversion of one molecule to another, new bonds are formed, old bonds broken, and other bonds experience changes, either transient or permanent, in their hybridization. Isotopic substitution can have a profound influence on the energetics of these bonding changes. Accordingly, studies of reactions in substrates with judicious isotopic substitution can often give important insight into the atomic details of the reaction mechanism (*1–7*).

For alternate, more sophisticated discussions of many aspects of isotope effects see:

1. K. B. Wiberg, *Chem. Rev.* **55,** 713 (1955).
2. R. P. Bell, "The Proton in Chemistry," Chapter IX. Cornell Univ. Press, Ithaca, New York, 1959.
3. L. Melander, "Isotope Effects on Reaction Rates." Ronald Press, New York, 1960.
4. E. A. Halevi, *Progr. Phys. Org. Chem.* **1,** 109 (1963).

Many of the properties of a chemical bond of interest to our discussion can be characterized by a force constant for the bond in question and the masses of the two atoms joined by the bond. The effect of isotopic substitution on the force constant is usually minor; however, the changes in mass can be a relatively major factor. For example, the substitution of deuterium (^{2}H) for hydrogen (^{1}H) doubles the mass of the atom in question. On the other hand, substitution of ^{13}C for ^{12}C increases the mass of the atom in question by only 8%. For this reason substitution of deuterium (or tritium) for hydrogen produces much larger effects than isotopic substitution of the heavier elements and isotope effect studies have, therefore, been principally concerned with hydrogen and its isotopes (*8*).

Division of isotope effects into two categories—primary and secondary—facilitates discussion. A primary kinetic isotope effect results when the bond to the isotopically substituted atom is broken in the transition state. A secondary isotope effect results when the bond to the isotopically substituted atom, though not broken, experiences significant changes in hybridization during the reaction.

The purpose of this chapter is to discuss briefly the underlying causes of primary and secondary kinetic effects for isotopes of hydrogen and to give a few examples of the application of these effects to problems of biochemical interest. Other examples are cited in Chapter 5, by Rose, in this volume.

II. Primary Isotope Effects

A. Origin and Magnitude

The zero point energy of a carbon–deuterium bond is less than the zero point energy of an analogous carbon–hydrogen bond. In many pro-

---

5. K. B. Wiberg, "Physical Organic Chemistry," pp. 351–364. Wiley, New York, 1964.

6. H. Simon and D. Palm, *Angew. Chem. Intern. Ed. Engl.* **5**, 920 (1966).

7. W. P. Jencks, "Catalysis in Chemistry and Enzymology," Chapter 4. McGraw-Hill, New York, 1969.

8. An interesting case of a $^{12}C/^{13}C$ isotope study is that observed (*9*) on the decarboxylation of oxalacetate catalyzed by manganese ion for which $k\,^{12}C/k\,^{13}C = 1.06$. In contrast, the enzyme-catalyzed decarboxylation shows no detectable isotope effect, demonstrating that loss of carbon dioxide is the rate-determining step of the nonenzymic reaction but that cleavage of the carbon–carbon bond is not the rate-determining step in the enzymic reaction.

9. S. Seltzer, G. A. Hamilton, and F. H. Westheimer, *JACS* **81**, 4018 (1959).

cesses in which a carbon–hydrogen bond is broken, the zero point energy of the bond is frozen out in the transition state. In such a transition state, therefore, both carbon–hydrogen and carbon–deuterium will have the same energy. Since the ground state energy of the carbon–deuterium bond is lower than the ground state energy of the analogous carbon–hydrogen bond, the difference in energy between the ground and transition state energy for the carbon–deuterium bond will be greater than for an analogous carbon–hydrogen bond. Consequently, the rate for the substrate containing deuterium will be slower than that for the substrate containing hydrogen and k_H/k_D will be greater than one.

A more quantitative estimate for k_H/k_D to be anticipated on the basis of the qualitative argument can be derived. The zero point energy of a bond is $\frac{1}{2}h\nu$, where h is Planck's constant and ν is the frequency of the vibration. For a simple harmonic oscillator, from Hooke's law, the frequency of the vibration, $\nu = \frac{1}{2\pi}\sqrt{k/m}$ where k is the force constant referred to earlier and m is the mass. Thus the zero point energy is $\epsilon = h/(4\pi)\sqrt{k/m}$ *(10)*. Since the force constant k is essentially the same for a carbon–hydrogen, a carbon–deuterium, or a carbon–tritium bond, the change in zero point energy will largely result from a change in mass and will be inversely related to the square root of the mass. Accordingly, the zero point energy of a carbon–deuterium bond should be about $1/\sqrt{2}$ that of a carbon–hydrogen bond. Actually, the stretching frequency for a carbon–hydrogen vibration is near 2900 cm^{-1} and that of a carbon–deuterium bond near 2100 cm^{-1} corresponding to zero point energies of 4.15 and 3.0 kcal/mole, respectively *(11, 12)*.

If, now, the hydrogen in question is transferred to some acceptor in the transition state, this stretching vibration can be frozen out. This will result in an activation energy for breaking a carbon–deuterium bond higher by about 1.15 kcal/mole (4.15–3.0) than for breaking an analogous carbon–hydrogen bond. For most reactions this difference in activation energy corresponds to a rate difference of about sevenfold. Figure 1 illustrates these differences schematically.

The concept of primary kinetic isotope effects as outlined above is something of an oversimplification and does not take account of the important concept of tunneling *(2)* in proton transfer reactions. For

10. The reduced mass $\mu = (m_1m_2)/(m_1 + m_2)$ should properly be used, but for vibration of a carbon–hydrogen bond the mass of the hydrogen is sufficiently small relative to carbon (and indeed the rest of the molecule to which it is attached) that this correction can be disregarded.

11. L. Bellamy, "The Infra Red Spectra of Complex Molecules," p. 14. Wiley, New York, 1958.

12. R. M. Silverstein and G. C. Bassler, "Spectrometric Identification of Organic Compounds," 2nd ed., p. 67. Wiley, New York, 1967.

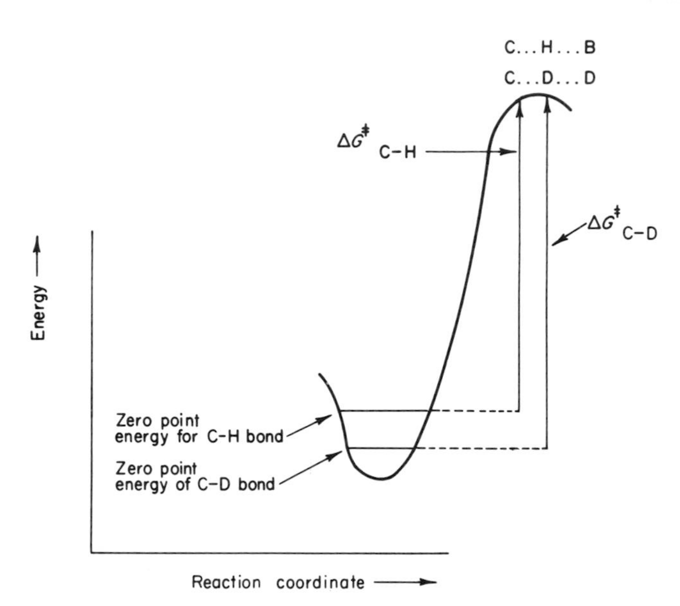

FIG. 1. Primary isotope effect.

example (*13*, *14*) $k_H/k_D = 24.1$ for removal of a proton from 2-nitropropane by 2,6-lutidine, a sterically hindered base. This unusually large isotope effect has been ascribed to tunneling. On the other hand, use of pyridine, which is not sterically hindered, as base shows $k_H/k_D = 9.8$. Values of k_H/k_D less than sevenfold may be anticipated if the carbon–hydrogen vibration of the ground state is not completely frozen out in the transition state.

As a general rule, an isotope effect of two to fifteen can be taken as evidence that the rate-determining step involves cleavage of the carbon–hydrogen bond in question. Conversely, the absence of an effect on the rate of a reaction when deuterium is substituted for hydrogen indicates that the bond in question is not broken in the rate-determining step. [It may, of course, be cleaved during the overall reaction (*15*).]

13. E. S. Lewis and L. H. Funderburk, *JACS* **89**, 2322 (1967).

14. E. S. Lewis and J. K. Robinson, *JACS* **90**, 4337 (1968).

15. If tritium is substituted for hydrogen, the isotope effect is increased relative to deuterium because of the increased mass of tritium. Use of assumptions similar to those used in evaluating k_H/k_D leads to a relationship between k_H/k_D and k_H/k_T (*16*):

$$k_H/k_T = (k_H/k_D)^{1.44}$$

or

$$\log k_H/k_T = 1.44 \log k_H/k_D$$

This relationship has been observed experimentally in several instances.

B. Examples

A clear example of a primary isotope effect in a simple reaction is that observed in the bromination of acetone at all but very low concentrations of bromine (*17*). In this reaction, the rate-determining step is the base-catalyzed removal of a proton to form an enolate ion which then reacts rapidly with bromine. The rate of this reaction as well as the rates of exchange of the hydrogen atoms of acetone with solvent when measured directly show $k_H/k_D \sim 6\text{–}10$.

$$CH_3\overset{O}{\overset{\|}{C}}CH_3 \;\; \overset{\ominus}{B} \longrightarrow \left[CH_3\overset{O}{\overset{\|}{C}}CH_2\cdots H\cdots B \right]^{\ominus} \xrightarrow[\text{step}]{\text{slow}} CH_3\overset{O}{\overset{\|}{C}}\overset{\ominus}{C}H_2 + BH \xrightarrow[Br_2]{\text{fast}} CH_3\overset{O}{\overset{\|}{C}}CH_2Br + Br^{\ominus}$$

Four examples of a primary kinetic isotope effect in hydrogen transfer processes of biochemical systems will be discussed: alcohol dehydrogenase, glucose-6-phosphate dehydrogenase, isomerase reactions catalyzed by coenzyme B_{12}, and isomerization of ribonuclease.

Alcohol dehydrogenase catalyzes the transfer of the pro-R hydrogen at C-1 of ethanol (*18*) to NAD^+ giving acetaldehyde and NADH with its newly acquired hydrogen in the pro-R position (*19*).

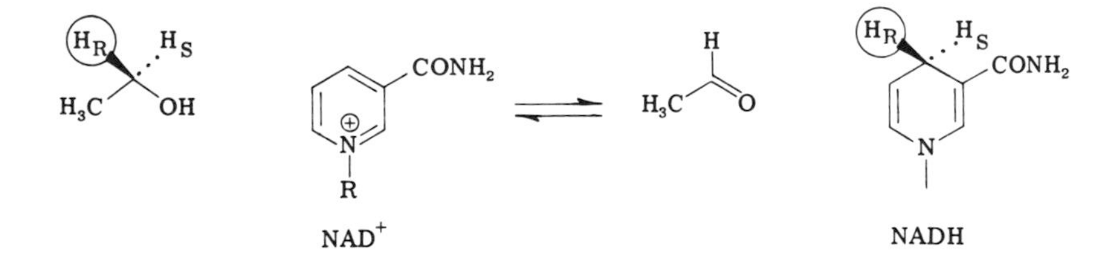

Deuterium substituted for hydrogen in the reaction NADH plus acetaldehyde catalyzed by horse liver alcohol dehydrogenase (*20, 21*) shows $k_H/k_D = 3.11$; for the reverse reaction $k_H/k_D = 2.28$. With yeast

16. C. G. Swain, E. C. Stivers, J. F. Renner, Jr., and L. J. Schaad, *JACS* **80**, 5885 (1958).
17. O. Reitz and J. Kopp, *Z. Physik. Chem.* **184A**, 429 (1939).
18. H. Weber, J. Seibl, and D. Arigoni, *Helv. Chim. Acta* **49**, 741 (1966).
19. For a review, see F. H. Westheimer, *Advan. Enzymol.* **24**, 441 (1962).
20. H. R. Mahler, R. H. Baker, Jr., and V. J. Shiner, Jr., *Biochemistry* **1**, 47 (1962).
21. R. H. Baker, Jr., *Biochemistry* **1**, 41 (1962).

alcohol dehydrogenase, a value of $k_H/k_D = 1.8$ was observed (*22*) for the reduction of acetaldehyde or the oxidation of ethanol. For 1-^{3}H ethanol values of $k_H/k_T = 3.2$ at 5–15°C and $k_H/k_T = 3.8$ at 20–35°C have been observed (*23, 24*) for liver alcohol dehydrogenase.

A simple interpretation of these experimental results is complicated by ambiguities of the kinetic analysis of the alcohol dehydrogenase system. For example, the Michaelis constant is also perturbed by substitution of deuterium for hydrogen ($K_H/K_D \sim 1.3$), an effect which suggests that formation of the enzyme–substrate complex itself results in some alteration in the bonding between carbon and hydrogen (*25*). However, the effects of isotopic substitution on the rate constants indicate a transition state for hydrogen transfer similar to that encountered, for example, in the Meerwein–Pondorf–Oppenauer equilibrium (*27*). One of many detailed possibilities is outlined utilizing the zinc of the dehydrogenase.

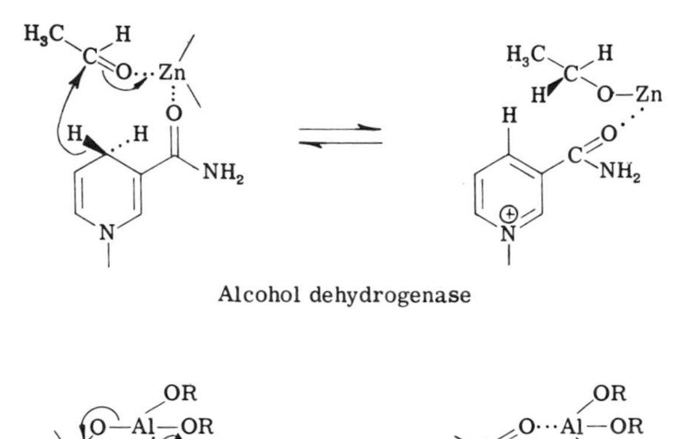

Alcohol dehydrogenase

Meerwein-Pondorf-Oppenauer equilibrium [*27*]

The oxidation of the aldehydic carbon of glucose 6-phosphate to a carboxyl group in 6-phosphogluconolactone is the first step in the important hexose monophosphate pathway of glucose metabolism. If deu-

22. H. R. Mahler and J. Douglas, *JACS* **79,** 1159 (1957).
23. D. Palm, *Z. Naturforsch.* **21b,** 540 (1966).
24. D. Palm, *Z. Naturforsch.* **21b,** 547 (1966).
25. The steric effect caused by substitution of deuterium for hydrogen is generally very small. One of the largest observed effects is in the reaction of 2-(α,α,α tri ^{2}H) methylpyridine with methyl iodide (*26*). The value $k_H/k_D = 1.030 \pm 0.003$ reflects the extreme case where the slight difference in the size of the methyl group causes a discernible rate change.
26. H. C. Brown and G. J. McDonald, *JACS* **88,** 2514 (1966).
27. R. B. Woodward, N. L. Wendler, and F. J. Brutschy, *JACS* **67,** 1425 (1945).

terium (*28*) or tritium (*29*) is substituted for hydrogen, the rate of oxidation by $NADP^+$ is significantly slowed for the isotopically substituted substrate.

In a study (*30*) of the similar oxidation reaction of glucose coupled to FAD and utilizing glucose oxidase from *Aspergillus niger,* Bright and Gibson found that with 1-^{1}H-D-glucose the principal rate-determining step in the overall reaction at high substrate concentrations is a first-order dissociation of product from the enzyme in the oxidative half-reaction. Use of 1-^{2}H-D-glucose as substrate, however, shifts the rate determining step to the transfer of deuterium from glucose to FAD in the reductive half-reaction. A value of $k_H/k_D = 15$ for this step gives good agreement for kinetic results observed with the natural and isotopically substituted substrates. The mechanism of hydrogen transfer shown was proposed to involve prior formation of a glucoside linkage between glucose and flavin followed by enzyme-base abstraction of hydrogen from C-1 of glucose as a proton and the concerted transfer of two electrons.

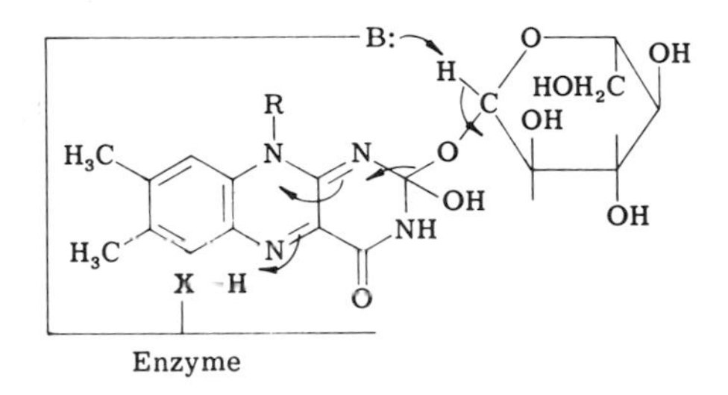

Coenzyme B_{12} catalyzes a variety of isomerizations (*31*) that can be summarized by

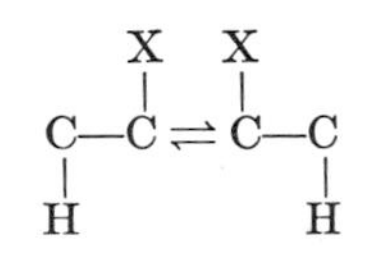

An example is the isomerization of methylmalonyl coenzyme A to succinyl coenzyme A. A study (*32*) of the effect of substituting deuterium

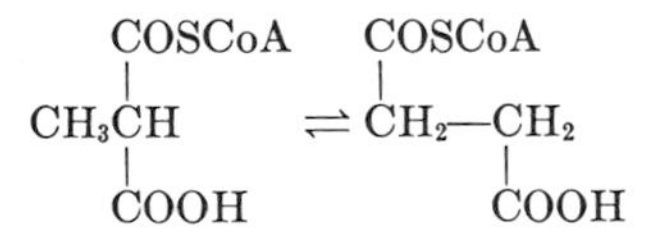

28. I. A. Rose, *JBC* **236,** 603 (1961).
29. J. C. Bartley and S. Abraham, *BBA* **148,** 563 (1967).
30. H. J. Bright and Q. H. Gibson, *JBC* **242,** 991 (1967).
31. H. P. C. Hogenkamp, *Ann. Rev. Biochem.* **37,** 225 (1968).
32. W. M. Miller and J. H. Richards, *JACS* **91,** 1498 (1969).

for hydrogen ($k_H/k_D = 3$) in the methyl group of methylmalonyl coenzyme A has shown that the rate-determining step involves transfer of one of the hydrogens of the methyl group, probably to C-5′ of the deoxyadenosine residue of coenzyme B_{12}. Though kinetic analysis of a reaction generally reveals isotope effects only in a single, rate-determining step a study of the partitioning of deuterium between products of differing isotopic composition can reveal isotope effects in steps occurring after the rate-determining one. In the case of the isomerization of methylmalonyl coenzyme A, the step in which a hydrogen is transferred, presumably between C-5′ of the deoxyadenosine residue of coenzyme B_{12} to C-3 of succinyl coenzyme, is found to have $k_H/k_D = 3$. Accordingly, a mechanism for the isomerization can be postulated as outlined:

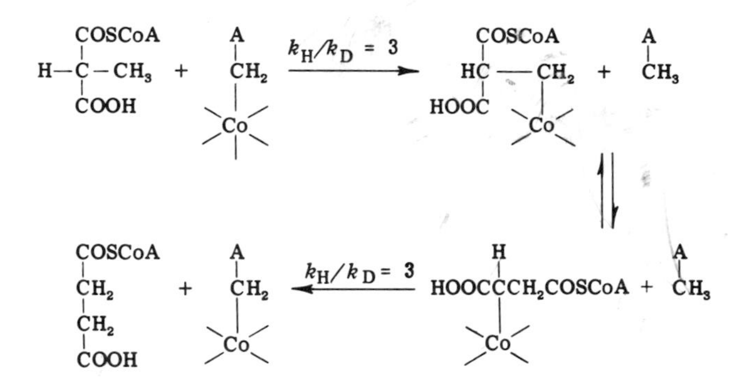

The primary isotope effects discussed so far only involve transfer of hydrogens that are not in rapid equilibrium with solvent. If the hydrogens being studied are however in easily exchangeable positions (for example, O—H and N—H) a change of solvent from H_2O to D_2O can have large effects on rate processes associated with transfers of this type.

A study (*33*) of the isomerization of ribonuclease, which occurs at 25° and at neutral pH values, by the temperature jump method in H_2O and

$$E_1H \overset{k_{12}}{\rightleftharpoons} E_2H \overset{K_{A_1}}{\rightleftharpoons} E_2 + H^{\oplus}$$

D_2O gave k_{H_2O}/k_{D_2O} for $k_{12} = 4.3$ and for $k_{21} = 6.4$. (For K_{A_1} the value of K_{H_2O}/K_{D_2O} is 5.0.) These values taken together with other aspects of the enzyme isomerization led to the proposal that the rate-determining step in the isomerization of ribonuclease involves a transfer of a proton probably from the carboxyl group of a glutamic acid residue to a histidine residue before loss of the proton by the enzyme.

33. T. C. French and G. C. Hammes, *JACS* **87**, 4469 (1965).

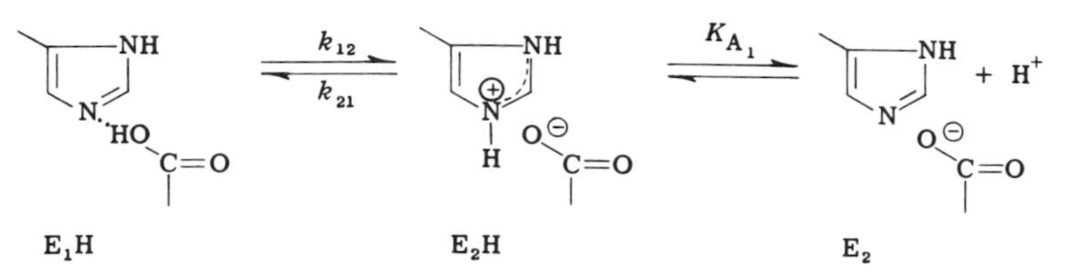

Other aspects of comparative studies of reactions in H_2O and D_2O should be noted. Changing the solvent from H_2O to D_2O can cause changes in the tertiary structure of enzymes which can in turn influence reaction rates. Also, equilibrium processes such as the relative acidities of groups can change appreciably between H_2O and D_2O. These phenomena are discussed by Jencks (*7*), for example.

III. Secondary Isotope Effects

A. Origin and Magnitude

Isotopic substitution of a hydrogen in a carbon–hydrogen bond that is not directly involved in a reaction can also cause changes in reaction rates that yield important insight into the mechanism. These secondary isotope effects have been particularly helpful in studying carbonium ion processes where changes in hybridization of the ionizing carbon can be diagnosed by isotopic substitution.

Consider the S_N1 ionization of a saturated carbinol derivative. In the ground state the carbon is tetrahedral and sp^3 hybridized. At the carbonium ion stage, the carbon has become planar, sp^2. The transition state probably strongly resembles the carbonium ion (*34*). The effect of substituting deuterium for hydrogen is thought to arise largely from the different responses these two isotopes have to such a change in hybridization (*35, 36*). This substitution causes little change in the stretching and one of the bending frequencies of the carbon–hydrogen bond; however, the other bending frequency changes from 1340 cm^{-1} in carbinols to 800 cm^{-1} in model compounds such as aldehydes or olefins with sp^2 hybridization. The change in this bending vibration to lower frequencies (and lower energy) as the hybridization changes from sp^3 to sp^2 means that the difference in zero point energy between a carbon–hydrogen and

34. G. S. Hammond, *JACS* **77,** 334 (1955).
35. A. Streitwieser, Jr., R. H. Jagow, R. C. Fahey, and S. Suzuki, *JACS* **80,** 2326 (1958).
36. E. R. Thornton, *Ann. Rev. Phys. Chem.* **17,** 349 (1966).

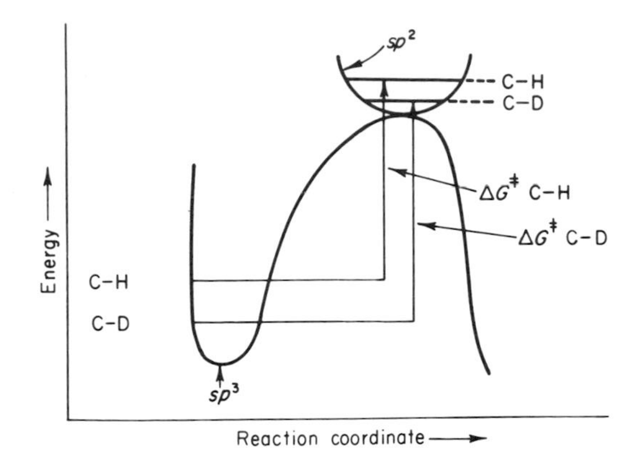

FIG. 2. Secondary isotope effect.

carbon–deuterium bond is lessened because the potential surface for the lower energy vibration is shallower (see Fig. 2). The expectation is, then, that the substrate with hydrogen will more easily form the sp^2 carbonium ion intermediate (and accordingly react faster) than will the substrate with deuterium; that is, k_H/k_D will be greater than one. The changes in vibrational frequencies from 1340 cm^{-1} to 800 cm^{-1} lead to an expected $k_H/k_D = 1.38$. A value of $k_H/k_D = 1.15$ is observed for the solvolysis of 1-^{2}H-cyclopentyl tosylate (*35*) and is a common value for a secondary isotope effect when deuterium is substituted for hydrogen on the α-carbon atom in a reaction involving a carbonium ion intermediate (*37*). In a case where the carbonium ion is fully formed (an equilibrium study of the formation of a carbonium ion from benzhydrol), a value of $K_H/K_D = 1.29$ has been observed (*38*). In contrast, if the transition state involves five groups situated around carbon, as the trigonal bipyramid of an S_N2 reaction, the substitution of deuterium for hydrogen has no effect on the rate because in both the ground and transition state the bending frequencies of the bonds to hydrogen are unchanged; that is, there is no "loosening" as in the case of a carbonium ion intermediate. For example, (*39*), the S_N2 reaction of isopropyl bromide with sodium ethoxide has $k_H/k_D = 1.00$.

Substitution of deuterium for hydrogen on the carbon adjacent to the one which becomes cationic also slows the rate of the reaction because, during formation of a carbonium ion, a loosening of this bond by hyper-

37. See Table VI in M. J. Nugent, R. E. Carter, and J. H. Richards, *JACS* **91**, 6145 (1969).
38. M. M. Mocek and R. Stewart, *Can. J. Chem.* **41**, 1641 (1963).
39. V. J. Shiner, *JACS* **74**, 5285 (1952).

conjugation with the vacant p orbital of the carbonium ion carbon occurs. Again, a hydrogen atom is better able to take advantage of this bond loosening than is a heavier deuterium atom. This effect is maximal when the dihedral angle between the vacant p orbital and the carbon–hydrogen bond in question is zero (maximum overlap) and vanishes when this angle becomes 90°.

A simplifying statement that collects the interpretations of isotope effects so far discussed is that the heavier isotope will be relatively more abundant in that species in which the potential well is steepest, that is, in that species in which the zero point vibration is highest.

B. Examples

Secondary isotope effects have been used recently to support carbonium ion type of intermediates in two processes of biochemical interest: the dehydration of L-malate catalyzed by fumarase (*40*) and the cleavage of glycosidic bonds of polysaccharide derivatives catalyzed by lysozyme (*41*).

In the dehydration of L-malate the α-secondary isotope effect, k_H/k_D, for isotopic substitution at C-2 is 1.09 (calculated from $k_H/k_T = 1.12$ observed experimentally), and k_H/k_D for isotopic substitution at C-3 in that position not removed in the elimination reaction (the β-secondary isotope effect) is 1.09 (again calculated from an experimentally determined value of $k_H/k_T = 1.12$). These results were interpreted as indicating that, as the hydroxyl group of L-malate leaves, considerable positive charge develops on the carbon to which it was attached. However, the carbonium ion is not as fully formed as in the case of solvolysis of cyclopentyl tosylate, which indicates the possibility of some nucleophilic participation either by a nucleophile on the enzyme surface or by one of the two carboxylate anions of the substrate.

Furthermore, the β-secondary isotope effect suggests that the conformation of the L-malate as the hydroxyl group departs should be one in which the carbon–hydrogen bond to the 3-(S) hydrogen (the hydrogen not eliminated) can overlap with the vacant p orbital on C-2. No primary kinetic isotope effect was observed in the dehydration of 3-(R)-3-^{2}H-2-(S)-malate (*42*).

40. D. E. Schmidt, Jr., W. G. Nigh, C. Tanzer, and J. H. Richards, *JACS* **91**, 5849 (1969).

41. F. W. Dahlquist, T. Rand-Meir, and M. A. Raftery, *Proc. Natl. Acad. Sci. U. S.* **61**, 1194 (1968); *Biochemistry* **8**, 4214 (1969).

42. H. Fisher, C. Frieden, J. S. M. McKee, and R. A. Alberty, *JACS* **77**, 4436 (1955).

Other evidence for the carbonium ion character of the transition state was provided by studies of fluorine-containing substrate analogs (*43a*).

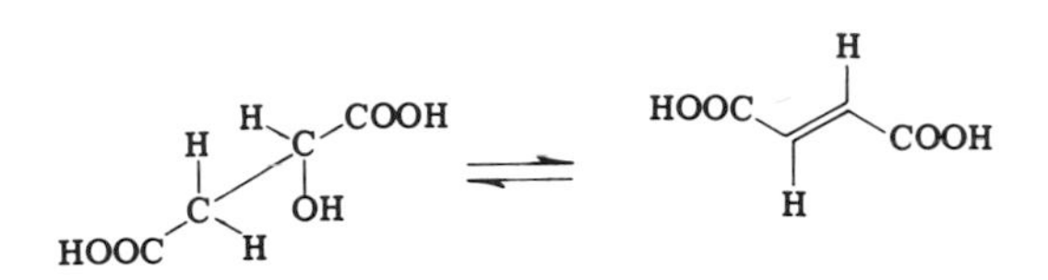

In his chapter, Rose (p. 307) has suggested that the sequence of steps 1 and 2 in the following scheme must be at least 70% toward equilibrium because oxygen exchange between malate and water is more rapid than

$$\mathrm{E} + \mathrm{Malate} \underset{1}{\rightleftharpoons} \mathrm{E_{malate}} \underset{2}{\overset{\mathrm{OH}^{\ominus}}{\rightleftharpoons}} \mathrm{E^{H}_{fumarate}} \underset{3}{\rightarrow} \mathrm{E^{H}} + \mathrm{fumarate}$$

carbon exchange between malate and fumarate (*43b*). Such an analysis overlooks the probable existence of a carbonium ion intermediate between E_{malate} and $E^{H}_{fumarate}$. Thus the complete scheme should be (*40*)

$$\mathrm{E} + \mathrm{M} \underset{1}{\rightleftharpoons} \mathrm{EM} \underset{2}{\overset{\mathrm{OH}^{\ominus}}{\rightleftharpoons}} \mathrm{EX} \underset{3}{\rightleftharpoons} \mathrm{EHF} \underset{4}{\rightarrow} \mathrm{EH} + \mathrm{F}$$

where X is the carbonium ion intermediate and k_{-2} is somewhat larger than k_3 to account for the rate of oxygen exchange between malate and water exceeding the rate of carbon exchange between malate and fumarate. For such a scheme, the observed kinetic isotope effect will be (if $k_1^{\mathrm{H}} \sim k_1^{\mathrm{T}}$, $k_{-1}^{\mathrm{H}} \sim k_{-1}^{\mathrm{T}}$, and $k_{-1} > k_2$)

$$\frac{d(F_{\mathrm{H}})/dt}{d(F_{\mathrm{T}})/dt} = \frac{(M_0^{\mathrm{H}})}{(M_0^{\mathrm{T}})} \frac{\kappa_3^{\mathrm{H}}}{\kappa_3^{\mathrm{T}}} \frac{\kappa_2^{\mathrm{H}}}{\kappa_2^{\mathrm{T}}} \frac{(\kappa_3^{\mathrm{T}} + \kappa_{-2}^{\mathrm{T}})}{(\kappa_3^{\mathrm{H}} + \kappa_2^{\mathrm{H}})}$$

We have presented (*40*) arguments and precedents for the approximation that, for secondary isotope effects in this system, $k_3^{\mathrm{T}} \sim k_3^{\mathrm{H}}$ and $k_{-2}^{\mathrm{T}} \sim k_{-2}^{\mathrm{H}}$ in which case the observed isotope effect measures to a good approximation $k_2^{\mathrm{H}}/k_2^{\mathrm{T}}$ and does *not* reflect quasi-equilibration between malate and fumarate bound to the active site of the enzyme.

Studies of the hydrolysis of phenyl 4-0-(2-acetamido-2-deoxy-β-D-glucopyranosyl)-β-D-glucopyranoside (I) have shown that substitution of deuterium by hydrogen at C-1 gives an α-secondary kinetic isotope effect of $k_{\mathrm{H}}/k_{\mathrm{D}} = 1.11$ over a wide pH range which supports the occurrence of a carbonium ion intermediate in this reaction. In contrast, the hydrolysis of phenyl β-D-glucopyranoside (II) catalyzed by β-galactosidase shows $k_{\mathrm{H}}/k_{\mathrm{D}} = 1.01$ suggesting that this enzyme effects hydrolysis by an S_N2-like displacement mechanism.

43a. W. G. Nigh and J. H. Richards, *JACS* **91,** 5847 (1969).
43b. J. N. Hansen, E. C. Dinovo, and P. D. Boyer, *JBC* **244,** 6270 (1969).

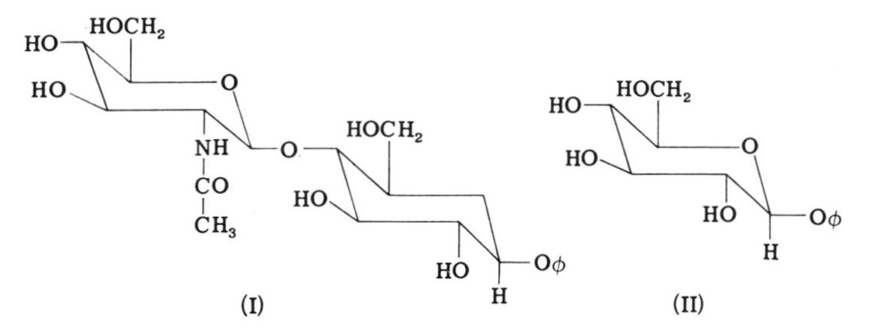

In summary, significant information about the nature of the transition state in many reactions catalyzed by enzymes can be obtained by studies in which deuterium (or tritium) is substituted for hydrogen in a bond which is either itself broken or is in the vicinity (α or β) of the center at which reaction occurs.

7

Schiff Base Intermediates in Enzyme Catalysis

ESMOND E. SNELL • SAMUEL J. DI MARI

I. Introduction

Among the common functional groups found in naturally occurring compounds is the carbonyl group, and many of the metabolic transformations that occur *in vivo* are readily understood in terms of the special properties it imparts to compounds that contain it (*1*). A general characteristic of carbonyl compounds is their ability to react with primary

1. Monographs that describe the chemistry of carbonyl compounds in substantial detail are now available [cf. C. D. Gutsche, "The Chemistry of Carbonyl Compounds." Prentice-Hall, Englewood Cliffs, New Jersey, 1967; S. Patai, ed., "The Chemistry of the Carbonyl Group." Wiley (Interscience), New York, 1966].

amines with loss of water to form Schiff bases (*azomethines* or *imines*). During the past three decades, the importance of such Schiff bases as intermediates in many of the enzymic transformations of both carbonyl compounds and amines (including the amino acids) has become apparent.

For purposes of this discussion, two categories of enzymes that operate via Schiff base mechanisms may be defined. In category I, the carbonyl (or potential carbonyl) group is present in the enzyme as part of a prosthetic group or coenzyme; the substrates are amines of various types. This category includes the numerous pyridoxal phosphate–dependent enzymes and a more recently discovered series of enzymes that contain covalently bound pyruvate (or other keto acids) as the prosthetic group. Enzymes of category II lack a carbonyl group but contain as part of the active site an amino group that reacts with a carbonyl-containing substrate. This category includes certain β-keto acid decarboxylases, various aldolases, etc. Enzymes of category I, but not those of category II, are usually inactivated by reduction in the absence of substrate with $NaBH_4$ or similar agents and by incubation with carbonyl reagents such as hydroxylamine, phenylhydrazine, and cyanide.

In this chapter we shall first consider some general properties of Schiff bases, then briefly consider reactions catalyzed by various subgroups of enzymes within categories I and II, together with present views concerning the mechanism of these reactions.

II. Schiff Base Formation: Nature and Catalytic Effects

A. General Characteristics

The Schiff base may be defined as the imine (III) arising from the condensation of an amine and a carbonyl-containing compound (Scheme 1). If the carbonyl compound (II) is an aldehyde, the resulting Schiff base (III) is an *aldimine;* if (II) is a ketone, (III) is a *ketimine* (*1a*). The extent of imine formation is markedly dependent upon the structure of the carbonyl compound and of the amine. Aldehydes react more rapidly and more completely than do ketones of similar structure. The con-

1a. Addition and elimination reactions presented here and in the following sections are shown as concerted processes to describe compactly the overall changes effected in the molecule as a result of the transformation discussed. This usage is not meant as a statement of the exact sequence of events by which one molecular species is transformed into another, i.e., by concerted as opposed to stepwise addition or elimination.

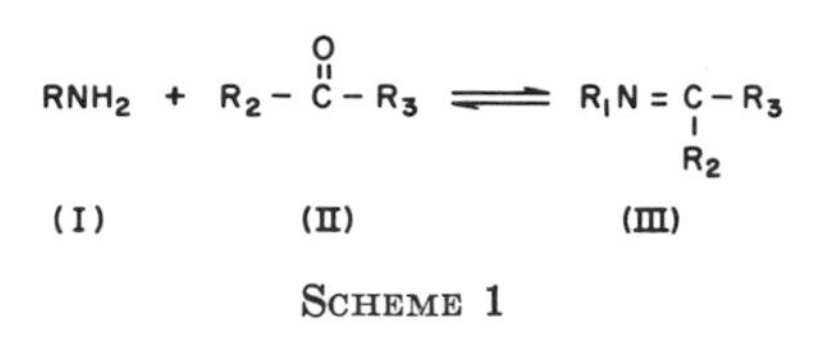

SCHEME 1

densation reaction is believed to be general acid catalyzed (Scheme 2); the initial step is the nucleophilic attack by the amino nitrogen upon the electron-deficient carbonyl group to form the aminocarbinol (IV) or (IV′). Such aminocarbinols are usually unstable and either revert to the starting materials or dehydrate to the imine (III) or (III′). This reaction

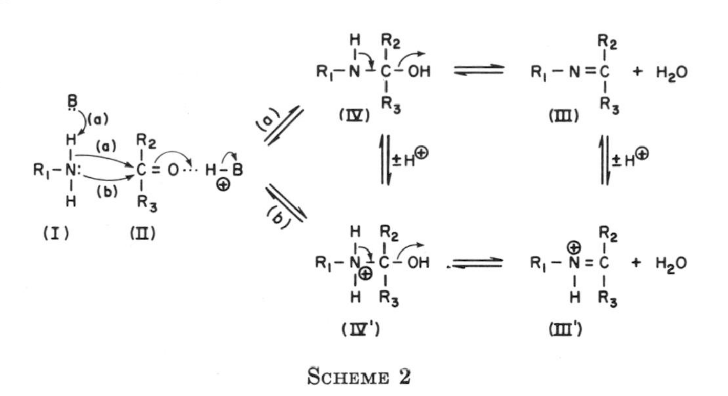

SCHEME 2

is readily reversible; as a consequence, the imine exists in aqueous solution as an equilibrium mixture with the aminocarbinol and its components (I) and (II).

The product imine (III) or (III′) is itself a reactive species and is prone to further reaction resulting in the addition of nucleophilic agents (e.g., H_2NR, HSR, and HOR) to the imine bond (e.g., Scheme 3). If a

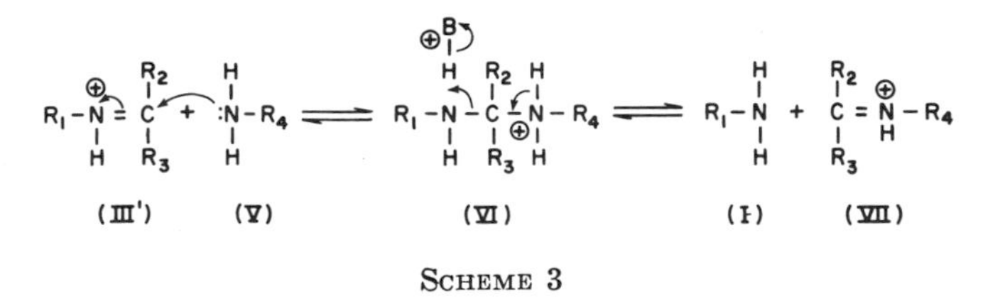

SCHEME 3

second amine [(V), Scheme 3] adds to this polarized bond, the labile adduct (VI) can collapse to either (III′) or (VII). This facile conversion of one imine to another is termed *transimination.* Addition reactions of this type are facilitated by the fact that the imine nitrogen is readily protonated (pK_a ~ 7.0) thereby increasing the positive character of the imine carbon. It is as a result of this ease of protonation on the imine nitrogen that the imine is more susceptible to nucleophilic addition reactions than the parent aldehyde or ketone (*2*).

The Schiff base may assume three different tautomeric forms by rearrangement of appropriate α-protons. A general base-catalyzed prototropic shift (*3–5*) (Scheme 4) converts the aldimine (VIII) to the keti-

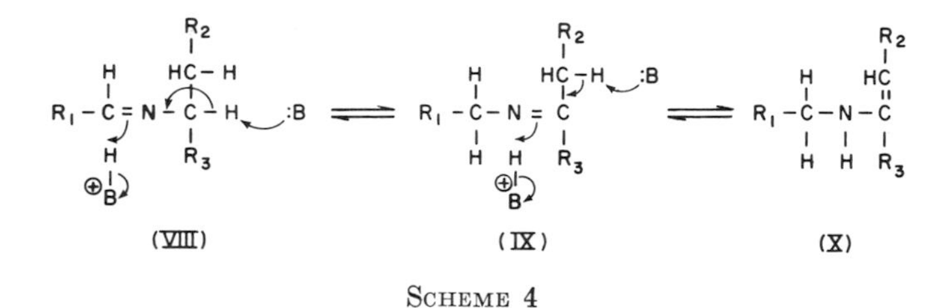

SCHEME 4

mine (IX). A further prototropic shift converts the ketimine (IX) to the *enamine* (X). The electronic distribution in a classic enamine is such (Scheme 5) that the β-carbon atom can serve as an efficient nucleophile,

$R_1-CH=CH-\ddot{N}H-R_2 \longleftrightarrow R_1-\overset{\ominus}{C}H-CH=\overset{\oplus}{N}H-R_2$

(XI) (XII)

SCHEME 5

a circumstance that facilitates β-addition reactions to the enamine (*6*). The negative character of the β-carbon arises as a result of the availability of the electron pair on nitrogen for resonance in the manner shown and is important in determining the course of enzymic reactions such as those catalyzed by the aldolases and transaldolases. However, if the enamine is situated in an electronic environment in which the nitrogen electrons are no longer free to participate exclusively in this type of resonance, e.g., because of protonation of the imine nitrogen or conjugation of the imine with an efficient electron sink, the nucleophilic character of the β-carbon can be altered. For example, the following major resonance forms (Scheme 6) may be written for enamines derived by appropriate

2. W. P. Jencks, "Catalysis in Chemistry and Enzymology," p. 73. McGraw-Hill, New York, 1969.
3. S. K. Hsü, C. K. Ingold, and C. L. Wilson, *JCS* p. 1778 (1935).
4. R. Pérez Ossorio and E. D. Hughes, *JCS* p. 426 (1952).
5. D. E. Metzler, M. Ikawa, and E. E. Snell, *JACS* **76**, 648 (1954).
6. H. O. House, "Modern Synthetic Reactions," pp. 198–199. Benjamin, New York, 1965.

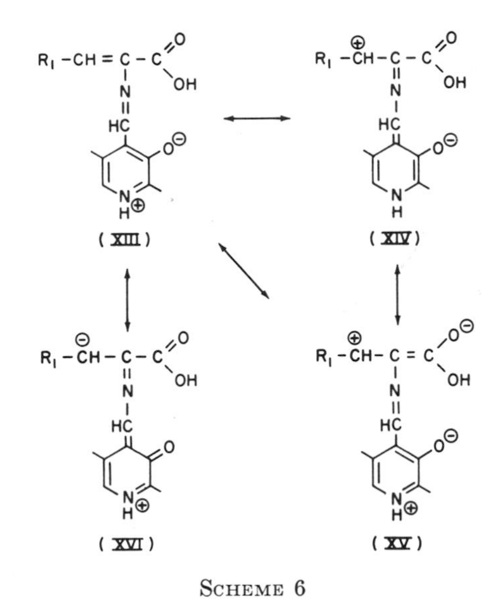

SCHEME 6

elimination reactions from pyridoxal (or pyridoxal phosphate) and certain amino acids. Forms (XIV) and (XV), in which the β-carbon atom has electrophilic character, appear to predominate as judged by the course of most reactions catalyzed by pyridoxal in both nonenzymic and enzymic systems (see Sections II,B and III,A).

In addition to permitting direct addition reactions of the types considered so far, imine formation also serves another important function: the weakening of each of the bonds to the α-carbon atom of the amine component of the imine. In this way the imine bond facilitates elimination reactions at this center. This generally directed bond-weakening effect became especially apparent from studies of reactions of pyridoxal with amino acids and will be discussed further under that heading.

B. Catalytic Effects of Schiff Base Formation in Nonenzymic Systems

1. *Reactions of Pyridoxal with Amino Acids*

Pyridoxal (PL) (XVII) or pyridoxal 5′-phosphate (PLP) (XVIII) catalyzes many reactions of amino acids. At equivalent concentrations, PLP is somewhat more effective than PL as a catalyst since the concentration of the free aldehyde necessary for catalysis is reduced in solutions of PL, but not in those of PLP, by formation of the internal hemiacetal (XIX). Many of these reactions are further catalyzed by metal ions

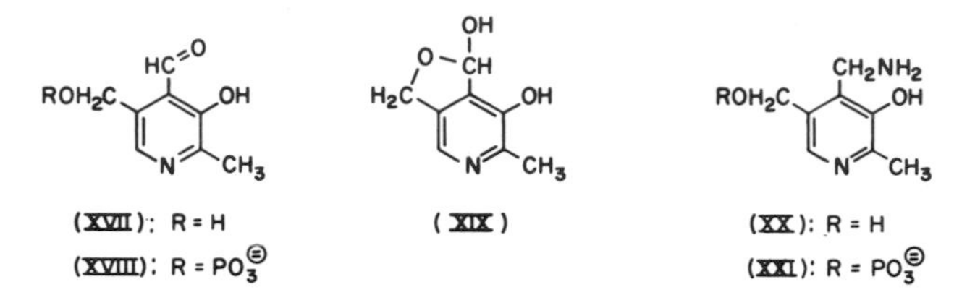

(*7, 8*), or less effectively by organic bases such as imidazole (*9, 10*). Reactions studied include racemization of alanine [Eq. (1)] (*11*), transamination [Eq. (2)] (*7, 12*), α,β-elimination reactions [Eq. (3)] (*13*), β-replacement reactions [Eq. (4)] (*5*), the reversible cleavage of α-amino-β-hydroxy acids [Eq. (5)] (*14, 15*), decarboxylation [Eq. (6)] (*16–18*), and decarboxylation-dependent transamination [Eq. (7)] (*16*).

$$\text{L-Alanine} \overset{\text{PL}}{\rightleftharpoons} \text{D-Alanine} \tag{1}$$

$$\text{RCHNH}_2\text{COOH} + \text{PL (or PLP)} \rightleftharpoons \text{PM (or PMP)} + \text{RCOCOOH} \tag{2}$$

[where R stands for the residual portion of most amino acids; PM, pyridoxamine (XX); and PMP, pyridoxamine phosphate (XXI)]

$$\text{R'YCHCHNH}_2\text{COOH} \xrightarrow{\text{PL}} \text{YH} + \text{NH}_3 + \text{R'CH}_2\text{COCOOH} \tag{3}$$

$$(\text{R'} = \text{H or CH}_3;\ \text{Y} = \text{OH},\ \text{OPO}_3^{2-},\ \text{O}\overset{\text{O}}{\overset{\|}{\text{C}}}\text{—R''},\ \text{SH, etc.})$$

$$\text{HOCH}_2\text{CHNH}_2\text{COOH} + \text{indole} \overset{\text{PL}}{\rightleftharpoons} \text{tryptophan} + \text{H}_2\text{O} \tag{4}$$

$$\text{R''CHOHCHNH}_2\text{COOH} \overset{\text{PL}}{\rightleftharpoons} \text{R''CHO} + \text{H}_2\text{NCH}_2\text{COOH} \tag{5}$$

$$(\text{R''} = \text{CH}_3,\ \text{C}_6\text{H}_5,\ \text{H, etc.})$$

$$\text{R}\overset{\text{CH}_3}{\overset{|}{\text{C}}}\text{NH}_2\text{COOH} \xrightarrow{\text{PL}} \text{R}\overset{\text{CH}_3}{\overset{|}{\text{C}}}\text{HNH}_2 + \text{CO}_2 \tag{6}$$

$$\text{R}\overset{\text{CH}_3}{\overset{|}{\text{C}}}\text{NH}_2\text{COOH} + \text{PL} \rightarrow \text{PM} + \text{R}\overset{\text{CH}_3}{\overset{|}{\text{C}}}\text{=O} + \text{CO}_2 \tag{7}$$

Each of these model reactions has its enzymic equivalent, and each of these enzymes contains pyridoxal phosphate as an essential coenzyme;

7. D. E. Metzler and E. E. Snell, *JACS* **74,** 979 (1952).
8. J. B. Longenecker and E. E. Snell, *JACS* **79,** 142 (1957).
9. T. C. Bruice and R. M. Topping, *JACS* **85,** 1480 and 1488 (1963).
10. J. W. Thanassi, A. R. Butler, and T. C. Bruice, *Biochemistry* **4,** 1463 (1965).
11. J. Olivard, D. E. Metzler, and E. E. Snell, *JBC* **199,** 669 (1952).
12. E. E. Snell, *JACS* **67,** 194 (1945).
13. D. E. Metzler and E. E. Snell, *JBC* **198,** 353 (1952).
14. D. E. Metzler, J. B. Longenecker, and E. E. Snell, *JACS* **75,** 2786 (1953).
15. D. E. Metzler, J. B. Longenecker, and E. E. Snell, *JACS* **76,** 639 (1954).
16. G. D. Kalyankar and E. E. Snell, *Biochemistry* **1,** 594 (1962).
17. E. Werle and W. Koch, *Biochem. Z.* **319,** 305 (1949).
18. J. W. Thanassi and J. S. Fruton, *Biochemistry* **1,** 975 (1962).

indeed, the participation of pyridoxal in some of the enzymic reactions [e.g., transamination (*12*), serine deamination (*13*)] was first postulated on the basis of these nonenzymic model reactions (*5, 19*). From the above facts one can conclude that pyridoxal must play a key catalytic role in the active site of each of these enzymes and that enzymic and nonenzymic reactions proceed by closely similar mechanisms. In fact, many current concepts concerning the mechanism of action of PLP enzymes find their principal experimental support in studies of the corresponding model systems, and for this reason a brief account of the model reactions appears justified. Several more detailed reviews of certain aspects of this topic are available (*20–23*).

A comparison of the ability of several aldehydes of modified structure to replace pyridoxal as a reactant in Eq. (2) showed that the phenolic group at position 3 and the formyl group at position 4 of pyridoxal were essential for transamination. The heterocyclic nitrogen was also necessary, apparently because it provided a strongly electronegative group so placed as to reduce the electron density about the formyl group in the 4 position. The methyl and hydroxymethyl groups at positions 2 and 5 were nonessential (*5*). Thus, the minimum structural features for the nonenzymic reaction were provided by 3-hydroxypyridine 4-aldehyde [subsequent studies have shown that this same compound also undergoes nonenzymic imidazole-catalyzed transamination with glutamate (*10*) and is also an excellent substrate for enzymic transamination by pyridoxamine pyruvate transaminase (*24*)]. Based on these considerations and the known propensity of imines of aromatic *o*-hydroxyaldehydes with amines to form metal chelates (*25*), it was postulated that each of the metal-catalyzed nonenzymic reactions resulted from initial formation of chelate complexes such as (XXII) (Scheme 7), where M is an appropriate di- or trivalent metal ion. Both the existence and the postulated structures of such chelate compounds have been proved by isolation and elemental analysis (*15, 26–30*) as well as by X-ray analysis of the crystalline compounds (*31, 32*).

19. J. B. Longenecker, M. Ikawa, and E. E. Snell, *JBC* **226,** 663 (1957).
20. A. E. Braunstein, "The Enzymes," 2nd ed., Vol. 2, p. 113, 1960. This article also contains a valuable historical account of the discovery of various pyridoxal phosphate-dependent enzymes.
21. E. E. Snell, *Vitamins Hormones* **16,** 77 (1958).
22. A. M. Perault, B. Pullman, and C. Valdemoro, *BBA* **46,** 555 (1961).
23. F. H. Westheimer, "The Enzymes," 2nd ed., Vol. 1, p. 259, 1959.
24. J. E. Ayling and E. E. Snell, *Biochemistry* **7,** 1626 (1968).
25. A. E. Martell and M. Calvin, "Chemistry of Metal Chelate Compounds," pp. 397–401. Prentice-Hall, Englewood Cliffs, New Jersey, 1952.
26. P. Fasella, H. Lis, N. Siliprandi, and C. Baglioni, *BBA* **23,** 417 (1957).
27. J. Baddiley, *Nature* **170,** 711 (1952).

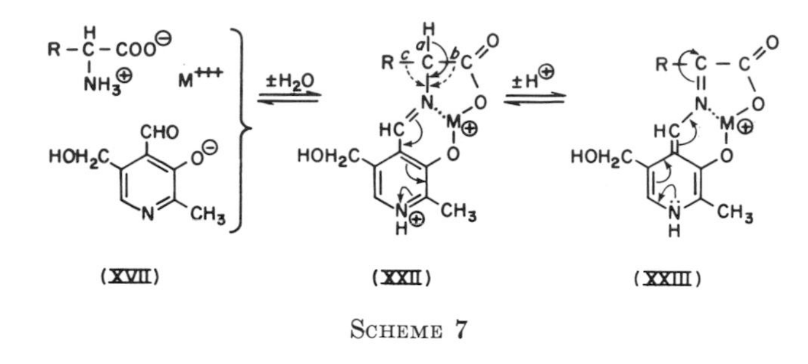

SCHEME 7

In such complexes, the considerable electron displacement from the bonds to the α-carbon atom of the amino acid residue toward the nitrogen of the aldimine linkage (cf. Schemes 4 and 5) is further intensified by: (a) conjugation of this imine bond with an efficient electron sink, the pyridine ring, (b) protonation of the pyridine nitrogen [p$K \cong 8.2$ (*33*)], and (c) the presence of the metal ion, which acts in the same manner as does protonation of the imine nitrogen. These combined effects result in the displacement of electrons from each of the three bonds to the α-carbon atom of the amino acid residue into the chelate complex (XXII) thus weakening these bonds (i.e., "activating" the amino acid) and permitting the reactions shown in Eqs. (1) to (7) to occur as a consequence of further reactions shown in Schemes 7 to 11. Racemization of amino acids [Eq. (1)] occurs as a consequence of reversible labilization of the α-hydrogen atom with loss of asymmetry via (XXIII) Scheme 7. Transamination [Eq. (2)] represents summation of the changes shown in Schemes 7 and 8. In this case the extra electron pair of (XXIII) localizes to form the ketimine (XXIV) (Scheme 8), re-

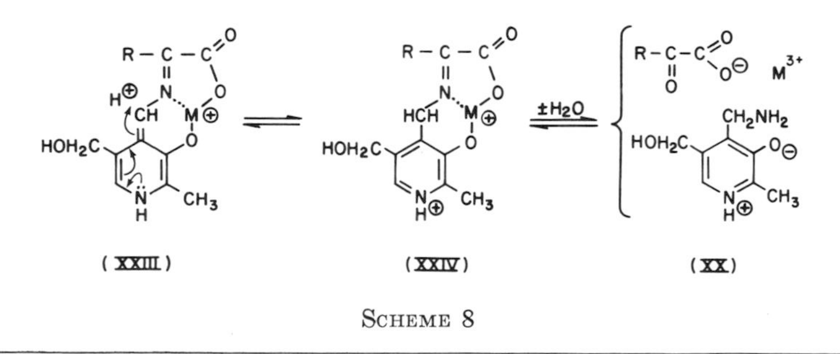

SCHEME 8

28. H. N. Christensen and S. Collins, *JBC* **220,** 279 (1956).
29. H. N. Christensen, *JACS* **81,** 6495 (1959).
30. L. Davis, F. Roddy, and D. E. Metzler, *JACS* **83,** 127 (1961).
31. E. Willstadter, T. A. Hamor, and J. L. Hoard, *JACS* **85,** 1205 (1963).
32. J. F. Cutfield, D. Hall, and T. N. Waters, *Chem. Commun.* p. 785 (1967).
33. D. E. Metzler and E. E. Snell, *JACS* **77,** 2431 (1955).

sulting in a reversible prototropic shift from aldimine to ketimine. Since both imines are in equilibrium with their components, the reaction shown in Eq. (2) results (*34*,a). α,β-Elimination reactions [Eq. (3)] and β-replacement reactions [Eq. (4)] similarly involve labilization of the α-hydrogen atom as shown in Scheme 7. These reactions occur only with amino acids having an electronegative group Y (e.g., OH, SH, and indole) on the β-carbon atom (Scheme 9). In these cases, the extra electron pair in (XXV) can be eliminated as an anion, Y^-, to yield (XXVI). If the latter hydrolyzes at this stage, the unsaturated amino

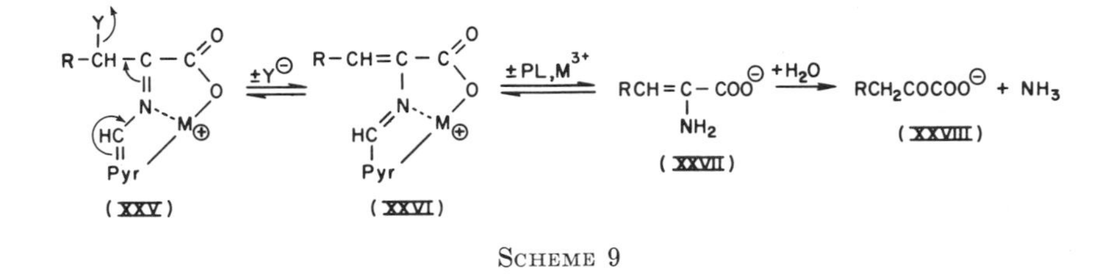

SCHEME 9

34. (a). In the ensuing discussion, chelated aldimines of type (XXII) (Scheme 7), quinonoid intermediates of the type (XXIII) (e.g., Schemes 7 and 8) resulting from the loss of any of the three groups about the α-carbon atom, and chelated ketimines of type (XXIV) (Scheme 8) which give pyridoxamine (PM, XX) or pyridoxamine 5′-phosphate (PMP, XXI) on hydrolysis will be abbreviated as shown in (XXIIa), (XXIIIa), and (XXIVa), respectively (Pyr stands for the substituted pyridine nucleus).

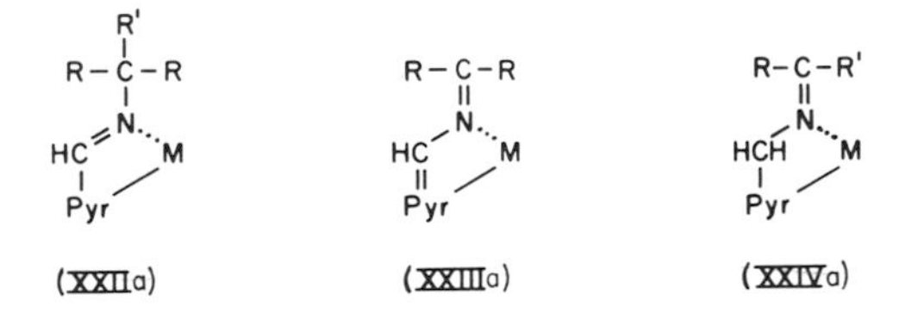

(b). In the discussion of analogous enzyme mechanisms the above abbreviations have been modified in the following manner:

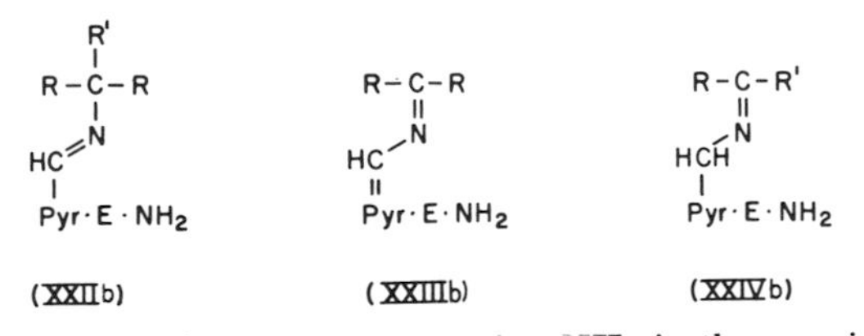

where $E \cdot NH_2$ represents the apoenzyme and $—NH_2$ is the ϵ-amino group of the lysine residue that forms a Schiff base with PLP in the holoenzyme. The free holoenzyme has been abbreviated in the manner shown below.

HC = N | | Pyr − E or Enz · PLP

acid (XXVII) formed decomposes spontaneously and irreversibly to the corresponding α-keto acid and ammonia, thus accounting for reactions shown in Eq. (3).

β-Replacement reactions can be explained in similar terms. If, for example, intermediate (XXVI) is generated from pyridoxal and serine (by the transformations shown in Schemes **7** and **9**) in the presence of a second nucleophile, indole, tryptophan is formed by the transition (XXVI) → (XXV) [= (XXIII)] → (XXII) and hydrolysis.

The cleavage of α-amino-β-hydroxy acids [Eq. (5), Scheme 10] and

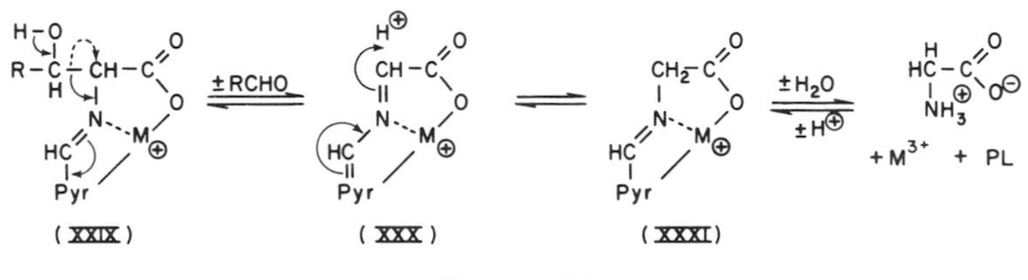

SCHEME 10

decarboxylation reactions [Eqs. (6) and (7), Scheme 11] differ mechanistically from the reactions considered so far only in that direct labilization of the appropriate —$\overset{\alpha}{\mathrm{C}}$—C— bond occurs (Schemes 10 and 11). That

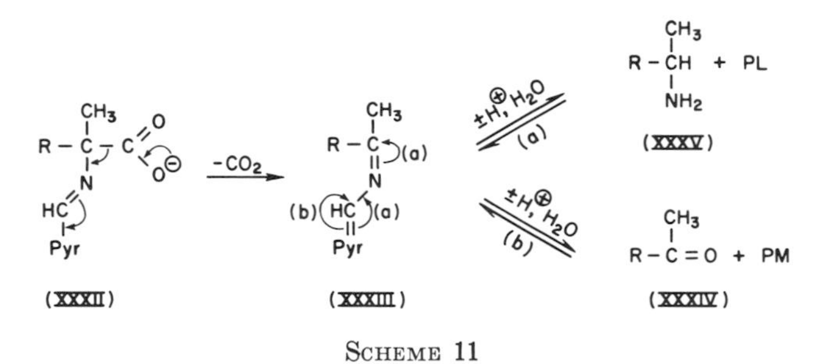

SCHEME 11

the reactions proceed in this way rather than indirectly via initial labilization of the —$\overset{\alpha}{\mathrm{C}}$—H was proved by model studies with α-substituted amino acids (*16, 19*). Perhaps the strongest evidence for quinonoid intermediates corresponding to (XXX) and (XXXIII) in these reactions is the demonstration (*16*) that the model decarboxylation reaction can yield as products those shown in both Eqs. (6) and (7); these products reflect the two possible alternate localizations of the extra electron pair present in (XXXIII) (reactions a or b, Scheme 11).

Catalysis of the reactions in Eqs. (1) to (5) by metal ions is explained in part by the fact that the extent of imine formation is enhanced by metal ions, especially in dilute solution and at low pH values (*35*). The chelated metal ion also (a) acts as does protonation (see Section II,A, pp. 337–338) to increase the electron-withdrawing character of the imine nitrogen, and (b) imposes a planar structure on the imine. Both factors contribute to the electron displacements and delocalization of charge necessary for cleavage of bonds to the α-carbon atom. The extent of catalysis has not been carefully quantitated; rate increases of 20- to 200-fold have been observed for the transamination reaction, depending upon the conditions (*8*). In nonaqueous solvents (*36*) and with appropriate substrates (*37*) metal ions are not catalytic; appropriate organic bases [e.g., imidazole (*9, 10*)] also catalyze certain of these reactions (although less efficiently than do metal ions) apparently by acting as general acid-base catalysts in the manner illustrated in Schemes 4 and 5. Decarboxylation reactions [Eqs. (6) and (7)] are partly inhibited by the same metal ions that catalyze other reactions, probably because chelate bonding of the carboxyl group inhibits its release as CO_2 (*16*).

2. *Catalysis by Schiff Bases Other Than Those of Pyridoxal or Pyridoxamine*

Catalysis of decarboxylation of oxalacetate by aniline citrate has found analytic use in biochemistry. This and similar decarboxylation reactions of β-keto acids apparently proceed as shown in Scheme 12 (*37a, 37b*). The reaction is closely related to enzymic decarboxylation of acetoacetate as shown by the elegant studies of Westheimer (*38–44*).

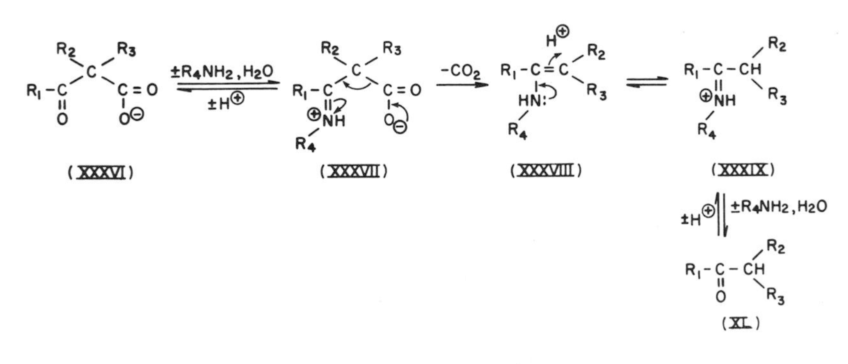

SCHEME 12

35. O. A. Gansow and R. H. Holm, *JACS* **91,** 573 (1969).
36. Y. Matsuo, *JACS* **79,** 2016 (1957).
37. C. Cennamo, *Boll. Soc. Ital. Biol. Sper.* **37,** 74 and 183 (1961).
37a. F. H. Westheimer and W. A. Jones, *JACS* **63,** 3283 (1941).
37b. J. P. Guthrie and F. H. Westheimer, *Federation Proc.* **26,** 562 (1967).
38. G. Hamilton and F. H. Westheimer, *JACS* **81,** 6332 (1959).
39. I. Fridovich and F. H. Westheimer, *JACS* **84,** 3208 (1962).

Decarboxylation of α-amino acids to the corresponding aldehydes occurs on boiling in aqueous solutions with α-keto acids [Eq. (8)] (*45*).

$$RCHNH_2COO^- + R'COCOO^- \rightarrow RCHO + CO_2 + R'CHNH_2COO^- \quad (8)$$

The reaction is analogous to decarboxylation-dependent transamination [Eq. (7)] between amino acids and pyridoxal and presumably occurs by a similar mechanism [cf. Scheme 11, (XXXII) → (XXXIV)] involving Schiff base formation between the two reactants (*16*).

III. Enzymes That Contain a Carbonyl Group (Category I)

A. Pyridoxal Phosphate Enzymes

1. *Interaction of Apoenzymes with Pyridoxal Phosphate*

Purified PLP enzymes differ greatly in their ease of resolution into coenzyme and apoenzyme. Aspartate aminotransferase, for example, is completely resolved by standing overnight at 0° in the presence of 60% saturated ammonium sulfate at pH 3.5 (PLP form) or at pH 5.0 (PMP form) (*46*), and the apoenzyme thus precipitated is completely reactivated on addition of the coenzyme. Resolution of some other enzymes requires dialysis against cysteine [e.g., arginine decarboxylase (*47*)], penicillamine [e.g., tryptophanase (*48*)] or other agents that react with PLP, with or without deforming buffers such as imidazole-citrate [e.g., phosphorylase (*49, 50*)]. A few [e.g., alanine aminotransferase (*51, 52*)] have not been fully resolved in a reversible manner by procedures thus far applied. Reactivation of various apoenzymes by PLP has not been sufficiently studied. The mechanism obviously is complex since complete re-

---

40. S. Warren, B. Zerner, and F. H. Westheimer, *Biochemistry* **5**, 817 (1966).
41. R. A. Laursen and F. H. Westheimer, *JACS* **88**, 3426 (1966).
42. F. H. Westheimer, *Proc. Chem. Soc.* p. 253 (1963).
43. W. Tagaki and F. H. Westheimer, *Biochemistry* **7**, 901 (1968).
44. W. Tagaki, J. P. Guthrie, and F. H. Westheimer, *Biochemistry* **7**, 905 (1968).
45. R. M. Herbst, *Advan. Enzymol.* **4**, 75 (1944).
46. H. Wada and E. E. Snell, *JBC* **237**, 127 (1962).
47. S. L. Blethen, E. A. Boeker, and E. E. Snell, *JBC* **243**, 1671 (1968).
48. Y. Morino and E. E. Snell, *JBC* **242**, 5591 (1967).
49. S. Shaltiel, J. L. Hedrick, and E. H. Fischer, *Biochemistry* **5**, 2108 (1966).
50. J. L. Hedrick, S. Shaltiel, and E. H. Fischer, *Biochemistry* **8**, 2422 (1969).
51. M. H. Saier and W. T. Jenkins, *JBC* **242**, 91 (1967).
52. T. Matsuzawa and H. L. Segal, *JBC* **243**, 5929 (1968).

activation may require several minutes or even longer (*53*) or may require addition of agents such as cysteine (*54*) to become complete. The subject is complicated by the fact that most PLP enzymes contain several subunits (Table I; see pp. 363–364). Dissociation into subunits is generally favored by resolution of the enzyme, and reassociation is favored by addition of PLP (*48, 55*). However, reactivation and reassociation of subunits are separate phenomena, since PLP analogs that permit reassociation of subunits do not necessarily regenerate enzymic activity (*48, 56*). Association with PLP frequently results in a readily measurable conformational change in the enzyme [cf. tryptophanase (*48*)], and the holoenzyme forms of most PLP enzymes are substantially more resistant toward denaturation than the corresponding apoenzymes (*48, 55*). Even in the single subunit enzyme, D-serine dehydratase, activation of the apoenzyme by PLP is a relatively slow process; resolution of the holoenzyme at concentrations near its dissociation constant for PLP requires several hours (*57*). These results suggest that activation of apoenzyme by PLP is at least a two-step process: a rapid, freely reversible association followed by a much slower and only sluggishly reversible conformational change. The latter change is not dependent upon imine formation (*57a*); indeed, direct fluorometric measurements show that activation of apoaspartate aminotransferase by PMP also involves two or more steps (*58*).

Studies with PLP analogs, summarized elsewhere (*21, 24, 59, 60*), have shown that every substituent of the PLP molecule contributes to binding of this coenzyme by its conjugate apoenzymes. The presence of the 5′-phosphate group is particularly important; the affinity of apoaspartate aminotransferase for free pyridoxal is less than 0.001 that for pyridoxal phosphate (*46*). From such studies it can be concluded that binding of PLP results from a mixture of ionic and other undefined forces (e.g., π-π or hydrophobic interactions). The imine bond contributes to the binding only in a secondary manner since it is broken on interaction with the substrate and since certain apoenzymes (e.g., certain aminotransferases) can be fully activated by pyridoxamine phosphate.

Similar studies show clearly that the structural requirements for coenzyme action with PLP-dependent enzymes resemble those for catalysis

53. A. Meister, H. A. Sober, and E. A. Peterson, *JBC* **206**, 89 (1954).
54. C. L. Sevilla and E. H. Fischer, *Biochemistry* **8**, 2161 (1969).
55. S. S. Tate and A. Meister, *Biochemistry* **7**, 3240 (1968).
56. S. S. Tate and A. Meister, *Biochemistry* **8**, 1056 (1969).
57. W. Dowhan and E. E. Snell, *JBC* (in press) (1970).
57a. M. Arrio-Dupont, *BBRC* **36**, 306 (1969).
58. J. E. Churchich and J. G. Farrelly, *JBC* **244**, 3685 (1969).
59. Y. Morino and E. E. Snell, *Proc. Natl. Acad. Sci. U. S.* **57**, 1692 (1967).
60. V. I. Ivanov and M. Ya. Karpeisky, *Advan. Enzymol.* **32**, 21 (1969).

of the model nonenzymic reactions discussed earlier. Neither the 2-methyl group nor the 5′-phosphate group (aside from its role in binding) of PLP is required for action of aspartate aminotransferase; the 3-hydroxyl group and the 4-formyl group are, however, essential. For some apoenzymes, 2-norpyridoxal 5′-phosphate, which lacks the 2-methyl group, is as effective or even more so than PLP itself (*47, 59, 60*).

2. *Structural and Spectral Properties*

In addition to the diversities just discussed, PLP enzymes also differ greatly with respect to their molecular weights, subunit composition, specific activities, and pH optima (Table I), and also in their spectral changes with pH. The absorption maxima at 410–435 mμ result from formation of an internal aldimine between the formyl group of PLP and the ε-amino group of a lysine residue on the protein, as first shown by Fischer for phosphorylase (*61*). This absorption maximum shifts substantially (to near 360 mμ) with increasing pH in some enzymes [e.g., aspartate aminotransferase (*62*), tryptophanase (*62a*)] but not in others [e.g., arginine decarboxylase (*47*), D-serine dehydratase (*57*)]; studies with model compounds (*63, 64*) have assigned the absorbance maximum near 415 mμ to one or more structures such as (XLI) ↔ (XLIII) and the absorbance at 360 mμ to similar structures [e.g., (XLIV)] lacking a proton. However, because of the several possible ionic and tautomeric forms, including those containing a proton on the pyridine nitrogen, the exact structures corresponding to these absorption maxima are not certain. The variation in position of the absorption maximum near 415 mμ from one enzyme to another also is not understood but presumably reflects subtle differences in coenzyme binding in each apoenzyme.

The reactivity of the imine bond (see Section II) is such that appropriate nucleophilic groups (e.g., —SH and —NH_2) from the protein can, in principle, add across it and shift the ultraviolet absorption band to still shorter wavelengths (ca. 330 mμ) similar to that shown by ketimines of pyridoxamine [(XLV), Scheme 13]. The same type of interaction between the imine form of the enzyme and its substrate leads via transimination (Scheme 3) to the Schiff base between enzyme and substrate. Whether the ε-amino group of the lysine residue released during this reaction remains in the vicinity of the bound substrate or moves out of the way during

61. E. H. Fischer, A. B. Kent, E. R. Snyder, and E. G. Krebs, *JACS* **80,** 2906 (1958).
62. W. T. Jenkins and I. W. Sizer, *JBC* **234,** 1179 (1959).
62a. Y. Morino and E. E. Snell, *JBC* **242,** 2800 (1967).
63. D. E. Metzler, *JACS* **79,** 485 (1957).
64. D. Heinert and A. E. Martell, *JACS* **85,** 183, 188, and 1334 (1963).

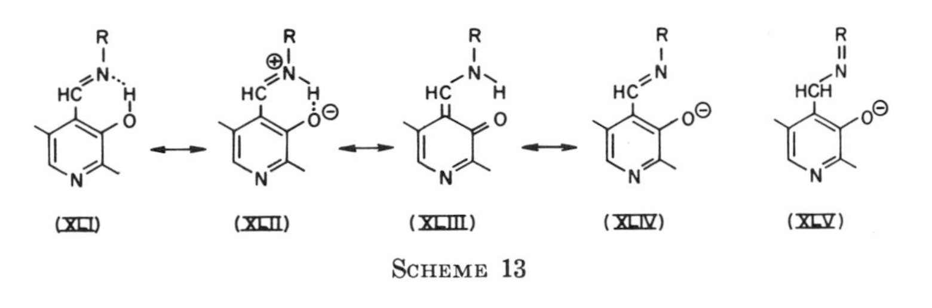

SCHEME 13

subsequent catalytic events is unknown; in the former case it might facilitate those events by acting as a base-acid catalyst. In either case, the sequence of amino acids surrounding the internal imine group is of interest and can be ascertained by sequencing peptides containing the covalently bound pyridoxyl residue isolated from partial digests of the $NaBH_4$-reduced enzyme. From the few enzymes so far studied in this way (Table II; see p. 366), it appears that the only common feature of these peptides is the lysine residue involved in interaction with PLP.

3. *Mechanism of the Enzymic Reactions*

Once the imine between enzyme and substrate is formed, the subsequent events during catalysis appear to proceed by mechanisms essentially similar to those presented for the related nonenzymic reactions (Section II,A). That is, each of the reactions is a consequence of cleavage of one of the three bonds to that substrate carbon atom adjacent to the imine nitrogen. A classification of the many known PLP-dependent reactions based upon this mode of action is given in Table III, pp. 367–368 (*65*); more extensive listings of individual transaminases (*66*), decarboxylases (*66*), and other PLP enzymes (*20, 67*) can be found elsewhere. Unlike the nonenzymic reactions, however, the enzymic reactions, with one probable exception [certain amine oxidases (*68*)], are not catalyzed by di- or trivalent metal ions, the functional role of the metal ion (see Section II,B,1, pp. 344–345) being played much more efficiently by the apoenzyme. In the following sections, only those few PLP enzymes that have undergone mechanistic investigations, or those for which no non-

65. Phosphorylase, the most plentiful PLP-dependent enzyme of mammals, does not act via a Schiff base mechanism, as shown by the fact that the borohydride-reduced enzyme still shows activity [E. G. Krebs and E. H. Fischer, *Advan. Enzymol.* **24**, 263 (1962)]. The mechanism by which PLP contributes to the activity of this enzyme is still unknown.

66. B. M. Guirard and E. E. Snell, *Comp. Biochem.* **15**, 138 (1964).

67. H. R. Mahler and E. H. Cordes, "Biological Chemistry," pp. 666–704. Harper, New York, 1966.

68. H. Yamada and K. T. Yasunobu, *JBC* **238**, 2669 (1963).

enzymic model reactions are known, will be discussed further. With the proviso that the enzymes exist in the imine form and interact with their substrates by transimination, tentative mechanisms for the racemization of amino acids, α,β-elimination and β-replacement reactions, decarboxylation reactions, and aldolase reactions can be derived from the mechanisms given in Section II,B,1 for the corresponding nonenzymic reactions simply by substituting a PLP enzyme [cf. (*34*,b) for abbreviations used] for the pyridoxal and metal ions shown in Schemes 7 to 11.

a. Transamination Reactions (*Class I-B, Table III*). The foregoing considerations can be illustrated by the mechanism suggested in 1962 (*69*) for enzymic transamination [Scheme (14)]. The PLP form of the

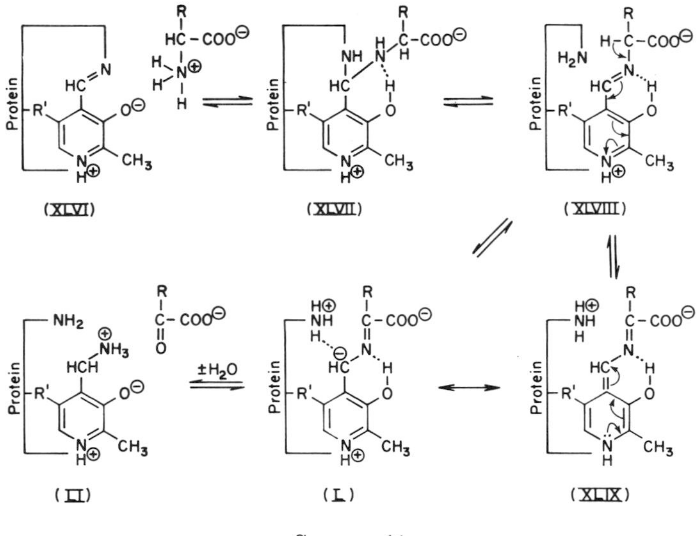

SCHEME 14

enzyme, which possesses specific affinity for its substrates, undergoes transimination with amino acid to yield the enzyme–substrate complex, liberating the ϵ-amino group of a lysine residue. This basic group (or another functionally similar group from the apoenzyme) enhances the previously discussed effects (Section II,A, p. 338) of the imine in conjugation with the pyridine ring, thus assisting in catalysis of the prototropic shift from aldimine to ketimine. The latter undergoes hydrolysis (presumably via an aminocarbinol) to the PMP form of the enzyme, thus completing catalysis of one-half of the overall reaction [Eq. (9)]. The process then proceeds in reverse with keto $acid_2$, yielding amino $acid_2$ [Eq. (10)] and

69. E. E. Snell, *Brookhaven Symp. Biol.* **15,** 32 (1962).

regenerating the PLP enzyme, thereby completing catalysis of the overall reaction [Eq. (11)].

$$\text{Amino acid}_1 + \text{PLP·E} \rightleftharpoons \text{keto acid}_1 + \text{PMP·E} \tag{9}$$

$$\text{Keto acid}_2 + \text{PMP·E} \rightleftharpoons \text{amino acid}_2 + \text{PLP·E} \tag{10}$$

$$\text{Sum:} \quad \text{Amino acid}_1 + \text{keto acid}_2 \rightleftharpoons \text{keto acid}_1 + \text{amino acid}_2 \tag{11}$$

In further support of this mechanism, the following observations predictable from it have been made with aspartate aminotransferase: (a) both the PLP and the PMP forms of the enzyme are readily isolated (*70–72*); (b) the two half-reactions [Eqs. (9) and (10)] can be observed spectrophotometrically at high concentrations of enzyme (*73*) or at high dilutions when the apotransaminase is used with PL as a poorly bound substitute for PLP (*46*); (c) borohydride reduction in the presence of substrate permits trapping of the predicted adduct between substrate and PLP as the corresponding acid-stable secondary amine (*74*); (d) addition of keto and amino acids at their equilibrium concentrations to high concentrations of enzyme, to produce steady state concentrations of all possible enzyme–substrate intermediates, results in spectral bands that can be separately attributed to the aldimine, the ketimine, and the intermediate quinonoid forms (*75*); (e) labilization of the α-hydrogen atom of its amino acid substrates is catalyzed by the enzyme and can occur independently of the overall reaction (*76*); and (f) the enzyme follows ping-pong, bi-bi kinetics (*77, 78*). When loosely bound PL replaces PLP in Eq. (9), the hydrogen atom added to the 4′-methylene group of the newly formed pyridoxamine occupies only one of the two stereochemically possible positions (*79*). This is also true for the closely related pyridoxamine pyruvate transaminase; with that enzyme the same proton removed from the α-amino acid appears in the methylene group of pyridoxamine (*80*). This latter observation strongly implies that the same basic group which assists in removing the α-proton during tautomerization of aldimine

70. W. T. Jenkins and L. D'Ari, *BBRC* **22**, 376 (1966).
71. H. Lis, P. Fasella, C. Turano, and P. Vecchini, *BBA* **45**, 529 (1960).
72. O. L. Polyanovskii, *Biokhimiya* **27**, 734 (1962).
73. W. T. Jenkins and L. D'Ari, *JBC* **241**, 2845 (1966).
74. E. A. Malakhova and Yu. M. Torchinskii, *Dokl. Akad. Nauk SSSR* **161**, 1224 (1965).
75. W. T. Jenkins, *Federation Proc.* **20**, 978 (1961).
76. W. T. Jenkins, *JBC* **239**, 1742 (1964).
77. C. P. Henson and W. W. Cleland, *Biochemistry* **3**, 338 (1964).
78. S. F. Velick and J. Vavra, *JBC* **237**, 2109 (1962).
79. H. C. Dunathan, L. Davis, P. G. Kury, and M. Kaplan, *Biochemistry* **7**, 4532 (1968).
80. J. E. Ayling, H. C. Dunathan, and E. E. Snell, *Biochemistry* **7**, 4537 (1968).

to ketimine assists in placing that same proton on the resulting pyridoxamine molecule. All of these findings indicate that Scheme 14 is correct in general outline as a minimal mechanism. Several additional steps (e.g., hydrolysis of the ketimine via a carbinolamine) might be added; indeed, temperature jump studies indicate at least eight reversible steps in the overall reaction (*81*). A much more detailed mechanism has been recently exhaustively discussed by Ivanov and Karpeisky (*60*).

b. α,β-Elimination and β-Replacement Reactions (*Class I-C, Table III*). Scheme 9, modified as specified earlier in this section for enzymic reactions, summarizes accurately present knowledge of the mechanism of these enzymic reactions. The most extensive investigations in this area have been those utilizing crystalline tryptophanase from *E. coli* (*62a, 82–84*). This enzyme catalyzes reaction 3 [Eq. (3)] where R′ is H and Y is the indolyl, or 4-, 5-, or 6-methyl indolyl radicals, —OH, —SH, —SCH_3, or —SCH_2CH_3. In the presence of added indole, the same enzyme synthesizes tryptophan from each of these same substrates. Thus, an enzyme-bound intermediate that can either decompose to pyruvate and ammonia or add indole to form tryptophan is formed from each amino acid substrate. This observation constitutes strong evidence for enzyme-bound α-aminoacrylate as the common intermediate. When high concentrations of any of these substrates are added to tryptophanase, an absorption peak near 505 mμ appears that disappears as these substrates are decomposed; however, when L-alanine, which is not altered by tryptophanase, is added, the same absorption appears and is stable. This peak is attributed to the quinonoid intermediate (XXV) (Scheme 9). This assignment is supported by the observation that in solutions of T_2O the enzyme catalyzes exchange of the α-hydrogen for tritium in alanine as well as in substrates of the enzyme; such exchange is sufficiently rapid so that it does not limit the rate of the β-elimination or replacement reactions (*62a*).

c. α,γ-Elimination and Replacement Reactions (*Class I-D, Table III*). Although small amounts of a keto acid thought to be α-ketobutyrate are formed when homoserine is heated with pyridoxal, its formation is almost obscured by more prominent products such as those formed by transamination (*5*). For this reason, no satisfactory nonenzymic models corresponding to enzymic γ-elimination or γ-replacement reactions have been studied. Enzymic reactions of this class are exemplified by the conversion

81. G. G. Hammes and J. L. Haslam, *Biochemistry* **8,** 1591 (1969).
82. W. A. Newton and E. E. Snell, *Proc. Natl. Acad. Sci. U. S.* **51,** 382 (1964).
83. W. A. Newton, Y. Morino, and E. E. Snell, *JBC* **240,** 1211 (1965).
84. Y. Morino and E. E. Snell, *JBC* **242,** 2793 (1967).

of homoserine to α-ketobutyrate (γ elimination), of *O*-phosphohomoserine to threonine (γ elimination–β substitution), and of *O*-succinylhomoserine (plus cysteine) to cystathionine (γ replacement). Several similar mechanisms have been proposed to explain these varied reactions; these may be consolidated as shown in Scheme 15. Conversion of homoserine (X =

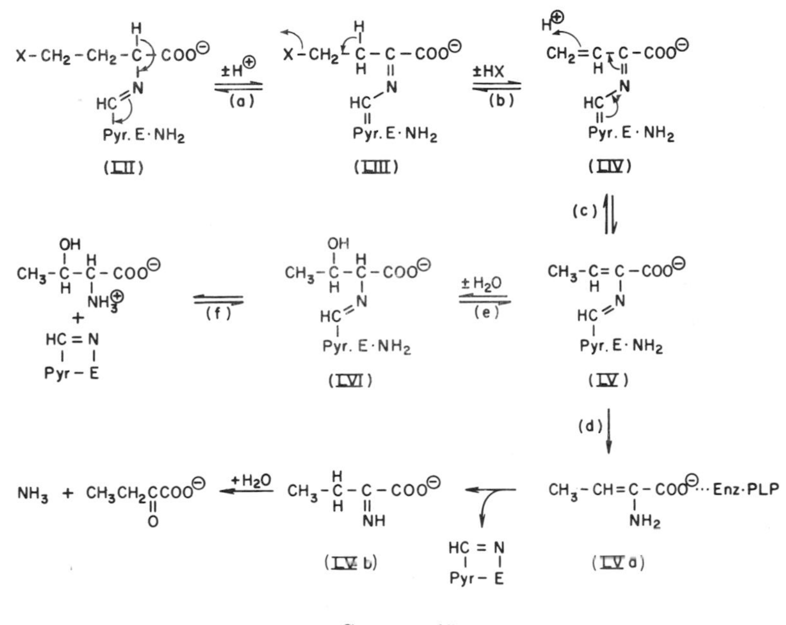

SCHEME 15

OH) to α-ketobutyrate, for example, could occur via reactions (a) to (d), Scheme 15. In the conversion of *O*-succinylhomoserine (X = —$OCOCH_2CH_2COOH$) to cystathionine (X = —SCH_2CHNH_2COOH), succinate is first eliminated [(LII) → (LIV)]; cysteine is then added to the β,γ-unsaturated intermediate (LIV) by reversal of the same reaction pathway [(LIV) → (LII)]. Conversion of *O*-phosphohomoserine to threonine would occur via (LII) → (LVI); addition of water to (LV) accounts for the observed incorporation of deuterium into the product threonine from D_2O (*85, 86*). In this general mechanism (Scheme 15), the primary step in each case is labilization of the α-hydrogen of the amino acid to yield the corresponding carbanionic or quinonoid intermediate (LIII). In support of this formulation, rapid exchange of tritium from T_2O into

85. M. M. Kaplan and M. Flavin, *JBC* **240**, 3928 (1965).
86. M. Flavin and C. Slaughter, *JBC* **235**, 1112 (1960).

cystathionine has been observed (*87*). In the formation of α-ketobutyrate from succinylhomoserine by a γ-cystathionine synthetase, one deuterium is incorporated from D_2O into the β position of α-ketobutyrate but is largely excluded from the γ position (*87*). This observation is consistent with the proposed mechanism if it is assumed that the proton which is removed from the β position of (LIII) is transferred as such to the γ-carbon atom [(LIV) $\rightarrow$ (LV)]; as a necessary corollary to this, it must also be assumed that this proton is initially removed from (LIII) by a general base group which is positioned within a sufficiently hydrophobic region of the active site to shield the proton from equilibration with solvent deuterium during the time required for its addition to the γ position of (LIV) (*88*). Deuterium is then incorporated into the β position of the product α-ketobutyrate during the conversion (LVa) $\rightarrow$ (LVb). Occurrence in this reaction of an α,β-unsaturated intermediate such as (LIV) with carbanionic character is indicated by trapping experiments with *N*-ethylmaleimide. When this reagent is included in the cystathionine synthetase system, one of the two possible enantiomers of the adduct, α-keto-3-[3′-(*N*′-ethyl-2′,5′-dioxopyrrolidyl)]butyric acid [compound (LVII)] is isolated, indicating that an intermediate is formed (LVa?) with sufficient nucleophilic character to react with the reagent (*89–91*).

When the same experiment was repeated in D_2O without the addition of trapping agent (*92*), optically active α-ketobutyrate was formed showing that the addition of a proton to the β-carbon of the species which is ultimately converted to α-ketobutyrate is stereospecific. If aminocrotonate were released into solution immediately upon the hydrolysis of (LV), racemic keto acid would be expected. Consequently, it was proposed that the addition of a proton to effect the conversion of aminocrotonate (LVa) to α-iminobutyrate (LVb) occurs while the intermediate is still fixed on the enzyme followed by nonenzymic hydrolysis of iminobutyrate to ketobutyrate (*91*).

d. Miscellaneous Enzymes of Class I-E, Table III. The mechanism of action of these enzymes has not been studied either with model systems or with purified enzymes. The reactions are most logically formulated as resulting from loss of the α-hydrogen, which leads to an α,γ-dicarbonyl structure resembling that of ethylacetoacetate (Scheme 16). This inter-

87. S. Guggenheim and M. Flavin, *BBA* **151**, 664 (1968).
88. A similar transfer of a single proton occurs from the α-carbon atom of alanine to the 4′-methylene group of pyridoxamine during the action of pyridoxamine pyruvate transaminase [cf. Ayling *et al.* (*80*)].
89. M. Flavin and C. Slaughter, *Biochemistry* **3**, 885 (1964).
90. M. Flavin and C. Slaughter, *Biochemistry* **5**, 1340 (1966).
91. M. Flavin and C. Slaughter, *JBC* **244**, 1434 (1969).
92. M. Krongelb, T. A. Smith, and R. H. Abeles, *BBA* **167**, 473 (1968).

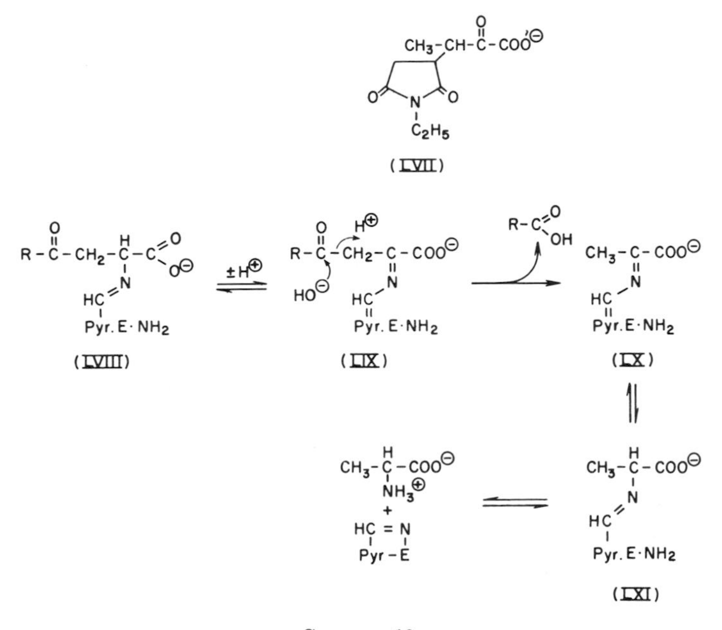

SCHEME 16

mediate could undergo cleavage adjacent to the γ-carbonyl substituent (cf. "acid hydrolysis" of acetoacetate) to yield CO_2 and alanine from aspartate or 3 hydroxyanthranilate and alanine from 3-hydroxykynurenine.

Certain amine oxidases appear to contain both PLP and copper (*68, 93–96*) and thus fall into the above category. Reversible resolution of the coenzyme and reconstitution of activity have not been accomplished unequivocally, and it is premature to suggest a mechanism. Model reactions have been studied briefly (*97, 98*), and a possible mechanism suggested (*98*).

e. Decarboxylation-Dependent Transaminases (*Class II-B, Table III*). A PLP enzyme that catalyzes the reaction shown in Eq. (12) has been isolated from a soil bacterium (*98a*).

93. H. Yamada and K. T. Yasunobu, *JBC* **237,** 3077 (1962).
94. H. Blaschko and F. Buffoni, *Proc. Roy. Soc.* **B163,** 45 (1965).
95. F. Buffoni and H. Blaschko, *Proc. Roy. Soc.* **B161,** 153 (1964).
96. B. Mondovi, M. T. Costa, A. Finazzi Agro, and G. Rotilio, *in* "Pyridoxal Catalysis: Enzymes and Model Systems" (E. E. Snell *et al.*, eds.), p. 403. Wiley (Interscience), New York, 1968.
97. M. Ikawa and E. E. Snell, *JACS* **76,** 4900 (1954).
98. G. A. Hamilton and A. Revesz, *JACS* **88,** 2069 (1966).
98a. G. B. Bailey and W. B. Dempsey, *Biochemistry* **6,** 1526 (1967).

$$\text{Pyruvate} + \text{isovaline} \rightarrow \text{L-alanine} + \text{butanone-2} + CO_2 \qquad (12)$$

The overall reaction was shown to proceed in two steps as shown in Eqs. (13) and (14). Equations (13) and (14) are specific enzymic equivalents

$$\text{Isovaline} + \text{PLP·E} \rightarrow \text{butanone-2} + CO_2 + \text{PMP·E} \qquad (13)$$

$$\text{PMP·E} + \text{pyruvate} \rightleftharpoons \text{L-alanine} + \text{PLP·E} \qquad (14)$$

of Eqs. (7) and (2), respectively; the reaction mechanism accordingly can be considered to be the sum of Scheme 11 [(XXXII) → (XXXIV), modified as in (*34*)] and Scheme 14 [(LI) → (XLVI)].

f. Enzymes Substituting RCO— for —COOH (Class II-C, Table III). The action of these enzymes resembles that of simple decarboxylases shown in Scheme 11 [modified as in (*34*)] with the modification that the leaving carboxyl group is replaced not by a proton [pathway (a)] but by an acyl group derived from an acyl-CoA, with expulsion of the extra electron pair as CoA (*99–101*).

B. Carbonyl-Containing Enzymes Other Than PLP Enzymes

Recently, at least five enzymes have been found that contain no PLP but contain a functionally essential carbonyl group, as judged by inhibition with carbonyl reagents. These are D-proline reductase (*102, 102a*), a histidine decarboxylase (*103–105*), adenosylmethionine decarboxylase (*106*), phenylalanine ammonia-lyase (deaminase) (*107*), and urocanase (*108*). Of these the first three contain covalently bound pyruvate as the prosthetic group; phenylalanine deaminase contains an unidentified residue (dehydroalanine?) that yields tritiated alanine on reduction with tritiated $NaBH_4$, and urocanase contains covalently bound α-ketobutyrate. Suggested mechanisms for two of these reactions are given below.

1. *Histidine Decarboxylase from Lactobacillus 30a*

Unlike mammalian histidine decarboxylases, which appear to be PLP enzymes, specific decarboxylases isolated from *Lactobacillus* 30a (*103–*

99. P. E. Braun and E. E. Snell, *JBC* **243,** 3775 (1968).
100. R. N. Brady, S. J. Di Mari, and E. E. Snell, *JBC* **244,** 491 (1969).
101. D. Shemin and C. S. Russell, *JACS* **75,** 4873 (1953).
102. D. Hodgins and R. H. Abeles, *JBC* **242,** 5158 (1967).
102a. D. S. Hodgins and R. H. Abeles, *ABB* **130,** 274 (1969).
103. J. Rosenthaler, B. M. Guirard, G. W. Chang, and E. E. Snell, *Proc. Natl. Acad. Sci. U. S.* **54,** 152 (1965).
104. G. W. Chang and E. E. Snell, *Biochemistry* **7,** 2005 and 2012 (1968).
105. W. D. Riley and E. E. Snell, *Biochemistry* **7,** 3520 (1968).
106. R. B. Wickner, C. W. Tabor, and H. Tabor, *Federation Proc.* **28,** 352 (1969).

105) and from a micrococcus (*109, 110*) contain no PLP. The lactobacillus enzyme contains approximately ten subunits, five of which contain a pyruvoylphenylalanyl residue at the N-terminal end. Largely in analogy with PLP-dependent decarboxylation reactions, the mechanism shown in Scheme 17 was suggested for its action. In support of the mech-

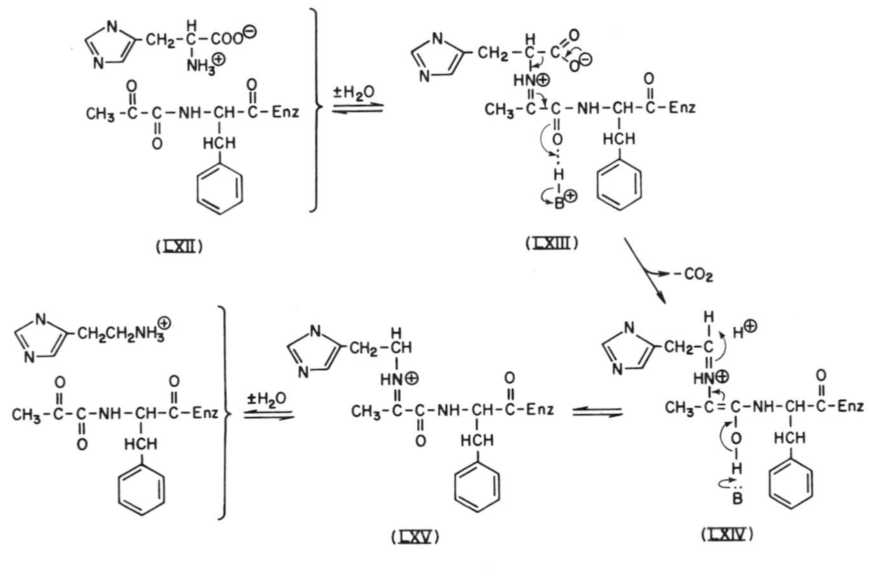

SCHEME 17

anism, reduction of the purified enzyme with $NaBH_4$ in the presence of labeled histidine followed by acid hydrolysis gave rise to two labeled products, which were identified as the reduction products of the Schiff bases formed by histidine and histamine, respectively, with pyruvate (*111*). This result demonstrates that the decarboxylase forms Schiff bases with these compounds, presumably as part of a catalytic cycle. An appropriate Lewis acid located near the carbonyl oxygen of the amide bond of the pyruvate residue might allow this oxygen to contribute to the action of the protonated imine bond by serving as an electron sink.

---

107. K. R. Hansen and E. A. Havir, *Federation Proc.* **28,** 600 (1969).
108. D. J. George and A. T. Phillips, *JBC* **245,** 528 (1970).
109. S. R. Mardashev, L. A. Siomina, N. S. Dabagov, and N. A. Gonchar, *in* "Pyridoxal Catalysis: Enzymes and Model Systems" (E. E. Snell *et al.*, eds.), p. 451. Wiley (Interscience), New York, 1968.
110. S. R. Mardashev, L. A. Semina, V. N. Prozorovskii, and A. M. Sokhina, *Biokhimiya* **32,** 761 (1967).
111. P. Recsei and E. E. Snell, *Biochemistry* **9,** 1492 (1970).

2. *Proline Reductase*

This enzyme catalyzes the reductive cleavage of D-proline (LXVI) to δ-aminovalerate (LXX). The suggested mechanism (*102, 102a*), not yet supported by chemical evidence, is shown in Scheme 18; the possibility is raised that a protonated carbinolamine may act in a manner equivalent to the corresponding imine.

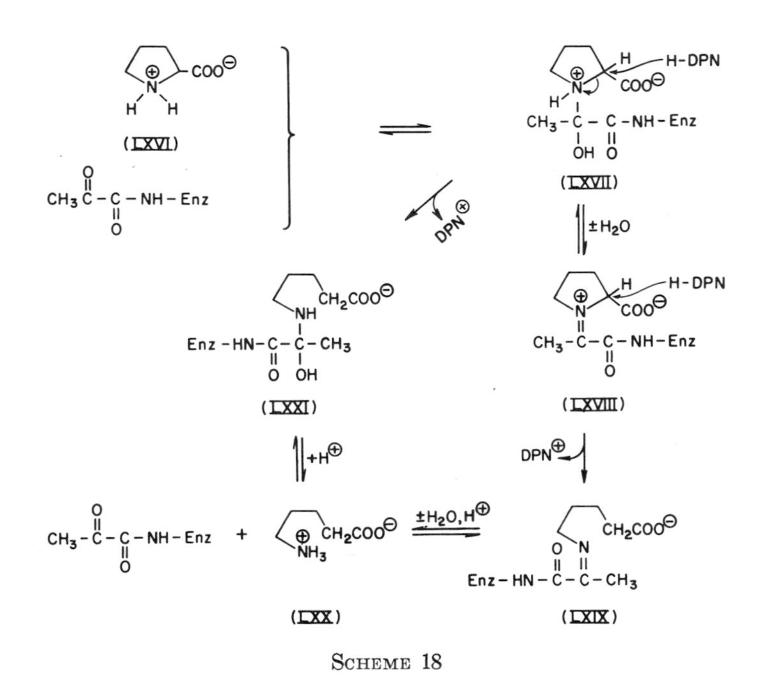

SCHEME 18

IV. Schiff Base Enzymes Lacking a Carbonyl Group (Category II)

Unlike the enzymes of category I discussed in Section III, those of category II do not contain a carbonyl group but act via intermediate imines formed between an amino group of the enzyme and a carbonyl group of the substrate. In practice this distinction has limited significance either as an aid to classification or from the mechanistic standpoint since enzymes of the two categories form a continuum ranging from those containing a covalently bound carbonyl-containing prosthetic group such as pyruvate at one extreme to those containing no carbonyl group at the other; an essential lysine residue of each of the latter enzymes participates in imine formation with a substrate containing the carbonyl group. The PLP enzymes, with their firmly bound coenzyme in imine linkage to an essential lysine residue, share characteristics of both extremes. The artificial character of the distinction is further emphasized by enzymes such as pyridoxamine pyruvate transaminase, which contains no carbonyl group (*112*) and hence belongs formally to category II. An ε-amino group of a lysine residue of this enzyme, like those of enzymes considered in this section, interacts with its substrate (pyridoxal) to form a Schiff base; subsequent catalytic events, however, are precisely analogous to those occurring during transamination reactions (*113*) involving PLP enzymes (Scheme 14).

Some representative enzymes of this category are listed together with some of their characteristics in Table IV, p. 369. Mechanisms proposed for several of these reactions are summarized below.

A. β-Keto Acid Decarboxylases

Two types of enzymes catalyzing the β-decarboxylation of β-keto acids are known. One type requires metal ions and the reactions do not involve imine formation (*114–117*); the second type involves imine formation between the substrate and a lysine residue of the enzyme (*40–42*). The mechanism originally suggested by Pedersen (*118*) for the amine-catalyzed nonenzymic reaction (Scheme 12) was shown by Westheimer (*38–44*) to apply to the action of crystalline acetoacetate decarboxylase from *Clostridium acetobutylicum*. As predicted by this mechanism, all of the ^{18}O present in the β-carbonyl group of acetoacetate was lost to the solvent during acetone formation (*38*), and interaction of enzyme with ^{14}C-labeled acetoacetate followed by $NaBH_4$ reduction led to inactivation of the enzyme and fixation of one mole of label per 50,000 g of protein (*39*). On acid hydrolysis, the labeled adduct, N^{ϵ}-(^{14}C-isopropyl)lysine was isolated, demonstrating that an ε-amino group of a lysine residue provided the catalytic amino function (*40–42*).

B. Aldolases and Transaldolases

Like the β-decarboxylases considered earlier, two classes of aldolases exist, one dependent upon metal ions for activity (*119*), the other inde-

112. H. Wada and E. E. Snell, *JBC* **237**, 133 (1962).
113. J. E. Ayling and E. E. Snell, *Biochemistry* **7**, 1616 (1968).
114. G. W. Kosicki and F. H. Westheimer, *Biochemistry* **7**, 4303 (1968).
115. D. Herbert, "Methods in Enzymology," Vol. 1, p. 753, 1955.
116. A. Schmitt, I. Bottke, and G. Siebert, *Z. Physiol. Chem.* **347**, 18 (1966).
117. M. F. Utter, "The Enzymes," 2nd ed., Vol. 5, p. 319, 1961.
118. K. J. Pedersen, *JACS* **60**, 595 (1938).
119. W. J. Rutter, *Federation Proc.* **23**, 1248 (1964).

pendent of metal ions but operating via a Schiff base mechanism (*120*). Fructose-1,6-diphosphate aldolase from rabbit muscle, for example, forms a Schiff base adduct with the substrate, dihydroxyacetone phosphate. Reduction of this enzyme–substrate complex with $NaBH_4$ yields a stable glycerophosphate-protein derivative with concomitant loss of activity; on acid hydrolysis, two moles of N^{ϵ}-(β-glyceryl)lysine can be isolated per mole of the inactivated enzyme (*120*).

Similar results have been obtained with transaldolase (*121–123*) which, in the absence of an aldehyde receptor, binds dihydroxyacetone so firmly in Schiff base linkage that this compound does not dissociate. Again the intermediate can be trapped by borohydride reduction (*123*).

These findings suggest the abbreviated mechanisms shown in Scheme 19 for the action of aldolases [(LXXII) → (LXXVII), $R_1 = PO_3^{2-}$] and transaldolases [(LXXII)→(LXXV), (LXXVIII)→(LXXX), R_1 = H].

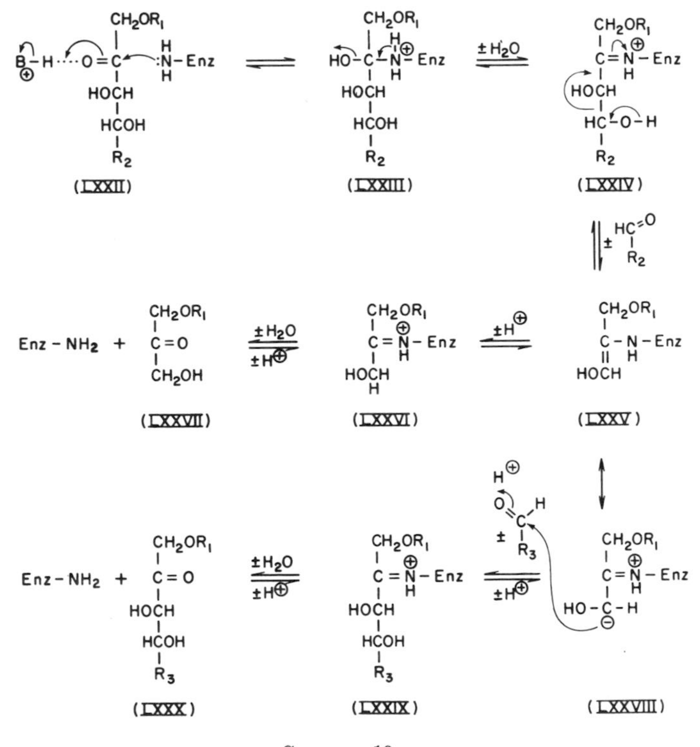

Scheme 19

120. E. Grazi, T. Cheng, and B. L. Horecker, *BBRC* **7**, 250 (1962). It had previously been shown that certain aldol reactions are catalyzed by amines [F. H. Westheimer and H. Cohen, *JACS* **60**, 90 (1938)] and J. C. Speck and A. A. Forist

C. δ-Aminolevulinic Acid Dehydratase

Biosynthesis of porphyrins proceeds through porphobilinogen (LXXXVII), which is formed by condensation of two molecules of δ-aminolevulinic acid (LXXXI). The mechanism for the reaction proposed by Nandi and Shemin (*124, 125*) is given in slightly modified form in Scheme 20. Following formation of an enzyme–substrate imine, the ini-

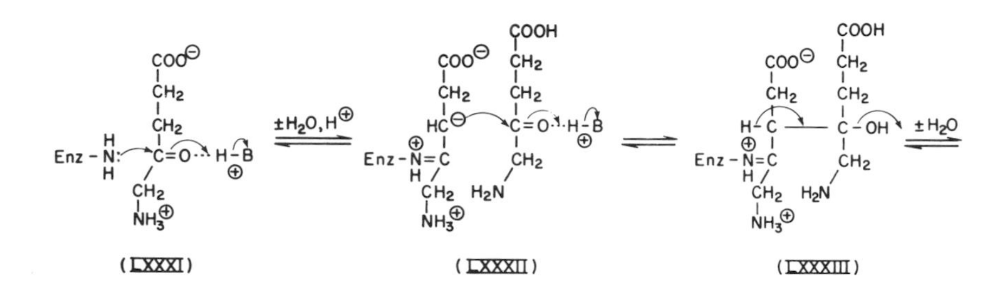

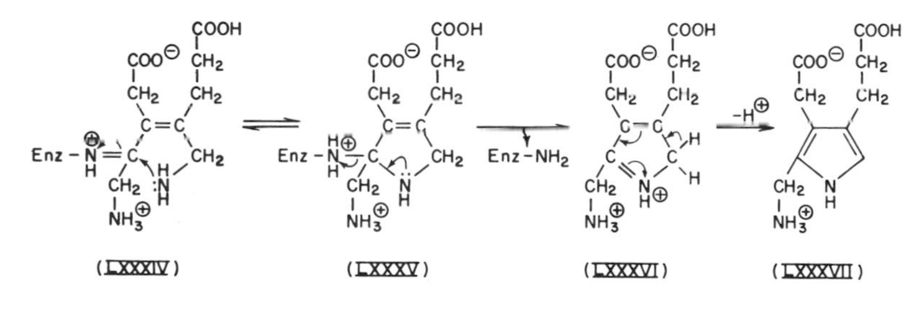

Scheme 20

tial condensation reaction (LXXXII) → (LXXXIII) yields an aldol intermediate; the mechanism is identical with that shown for the similar condensation in Scheme 19 [(LXXVIII) → (LXXIX)]. Following dehy-

---

[*JACS* **79**, 4659 (1957)] had suggested the possibility that the active group of aldolase might be an amine.

121. R. Venkataraman and E. Racker, *JBC* **236**, 1876 (1961).

122. S. Pontremoli, B. D. Prandini, A. Bonsignore, and B. L. Horecker, *Proc. Natl. Acad. Sci. U. S.* **47**, 1942 (1961).

123. B. L. Horecker, S. Pontremoli, C. Ricci, and T. Cheng, *Proc. Natl. Acad. Sci. U. S.* **47**, 1949 (1961).

124. D. Shemin and D. L. Nandi, *Federation Proc.* **26**, 390 (1967).

125. D. L. Nandi and D. Shemin, *JBC* **243**, 1236 (1968).

dration of the aldol, ring closure and displacement of enzyme occur [(LXXXIV) → (LXXXVI)] by a reaction essentially identical with a transimination reaction (Scheme 3), resulting in the formation of a new intramolecular cationic Schiff base (LXXXVI) which provides the electrophilic center required for labilization of a hydrogen atom and formation of the product (LXXXVII). The Schiff base mechanism is supported by the observation that the enzyme is inactivated in the presence of labeled δ-aminolevulinate by reduction with $NaBH_4$ with concomitant fixation of label to protein.

V. Concluding Remarks

The preceding discussion shows the versatility of the imine bond in promoting an extraordinary variety of apparently different reactions of biochemical importance. Imines of many carbonyl compounds promote such reactions, even in the absence of specific enzyme proteins, and their effectiveness is especially pronounced when the imine is conjugated with an electron sink such as that provided by PLP. Such nonenzymic reactions constitute useful models for the enzymic systems. The protein portion of the enzyme, however, by supplying a specific affinity for the substrate and a variety of additional appropriately placed catalytic functional groups (principally acid-base groups) enhances the rates of the nonenzymic reactions by factors as high as 10^7 or more and imparts reaction specificity to the system. The recent discovery of pyruvate and α-ketobutyrate as catalytic residues in proteins indicates that many additional carbonyl-dependent reactions may be discovered. The role of the carbonyl group in these reactions can now be reasonably well understood as an expression of the characteristics of the intermediate imines. The principal task of the enzyme chemist here, as with other enzymes, will now be to determine the sequence and three-dimensional structure of a sufficient number of these enzymes to permit a detailed understanding of the placement and role of additional functional groups which contribute to their substrate specificity and catalytic action.

TABLE I

SOME COMPARATIVE CHARACTERISTICS OF VARIOUS HOMOGENEOUS PLP ENZYMES

Enzyme[a]	Source	Specific activity[b]	MW ($\times 10^{-4}$)	Subunits	Combined PLP's	Spectral bands >280 mμ[c]		Ref.
I. Aminotransferases (pH optima, 7.7–8.5)								
L-Alanine (EC 2.6.1.2)	Pig heart	340	11.5	—	2	325	$425_{5.5}$	(*51*)
D-Aspartate (EC 2.6.1.10)	*B. subtilis*	90	5.3	—	1	330	$415_{5.2}$	(*126*)
L-Aspartate (EC 2.6.1.1)	Ox heart muscle	98	9.6	2	2	365	$425_{6.2}$	(*127, 128, 128a*)
	Pig heart muscle	31	7.8	2	2	362	$430_{4.6}$	(*129, 130*)
L-Histidinol phosphate (EC 2.6.1.9)	*S. typhimurium*	24[d]	5.9	2	1	338_{13}		(*131, 132*)
L-Lysine[h]	*A. liquidum*	19[d]	11.6	—	2	340	$415_{7.4}$	(*133*)
Pyridoxamine pyruvate	*Pseudomonas* MA-1	23	14.8	4	2[e]	325	$410^{e}_{7.0}$	(*112, 134*)
II. Racemases (pH optima, 7.5–10.0)								
Alanine (EC 5.1.1.1)	*P. putida*	3680	6.0	—	1		$420^{d}_{7.4}$	(*135*)
Arginine	*P. graveolens*	1250	16.7	—	4(?)		420	(*136*)
III. Enzymes Catalyzing α,β-Elimination (pH optima, 7.2–10.5)								
D-Serine dehydratase (EC 4.2.1.14)	*E. coli*	300	4.5	1	1	330	$405\text{–}415_{6.0}$	(*57, 137, 138*)
L-Threonine dehydratase (biosynthetic) (EC 4.2.1.16)	*S. typhimurium*	450[d]	19.4	4	2		$410\text{–}420_{7.4}$	(*139*)
L-Threonine dehydratase (degradative)								
A. AMP dependent	*E. coli*	880	14.7	4(?)	4		$415_{6.8}$	(*140, 140a*)
B. ADP dependent	*C. tetanomorphum*	395[d]	16.0(?)	8(?)	—		$415_{7.0}$	(*141, 142, 142a*)
Tryptophanase	*E. coli*	26	22.0	4	4	330	$425_{7.0}$	(*83, 143*)
	B. alvei	13[d]	22.0	—	2		$410\text{–}425_{7.0}$	(*144, 145*)
β-Tyrosinase	*E. intermedia*	2	17.0	—	2	335	$425_{7.0}$	(*146*)

TABLE I (*Continued*)

Enzyme[a]	Source	Specific activity[b]	MW ($\times 10^{-4}$)	Subunits	Combined PLP's	Spectral bands >280 mμ[c]		Ref.
IV. Enzymes Catalyzing β-Substitution (pH optima, 7.0–8.0)								
O-Acetylserine sulfhydrylase-A[f]	*S. typhimurium*	1100	6.8	2	2		$412_{7.6}$	(*147*)
Tryptophan synthetase, B protein[g] (EC 4.2.1.20)	*E. coli*	135[d]	9.0–10.8	2	2	332	$407_{7.5}$	(*148, 148a*)
V. Enzymes Catalyzing γ-Elimination and Substitution (pH optima, 8.0)								
L-Homoserine dehydratase (EC 4.2.1.15)	Rat liver	6[d]	19.0	—	4(?)		$425_{7.5}$	(*149*)
Cystathionine γ-synthetase	*S. typhimurium*	18[d]	16.0	4	4		$422_{7.3}$	(*150*)
VI. Decarboxylases (Carboxy-lyases) (pH optima, 3.8–5.8)								
L-Arginine (EC 4.1.1.19)	*E. coli*	410	85.0	10	10		$420_{6.6}$	(*47, 151*)
L-Glutamate (EC 4.1.1.15)	*E. coli*	110	31.0	6	6	$340_{7.0}$	$420_{4.5}$	(*152, 153*)
L-Lysine (EC 4.1.1.18)	*B. cadaveris*	86	100.0	—	10		$425_{6.2}$	(*154*)
L-Aspartate 4-decarboxylase (EC 4.1.1.12)	*P. dacunhae*	84	67.5	12	12		$360_{6.8}$	(*155, 155a*)
VII. Hydroxymethyltransferases (Aldolases)								
Serine (EC 2.1.2.1)	Rabbit liver	14[d]	33.1	—	4		$430_{7.3}$	(*156, 157*)

[a] Where applicable, the trivial name recommended by the International Union of Biochemistry is given for each enzyme together with its Enzyme Commission classification number.

[b] Units represent micromoles of substrate transformed per minute per milligram of protein.

[c] Subscript number indicates pH at which the cited spectrum was recorded.

[d] Calculated from data presented in the cited works.

[e] The substrate for this enzyme is pyridoxal; no pyridoxal phosphate is present. The spectral band at 410 mμ appears only in the presence of pyridoxal.

[f] This enzyme interacts with serine transacetylase to form a complex known by the trivial name *cysteine synthetase*.

[g] The "B protein" is the pyridoxal phosphate containing component of tryptophan synthetase.

[h] This enzyme acts upon the ε-amino group of lysine.

126. M. Martinez-Carrion and W. T. Jenkins, *JBC* **240,** 3538 and 3547 (1965).
127. V. Scardi, *in* "Pyridoxal Catalysis: Enzymes and Model Systems," (E. E. Snell *et al.*, eds.), p. 179. Wiley (Interscience), New York, 1968.
128. G. Marino, A. M. Greco, V. Scardi, and R. Zito, *BJ* **99,** 589 (1966).
128a. G. Marino, V. Scardi, and R. Zito, *BJ* **99,** 595 (1966).
129. B. E. C. Banks, S. Doonan, and C. A. Vernon, *in* "Pyridoxal Catalysis: Enzymes and Model Systems" (E. E. Snell *et al.*, eds.), p. 305. Wiley (Interscience), New York, 1968.
130. W. T. Jenkins, D. A. Yphantis, and I. W. Sizer, *JBC* **234,** 51 (1959).
131. R. G. Martin and R. F. Goldberger, *JBC* **242,** 1168 (1967).
132. R. G. Martin, M. J. Voll, and E. Appella, *JBC* **242,** 1175 (1967).
133. K. Soda and H. Misono, *Biochemistry* **7,** 4110 (1968).
134. H. Kolb, R. D. Cole, and E. E. Snell, *Biochemistry* **7,** 2946 (1968).
135. G. Rosso, K. Takashima, and E. Adams, *BBRC* **34,** 134 (1969).
136. T. Yorifuji, K. Ogata, and K. Soda, *BBRC* **34,** 760 (1969).
137. D. Dupourque, W. A. Newton, and E. E. Snell, *JBC* **241,** 1233 (1966).
138. R. Labow and W. G. Robinson, *JBC* **241,** 1239 (1966).
139. R. O. Burns and M. H. Zarlengo, *JBC* **243,** 178 and 186 (1968).
140. Y. Shizuta, A. Nakazawa, M. Tokushigi, and O. Hayaishi, *JBC* **244,** 1883 (1969).
140a. P. D. Whanger, A. T. Phillips, K. W. Rabinowitz, J. R. Piperno, J. D. Shada, and W. A. Wood, *JBC* **243,** 167 (1968).
141. H. R. Whiteley and M. Tahara, *JBC* **241,** 4881 (1966).
142. H. R. Whiteley, *JBC* **241,** 4890 (1966).
142a. A. Nakazawa and O. Hayaishi, *JBC* **242,** 1146 (1967).
143. Y. Morino and E. E. Snell, *JBC* **242,** 5602 (1967).
144. J. A. Hoch and R. D. DeMoss, *Biochemistry* **5,** 3137 (1966).
145. J. A. Hoch, F. J. Simpson, and R. D. DeMoss, *Biochemistry* **5,** 2229 (1966).
146. H. Yamada, H. Kumagai, H. Matsui, H. Ohgishi, and K. Ogata, *BBRC* **33,** 10 (1968).
147. M. A. Becker, N. M. Kredich, and G. M. Tomkins, *JBC* **244,** 2418 (1969).
148. D. A. Wilson and I. P. Crawford, *JBC* **240,** 4801 (1965).
148a. G. M. Hathaway, S. Kida, and I. P. Crawford, *Biochemistry* **8,** 989 (1969).
149. Y. Matsuo and D. M. Greenberg, *JBC* **230,** 545 (1958).
150. M. M. Kaplan and M. Flavin, *JBC* **241,** 5781 (1966).
151. E. A. Boeker and E. E. Snell, *JBC* **243,** 1678 (1968).
152. P. H. Strausbach and E. H. Fischer, *Biochemistry* **9,** 226 (1970).
153. P. H. Strausbach and E. H. Fischer, *Biochemistry* **9,** 233 (1970).
153a. K. Soda and M. Moriguchi, *BBRC* **34,** 34 (1969).
154. S. S. Tate and A. Meister, *Biochemistry* **9,** 2626 (1970).
155. T. Kakimoto, J. Kato, T. Shibatani, N. Nishimura, and I. Chibata, *JBC* **244,** 353 (1969).
155a. W. F. Bowers, V. B. Czubaroff, and R. H. Hashemeyer, *Biochemistry,* **9,** 2620 (1970).
156. L. Schirch and M. Mason, *JBC* **237,** 2578 (1962).
157. L. Schirch and M. Mason, *JBC* **238,** 1032 (1963).

TABLE II
PEPTIDE SEQUENCES NEAR THE BOUND PLP OF VARIOUS BOROHYDRIDE-REDUCED PLP ENZYMES[a]

Enzyme	Sequence	Ref.
Extramitochondrial AAT	Ser·Lys(Pxy)·Asn·Phe	(*158*)
Mitochondrial AAT	Ala·Lys(Pxy)·Asn·Met	(*158*)
Glutamate decarboxylase	Ser·Ile·Ser·Ala·Ser·Gly·His·Lys(Pxy)·Phe	(*153*)
Phosphorylase	Met·Lys(Pxy)·Phe·Met	(*159*)
Pyridoxamine pyruvate transaminase[b]	Val·Thr·Gly·Pro·Asp·Lys(Pxy)·Cys(SH)·Leu	(*160*)
Tryptophanase	Ser·Ala·Lys·Lys(Pxy)·Asp·Ala·Met·Val·Pro·Met	(*161*)
Tryptophan synthetase (B protein)	Glu·Asp·Leu·Leu·(Gly_2,Ala,His_2)·Lys(Pxy)·Thr	(*162*)

[a] Abbreviations used in the table are: standard abbreviations for the amino acids; AAT, aspartate aminotransferase; Pxy, pyridoxyl or phosphopyridoxyl residue.
[b] Pyridoxal (a substrate) is bound in the manner shown; the enzyme contains no PLP.

158. Y. Morino and T. Watanabe, *Biochemistry* **8**, 3412 (1969).
159. E. H. Fischer, private communication.
160. H. Kolb, J. M. Hodsdon, R. D. Cole, and E. E. Snell, to be published.
161. H. Kagamiyama, Y. Morino, and E. E. Snell, *JBC* **245,** 2819 (1970).
162. I. Crawford, private communication.

TABLE III

CLASSIFICATION OF PLP ENZYMES BY NATURE OF THE PRIMARY BOND LABILIZED[a,b]

Enzyme	Reaction
I. Enzymes Labilizing a —C—H Bond [Bond *a* in (XXII)]	
A. Racemases [Eq. (1), Section II, text], e.g.,	
1. L-Alanine	L-Ala ⇌ D-Ala
2. Diaminopimelate	L,L-DAP ⇌ *meso*-DAP
B. Transaminases (aminotransferases) [Eq. (2), Section II, text], e.g.,	
1. L-Aspartate	L-Asp + α-KG ⇌ L-Glu + OA
2. β-Alanine:pyruvate	β-Ala + Pyr ⇌ L-Ala + malonic semialdehyde
3. γ-Aminobutyrate	γ-ABA + α-KG ⇌ L-Glu + succinic semialdehyde
4. Histidinol phosphate	L-Hist-P + α-KG ⇌ L-Glu + imidazole acetol-P
C. Enzymes catalyzing α,β-elimination and β-replacement reactions [Eqs. (3) and (4), of text], e.g.,	
1. Tryptophanase	L-Trp + H_2O → Indole + Pyr + NH_3
2. D-Serine dehydratase	D-Ser → Pyr + NH_3
3. Cysteine synthetase	*O*-Ac-Ser + H_2S → AcOH + L-Cys
4. Tryptophan synthetase	Indoleglycerol-P + Ser → Trp + 3-P-glyceraldehyde
5. Cystathionine synthetase	L-Ser + L-Homocys → Cystathionine + H_2O
D. Enzymes catalyzing α,γ-elimination and replacement reactions, e.g.,	
1. *O*-Succinylhomoserine deaminase	*O*-Succ-Homoser → Succinate + α-KB + NH_3
2. Cystathionase (animal)	L-cystathionine → L-Cys + α-KB + NH_3
3. Cystathionine synthetase (plant)	*O*-Ac-Homoser + L-Cys → AcOH + L-cystathionine
E. Miscellaneous enzymes	
1. Aspartate β-decarboxylase	L-Asp → CO_2 + L-Ala
2. 3-Hydroxykynureninase	3-HK + H_2O → 3-Hydroxyanthranilate + L-Ala
3. Amine oxidase	Histamine + O_2 + H_2O → Imidazole acetaldehyde + NH_3 + H_2O_2
II. Enzymes Labilizing a —C—C— Bond: Decarboxylases [Bond *b* in (XXII)][c]	
A. Simple decarboxylases [Eq. (6), Section II, text], e.g.,	
1. L-Arginine	L-Arg → Agmatine + CO_2
2. L-Glutamate	L-Glu → γ-ABA + CO_2
3. *meso*-Diaminopimelate	*m*-DAP → L-Lys + CO_2
B. Decarboxylation-dependent transaminases [partial reaction: Eq. (7), Section II, text], e.g.,	
1. α-Aminoisobutyrate	α-AIB + Pyr → L-Ala + acetone + CO_2
C. Enzymes substituting RCO— for —COOH	
1. δ-Aminolevulinate synthetase	Succinyl-CoA + Gly → δ-AL + CO_2
2. 3-Ketosphinganine synthetase	Palmityl-CoA + Ser → 3-KS + CO_2

TABLE III (*Continued*)

III. Enzymes Labilizing a —C—C— Bond: Aldolases [Bond *c* in (XXII)][c]	
1. Threonine	L-Thr ⇌ Gly + acetaldehyde
2. Serine transhydroxymethylase	L-Ser + FH_4 ⇌ Gly + $CH_2{=}FH_4$
3. 1-Phosphodihydrosphingosine	1-P-DHS → Palmitaldehyde + *O*-P-ethanolamine

[a] More extensive lists of PLP enzymes are provided in Braunstein (*20*), Guirard and Snell (*66*), and Mahler and Cordes (*67*).

[b] The abbreviations used in this table are: the accepted abbreviations for amino acids; α-KG, α-ketoglutarate; OA, oxalacetate; Pyr, pyruvate; α-KB, α-ketobutyrate; γ-ABA, γ-aminobutyrate; Homoser, homoserine; Homocys, homocysteine; FH_4, tetrahydrofolate; $CH_2{=}FH_4$, 5,10-methylenetetrahydrofolate; δ-AL, δ-aminolevulinate; 3-KS, 3-ketosphinganine; 1-P-DHS, 1-phosphodihydrosphinganine.

[c] The physiological role of these enzymes lies in degradation of their substrates. The aldolase reactions are freely reversible; the decarboxylase reactions are irreversible for practical purposes. The reverse reactions (from right to left) in each case may be considered to result from labilization of an α-hydrogen atom, which emphasizes the close relationship of these reactions with those of Class I.

TABLE IV
SOME COMPARATIVE PROPERTIES OF SCHIFF BASE ENZYMES THAT CONTAIN NO CARBONYL GROUP

Enzyme[a]	Source	Specific activity[b]	MW ($\times 10^{-4}$)	Subunits	Binding sites	Ref.
	I. Decarboxylases (Carboxy-lyases)					
Acetoacetate (EC 4.1.1.4)	*C. acetobutylicum*	38	26.0	8(?)	—	(*163, 164*)
	II. Aldolases (Aldehyde-lyases)					
Deoxyriboaldolase (EC 4.1.2.4)	*L. plantarum*	620[c]	6.1	1(?)	1	(*165, 166*)
Fructose diphosphate	Bovine liver	20[c]	15.9	—	—	(*167*)
(EC 4.1.2.13)	*Boa constrictor constrictors*	12	15.3	3–4	3	(*168*)
	Rabbit skeletal muscle	10[c]	16.0	4	2	(*169–172*)
4-Hydroxy-2-keto-	Rat liver	7.7[c]	12.0	—	1–2	(*173*)
glutarate	Beef liver	7.2	12.0	—	1	(*174*)
Phospho-2-keto-3-deoxygluconate (EC 4.1.2.14)	*P. fluorescens*	312[c]	8.7	2	2	(*175, 176*)
	III. Transaldolases					
Transaldolase (EC 2.2.1.2)	*C. utilis*	48	5.7	—	1	(*122, 177*)
	IV. Dehydratases					
δ-Aminolevulinate (EC 4.2.1.24)	*R. spheroides*	2.0[c]	25.0(?)[d]	2(?)[e]	—	(*178, 179*)
2-Keto-3-deoxy-L-arabonate (EC 4.2.1.-)	*P. saccharophila*	54	8.5	—	1	(*180, 181*)

[a] Whenever possible, the trivial name recommended by the International Union of Biochemistry as well as the assigned classification number for each enzyme listed have been used.

[b] Units represent micromoles of substrate transformed per minute per milligram of protein.

[c] Calculated from data presented in the references cited.

[d] This enzyme appears to exist as an equilibrium mixture of three species: MW 250,000, 500,000, and 750,000. It is uncertain whether the two higher molecular weight proteins occur naturally or are aggregates of the 250,000 species arising from isolation procedures.

[e] Subunits/250,000 molecular weight.

163. B. Zerner, S. M. Coutts, F. Lederer, H. H. Waters, and F. H. Westheimer, *Biochemistry* **5**, 813 (1966).

164. F. Lederer, S. M. Coutts, R. A. Laursen, and F. H. Westheimer, *Biochemistry* **5**, 823 (1966).

165. P. Hoffee, O. M. Rosen, and B. L. Horecker, *JBC* **240,** 1512 (1965).
166. O. M. Rosen, P. Hoffee, and B. L. Horecker, *JBC* **240,** 1517 (1965).
167. R. J. Peanasky and H. A. Lardy, *JBC* **233,** 365 and 371 (1958).
168. E. Schwartz and B. L. Horecker, *ABB* **115,** 407 (1966).
169. J. F. Taylor, A. A. Green, and G. T. Cori, *JBC* **173,** 591 (1948).
170. J. F. Taylor and C. Lowry, *BBA* **20,** 109 (1956).
171. C. Y. Lai, O. Tchola, T. Cheng, and B. L. Horecker, *JBC* **240,** 1347 (1965).
172. C. L. Sia and B. L. Horecker, *ABB* **123,** 186 (1968).
173. R. G. Rosso and E. Adams, *JBC* **242,** 5524 (1967).
174. R. D. Kobes and E. E. Dekker, *BBRC* **25,** 329 (1966).
175. H. P. Meloche and W. A. Wood, *JBC* **239,** 3515 (1964).
176. H. P. Meloche, J. I. Ingram, and W. A. Wood, "Methods in Enzymology," Vol. 9, p. 520, 1966.
177. B. L. Horecker, P. T. Rowley, E. Grazi, T. Cheng, and O. Tchola, *Biochem. Z.* **338,** 36 (1963)
178. D. L. Nandi, K. F. Baker-Cohen, and D. Shemin, *JBC* **243,** 1224 (1968).
179. D. L. Nandi and D. Shemin, *JBC* **243,** 1231 (1968).
180. A. C. Stoolmiller and R. H. Abeles, *JBC* **241,** 5764 (1966).
181. D. Portsmouth, A. C. Stoolmiller, and R. H. Abeles, *JBC* **242,** 2751 (1967).

8

*Some Physical Probes of Enzyme Structure in Solution**

SERGE N. TIMASHEFF

I. Introduction

The knowledge of the three-dimensional structures of a number of enzymes and other proteins, derived from X-ray crystallographic studies has greatly advanced our understanding in structural terms of the prop-

* The studies from our laboratories that are included in this chapter were supported in part by the National Institutes of Health Research Grant GM-14603.

erties of these macromolecules. Frequently, however, it becomes necessary to establish structural features of these molecules in solution and, in particular, to monitor changes in structure which may occur as a result of changes in environment or of interactions with various ligands such as substrates, inhibitors, or activators. Solution studies are obviously not capable of approaching the degree of resolution afforded by crystallographic techniques; they can, however, lead to the description of the general structural features of a biological macromolecule and pinpoint some particular details about the location and interactions of various specific groups on proteins. When we study enzymic reactions, we normally deal with systems in dilute aqueous solutions, in which the enzymes exist in a dynamic state as they undergo small conformational changes accompanied by changes in their interaction properties and in chemical reactivity. At present, the evidence seems to indicate that the structures of most proteins in the crystal state and in solution are not significantly different (*1*); furthermore, studies on enzyme–substrate and enzyme–inhibitor interactions in crystals have also made possible a test of various conformational changes which might accompany binding of these ligands (*2*, *3*). At present, the vast majority of enzymic systems are not amenable to direct crystallographic examination of conformation and the needed information must be obtained from solution studies. The X-ray crystallography of enzymes is treated by Eisenberg in Chapter 1, Volume I; therefore, it will not be discussed in the present chapter, except that it will be referred to when the crystallographic results are necessary for the interpretation of results obtained with solutions.

The problems which will be discussed in this chapter are concerned with the determination of enzyme secondary structure in solution, with the probing of side chain location and order, i.e., tertiary structure, and with perturbations of this order by the binding of ligands to specific sites on enzymes. The methods used in such studies are essentially of three categories: spectroscopic, such as ultraviolet and infrared absorption, optical activity, fluorescence, and nuclear magnetic resonance; thermodynamic, such as titration of specific ionizable groups, frequently with the use of their absorption properties in various spectral regions; and geometric, such as small angle X-ray scattering and various hydrodynamic techniques. No attempt will be made at an exhaustive presentation of the field but rather the salient features of various approaches will be presented, followed by a description of a few specific examples which

1. J. A. Rupley, *in* "Structure and Stability of Biological Macromolecules" (S. N. Timasheff and G. D. Fasman, eds.), p. 121. Marcel Dekker, New York, 1969.
2. F. A. Quiocho and F. M. Richards, *Biochemistry* **5**, 4062 (1966).
3. L. Stryer, *Ann. Rev. Biochem.* **37**, 25 (1968).

demonstrate what can be accomplished by any given technique in the present state of the art.

II. Secondary and Tertiary Structures

Recent years have witnessed an explosion of solution studies aimed at an understanding of the conformations of enzymes and other proteins. In these studies, the emphasis has been on spectroscopic techniques. In these techniques the fundamental observations are based on the interaction of electromagnetic radiation of a given frequency with some specific transitions of structural elements within the molecules. The nature of the investigated transition in each case is determined by the coincidence of the energy of the radiation with that of the transition. Thus, in the infrared spectral region, the radiation interacts with the modes of vibration of specific bonds; in the ultraviolet region, the interactions are with the electric and magnetic transitions of electrons in various groups on the protein. In complicated molecules, such as enzymes, the transitions are not independent of each other but usually they are coupled as a result of interactions between groups. Therefore, it is reasonable to expect that their energies, and thus band positions, will be affected by the conformations in which the specific groups are present within an enzyme. Secondary structure concerns the location in three-dimensional space and mutual interactions of the fundamental repeating units of the protein backbone, the peptide groups. This chapter will deal with the effects of conformation on the absorption of radiation by peptide groups in various spectral regions. Considerations of tertiary structure will involve similar effects on the spectra of side chain residues, more specifically, of differences in their environment, orientation with respect to other groups and interactions with the backbone chain and other structural elements of the protein molecule.

Structural analysis by spectroscopic techniques is complicated by a number of factors, the most prominent one being poor band resolution and the resulting band overlaps. The position of an absorption maximum does not represent necessarily the center of a band characteristic of a specific transition, and thus, conformation, but it is the result rather of the additivity of absorptions of closely located transitions which correspond to the various constituent conformations of a protein. The interpretation of any piece of spectroscopic data, therefore, must be undertaken with extreme caution; in fact, the most promising way of approaching solution structure is by the parallel application of several unrelated techniques, with subsequent interpretation of the data in terms of a conformation that can fully account for all the observations made.

III. Infrared Spectroscopy

The infrared absorption spectra of polypeptides and fibrous proteins have been the object of extensive studies over a number of years. These were culminated by Miyazawa's theoretical analysis of the amassed data on the positions and shifts of the amide I and II bands in terms of a model based on a weakly coupled oscillator (*4–8*). As a result, it is possible now to relate small differences in band position to specific differences in the conformations of polypeptide chains. In protein spectra, the amide I band, located between 1600 and 1700 cm^{-1}, results mostly from C=O stretching, with small contributions from C—N stretching and N—H in-plane bending. The amide II band is located between 1500 and 1600 cm^{-1}; it corresponds to N—H bending and C—N stretching. In a polypeptide chain or protein, the exact positions of these bands are a function of chain conformation, namely, of the interactions between neighboring —C(=O)—N(H)— groups. These interactions may be of two kinds: (1) the vibrations of adjacent peptide groups may be coupled through the α-carbon atom; (2) they may be coupled across hydrogen bonds. In any given conformation, the transition moment to an excited state is a function of the coupling constants D_i multiplied by a phase factor ($\cos \theta_i$), where θ_i is the phase angle between the motions of the interacting oscillators. As a result, a spectral band with an unperturbed frequency ν_0 becomes split into several bands, whose frequencies ν are given by $\nu = \nu_0 + \Sigma_i D_i \cos \theta_i$. Since both D_i and θ_i are functions of the geometry and mutual organization in space of peptide bonds, the numbers and positions of bands observed for any transition in polypeptide chains are also a function of their conformations.

In synthetic polypeptides band assignment was greatly facilitated by dichroic measurements on oriented films; in globular proteins and en-

4. T. Miyazawa, *J. Chem. Phys.* **32**, 1647 (1960).

5. T. Miyazawa and E. R. Blout, *JACS* **83**, 712 (1961).

6. T. Miyazawa, *in* "Poly-α-Amino Acids" (G. D. Fasman, ed.), p. 69. Marcel Dekker, New York, 1967.

7. S. Krimm, *JMB* **4**, 528 (1962).

8. H. Susi, *in* "Structure and Stability of Biological Macromolecules" (S. N. Timasheff and G. D. Fasman, eds.), p. 575. Marcel Dekker, New York, 1969.

TABLE I
AMIDE I BAND FREQUENCIES FOR PROTEINS IN VARIOUS CONFORMATIONS

Conformation	In H_2O (cm^{-1})	In D_2O (cm^{-1})	Films or crystals (cm^{-1})
α Helix	1652	1650	1652
Unordered	1656	1643	1658
APCP	1632(s)[a]	1632(s)	1632(s)
	1690(w)	1675(s)	1685(w)
Parallel β			1632 (calc)[b]
ν_0	1661	1654	1658

[a] s and w refer to strong and weak bands, respectively,
[b] From Krimm (*7*).

zymes in solution the advantages of orientation are not available and it becomes necessary to rely on band positions alone. The positions in solution of the amide I bands which correspond to various conformations have been established by studies of a number of model polypeptides and proteins in aqueous solution (*9, 10*). These are listed in Table I. Infrared spectroscopy of proteins in solution is complicated by the fact that the usual solvent H_2O has a strong broad absorption band centered at 1650 cm^{-1}; as a result, routine solution measurements must be done in D_2O, which is essentially transparent in the spectral region of interest. This introduces uncertainties which stem from the facts that (1) deuteration of a protein may alter its conformation, and (2) it may result in band shifts in the infrared spectrum. While the first uncertainty must be eliminated by auxiliary experiments, the second one can be checked by comparing the solution spectra of proteins in D_2O with spectra of solid (crystalline) proteins suspended in some inert medium, such as Nujol, or cast as films out of aqueous solution. Examination of Table I leads to two important conclusions: (1) the band positions in three typical conformations, α-helical, antiparallel chain pleated sheet (APCP), or β, and unordered, are quite distinct; (2) the band positions in the α-helical and APCP conformations are essentially invariant whether the protein spectra are obtained on suspensions of crystals or on solutions of the proteins either in H_2O or D_2O. It is only in the unordered conformation that deuteration of the protein results in a band shift to lower frequency, probably reflecting interactions with solvent. Of particular importance is the presence of a second, weaker band at 1675–1690 cm^{-1} in the spectrum of the APCP structure; this band can be used to advantage diagnostically for the presence of that particular conformation in a protein. The parallel

9. S. N. Timasheff and H. Susi, *JBC* **241,** 249 (1966).
10. H. Susi, S. N. Timasheff, and L. Stevens, *JBC* **242,** 5460 (1967).

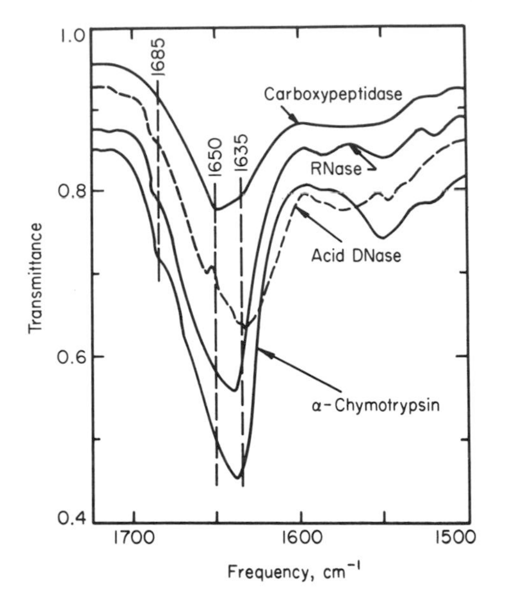

FIG. 1. Infrared spectra in the amide I band region of various enzymes in D_2O solution (arbitrarily displaced along the transmittance scale) (*11, 12*).

β structure, while exhibiting a principal band in the same position as the antiparallel one, does not have the higher frequency absorption (*7*).

The conformations of a number of enzymes have been examined in the region of the amide I band absorption (*11–15*). Some typical spectra are shown on Fig. 1. Some of the significant features of these spectra merit comment. First, it is evident that the positions and shapes of the spectral bands differ for the various enzymes. Analysis of spectral details have permitted the assignment of some structural features. Both ribonuclease and α-chymotrypsin have absorption maxima at 1637–1640 cm^{-1} with marked shoulders at 1685 cm^{-1}. The shoulder at 1685 cm^{-1} is a definite indication of the presence of APCP conformation within these enzymes, in full agreement with the X-ray structural results (*16–18*); the shoulders

11. S. N. Timasheff, H. Susi, and L. Stevens, *JBC* **242,** 5467 (1967).
12. S. N. Timasheff, H. Susi, R. Townend, L. Stevens, M. J. Gorbunoff, and T. F. Kumosinski, *in* "Conformation of Biopolymers" (G. N. Ramachandran, ed.), Vol. 1, p. 173. Academic Press, New York, 1967.
13. S. N. Timasheff and L. Stevens, *Trans. Bose Res. Inst.* (*Calcutta*) **31,** 50 (1969).
14. K. Hamaguchi, *J. Biochem.* (*Tokyo*) **56,** 441 (1964).
15. P. Y. Chang, *Proc. Natl. Acad. Sci. U. S.* **55,** 1535 (1966).
16. B. W. Matthews, P. B. Sigler, R. Henderson, and D. M. Blow, *Nature* **214,** 652 (1967).
16a. J. J. Birktoft, B. W. Matthews, and D. M. Blow, *BBRC* **36,** 131 (1969).
17. G. Kartha, J. Bello, and D. Harker, *Nature* **213,** 862 (1967).

around 1650 cm^{-1} are consistent with the presence of a limited amount of α helix, although they could not be used as diagnostic evidence for the presence of that conformation. Carboxypeptidase A, to the contrary, has a maximum at 1650 cm^{-1}, a pronounced shoulder at 1637 cm^{-1}, and no discernible band at 1685 cm^{-1}. This is consistent with the presence of both α-helical and β conformations, with a larger amount of the former. X-ray diffraction studies (*19*) have shown that this enzyme contains about 25–30% of α helix and about 18% of β conformation; the latter, however, is of both the parallel and antiparallel types, with a predominance of the former; from Table I, a band at 1685 cm^{-1} could have been expected, as a contribution of the antiparallel pleated sheet structure. This band, however, is quite weak and it is very probable that its intensity in carboxypeptidase A is too low to be resolved from the steeply falling edge of the strong lower frequency band. Another interesting example is afforded by the allosteric enzyme, acid deoxyribonuclease. This enzyme is known to be a dimer of two identical chains (*20, 21*). Its secondary structure, however, is not known. Its circular dichroism and optical rotatory dispersion spectra (*22*) are characterized by weak overlapping bands, which suggest the presence of a mixture of conformations, with little α helix. Its infrared spectrum, shown in Fig. 1, is similar to that of ribonuclease; the 1685-cm^{-1} shoulder suggests that parts of the polypeptide chain in this enzyme are folded in an antiparallel pleated sheet structure. These examples indicate the extent of information on the secondary structure of enzymes which may be derived from infrared spectroscopy. It is evident that the information is very limited and only of a general nature. At best, this method can be used to corroborate or interpret results obtained by other techniques such as circular dichroism or optical rotatory dispersion as well as to ascertain whether significant conformational differences exist between a given enzyme in the crystal and solution states. Where the crystal structure is available, it is consistent with the general conclusions drawn from the analysis of the infrared spectra. The particular advantage of the infrared technique is that the spectra are not complicated by overlapping side chain absorption, a situation which gives rise to difficulties in the ultraviolet region.

While infrared spectroscopy cannot give more than a general indication

---

18. H. W. Wyckoff, K. D. Hardman, N. M. Allewell, T. Inagami, L. N. Johnson, and F. M. Richards, *JBC* **242,** 3984 (1967).
19. W. N. Lipscomb, J. A. Hartsuck, G. N. Reeke, Jr., F. A. Quiocho, P. H. Bethge, M. L. Ludwig, T. A. Steitz, H. Muirhead, and J. C. Coppola, *Brookhaven Symp. Biol.* **21,** 24 (1969).
20. G. Bernardi, *Advan. Enzymol.* **31,** 1 (1968).
21. G. Bernardi, E. Appella, and R. Zito, *Biochemistry* **4,** 1725 (1965).
22. G. Bernardi and S. N. Timasheff, *ABB* (in press) (1970).

of conformational features present within a protein, it can be used to advantage in studying changes in conformation induced by changes in medium or by interactions. A typical example is given by bovine carbonic anhydrase (*11*). The infrared spectrum of this enzyme at neutral pH displays an amide I band with a maximum at 1637 cm^{-1}. When the pH is lowered, the amide I band maximum shifts from 1637 to 1647 cm^{-1}. Such a shift is consistent with a transition from a structure containing a significant amount of β conformation to one richer in α helix, as suggested by optical rotatory dispersion and circular dichroism observations (*23*); the latter spectra, however, are so complex at neutral pH that they essentially defy conformational analysis. When the pH is raised to 12.0, the amide I band in the infrared spectrum of bovine carbonic anhydrase remains identical to that observed at pH 7.4, suggesting that no major changes in secondary structure occur, while a similar pH shift is accompanied by gross changes of the optical rotatory dispersion and circular dichroism spectra. These infrared results permit a better understanding of the observations that the availability of tyrosine residues to titration remains unaltered up to pH 12 (*24*), as well as of the reactivity of these residues with cyanuric fluoride (*25*), which is most probably related to small changes in the tertiary structure and the charge configuration of the enzyme. In the case of the human enzyme, a similar increase in pH is accompanied by significant changes in the optical rotatory dispersion and circular dichroism spectra (*23, 26–28*), which could not be interpreted in terms of secondary structure because of the presence of overwhelming aromatic residue transition bands. In view of the infrared results, it is now possible to assign these changes in the ultraviolet region almost exclusively to changes in side chain transitions, that is, to changes in the tertiary structure.

The examples given above show the extent to which infrared absorption in the region of the amide I band may be useful in the general structural characterization of an enzyme and in the analysis of conformational changes. Other information of similar nature may be also obtained from other bands characteristic of peptide transitions. For example, the amide V band gives better resolution; unfortunately, it is very weak. This band is located at much lower frequencies and is related to out-of-plane NH

23. S. Beychok, J. McD. Armstrong, C. Lindblow, and J. T. Edsall, *JBC* **241,** 5150 (1966).
24. A. Nillson and S. Lindskog, *European J. Biochem.* **2,** 309 (1967).
25. M. J. Gorbunoff, *Federation Proc.* **28,** 854 (1969).
26. D. V. Myers and J. T. Edsall, *Proc. Natl. Acad. Sci. U. S.* **53,** 169 (1965).
27. J. E. Coleman, *Biochemistry* **4,** 2644 (1965).
28. A. Rosenberg, *JBC* **241,** 5126 (1966).

bending (*6*, *29*). Fasman and Miyazawa (*30*) have found that for synthetic polypeptides the characteristic band positions are α helix, 610–620 cm^{-1}; APCP (β), 695–705 cm^{-1}; and unordered, 650 cm^{-1}. These bands are weak relative to amide I and II, and N-deuteration results in considerable shifts. While they have not been applied yet to structural studies of enzymes, their good resolution should render them useful for such studies, in particular as highly sensitive spectrophotometers become available.

IV. Ultraviolet Absorption of Peptide Groups

Amide groups undergo electronic transitions in the ultraviolet spectral region below 240 nm. The nature and origin of these transitions has been reviewed recently (*31*) and will not be discussed here. Since, in coupled systems such as a polypeptide chain, the interactions between the transition moments of neighboring groups are functions of intergroup distance and orientation, the exact positions and intensities of the bands are a function of conformation. As a result, it is possible to use the ultraviolet absorption spectra of proteins in the region between 190 and 230 nm to analyze their conformations. The theoretical analysis leading to band identification is similar (*31–35*) to that described above for infrared spectroscopy and will not be discussed here. The absorption spectra characteristic of the α-helix, APCP, and random conformations are shown in Fig. 2 (*36*). While the spectra of the unordered, or random, and β conformations are very similar, formation of the α helix is accompanied by strong hypochromism. In analyzing the ultraviolet spectrum of a protein in terms of conformation, it must be first corrected for the contribution of aromatic side chain absorption, since these groups also undergo transitions in the same spectral region. The resultant curve is

29. T. Miyazawa, K. Fukushima, and S. Sugano, *in* "Conformation of Biopolymers" (G. N. Ramachandran, ed.), Vol. 2, p. 557. Academic Press, London, 1967.

30. G. D. Fasman and T. Miyazawa, in preparation.

31. W. B. Gratzer, *in* "Poly-α-Amino Acids" (G. D. Fasman, ed.), p. 177. Marcel Dekker, New York, 1967.

32. R. W. Woody, Doctoral Dissertation, University of California, Berkeley, California, 1962.

33. E. S. Pysh, *Proc. Natl. Acad. Sci. U. S.* **56,** 825 (1966).

34. G. Holzwarth and P. Doty, *JACS* **87,** 218 (1965).

35. R. W. Woody, *Biopolymers* **8,** 669 (1969).

36. K. Rosenheck and P. Doty, *Proc. Natl. Acad. Sci. U. S.* **47,** 1775 (1961).

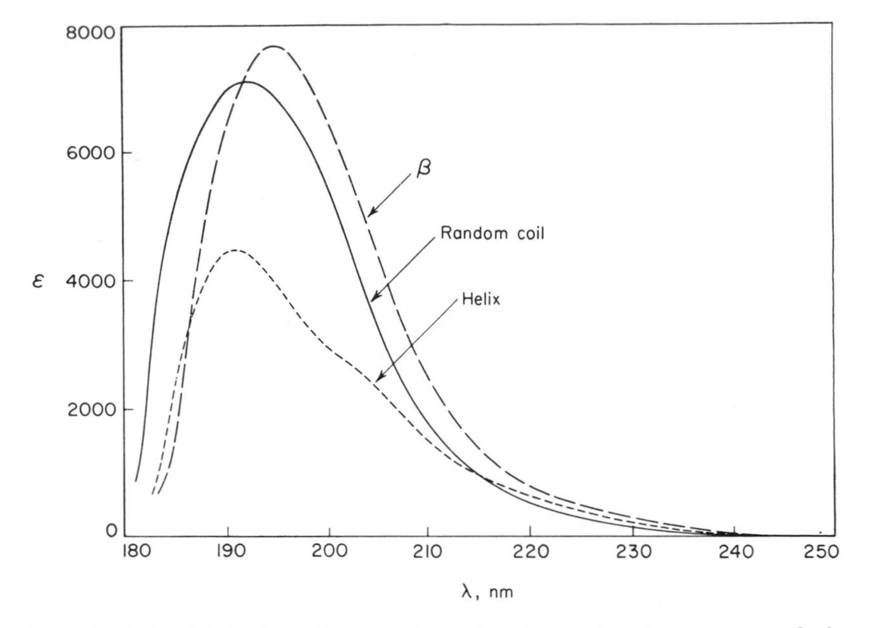

FIG. 2. Ultraviolet absorption spectra of poly-L-lysine in aqueous solution: random coil, pH 6.0, 25°C; α helix, pH 10.8, 25°C; β conformation, pH 10.8, 52°C [from Rosenheck and Doty (*36*)].

then fitted by a proper combination of the constituent conformational spectra. Such spectral analyses of protein conformation have been rather disappointing up to now, since it has not been possible to arrive at perfect fits in terms of the known spectra of polypeptides in the α-helical, APCP, and random conformations (*31*, *36*). This situation most probably reflects the fact that proteins contain a number of conformations other than the standard three. Certainly the regions devoid of long-range order within a protein structure cannot be identified with the random conformation of synthetic polypeptides in solution. These regions are highly organized and their structures are constrained by various interactions within the protein. The individual residues of randomly coiled polypeptides in solution, however, are in a state of Brownian motion, i.e., their mutual positions in space are constantly changing. Indeed, while the structures of individual molecules of a given protein are identical within the limits of thermal fluctuations in a randomly coiled synthetic polypeptide in solution, the probability is high that at any given instant no two molecules have identical conformations. The implications of these differences between randomly coiled synthetic polypeptides and the regions of protein structure devoid of long-range order will be discussed later in the examination of the pertinent circular dichroism spectra.

V. Circular Dichroism and Optical Rotatory Dispersion

While subject to some of the same limitations as ultraviolet absorption, as well as to greater theoretical uncertainty (*37*), the use of the optical rotatory properties of proteins has met with somewhat greater success. The optical rotatory properties of peptide bands stem from the same transitions as the ultraviolet absorption bands and convey essentially the same information. However, while absorption reflects electric moments only, optical activity results from a product of electric and magnetic moments (*38*). As a result, a transition characterized by weak absorption may have strong rotation, and vice versa. Thus, ideally, the two techniques should be complementary. Optical activity may be examined in two different ways: (1) by optical rotatory dispersion (ORD), i.e., the dependence of optical rotation on wavelength; and (2) by circular dichroism (CD), i.e., the difference between the absorption of left and right circularly polarized light. The advantages of CD are that it yields discrete spectral bands; furthermore, since the bands may be positive or negative, greater band resolution is afforded than in usual ultraviolet absorption which measures the sum of the absorptions of these two components of light. Optical rotation, being a dispersion phenomenon, gives rise to infinite bands and, thus, to strong overlaps. While this leads to difficulty in the resolution of bands in the region of transitions, this very overlap makes it possible to measure the optical rotation far from the wavelengths of particular transitions and may be used to advantage to follow changes in conformation without any specific knowledge of the structural features involved; this may be done by following either the variation in rotation at a single wavelength or the variation in the characteristic parameters of empirical relations, such as the Moffitt–Yang or Schechter–Blout equations (*39, 40*), which describe the dependence of optical rotation on wavelength far from the region of transition. Circular dichroism and ORD are different manifestations of the same phenomenon and are related mathematically by the Kronig–Kramers transform (*41, 42*). This interconversion is very useful in checking the internal consistency of measurements and as a help in band resolution (*12, 43–46*).

37. I. Tinoco, Jr., A. Halpern, and W. T. Simpson, *in* "Polyamino Acids, Polypeptides and Proteins," (M. A. Stahmann, ed.) p. 147. Univ. of Wisconsin Press, Madison, Wisconsin, 1962.

38. W. Kauzmann, "Quantum Chemistry," Chapters 15 and 16. Academic Press, New York, 1957.

39. W. Moffitt and J. T. Yang, *Proc. Natl. Acad. Sci. U. S.* **42,** 596 (1956).

40. E. Schechter and E. R. Blout, *Proc. Natl. Acad. Sci. U. S.* **51,** 695 (1964).

41. A. Moscowitz, *in* "Optical Rotatory Dispersion" (K. Djerassi, ed.), p. 150. McGraw-Hill, New York, 1960.

A. Secondary Structure

With the availability of highly sensitive instrumentation, which first permitted ORD (and later CD) measurements in the far ultraviolet, a vast amount of experience has been accumulated on the use of these techniques in the determination of protein and enzyme structure. We will confine our discussion strictly to the use of CD. The CD spectra of polypeptides in the α-helical and random conformations in solution were first reported by Holzwarth and Doty (*34*); the solution spectrum of the APCP structure was observed simultaneously on poly-L-lysine (*47, 48*) and silk fibroin (*49*). The early work has been reviewed recently (*50*). Spectra typical of the three conformations in aqueous solution are shown in Fig. 3. Evidently these conformations are quite distinguishable in the case of a polypeptide in solution. The α-helix has two negative bands at 208 and 221 nm and a positive band at 191 nm; the APCP structure has a negative band at 217 nm and a positive band at 195 nm; the random conformation is characterized by a negative band at 196 nm, weak positive absorption at 218 nm, and very weak negative absorption at 235 nm. More recent studies have shown that the true situation in a protein may be much more complicated than a simple combination of these three spectra. Studies on polypeptide films have revealed that both the β and random (or unordered) conformations may give rise to more than one type of CD spectrum. Thus, Fasman and Potter (*51*) found in ORD studies that the APCP conformation may generate two general families of spectra, which they called I-β and II-β. This finding was confirmed by CD experiments (*52, 52a*). Typical spectra are shown in Fig. 3. They indicate that the nature of side chains and probably side chain interactions con-

---

42. A. Moscowitz, *Advan. Chem. Phys.* **4,** 67 (1962).

43. R. Townend, T. F. Kumosinski, and S. N. Timasheff, *JBC* **242,** 4538 (1967).

44. J. P. Carver, E. Schechter, and E. R. Blout, *JACS* **88,** 2550, 2562 (1966).

45. S. Beychok, *in* "Poly-α-Amino Acids" (G. D. Fasman, ed.), p. 293. Marcel Dekker, New York, 1967.

46. J. E. Coleman, *JBC* **243,** 4574 (1968).

47. R. Townend, T. F. Kumosinski, S. N. Timasheff, G. D. Fasman, and B. Davidson, *BBRC* **23,** 163 (1966).

48. P. K. Sarkar and P. Doty, *Proc. Natl. Acad. Sci. U. S.* **55,** 981 (1966).

49. E. Iizuka and J. T. Yang, *Proc. Natl. Acad. Sci. U. S.* **55,** 1175 (1966).

50. S. N. Timasheff and M. J. Gordunoff, *Ann. Rev. Biochem.* **36,** 13 (1967).

51. G. D. Fasman and J. Potter, *BBRC* **27,** 209 (1967).

52. L. Stevens, R. Townend, S. N. Timasheff, G. D. Fasman, and J. Potter, *Biochemistry* **7,** 3717 (1968).

52a. G. D. Fasman, R. Hoving, and S. N. Timasheff, *Biochemistry* (in press) (1970).

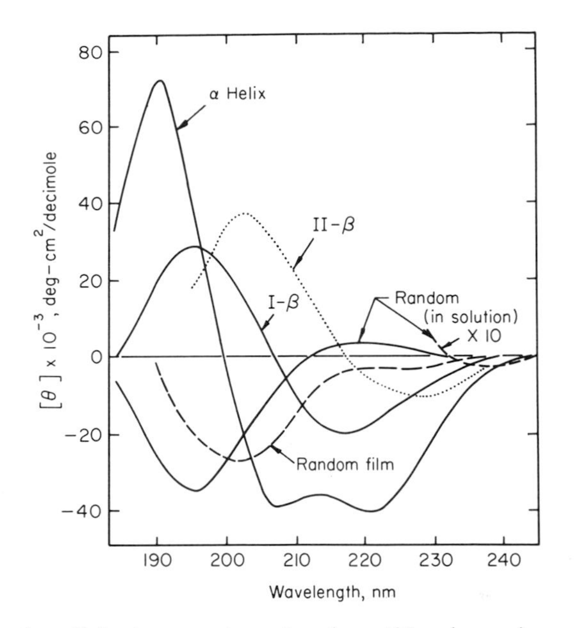

FIG. 3. Circular dichroism spectra of polypeptides in various conformations. α Helix: poly-L-lysine, pH 11.2, 25°C; I-β: poly-L-lysine, pH 11.2, after heating 20 min at 51°C; II-β: poly-*S*-carbobenzoxy-methyl-L-cysteine film cast from trifluoroacetic acid solution; random (in solution): poly-L-lysine, pH 7.5; random film: poly-L-lysine film cast from an aqueous pH 7.5 solution. [θ] is the ellipticity uncorrected for the refractive index of the solvent (*41*, *47*, *52*).

trol both band position and intensity. A similar situation exists in the case of the unordered conformation. Whereas, in solution, polypeptides in this conformation give spectra similar to that of poly-L-lysine, the spectrum changes drastically when an unordered polypeptide is cast as a film, as shown in Fig. 3; the strong negative band shifts to a higher wavelength, and the spectral features above 210 nm are replaced by a broad negative shoulder centered around 220 nm. This observation may be particularly pertinent to protein and enzyme structural analysis for two reasons:

(1) the unordered sections of globular proteins are not truly random chains, such as polypeptides in solution; in fact, they are organized in a definite geometry and subject to constraints imposed by the general folding of the molecule; in this respect, they are much better modeled by polypeptides in the state of hydrated films, in which the chain segments are also constrained and essentially immobile, than by random flight polymers.

(2) Denatured proteins give CD spectra much more reminiscent of

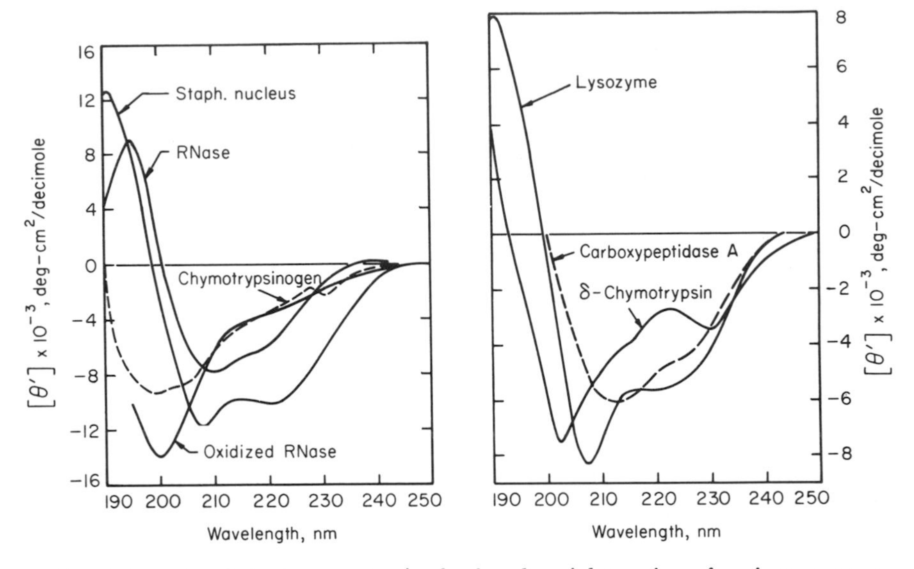

FIG. 4. Circular dichroism spectra in the far ultraviolet region of various enzymes. $[\theta']$ is the ellipticity corrected for the refractive index of the solvent (*12, 13, 22, 52a, 71, 72*).

those of polypeptides in film form than in solution (*52a*); an example is given by the spectrum of oxidized ribonuclease shown in Fig. 4.

Further complications arise from the facts that band intensities and positions are sensitive to the nature of the environment (*49, 53*), to the extent over which the long-range order persists and to the state of association of polypeptides (*54*). It should also be pointed out that amide groups included in structures other than α-helical or β may give rise to CD spectra similar to those shown in Fig. 3 (*55–59*). The positions of CD bands associated with the various conformations are summarized in Table II.

Taking cognizance of the limitations of CD, it seems worthwhile to examine the extent to which enzyme structure may be probed by this technique by looking at some specific examples. In Fig. 4 are shown the

53. R. W. Woody, *J. Chem. Phys.* **49,** 4797 (1968).
54. G. G. Hammes and S. E. Schullery, *Biochemistry* **7,** 3882 (1968).
55. J. T. Yang, *in* "Conformation of Biopolymers" (G. N. Ramachandran, ed.), Vol. 1, p. 157. Academic Press, New York, 1967.
56. E. B. Nielsen and J. A. Schellman, *J. Phys. Chem.* **71,** 2297 (1967).
57. D. Balasubramanian and D. B. Wetlaufer, *in* "Conformation of Biopolymers" (G. N. Ramachandran, ed.), Vol. 1, p. 147. Academic Press, New York, 1967.
58. N. J. Greenfield and G. D. Fasman, *Biopolymers* **7,** 595 (1969).
59. S. Laiken, M. Printz, and L. C. Craig, *JBC* **244,** 4454 (1969).

TABLE II
CIRCULAR DICHROISM BAND POSITIONS FOR VARIOUS STRUCTURES

Structure	Band positions (nm) and sign
α Helix	221(−); 208(−); 191(+)
I-β	217(−); 195–198(+)
II-β	227(−); 198–203(+)
Random (solution)	235(−); 217(+); 196(−)
Unordered (film)	202(−); 220–230 sh(−)

ultraviolet CD spectra of ribonuclease, oxidized ribonuclease, lysozyme, carboxypeptidase A, δ-chymotrypsin, chymotrypsinogen, and staphylococcal nuclease. Just as in the case of the infrared region, these spectra are distinct from each other and are, in general, consistent with structures known from X-ray crystallography (*17–19, 60*). Greenfield and Fasman (*61, 62*) have made calculations of ORD and CD spectra of enzymes in terms of various compositions of α-helical, APCP, and random structures, using poly-L-lysine as a standard. For the proteins shown in Fig. 4, in the case of CD, they were able to account for the spectra between 205 and 240 nm in terms of the following compositions.

Ribonuclease: 9.3% α helix, 32.6% APCP, 58.1% random
Lysozyme: 28.5% α helix, 11.1% APCP, 60.4% random
Carboxypeptidase A: 13% α helix, 30.6% APCP, 56.4% random

In the first two cases the agreement is good with structures determined by X-ray crystallography (*17, 18, 60*). In the case of carboxypeptidase A, the CD analysis overestimates the contents of APCP structure almost by a factor of two, while underestimating the α-helical contents by an equal amount (*19*). This enzyme, however, is known to contain parallel β structure as well as antiparallel; this has not been taken into account specifically in the calculations. It is known, however, that the CD spectra of the two β structures should be almost identical (*35*). In the case of the staphylococcal nuclease, the spectrum shown in Fig. 4 is very typical for the presence of a significant amount of α helix (about 20%, when poly-L-lysine is used as a standard) (*22, 62a*). This value is in reasonable agreement with the results of X-ray diffraction studies (*63*). Discrep-

60. C. C. F. Blake, D. F. Koenig, G. A. Mair, A. C. T. North, D. C. Phillips, and V. R. Sarma, *Nature* **206,** 757 (1965).

61. N. Greenfield and G. D. Fasman, *Biochemistry* **8,** 4108 (1969).

62. N. Greenfield, B. Davidson, and G. D. Fasman, *Biochemistry* **6,** 1630 (1967).

62a. H. Taniuchi and C. B. Anfinsen, *JBC* **243,** 4778 (1968).

63. A. Arnone, C. J. Bier, F. A. Cotton, E. E. Hazen, Jr., D. C. Richardson, and J. S. Richardson, *Proc. Natl. Acad. Sci. U. S.* **64,** 420 (1969).

ancies which occur on occasion between estimates of the α-helical contents based on CD measurements and X-ray diffraction may be due to a variety of causes: First, the various factors discussed above may contribute to a wrong estimate. Second, in proteins, there may be individual residues with bond directions and angles identical to those found in ordered structures but which are not parts of such long-range structures. These will have transition moments and, thus, optical rotatory properties qualitatively similar to those of the given conformations, while in the structure deduced from X-ray diffraction they will lie in regions devoid of long-range order.

These examples indicate the degree of success that may be expected at present in the determination of the absolute secondary structure of enzymes by circular dichroism. While having serious limitations when used as a tool for the determination of the secondary structure of a protein, circular dichroism may be very useful in comparative studies of selected structures and of conformational changes which accompany enzyme activation or the binding of substrates. In such studies, it is frequently not essential to identify the exact nature of the conformational change which occurs; it may be sufficient to know the manner in which a specific band varies with a change in experimental conditions. This will be illustrated below by the specific cases of ribonuclease, carbonic anhydrase, and chymotrypsin, including its activation from zymogen.

B. Tertiary Structure

While CD spectra in the far ultraviolet reflect principally peptide bond transitions, and thus the secondary structure of a protein, important information on enzyme structure may be derived also from transitions of the side chains. These may reflect features of both the secondary and tertiary structures. Circular dichroism bands characteristic of amino acid side chains span the spectral region below 330 nm. Above 240–250 nm, there is no overlap with peptide bond transitions so that this region may be used to follow changes in side chain conformation and environment. Below 250 nm, the side chain transitions are superimposed on usually stronger peptide bands and therefore are not easy to distinguish, although, in some particular cases, they may become prominent features of the spectrum. The principal contributions are made by tryptophan and tyrosine residues, as well as by cystine disulfide bridges, although phenylalanine and histidine may also make significant contributions. The side chain CD spectra are usually complex as a result of the overlap of closely positioned bands. Furthermore, band positions, signs, and intensities may vary for any given type of residue as a function of its environment. Factors which determine these characteristics of given bands include the nature of the ordered or unordered polypeptide fold to which the side chain is attached, the nature of its environment (polarity and asymmetry), the exact conformation of the side chain, the proximity of charged groups, the proximity of other groups undergoing electronic transitions with which the group in question may interact, and ionization of the residue in the case of tyrosine and histidine.

The positions of the CD bands of various amino acids were examined first by Legrand and Viennet (*64*), they have been summarized recently by Beychok (*45, 65*). Incorporation of the amino acids into polypeptides results in band shifts which vary strongly with conformation. A typical example of the type of band shift which may occur is found in the results of Beychok and Fasman on poly-L-tyrosine (*66*) shown in Fig. 5. It is evident that a shift from the α-helical to the random conformation is accompanied by major changes in the spectrum: Not only are band inten-

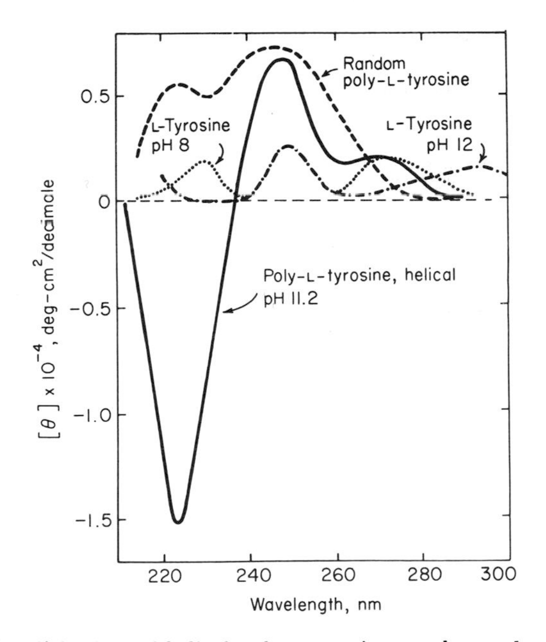

FIG. 5. Circular dichroism of helical poly-L-tyrosine, random poly-L-tyrosine (both at pH 11.2), and L-tyrosine at the pH values indicated [from Beychok and Fasman (*66*)].

64. M. Legrand and R. Viennet, *Bull. Soc. Chim. France* No. 3, 679 (1965).
65. S. Beychok, *Science* **154,** 1288 (1966).
66. S. Beychok and G. D. Fasman, *Biochemistry* **3,** 1675 (1964).

sities and positions altered, but even the sign of the absorption may change, as, for example, at 222 nm. Phenylalanine and its derivatives were found by Horwitz *et al.* (*67*) to give complex spectra between 250 and 268 nm; the band positions were essentially independent of structure and solvent, but their signs varied, being reversed both as a function of aggregation and residue conformation. A similar situation has been found to be true in the case of tryptophan. Peggion *et al.* (*68*) have determined the CD spectra of polytryptophan and of copolymers of tryptophan with γ-ethyl-L-glutamate, dissolved in ethylene glycol monomethyl ether. The spectra are shown in Fig. 6. In the near ultraviolet, positive bands at 293, 285, 278, and 266 nm reflect aromatic transitions. The progression of the curves below 240 nm from one dominated by a positive band at 230 nm and a negative one at 220 nm, characteristic of poly-L-tryptophan, to a spectrum characteristic of an α-helical structure, shows the overwhelming contribution which may be made by aromatic transitions in the far ultraviolet region as well. An example of such complex enzyme spectra is found in carbonic anhydrase (*23*, *46*). It is known that

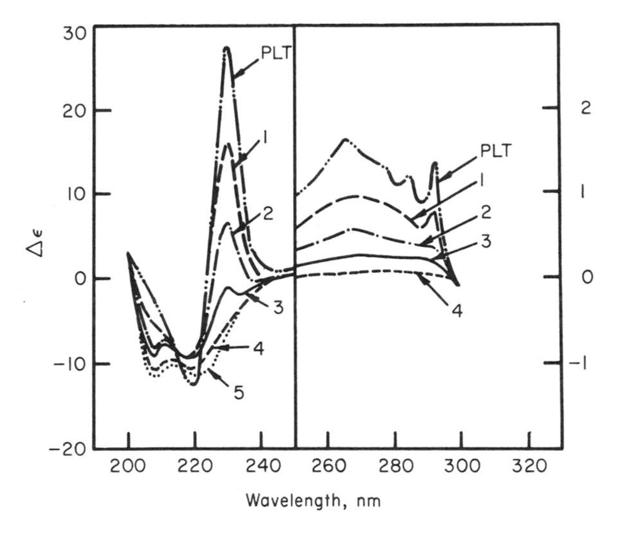

FIG. 6. Circular dichroism spectra of poly-L-tryptophan (PLT) and of copolymers of L-tryptophan and γ-ethyl-L-glutamate in ethylene glycol monomethyl ether solutions. The numbers on the curves refer to the following mole fractions of L-tryptophan in the copolymers: 1, 0.855; 2, 0.679; 3, 0.502; 4, 0.315; and 5, 0.159. Δε is the difference between the absorptions of left and right circularly polarized light [from Peggion *et al.* (*68*)].

67. J. Horwitz, E. H. Strickland, and C. Billups, *JACS* **91**, 184 (1969).
68. E. Peggion, A. Cosani, A. S. Verdini, A. Del Pra, and M. Mammi, *Biopolymers* **6**, 1477 (1968).

the tryptophan copolypeptides are in the α-helical conformation (*68, 69*). The spectrum of poly-L-tryptophan cast as a film has the same bands above 250 nm (*52*). The 230-nm band is negative, however, or of opposite sign to that found in solution (*68*); furthermore, the detailed structure of the spectrum is a function of film orientation. This indicates the strong effects which may result from the imposition of constraints on the random motion of polypeptide chains; such constraints are almost absent in solution but are prevalent both in films and in the interior of protein molecules.

Quite recently, Strickland *et al.* (*70*) have analyzed in detail the CD spectra of tryptophan derivatives in solvents of various polarities at room temperature and at 77°K. They were able to identify a number of overlapping dichroic bands which were assigned to various transitions. The low temperature CD spectra were almost twenty times as intense as those obtained at room temperature; the absorption weakness at 298°K reflects conformational mobility with cancellation of bands that correspond to different conformations. In the light of this observation, Strickland *et al.* analyzed the near ultraviolet CD spectra of chymotrypsinogen obtained at 298° and 77°K. The room temperature spectrum is shown in Fig. 7 (*71, 72*). The bands above 285 nm have been assigned to tryptophan transitions on the basis of comparison with model compounds (*70, 71*), as well as the observation that acetylation of three of the four tyrosines of this protein produced no changes in the CD spectrum as shown in Fig. 7 (*71*). Cooling of the chymotrypsinogen solution to 77°K resulted in only a slight change in the spectral intensity (*70*). The last observation, together with the fact that the optically active bands of tryptophan within chymotrypsinogen are red shifted with respect to the model compounds, led Strickland *et al.* (*70*) to the conclusion that in this zymogen the tryptophan residues are buried within the protein and thus have a relatively rigid position in space even at room temperature.

In the case of cystine residues, the situation is still more complicated. Not only is the CD spectrum a function of the various structural and environmental factors listed above, but also it is strongly affected by the dihedral angle about the disulfide bridge. Beychok (*45*) and Coleman and Blout (*73, 74*) have investigated a number of disulfide-containing

69. G. D. Fasman, M. Landsberg, and M. Buchwald, *Can. J. Chem.* **43,** 1588 (1965).
70. E. H. Strickland, J. Horwitz, and C. Billups, *Biochemistry* **8,** 3205 (1969).
71. M. J. Gorbunoff, *Biochemistry* **8,** 2591 (1969).
72. M. J. Gorbunoff, in preparation.
73. D. L. Coleman and E. R. Blout, *in* "Conformation of Biopolymers" (G. N. Ramachandran, ed.), Vol. 1, p. 123. Academic Press, New York, 1967.
74. D. L. Coleman and E. R. Blout, *JACS* **90,** 2405 (1968).

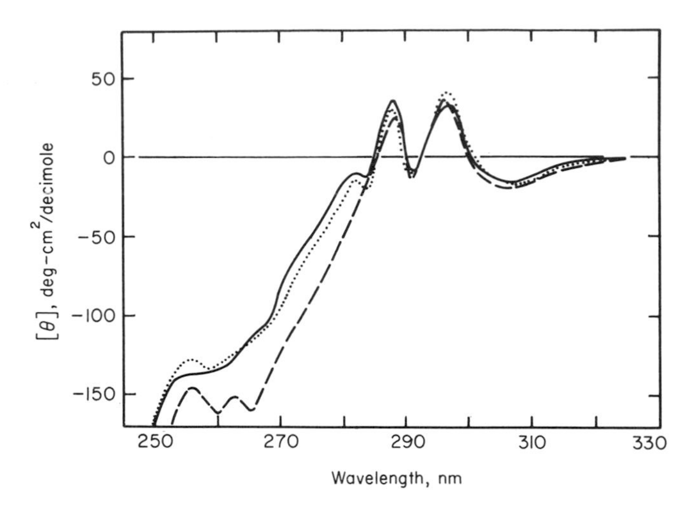

FIG. 7. Circular dichroism spectra in the near ultraviolet region of native chymotrypsinogen (*71*) in 0.001 *M* HCl(——), δ-chymotrypsin in 0.001 *M* HCl (*72*) (– –), and chymotrypsinogen with 3.2 tyrosines O-acetylated in pH 7.5 borate buffer (*71*) (......).

compounds with the general conclusion that the principal transitions result in bands at around 200 nm and 260 nm, the signs being a function of the exact disulfide configuration. Beychok (*45*) has shown, furthermore, that with the proper configuration a disulfide bridge may give CD extrema at wavelengths as high as 350 nm. The contribution from disulfide rotation may be quite large. Coleman and Blout (*74*) estimated that for a polypeptide containing 25–40% α helix, 10% of the optical rotation near 210 nm may result from disulfide transitions if the chain contains two S—S bonds per hundred residues, a situation not unlike that found in many enzymes. They have also concluded that extensive cancellation of contributions from the 200-nm transition may occur in proteins because of variations of the dihedral angles of the S—S bonds.

From the preceding discussion, it is quite evident that side chain transitions give rise to a number of overlapping bands, which vary greatly in sign, position, and intensity, rendering the analysis of a protein spectrum extremely difficult. The positions of bands associated at present with various amino acid side chains are summarized in Table III. As a result, up to now, there has been only a limited success in the determination of tertiary structure from the side chain CD bands. Conversely, it is hoped that, as more data become available from properly controlled experiments on well-characterized model systems, the very complexity of the spectral patterns should become a useful tool in the determination of structural

TABLE III
IDENTIFIED CD BANDS OF CHROMOPHORIC RESIDUES

Residue	Band position (nm)	Residue	Band position (nm)
Trytophan	298–301	Tyrosine	275
	294–297	(unionized)	226
	290–293		
	287–290	(ionized)	292
	283–286		235
	277–278		
	265–270		
	222–225		
Cystine	260	Phenylalanine	268
	200		266.5
	(position varies with S—S dihedral angle)		264–265
			262
			259–260

details and, in particular, of changes in the tertiary structures of enzymes.

C. Typical Cases

1. *Chymotrypsin*

One of the first indications that a structural reorganization takes place when chymotrypsinogen is converted to the enzyme was the observation of Neurath *et al.* (*75*) that this reaction was accompanied by a change in $[\alpha]_D$ from —79.8 to —60.3. That this did not reflect a major change in α-helical contents is evident from the observations of Raval and Schellmann (*76*) and Fasman *et al.* (*77*) that the Moffitt–Yang b_0 parameter did not change within experimental error during activation at pH 6.8–7.0. Extending the optical rotation dispersion studies into the far ultraviolet region, Raval and Schellman (*76*), Biltonen *et al.* (*78*), and Fasman *et al.* (*77*) observed that activation of the zymogen to the enzyme was accompanied by spectral changes in the region between 220 and 235 nm, where an apparently negative Cotton effect, centered at about 230 nm, became much more pronounced in the enzyme than in the precursor. The relation of this spectral change to molecular events

75. H. Neurath, J. A. Rupley, and W. F. Dreyer, *ABB* **65**, 243 (1956).
76. D. N. Raval and J. A. Schellman, *BBA* **107**, 463 (1965).
77. G. D. Fasman, R. J. Foster, and S. Beychok, *JMB* **19**, 240 (1966).
78. R. Biltonen, R. Lumry, V. Madison, and H. Parker, *Proc. Natl. Acad. Sci. U. S.* **54**, 1018 and 1412 (1965).

was clarified later in the elegant studies of Hess and co-workers (*79, 80*), who examined with a variety of techniques the problem of the detailed structural changes which accompany the activation of chymotrypsin. This section concerns solely ORD and CD analysis.

The far ultraviolet CD spectra of δ-chymotrypsin and chymotrypsinogen at pH 7.5 are shown on Fig. 4 (*60, 71, 72*). Both are dominated by broad negative absorption between 190 and 240 nm with a maximal amplitude between 197 and 207 nm. Between 220 and 235 nm both spectra contain a negative maximum near 230 nm and a negative minimum (located at 225 nm in the zymogen and 220 nm in the enzyme). The ellipticities of the extrema, however, are not identical. For the active enzyme, the ellipticity is more negative than for the zymogen, and the difference between $[\theta']_{230}$ and $[\theta']_{220-225}$ is more pronounced. At lower wavelengths, the amplitude of the CD is more negative in the precursor than in δ-chymotrypsin, suggesting that the enzyme structure contains somewhat more order than is found in chymotrypsinogen. The small differences in CD are also present in the spectral region of side chain absorption. Between 250 and 285 nm, the CD of δ-chymotrypsin is more negative than that of the precursor. Above 285 nm the CD spectra are essentially identical, indicating that activation has little effect on the environment of the tryptophan residues. Thus, a subtle conformational difference is revealed between chymotrypsinogen and the active enzyme. Hess and coworkers (*79, 80*) related these spectral changes to alterations in the conformation of specific amino acid residues by the following analysis.

The chemical difference between chymotrypsinogen and δ-chymotrypsin is known to be the absence of a dipeptide between residues 13 and 16 in the enzyme. Further peptide bond cleavage leads to the formation of α-chymotrypsin. Since the CD spectra of α- and δ-chymotrypsin are essentially identical in the 220–240 nm spectral region, the entire conformational difference between chymotrypsinogen and the enzymes must be related to the removal of the Ser 14–Arg 15 dipeptide. Crystallographic studies have shown that in α-chymotrypsin, at neutral pH, the α-amino group of Ile residue 16 and the carboxyl group of Asp 194 form an ion pair and then point toward the nonpolar inside of the protein molecule (*81*). Thus, from a comparison of chymotrypsin and chymotrypsinogen X-ray data (*82*), it appears that activation of the enzyme is accom-

79. J. McCann, G. D. Fasman, and G. P. Hess, *JMB* **39**, 551 (1969).
80. G. P. Hess, *Brookhaven Symp. Biol.* **21**, 155 (1969).
81. B. W. Matthews, P. B. Sigler, R. Henderson, and D. M. Blow, *Nature* **214**, 5089 (1967).
82. J. Kraut, H. T. Wright, M. Kellerman, and S. T. Freer, *Proc. Natl. Acad. Sci. U. S.* **58**, 304 (1967).

panied by a displacement of Ile 16 by a distance of 10 to 15 Å (*79*). In the absence of the positive charge on Ile 16, Asp 194 must point into the solvent (*83*), since the presence of a charged group within the low polarity interior of a protein molecule would result in an unfavorable free energy contribution of sufficient magnitude to destabilize the native structure of the enzyme (*84*). Thus, movement of Ile 16 on activation implies movement of Asp 194. Movement of Asp 194, in its turn, requires movement of Ser 195 (*80*), which is required for the catalytic activity of the enzyme. In the active enzyme it appears to be hydrogen bonded to His 57 which is also required for activity.

That these movements of the amino acid residues around the active site of chymotrypsin are indeed the cause of the change in the optical rotatory properties between 220 and 240 nm was shown by the following conformational experiments (*79*, *80*). Raising the pH from neutrality to 10, gradually transformed the CD spectrum of active chymotrypsin to one identical with that of chymotrypsinogen (see Fig. 8a). This change occurs with an apparent pK of 8.5, in agreement with the earlier observation that a change in specific rotation at 313 nm also occurs with an increase in pH and can be accounted for by the ionization of a single group with an apparent pK of 8.5 (*85*). Hess and co-workers (*79*, *80*) identified the group involved in the optical rotatory change in the following way. The amino groups of chymotrypsinogen were fully acetylated and the product was activated by trypsin to yield fully active acetylated δ-chymotrypsin. This product must have a single amino group, that of Ile 16. Circular dichroism experiments were then carried out as before, and the results are shown in Fig. 8a. The spectrum of active acetylated δ-chymotrypsin at neutral pH is identical with that of the normal enzyme at the same conditions; that of acetylated chymotrypsinogen, whether at neutral or alkaline pH, is the same as that of the unmodified precursor. An increase in pH shifted the CD spectrum of the acetylated enzyme to that of the zymogen. Further acetylation of the amino group of Ile 16 in acetylated δ-chromotrypsin also displaced the CD spectrum toward that of the zymogen even at neutral pH, as shown in Fig. 8a. These experiments clearly demonstrate the participation of Ile 16 in the observed shifts of the CD band; protonation of its α-amino group is required for the spectrum typical of the active enzyme. Modification of Ser 195 by formation of DIP-chymotrypsin (*79*) resulted in CD spectra which were identical with that of the active enzyme whether measured

83. P. B. Sigler, D. M. Blow, B. W. Matthews, and R. Henderson, *JMB* **35**, 143 (1968).
84. C. Tanford and J. G. Kirkwood, *JACS* **79**, 5333 (1957).
85. H. L. Oppenheimer, B. Labouesse, and G. P. Hess, *JBC* **241**, 2720 (1966).

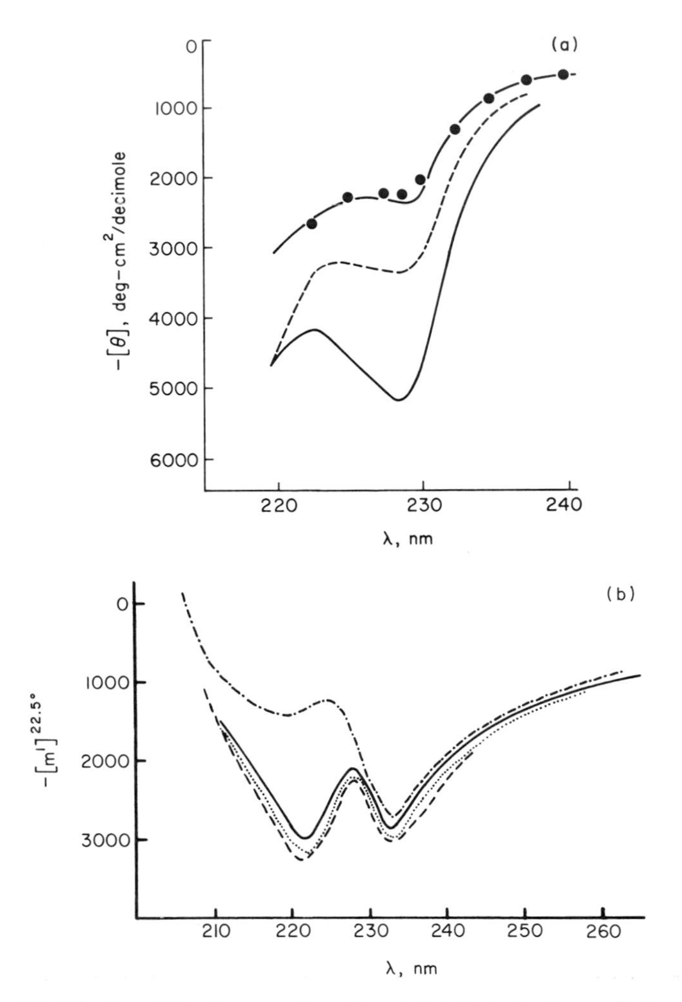

FIG. 8. (a) Circular dichroism spectra of acetylated derivatives of chymotrypsin. Lower curve, acetylated δ-chymotrypsin at pH 6.7; middle curve, acetylated δ-chymotrypsin at pH 10; upper curve, acetylated chymotrypsinogen in the pH region 6.7–10; (●) reacetylated acetylated δ-chymotrypsin in the pH region 6.7–10. (b) Ultraviolet optical rotatory dispersion curves for acetylated chymotrypsinogen and acetylated δ-chymotrypsin at pH 7.0 and pH 10.5: (–·–) acetylated δ-chymotrypsin at pH 7.0; (——) acetylated chymotrypsinogen at pH 7.0; (···) acetylated δ-chymotrypsin at pH 10.5; (---) acetylated chymotrypsinogen at pH 10.5.

at neutral or alkaline pH; this suggested that the presence of the bulky group on Ser 195 prevents the conformational change accompanied by the break of the ion pair, by steric interference with the motion of Asp 194 into solvent. The presence of the charged carboxyl inside the protein molecule requires the continued forced protonation of the Ile 16 α-amino group even above pH 10. This was demonstrated to be true indeed by titration studies which indicate that in DIP-chymotrypsin one less group is deprotonated at high pH than in the parent enzyme (*85*).

The chymotrypsin studies are probably the best example available at present of the manner in which CD may be used as a probe of conformational changes which accompany the activation of the zymogen to an enzyme, as well as changes in solvent properties (here variation of pH). It should be stressed, however, that, while some of the side chains involved in the conformational transition have been identified, the actual nature of the electronic transitions affected is not known, nor are the details of the conformational changes known on the atomic level. The CD spectral changes observed could conceivably reflect changes in the conformations of the amino acid residues identified as being involved in the interactions; in particular, a histidine residue may undergo a spectral transition within the frequency range in question (*45*); the exact position and amplitude of its CD band may be affected by changes in charge environment, resulting from the motions of Ile 16 and Asp 194. On the other hand, it is quite possible that the spectral details between 220 and 235 nm reflect the transition of a tyrosine residue (*86*). This transition could be affected by changes in the polarity or charge distribution of the environment as the active site residues change their coordinates during activation. Indeed, it is quite suggestive that activation of chymotrypsinogen to chymotrypsin generates a tyrosine difference spectrum (*87*), indicating a change in the environment of these residues; furthermore, only three tyrosine residues are available to acetylation with *N*-acetlylimidazole in chymotrypsinogen, while all four residues are accessible in α-chymotrypsin (*71*); this difference by one accessible tyrosine residue is maintained when cyanuric fluoride is used as the probe; α-, γ-, and δ-chymotrypsins have maximal reactivities of three residues with this reagent (*72, 88*), while only two groups can be induced to react in chymotrypsinogen (*71*).

A further complication in the identification of the transition in question stems from ignorance of whether the CD spectral changes between 220 and 240 nm are related to a negative band centered around 230 nm, a positive band centered at about 220 nm, or both. This results from the superposition of these weak bands on a rapidly changing negative band located at lower wavelengths. It is possible to seek an answer to this

86. M. N. Pflumm and S. Beychok, *JBC* **244,** 3973 (1969).
87. P. Benmouyal and C. G. Trowbridge, *ABB* **115,** 67 (1966).
88. Y. Hachimori, K. Kurihara, H. Horimishi, A. Matsushima, and K. Shibata, *BBA* **105,** 167 (1965).

question in the comparison of the CD and ORD spectra, which must be related by the Kronig–Kramers transform. The ORD curves of acetylated chymotrypsinogen and δ-chymotrypsin at conditions identical with those of Fig. 8a are shown in Fig. 8b (*79*). The ORD curves are identical above 230 nm; they display a trough of equal amplitude at 233 nm. Below 230 nm, a peak appears at 228 nm and a second trough at 222 nm. In the active form of the enzyme, the rotation between 230 and 210 nm is less negative than in the inactive conformation. Comparison between the presently available ORD and CD data, however, does not permit us to arrive at an unequivocal assignment of the sign, position, and intensity of this very interesting band.

2. *Ribonuclease*

The difficulties and uncertainties encountered in CD band assignment and the methods which must be employed are extremely well illustrated by the case of ribonuclease. This task would appear, *a priori,* to be quite simple since this enzyme is devoid of tryptophan residues. Yet, even in this case, it required extensive experiments to arrive at correct band assignments. The CD spectrum of ribonuclease A is shown in Figs. 4 and 9. In the far ultraviolet region it is characterized by negative absorption between 197 and 235 nm, with a maximum at 210 nm and a shoulder at 217 nm. Above 235 nm, a positive band appears at 240 nm, followed by negative absorption between 250 and 300 nm, with a maximum at 275 nm.

For purposes of conformational analysis, the spectral regions above and below 235 nm should be discussed separately. Pflumm and Beychok (*86*) decomposed the far ultraviolet spectral region into contributions from α-helical, antiparallel β, and random conformations; they used as standards spectra of poly-L-glutamic acid, poly-L-lysine, and poly-L-glutamate. They were able to fit the experimental spectrum in terms of a conformational composition of 11.5% α helix, 33% β structure, and 39% random coil, with the remainder being unassigned. This is in good agreement with the fit found by Greenfield and Fasman (*61*). In these calculations, they could keep the polypeptide band positions, half-widths and maximum amplitudes for the 222 and 207 nm α-helical bands and the 217 and 195 nm β-structure bands. Adjustments had to be made in the position (3 nm red shift) and intensity (about 50% of expected) of the 192-nm α-helical band and in the intensity (about 70% of expected) of the 198-nm random coil band. A small positive band at 226 nm had to be included to make a reasonable fit. Pflumm and Beychok (*86*) proposed as a possible explanation for the band adjustments the presence of short

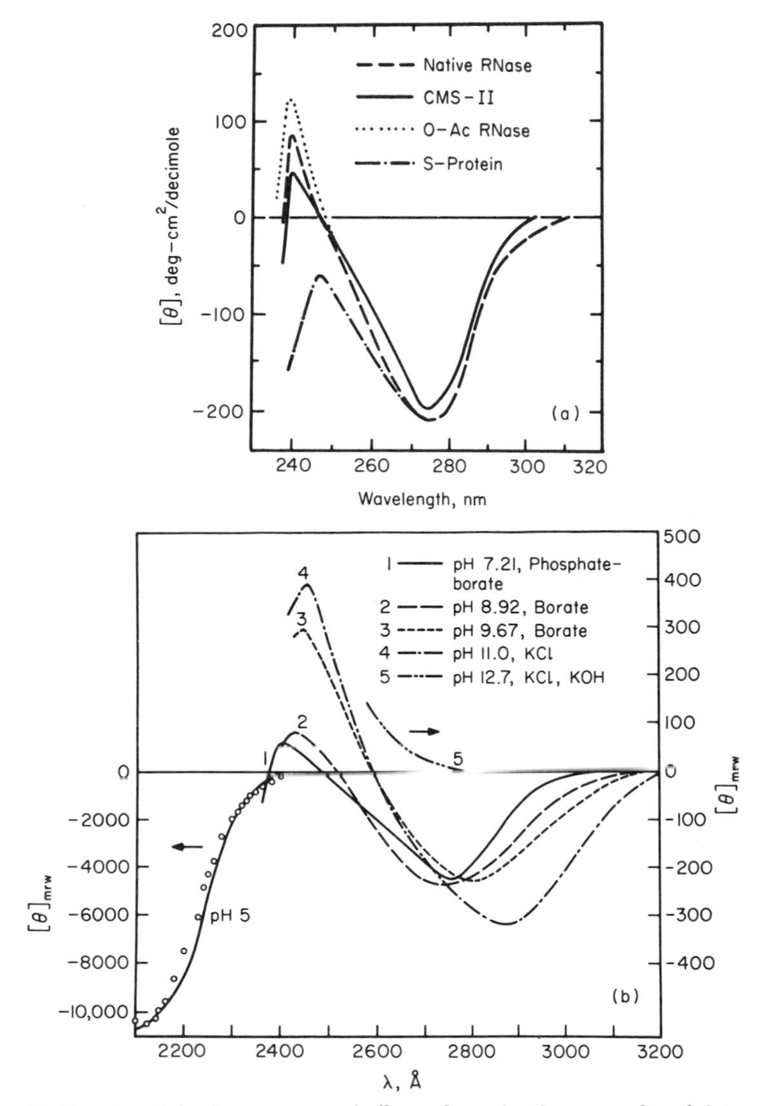

FIG. 9. Circular dichroism spectra of ribonuclease in the near ultraviolet region: (a) Ribonuclease A and various derivatives (*86*, *92*, *95*); (b) ribonuclease A at various pH values ($[\theta]_{mrw}$ is the ellipticity per mean residue weight) [from Pflumm and Beychok (*95*)].

or distorted helices and the observation that ribonuclease, which had been reoxidized in urea and is unordered, also yields an ellipticity value at 198 nm which is about two-thirds of that exhibited by the randomly coiled polymers. The positive band at 226 nm has been attributed to tyrosine transitions. An alternate interpretation of the spectrum has been proposed by Schellman and Lowe (*89*). They suggest that the CD spectrum between 200 and 235 nm can be accounted for in terms of a combination of β structure and α helix, if the α-helical Cotton effect in ribonuclease is blue shifted by 5 nm. This would require a change in the exciton energy of the helical n,π^* transition of about 3 kcal/mole of residues, a situation which they find reasonable for proteins which contain short and/or distorted α helices. A contribution from tyrosine transitions in the region between 215 and 230 nm was considered unlikely by Schellman and Lowe since titration of ribonuclease to pH 11.3 has no effect on the CD in that wavelength interval. Pflumm and Beychok (*86*), however, point out that in the case of *N*-acetyltyrosineamide, the intensities of the CD spectra at pH 7 and 12 do not differ sufficiently between 215 and 230 nm to be detectable in a protein such as ribonuclease. With the elimination of this argument against the Pflumm and Beychok band assignment at 226 nm, it would seem that their conformational analysis is reasonable, in particular, since it is consistent with the findings of X-ray diffraction studies on ribonuclease (*17, 18*). One last comment seems in order. In fitting the experimental data, Pflumm and Beychok had to assign insufficient contents of random coil conformation, leaving about 18% of the structure unaccounted for. It must be repeated again that the unordered portion of a protein molecule is not equivalent to a polymer coil in the "random flight" sense, but it is highly constrained and has little freedom of motion. Thus, it would be surprising if its CD spectrum were to average out to that of a truly randomly coiled polypeptide chain, and ribonuclease seems to be a good case in point.

Let us turn now to the structural information inherent in the positive band at 240 nm. The assignment of this band has been the subject of several studies (*65, 86, 90–92*). It can result from either a tyrosyl or a disulfide transition. Chemical modification and variation of pH permitted assigning this band to a tyrosine transiton. The evidence in favor of this assignment is manyfold: Pflumm and Beychok (*86*) have found that the intensity increases and the band is red shifted when the pH is raised from 8.9 to 11, as would be expected from tyrosine ionization; the intensity at 240 nm increases upon acetylation of the three accessible tyrosines as shown in Fig. 9a; nitration with tetranitromethane results in diminished intensity. Simons and Blout (*92*) have examined the pH dependence of this band as well as the differences in the CD spectra of

89. J. A. Schellman and M. J. Lowe, *JACS* **90,** 1070 (1968).
90. R. J. Simpson and B. L. Vallee, *Biochemistry* **5,** 2531 (1966).
91. N. S. Simmons and A. N. Glazer, *JACS* **89,** 5040 (1967).
92. E. R. Simons and E. R. Blout, *JBC* **243,** 218 (1968).

ribonuclease S, the S protein and ribonuclease S reconstituted by addition of the S peptide to the S protein. Acid denaturation eliminates this band in ribonuclease A at pH 2 and in ribonuclease S at pH 3.5. As shown on Fig. 9a, the S protein does not have this positive band; this derivative differs from the parent enzyme in that it contains four available tyrosines rather than three. This evidence again implicates a tyrosine residue as being responsible for the 240-nm band and specifically the group whose environment changes on removal of the S peptide, i.e., Tyr 25. The chemical modification data (*86*), on the other hand, point to the participation of a solvent accessible tyrosine. It would appear, therefore, that the band observed at 240 nm contains contributions from transitions of both at least one accessible and one inaccessible tyrosine residues of ribonuclease.

The 240-nm band of ribonuclease has also been found to be an extremely sensitive probe of minor conformational changes. Following the discovery by Anfinsen and co-workers (*93, 94*) that when ribonuclease is reduced and reoxidized its enzymic activity is essentially regained, Pflumm and Beychok carried out a detailed investigation of this process by CD (*95*). They found that the CD spectrum of the reoxidized enzyme differs from that of the native enzyme in a decrease of positive intensity at 240 nm. Fractionation on a column resulted in several components, one of which, CMS-II, exhibited full enzymic activity. Its CD spectrum, shown in Fig. 9a, however, differed again from that of the native enzyme in the 240-nm region, while it was essentially identical with that of the parent enzyme at other wavelengths. Thus, it appears that an enzyme, ribonuclease, may have two fully active forms which differ in small conformational features, the exact nature of which is not yet known.

Another interesting use of this band has been made by Simons *et al.* (*96*) who found that when ribonuclease A was heated at pH 6.46, the intensity at 240 nm decreased in a bimodal fashion as shown in Fig. 10. The first change is gradual and noncooperative; it is followed by a cooperative transition which corresponds to the thermal denaturation of the enzyme. It would appear, therefore, that when ribonuclease is heated from 15° to 50°C, a small local change in conformation occurs. Comparison with similar data on the S protein, which does not display this bimodality, and with the above-described observations, has led the authors to speculate that the first slow change in ellipticity corresponds

93. C. B. Anfinsen and E. Haber, *JBC* **236,** 1361 (1961).

94. C. J. Epstein, R. F. Goldberger, D. M. Young, and C. B. Anfinsen, *ABB* Suppl. 1, 223 (1962).

95. M. N. Pflumm and S. Beychok, *JBC* **244,** 3982 (1969).

96. E. R. Simons, E. G. Schneider, and E. R. Blout, *JBC* **244,** 4023 (1969).

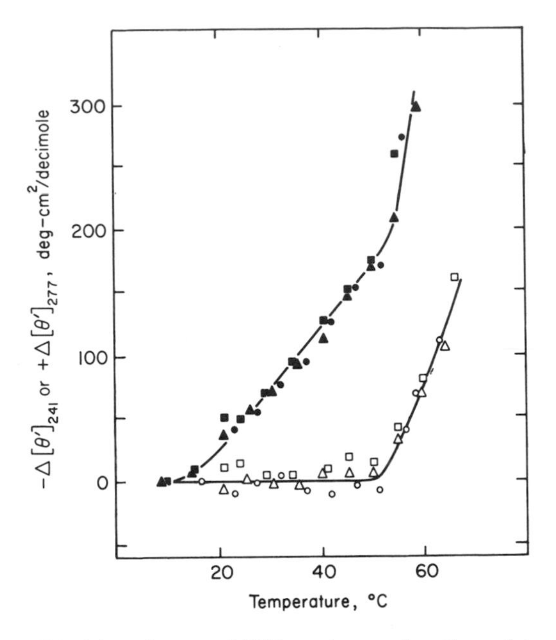

FIG. 10. Molar ellipticity change of RNase A as a function of temperature: (■) 0.1 *M* phosphate at 241 nm, (□) 277 nm; (●) 0.1 *M* acetate at 241 nm, (○) 277 nm; (▲) 0.1 *M* sulfate at 241 nm, and (△) 277 nm [from Simons and Blout (*92*)].

to an initial detachment of noncovalent bonds between the S-peptide and S-protein portions of the enzyme, with the release to solvent of one of the abnormal tyrosines. It is particularly interesting that the 275-nm band, which will be discussed now, does not undergo the first transition but changes sharply only in the general unfolding of the secondary structure. It would appear reasonable to infer from this that Tyr 25 makes no contribution to the higher wavelength band and that variations in the two bands may serve as probes of two distinct conformational changes.

Assignment of the negative band at 275 nm again is not immediate. In the absence of tryptophan, this band may result from either tyrosine or cystine transitions (*97*). Simpson and Vallee (*90*) and Beychok (*65*) found that when pH is raised to 11, the band is red shifted to 288 nm and its ellipticity increases by 50%, as shown in Fig. 9b. This suggests the participation of exposed and ionizable tyrosines. Acetylation of tyrosine hydroxyl groups, however, results in only a 10% decrease in band intensity, implicating either buried tyrosines or disulfides as

97. S. Beychok, *Proc. Natl. Acad. Sci. U. S.* **53,** 999 (1965).

responsible for the remainder of the band. That disulfides do make a contribution is supported by the observation that a small negative CD band persists at 275 nm even in 8 *M* urea at pH 11, conditions at which tyrosines should be totally randomized (*86*). Nitration experiments cast further light on the nature of the groups involved. Beaven and Gratzer (*98*) found that nitration of 1.8 tyrosines/mole of protein resulted in the appearance of a dichroic band at 360 nm with no changes in intensity at 275 nm, suggesting that the new band at 360 nm results from previously optically inactive exposed residues. Nitration of all three exposed groups (*95*) is accompanied by a decrease in optical activity at 275 nm, as well as in the far ultraviolet, indicating some conformational change. It would appear, therefore, that the 275 nm band stems from the summation of transitions of disulfides, some buried tyrosines, and possibly one or two of the exposed tyrosines.

Summarizing all the available information on the behavior of the 240- and 275-nm CD bands leads to the conclusion that both the exposed and buried tyrosines of ribonuclease are in different environments. Thus, Tyr 25 seems to be optically inactive, while the other buried residues contribute to the CD; on the other hand, only the least accessible of the three exposed residues seems to be optically active since nitration of two groups does not change the spectrum at 275 nm; that one of the exposed groups is present in asymmetric environment is required by the observation of the pH dependence of the 275-nm band below the onset of irreversible changes. Further support for this conclusion may be drawn from solvent perturbation spectroscopy (*99*) and chemical modification studies (*100*). Both studies indicate that one of the exposed residues is significantly less accessible to the environment than the other two.

3. *Carbonic Anhydrase*

The third specific system which we shall describe, that of carbonic anhydrase, will serve to illustrate two problems. First, it is probably the enzyme with the most complicated CD spectrum that has been subjected to detailed analysis; second, it has been the object of a detailed study of the effect of binding of an inhibitor on the CD spectrum.

The CD spectra of a number of isozymes of carbonic anhydrase from various sources have been reported by several authors (*12, 13, 23, 46, 101*). All are similar in their general complexity; the small differences in

98. G. H. Beaven and W. B. Gratzer, *BBA* **168**, 456 (1968).
99. T. T. Herskovits and M. Laskowski, Jr., *JBC* **235**, PC56 (1960).
100. M. J. Gorbunoff, *Biochemistry* **6**, 1606 (1967).
101. M. J. Gorbunoff, *ABB* **138**, 684 (1970).

their amino acid compositions, however, do result in significant variations. This suggests that the various enzymes—whether human, bovine, or monkey—have similar general structural features but differ mainly in details of tertiary structure. The CD spectra of human carbonic anhydrase B and monkey enzyme C at pH 7.5 are shown in Fig. 11. The positions of the various extrema are indicated in the figure. Beychok *et al.* (*23*) carried out a detailed study of the pH dependence of the ORD and CD of the human enzymes, while Rosenberg (*28*) reported

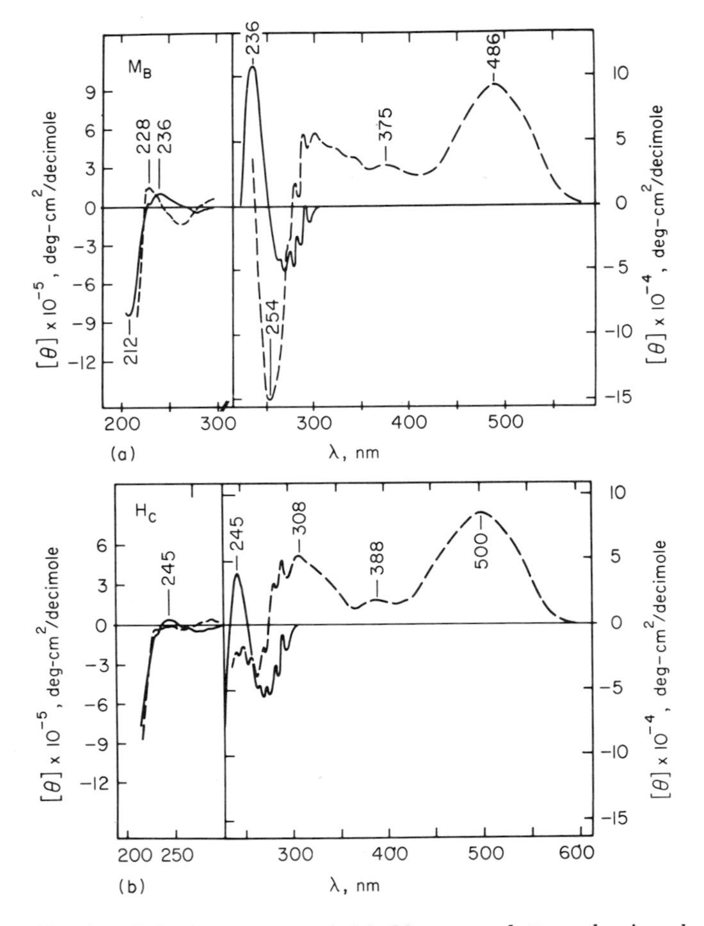

Fig. 11. Circular dichroism spectra of (a) *Macaca mulatta* carbonic anhydrase B (M_B) and (b) human carbonic anhydrase C (H_C) and their 1:1 complexes with Neoprontosil. (——) Isozyme and (- - -) isozyme plus equimolar Neoprontosil. All solutions contained 5×10^{-5} *M* enzyme ($+5 \times 10^{-5}$ *M* azosulfonamide in the case of the complexes), 0.025 *M* tris, pH 7.5, 25°. The ellipticities are calculated per decimole of protein, rather than mean amino acid residue [from Coleman (*46*)].

similar ORD results for the human and bovine species. It was found that the extrema located at 275 and 236 nm shift to higher wavelengths when the pH is raised to above 10, suggesting that those bands correspond to tyrosine transitions. The higher wavelength extrema are certainly due to tryptophans (*70*); carbonic anhydrase contains no disulfide bridges. In the lower wavelength region, the spectra are difficult to interpret. They are in general weak and the band positions do not correspond to those of any given ordered structure, such as the α helix or β conformation. In this spectral region, as well, it appears that aromatic side chain transitions predominate. Beychok *et al.* (*23*) have examined the effects on CD of both acid and alkaline denaturation. In both cases, many of the spectral details, in particular, in the region between 250 and 300 nm, disappear suggesting the loosening of enzyme structure, with a strong enhancement of the freedom of rotation of the aromatic residues (*70*). The CD and ORD spectra, in the region below 240 nm, would suggest an increase in the α-helical contents on acid denaturation, in agreement with the infrared spectra (*11*). The ORD spectra at alkaline pH values are consistent with the presence of significant amounts of α-helical structure. Since the infrared spectra of the bovine enzyme at neutral pH and at pH 12 are essentially identical, one might speculate that the high pH CD and ORD spectra describe a secondary structure not very different from that of the native enzyme. X-ray crystallographic studies at 5.5 Å resolution (*102*) indicate that human carbonic anhydrase C may contain as much as 30% α helix; examination of the manner in which the polypeptide chain is folded into the globular structure indicates some regions in which the chains run parallel to each other, suggesting the presence of β conformation.

Coleman (*46*) has carried out a particularly interesting CD study on the interactions of various carbonic anhydrases with a nonoptically active inhibitor, 4′-sulfamylphenyl-2-azo-7-acetamido-1-hydroxynaph-

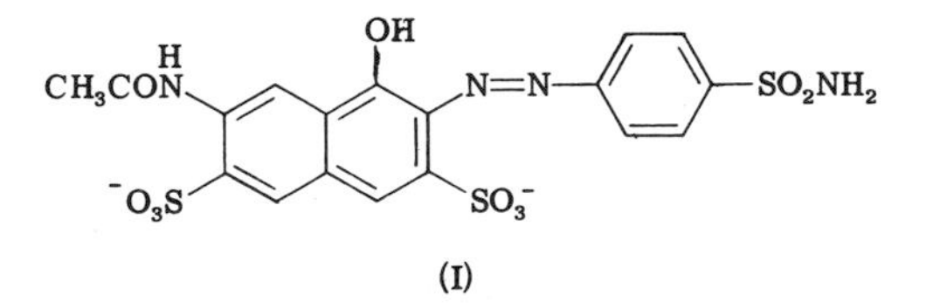

(I)

thalene-3,6-disulfonate (Neoprontosil). This compound is known to complex with the various carbonic anhydrases and to inhibit their activity.

102. K. Fridborg, K. K. Kannan, A. Liljas, J. Lundkin, B. Strandberg, R. Strandberg, B. Tilander, and G. Wirén, *JMB* **25,** 505 (1967).

X-ray studies have shown that the inhibitor is inserted into a crevice leading to the zinc atom (*102*).

The CD spectra of monkey carbonic anhydrase B and human enzyme C in the free and complexed states are shown in Fig. 11. (The spectra of the human and monkey enzymes C are essentially identical.) It is evident that complexing to the enzyme of Neoprontosil, which by itself is devoid of optical activity, results in dramatic CD spectral changes. It is particularly striking that the positions of the new bands are different for the two enzymes, which are very similar in amino acid composition. This suggests that there are considerable structural differences around the active sites of the two isozymes; particularly striking is the presence of a large negative band at 254 nm in the monkey carbonic anhydrase B complex and its absence in human carbonic anhydrase C complex. These structural differences have been found by Coleman (*46*) to be paralleled by differences in the esterase activities of the enzymes. Therefore, it appears that the azosulfonamide acts as a highly sensitive probe of the active-site topography.

The azosulfonamide is optically inactive. Its combination with the protein, however, induces strong optical rotatory power which is of a magnitude typical of inherently dissymmetric chromophores. It seems, therefore, that the inhibitor becomes highly immobilized in an asymmetric environment within the protein molecule. Coleman (*46*) concludes that the induced optical activity must result from coupling with transitions of the protein or from dissymmetric electrostatic perturbations produced by charged groups or dipoles of the protein. In fact, the observed CD spectral shifts are consistent with the location of this molecule within a hydrophobic cavity (*103, 104*) in the active center of carbonic anhydrase.

The extreme complexity of overlapping transitions which result from the interaction of the inhibitor with the enzyme is demonstrated in Fig. 12. Here the CD spectrum between 235 and 500 nm of the bovine enzyme B and its complex have been decomposed into sets of Gaussian bands. Very strikingly, seven bands are needed to account for the enzyme spectrum, while an additional six bands arise on binding of the inhibitor. A particularly interesting feature is that four weak bands in the aromatic region of the enzyme spectrum (at 267, 275, 283, and 296 nm) are found intact in the spectrum of the complex. Therefore, the groups responsible for these transitions remain unaffected by the presence of the sulfonamide. These bands correspond to both tyrosine and tryptophan

103. M. E. Riepe and J. H. Wang, *JACS* **89,** 4229 (1967).
104. R. F. Chen and J. C. Kernohan, *JBC* **242,** 5813 (1967).

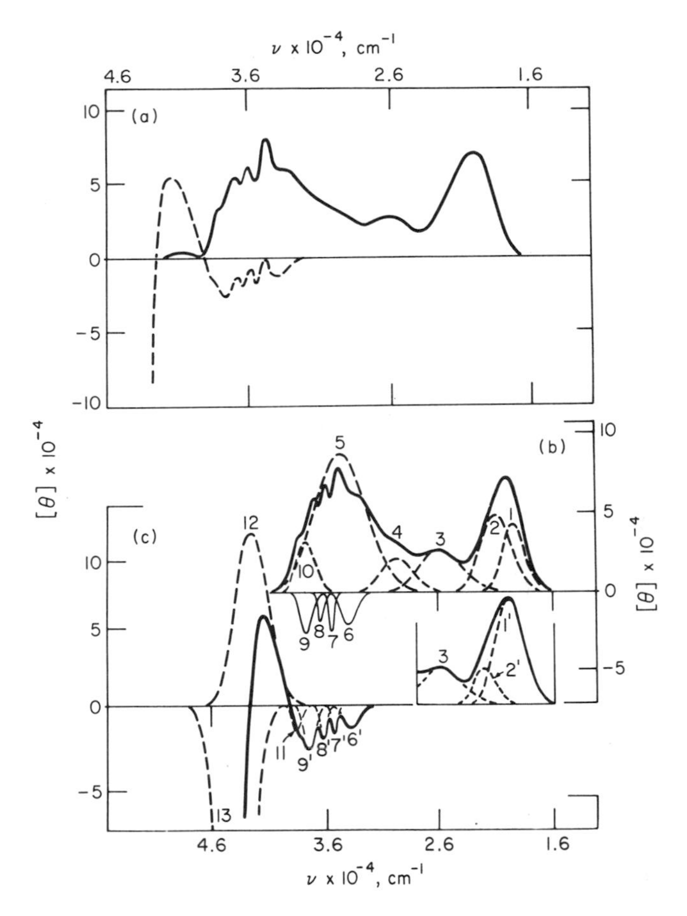

FIG. 12. Circular dichroism spectra of bovine carbonic anhydrase B and its 1:1 complex with Neoprontosil. Resolution into a set of overlapping Gaussian ellipticity bands: (a) (——) observed CD spectrum of the complex; (- - -) observed CD spectrum of the native enzyme. (b) (complex) (——) Resultant envelope for the ten Gaussian bands labeled from 1 to 10; (---) bands attributable to the sulfonamide; (——) bands attributable to the protein. The subset curve, bands 1′ and 2′, shows an alternate fitting of the lowest energy region. (c) (native enzyme) (——) Resultant envelope for the seven Gaussian bands labeled on the figure. In (b) and (c), the envelopes correspond within the limit of error to the experimental CD curves. The ellipticities are calculated per decimole of protein [from Coleman (*46*)].

transitions which, by inference, should be located in a region of the enzyme that is distant from the active center. A further observation of great interest is the fact that the CD spectra of the enzyme–inhibitor complexes are different whether the metal atom on the active site is Zn(II) or Co(II) (*46*). This may result from differences in binding affinity or from slight conformational differences in the active center in the presence of the different metals.

4. *Pyrocatechase*

An example of the effect of a bound metal atom on the CD spectrum of an enzyme is found in the recent study on pyrocatechase (*105*). Pyrocatechase is an enzyme which requires ferric iron for enzymic activity. Nakazawa *et al.* (*105*) examined the effects of iron binding and interaction with substrate on the CD spectrum of this enzyme, with pertinent results shown in Fig. 13. It is very striking that in both cases the qualitative features of the spectrum in the aromatic region remain essentially unchanged but the band intensities change drastically. The spectrum in the near ultraviolet is positive with five maxima at 292, 285, 278, 265, and 255 nm. Removal of iron causes a decrease in all the bands; addition of catechol results in a decrease in band intensity above 265 nm and an enhancement in positive ellipticity below that wavelength. While removal of iron seems to affect equivalently all the aromatic transitions without any changes in secondary structure, binding of the substrate is reflected differently in the bands typical of tryptophan (285 and 292 nm) and those characteristic of the tyrosine phenolic ring. Thus, when substrate is bound, positive CD extrema appear at 253 and 275 nm; these are known to be related to tyrosines, and in the case of this particular enzyme the 255-nm band has been unequivocally related to a phenolic transition. Catechol itself is optically inactive (*105*). It is interesting to speculate whether its immobilization in the active site of the enzyme does not impart to it optical activity, as a result of effects similar to those found with carbonic anhydrase. The iron enzyme also displays dichroism in the visible region, with a negative maximum at 327 nm and a broad negative band centered at 500 nm. Removal of iron eliminates these bands totally with no changes in secondary structure. When the iron-containing enzyme was heated, the ellipticities at 500, 327, 292, and 285 nm changed linearly with a decrease in enzymic activity, again without any changes in secondary structure, as evidenced by a lack of change of the CD spectrum below 235 nm. In the absence of iron, no effect could be detected upon addition of the substrate. It would appear, therefore, that the binding both of the metal ion and of the substrate induce subtle conformational changes in the active site which are directly related to the enzymic activity. In neither case is a major change in secondary structure necessary.

105. A. Nakazawa, T. Nakazawa, S. Kotani, M. Nozaki, and O. Hayashi, *JBC* **244,** 1527 (1969).

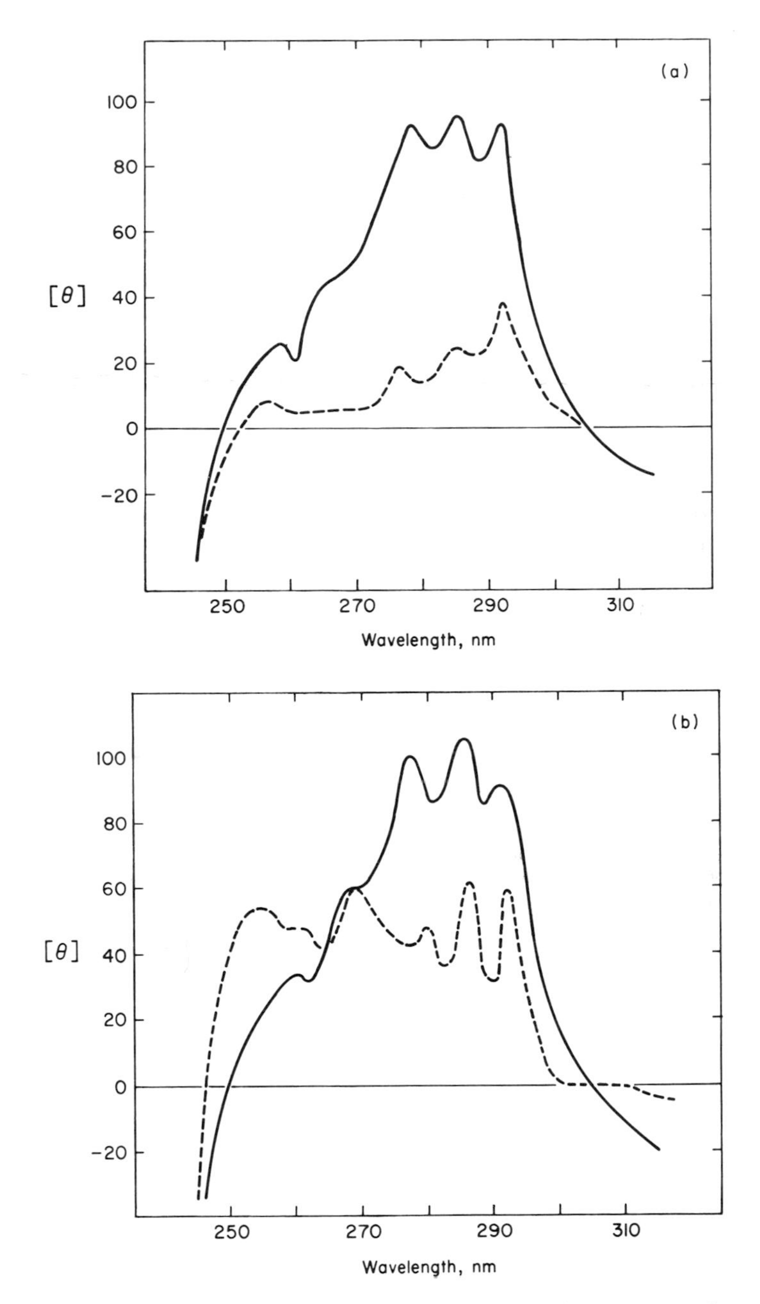

FIG. 13. (a) Circular dichroism of the holo- (——) and apopyrocatechase (---) in 0.05 M tris Cl, pH 8.0 [from Nakazawa *et al.* (*105*)]. (b) The effect of pH on the ultraviolet circular dichroism of pyrocatechase: (——) before the addition of catechol; (---) after the addition of catechol [from Nakazawa *et al.* (*105*)].

D. Conclusion

These examples of CD studies on particular enzymes indicate the extent to which this technique is capable at present of probing the structure of enzymes in solution. It is evident that, while much qualitative information of a general nature may be obtained, little actual structural information follows. At best, it is possible to determine which types of residues in the three-dimensional structure of the enzyme are close to the active site or are affected by interaction of the enzyme with substrate or inhibitor as well as by changes in conformation. In all cases, where more detailed conclusions could be drawn, information had to be drawn from other sources. For example, the detailed interpretation of structural changes responsible for the changes in the CD spectrum between 220 and 235 nm when chymotrypsinogen is activated to chymotrypsin could certainly not have been carried out on the basis of CD data alone. In fact, CD turned out to be an excellent probe for following the changes; these experiments, however, were done in the blind from the point of view of CD since the exact nature of the transitions followed is still unknown.

VI. Ultraviolet Difference Spectroscopy

Because of the strong overlapping of bands ultraviolet absorption spectroscopy of proteins in solution can yield only gross general structural information. A much more sensitive way of detecting small discrete changes in the environment of particular absorbing groups is afforded by the differential technique. In this approach, two solutions of the same protein in different environments are compared directly at identical concentrations; since one solution serves as the sample and the other as the blank, all the common features of their spectra cancel out and only those transitions which had been displaced with respect to each other because of alterations in the environment are manifested by positive or negative differential bands. This enzyme structural probe is based on the fact that the transition moments which give rise to the absorption bands of chromophores are strong functions of their interactions with the environment. Electrical transition moments may be perturbed by a number of factors; these include the polarity of the environment, the polarizability of the solvent, the presence of charges or dipoles in the vicinity of the chromophore, and the formation of hydrogen bonds (for example, in the case of the tyrosine phenolic radical). Such interactions lead to shifts both in the positions and intensities of the bands. For example, Yanari and Bovey (*106*) have shown that the spectra of indole, phenol, and benzene undergo a red shift when the refractive index of the solvent is increased. In general, transfer of aromatic side chains from an aqueous to a hydrophobic medium results in a red shift and an exaltation of the absorption. The general principles of difference spectroscopy and the types of information that can be obtained have been adequately reviewed (*107–109a*) and will not be discussed in detail here.

Ultraviolet difference spectroscopy of enzymes has gained its widest applications in studies aimed at the probing of general conformational changes which occur, for example, when the pH is varied over a wide range (acid difference spectra), and in investigations of the dissociation behavior of ionizable chromophores, in particular of tyrosines. While the first type of application is outside the scope of this chapter, the second will be discussed in a later section. Suffice it only to demonstrate here by a few specific examples the type and degree of insight that may be obtained into changes in the environment of particular chromophoric groups which occur when a zymogen is activated to an enzyme or when the enzyme interacts with ligands such as substrates or inhibitors.

The specific residues which have been used most extensively as markers of conformational details are tryptophans and tyrosines. When their environment is perturbed, these residues yield difference absorption bands in the near ultraviolet region between 270 and 300 nm. Specifically, the unionized tyrosine difference absorption bands occur at about 278 and 287 nm, while the tryptophan bands are at about 284 and 292 nm, with possibly a weak band close to 275 nm (*107*, *110*, *111*). Thus, below the pH of tyrosine ionization, the presence of a difference band at 292 nm indicates the involvement of tryptophans, while similar bands at lower wavelengths may result from either tyrosines or tryptophans and require for their complete interpretation independent information such as knowledge of the exact amino acid composition or of the manner in which the difference spectra are affected by chemical modifications.

106. S. Yanari and F. A. Bovey, *JBC* **235**, 2818 (1960).
107. D. B. Wetlaufer, *Advan. Protein Chem.* **17**, 303 (1962).
108. M. Laskowski, Jr., *Federation Proc.* **25**, 20 (1966).
109. T. T. Herskovits, "Methods in Enzymology," Vol. 11, p. 748, 1967.
109a. J. W. Donovan, *in* "Physical Principles and Techniques of Protein Chemistry" (S. J. Leach, ed.), Vol. 1, p. 102. Academic Press, New York, 1969.
110. E. J. Williams and M. Laskowski, Jr., *JBC* **240**, 3580 (1965).
111. T. T. Herskovits and Sr. M. Sorensen, *Biochemistry* **7**, 2523 (1968).

A. Solvent Perturbation Spectroscopy

A very ingenious approach to the detailed probing of enzyme topography is that of solvent perturbation difference spectroscopy developed by Laskowski and his school (*99, 108, 110, 112, 113*). The method is based on the very simple principle that spectral bands undergo small shifts when the polarity of the environment is changed. Thus, if an absorbing group is present in the surface of the enzyme in contact with aqueous solvent, addition of a nonaqueous component to the solution results in slight alterations in its spectrum since its transition moments are affected by interactions with the new solvent component. To the contrary, if the same group is present inside the hydrophobic interior of the protein molecule, addition of the new solvent component, if it does not alter the conformation of the protein, should not affect its absorption spectrum. Operationally, a difference spectrum is measured between a solution of the protein in aqueous medium containing 20% of the inert perturbant (e.g., ethylene glycol, glycerol, sucrose, methanol, and polyethylene glycol) and the same solution without the perturbant. A difference spectrum such as the one shown in Fig. 14 is obtained (*114*). The existence of such a difference spectrum indicates that absorbing groups are in sufficient contact with solvent for their transition moments to be perturbed by the additive. The extent of group exposure is then calculated from the ratio of the differential peak intensities with those obtained with a fully unfolded protein. For example, the 292-nm peak present in the chymotrypsinogen difference spectrum, when polyethylene glycol is used as perturbant, has a height ($\Delta\epsilon/\epsilon_{282}$) of 0.026 and that of the hypothetical denatured protein in water has a height of 0.063 (*114*). Since the last value corresponds to perturbation of all absorbing groups, i.e., contact of all groups with solvent, the ratio of these two numbers yields the extent of exposure of the pertinent groups, namely, $0.026/0.063 = 0.41$. Since the peak at 292 nm corresponds to tryptophan residues (*110*), this means that chymotrypsinogen at pH 4.4 has 41% of its tryptophans exposed to solvent if contact with polyethylene glycol is used as criterion.

If perturbants of different sizes are used, this technique becomes a probe of the surface topography of the enzyme molecules. Laskowski

112. M. Laskowski, Jr., R. H. Cramer, T. T. Herskovits, and C. C. Wang, *Federation Proc.* **19**, 343 (1960).

113. T. T. Herskovits and M. Laskowski, Jr., *JBC* **237**, 2481 (1962).

114. E. J. Williams, T. T. Herskovits, and M. Laskowski, Jr., *JBC* **240**, 3574 (1965).

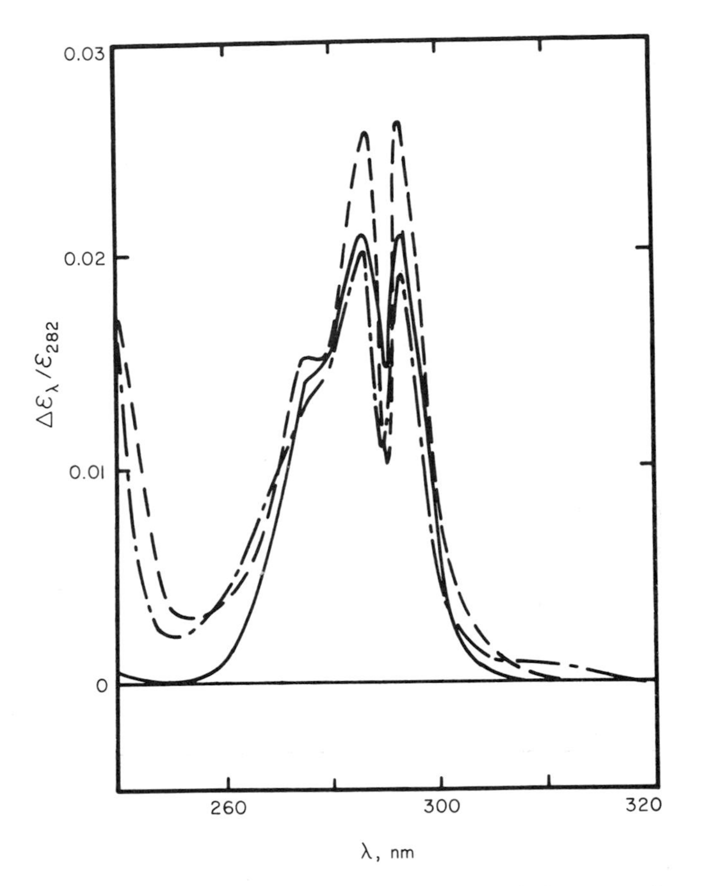

FIG. 14. The solvent perturbation difference spectra of chymotrypsinogen (---), α-chymotrypsin (——), and diisopropylphosphoryl-α-chymotrypsin (-—-) in 0.05 *M* sodium acetate–acetic acid buffer, pH 4.4, 0.1 *M* NaCl. The perturbant is 20% (v/v) polyethylene glycol; protein concentration = 0.13% [from Williams *et al.* (*114*)].

(*108*) has discussed two distinct situations that must be considered. The first concerns the presence of crevices on the protein surface, the second deals with partially exposed groups. Crevices may be detected by using perturbants with various van der Waals radii. The principle is illustrated in Fig. 15. It is seen that if some of the chromophores are located within narrow crevices such chromophores can be perturbed only by perturbants small enough to enter the crevice; they are not affected and hence remain undetected by larger ones. On the other hand, chromophores located on the surface or within wide crevices can be detected by a greater variety of perturbants of various sizes.

Another situation which must be considered is the existence of the

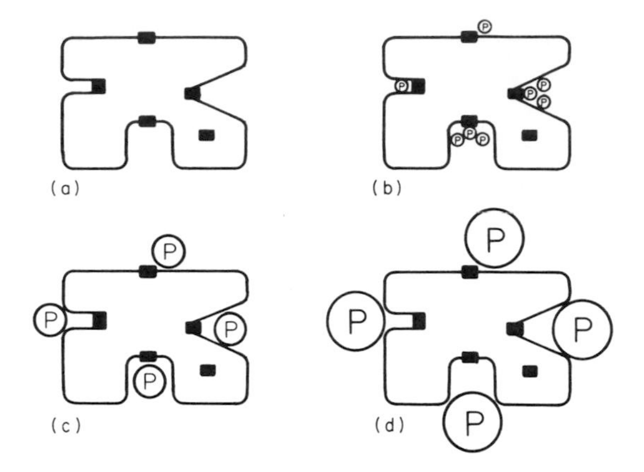

FIG. 15. Schematic diagram showing the use of perturbants of different sizes to detect crevices within proteins. The black rectangles denote chromophores; the circles, P, denote perturbant molecules. (a) Reference protein solution; (b) smallest perturbant approaches all four exposed residues, giving 80% exposure; (c) intermediate size perturbant approaches three residues, giving 60% exposure; and (d) large perturbant approaches one residue, giving 20% exposure.

chromophoric groups in a partially exposed state. The detection of 50% exposure of a certain type of group with a variety of perturbants may have several interpretations. This may mean that all the groups are half-exposed (e.g., with only one side in contact with solvent), that half of the groups are fully exposed and half are completely buried, or that there is a combination of degrees of exposure which average out to 50%. Resolution of this question usually requires the recourse to other experiments such as chemical modification. For example, in the case of chymotrypsinogen, oxidation of three tryptophans with *N*-bromosuccinimide eliminated the solvent perturbation difference spectrum and no further change was evident on oxidation of the remaining residues (*110*). It seems reasonable to conclude then that in chymotrypsinogen three tryptophans are fully exposed and five are completely buried. In ribonuclease, solvent perturbation indicates that about half of the tyrosine residues are exposed to solvent (*112*, *115*); chemical modification studies with cyanuric fluoride (*100*) showed that the three normally titrating residues are accessible to different extents to this reagent; in fact, two groups react quite readily, while the third group requires some loosening of the enzyme. This conclusion was confirmed by detailed solvent perturbation spectroscopy studies (*115*) in which the spectral data could be accounted for best in terms of two fully exposed and two partially buried tyrosine residues.

115. T. T. Herskovits and M. Laskowski, Jr., *JBC* **243**, 2123 (1968).

At present, the topography of a number of enzymes has been probed using this technique. It has been established, as a general rule, that even though they are hydrophobic in character at least part of the tyrosines and tryptophans are exposed to solvent, the values usually clustering about 50%. In more detailed studies on some enzymes it has been possible to show that the tyrosines and tryptophans of glutamic dehydrogenase (*116*) and the tryptophans of α-lactalbumin (*117*) are located within crevices. On the other hand, none of the tyrosines of ribonuclease (*112*, *115*) nor of the tryptophans of chymotrypsinogen (*114*) and lysozyme (*114*, *118*) is located within crevices. In pepsin (*119*) and aldolase (*119*, *120*) both types of residues exhibit various degrees of exposure but are not located within crevices. Other residues within enzymes that have been examined with a single perturbant only, thus precluding conclusions concerning the existence of crevices, include the tyrosines of taka-amylase (*121*), lactate dehydrogenase (*122*), glyceraldehyde-3-phosphate dehydrogenase (*122*), and human carbonic anhydrase B (*123*) and the tryptophans of carboxypeptidase (*124*), human carbonic anhydrase B (*123*) and trypsin (*125*). In the last case, the interesting observation was made that complexing of trypsin with pancreatic inhibitor results in little change in the degree of exposure of the tryptophans of either protein, indicating that these groups are removed from the site of enzyme–inhibitor complexing (*125*).

B. Typical Cases

1. *Staphylococcal Nuclease*

A particularly striking example of the effect of ligand binding on group exposure is found in staphylococcal nuclease (*126*). This enzyme contains seven tyrosines and one tryptophan. Cuatrecases *et al.* (*126*) examined its solvent perturbation difference spectra in the absence and

116. D. Cross and H. Fisher, *Biochemistry* **5,** 880 (1966).
117. M. J. Kronman and L. G. Holmes, *Biochemistry* **4,** 526 (1965).
118. K. Hayashi, T. Imoto, and M. Funatsu, *J. Biochem.* (*Tokyo*) **55,** 516 (1964).
119. T. T. Herskovits and Sr. M. Sorensen, *Biochemistry* **7,** 2533 (1968).
120. J. W. Donovan, *Biochemistry* **3,** 67 (1964).
121. T. Friedman and C. J. Epstein, *JBC* **242,** 5131 (1967).
122. S. Libor, E. Elodi, and Z. Nagy, *BBA* **110,** 484 (1965).
123. R. H. Stellwagen, L. Riddiford, and J. T. Edsall, *Abstr., 145th ACS Meeting, New York, 1963* p. 66C.
124. H. Fujioka and R. Imahori, *J. Biochem.* (*Tokyo*) **53,** 244 (1963).
125. H. Edelhoch and R. F. Steiner, *JBC* **240,** 2877 (1965).
126. P. Cuatrecases, S. Fuchs, and C. B. Anfinsen, *JBC* **242,** 4759 (1967).

presence of the competitive inhibitor, deoxythymidine 3′,5′-diphosphate. Using glycerol and ethylene glycol as perturbants, it was found that in the free enzyme 4–5 tyrosines are exposed to solvent, while the tryptophan is completely buried. Binding of the inhibitor leads to dramatic spectral changes: Only one tyrosine remains exposed to solvent, while the inhibitor itself gives no solvent perturbation spectrum. These results indicate that attachment of the diphosphate to the active site of the enzyme causes it to become surrounded with protein and to be taken out of contact with the solvent. This must be accompanied either by a conformational change which buries the tyrosines or by direct interaction between the inhibitor and 3–4 tyrosines in such a way that the latter become shielded from contact with solvent.

2. *Chymotrypsin*

The various ramifications of difference spectroscopy have been used quite successfully to probe the structural events that accompany the zymogen → enzyme activation and enzyme–substrate interactions in the chymotrypsin system.

The activation of chymotrypsinogen to chymotrypsin generates the difference spectrum shown in Fig. 16 (*87*, *126a*). The appearance of

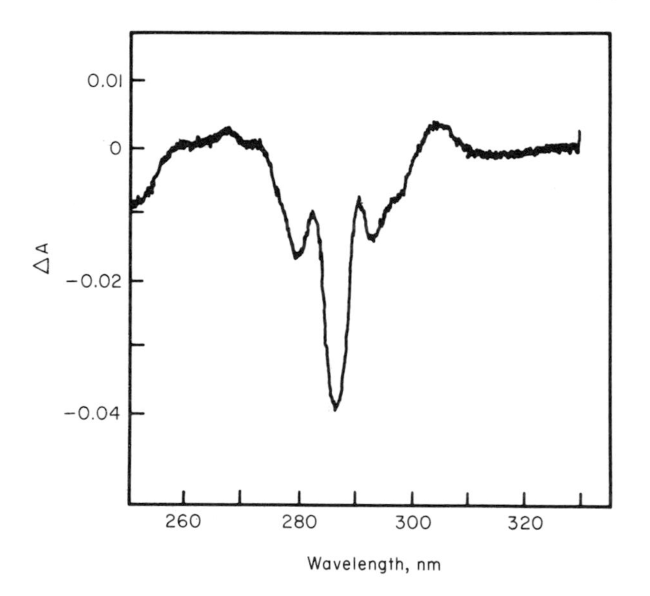

FIG. 16. Difference spectrum obtained on fast activation of chymotrypsinogen to chymotrypsin [from Benmouyal and Trowbridge (*87*)].

126a. C. H. Chervenka, *BBA* **26,** 222 (1957); **31,** 85 (1959).

differential absorption bands at 292, 287, and 278 nm indicates that both tryptophan and tyrosine transitions are affected; thus, the environment of both types of residues changes on activation.

Williams *et al.* (*114*) and Oppenheimer *et al.* (*127*) have carried out solvent perturbation difference spectroscopy studies on chymotrypsinogen, α-chymotrypsin, and diisopropylphosphoryl-α-chymotrypsin. The difference spectra obtained when polyethylene glycol is used as perturbant are shown on Fig. 14. Both tyrosine and tryptophan residues are exposed to solvent in all three proteins. The results obtained with several perturbants of different sizes for the 292-nm difference band characteristic of tryptophans are summarized in Table IV. Williams *et al.* (*114*) have concluded from these data that the exposure of tryptophans is not changed significantly in the activation process. The observed decrease in exposure to large, strongly interacting perturbants, dimethyl sulfoxide and polyethylene glycol, suggests a small local conformational change. The differences between DIP-α-chymotrypsin and α-chymotrypsin are very small. They are of a sufficient magnitude, however, to suggest that subtle changes in the environment of a tryptophan residue occur on diisopropylphosphorylation of Ser 195. In fact, Oppenheimer *et al.* (*127*) concluded that the differences in the spectra of α-chymotrypsin and the DIP derivative result from a tryptophyl residue which is buried in both the enzyme and the derivative.

That the environment of tyrosines is altered on activation is evident from the activation difference spectrum (Fig. 16) (*87*), which has a

TABLE IV
RELATIVE EXPOSURE OF TRYPTOPHYL RESIDUES OF CHYMOTRYPSINOGEN, α-CHYMOTRYPSIN, AND DIISOPROPYLPHOSPHORYL-α-CHYMOTRYPSIN (*114*)

	$(\Delta\epsilon/\epsilon)_{rel}$[a]		
Perturbant 20% (v/v)	Chymotrypsinogen	α-Chymotrypsin	Diisopropylphosphoryl-α-chymotrypsin
Sucrose	0.49 (0.55)	0.47 (0.52)	0.44 (0.48)
Glycerol	0.43 (0.35)	0.40 (0.33)	0.38 (0.31)
Ethylene glycol	0.44 (0.35)	0.41 (0.33)	0.43 (0.35)
Dimethyl sulfoxide	0.39 (0.31)	0.30 (0.23)	0.28 (0.22)
Polyethylene glycol	0.41 (0.30)	0.33 (0.24)	0.30 (0.22)

[a] Exposure estimated relative to the hypothetical denatured proteins in water; values in parentheses are relative to the model mixture analogs.

127. H. L. Oppenheimer, J. Mercouroff, and G. P. Hess, *BBA* 71, 78 (1963).

strong band at 287 nm. This is supported by the difference between solvent perturbation difference spectra of the zymogen and the enzyme (Fig. 14) (*114*), as well as by the results of chemical modification experiments (*71*). In the case of the tyrosines, however, none of the spectral data can be interpreted directly in terms of changes in tyrosine exposure. On activation, Tyr 146 becomes the C-terminal group of the B chain. This should have two effects on the chromophore (*71*): first, it should become more available to solvent and thus be immersed in an environment of different polarity; second, the appearance of an α-carboxyl group on the same residue should alter significantly the charge environment of the phenolic ring. Wetlaufer *et al.* (*128*) have shown that ionization of the α-carboxyl group of an aromatic amino acid perturbs its spectrum. The change in tyrosine absorption on removal of residues 147 and 148 may, therefore, reflect in great part the introduction of a new negative charge in the vicinity of the side chain of Tyr 146.

Difference spectroscopy has also been used to probe chymotrypsin–substrate interactions. Benmouyal and Trowbridge (*87*) determined the difference spectra between α-chymotrypsin and its complexes with several ligands. The difference spectra obtained when *p*-toluenesulfonylarginine methyl ester and acetyl phenylalanine ethyl ester were used as ligands are very complicated and quite different in character. Since neither ligand absorbs at wavelengths above 275 nm, the difference spectra must reflect changes in the environment of enzyme chromophoric residues. Since both the positions and signs of the difference bands are different for the two substrates, it would appear that their binding to the enzyme induces nonidentical local structural perturbations. Similar effects have been observed by Burr and Koshland (*129*) in experiments on the binding of substrate to α-chymotrypsin which had been chemically modified by the incorporation of a chromophoric reporter group (see Chapter 7 by Koshland, Volume I).

An elegant application of solvent perturbation spectroscopy to enzyme–substrate interaction is the study of Mercouroff and Hess (*130*) on *trans*-cinnamoyl-chymotrypsin (CIN-CT). The difference spectrum between CIN-CT and α-chymotrypsin at pH 2.0, 21°C, gives a peak with a maximum at 290 nm. At 51°C, the same difference spectrum peaks at 281 nm, i.e., at the same positions as that of the model com-

128. D. B. Wetlaufer, J. T. Edsall, and B. R. Hollingworth, *JBC* **233,** 1421 (1958).

129. M. Burr and D. E. Koshland, Jr., *Proc. Natl. Acad. Sci. U. S.* **52,** 1017 (1964).

130. J. Mercouroff and G. P. Hess, *BBRC* **11,** 283 (1963).

pound *O*-CIN-*N*-acetyl serineamide, in which the cinnamoyl group is exposed to solvent. This spectral shift suggests that in the enzyme–substrate compound the CIN group is not in contact with solvent at room temperature, becoming exposed only after thermal unfolding. This conclusion was confirmed by solvent perturbation spectroscopy experiments. When 20% ethylene glycol or 20% glycerol were used as perturbants, the solvent perturbation difference spectrum of native CIN-CT was identical with that of native α-chymotrypsin; thus, the cinnamoyl group was not coming in contact with solvent. Disruption of the tertiary structure of the CIN–enzyme compound by peptic hydrolysis was accompanied by the generation of a solvent perturbation difference spectrum, i.e., by the exposure to solvent of the bound substrate. These observations indicate that binding of the cinnamoyl group to the active site of the enzyme results in its complete burial within the native enzyme. Had it been present either in a crevice or in a partly exposed state, the small perturbants used would have generated a typical cinnamoyl difference spectrum. The removal from contact with solvent of the substrate chromophoric group when it is bound to the enzyme is consistent with the occurrence of a local conformational change. Such a structural change had been detected previously by Havsteen and co-workers (*131, 132*) who found by optical rotatory dispersion and difference spectroscopy large differences between the thermodynamic parameters of structural transitions of the enzyme and enzyme–substrate compounds, when they probed the systems α-chymotrypsin, monoacetyl-α-chymotrypsin, and diisopropylphosphoryl-α-chymotrypsin.

Comparison of the results obtained with two probing techniques (circular dichroism and difference spectroscopy) permits more detailed conclusions to be drawn on the state of particular amino acids within the enzyme. When chymotrypsinogen is activated to chymotrypsin or when the enzyme reacts with substrates, the environments of both tyrosine and tryptophan residues change. Yet, the activation process does not affect the near ultraviolet tryptophan circular dichroism spectrum (*71, 72*) (see Fig. 7). Chymotrypsinogen has eight tryptophans. *N*-Bromosuccinimide oxidation (*110*) permits their grouping into two classes: three residues are fully exposed and five are essentially buried. The optical activity must therefore reside fully in the buried tryptophans. These must be more constrained in space than the exposed ones, whose freedom of motion may be sufficient to mutually cancel CD absorption bands (*70*).

131. B. H. Havsteen and G. P. Hess, *JACS* **85**, 791 (1963).
132. B. Havsteen, B. Labouesse, and G. P. Hess, *JACS* **85**, 796 (1963).

VII. Fluorescence Spectroscopy

One of the most useful and versatile probes of enzyme conformation in solution is fluorescence spectroscopy. Since its application by Weber (*133–138*) to protein studies, this technique has undergone extensive development and has found wide uses in the probing of various structural problems which arise in enzyme chemistry. The scope and applicability of this technique has been discussed in a number of reviews (*133, 136, 139–143a*). Briefly, the types of structural information which may be obtained include the extent of flexibility of a protein and the hydrodynamic volume of its subunits; the degree of polarity of particular regions, e.g., the active site of an enzyme; the distances between specific groups on a protein as well as complexed ligands and specific interactions which occur between them; the quarternary structure of a protein and the rate of very rapid conformational transitions. The various kinds of structural information are obtained from determinations of the excitation and emission spectra of fluorescent groups attached to proteins, in particular, from the band positions and intensities under various circumstances, the transfer of energy from one emitter to another, the effect of the protein environment on quantum yield, including quenching of fluorescence by vicinal groups such as ionizable amino acid residues, and the rate of decay of the parallel and perpendicular components of fluorescent radiation.

The fluorescent chromophores which are used in enzyme studies are of three types: intrinsic, i.e., the tryptophan, tyrosine and phenylalanine

133. G. Weber, *in* "Light and Life" (W. D. McElroy and B. Glass, eds.), p. 82. Johns Hopkins Press, Baltimore, Maryland, 1961.

134. G. Weber, *BJ* **51,** 145 (1952).

135. G. Weber, *BJ* **51,** 155 (1952).

136. G. Weber, *Advan. Protein. Chem.* **8,** 415 (1953).

137. G. Weber and D. J. R. Lawrence, *BJ* **56,** 31P (1954).

138. G. Weber and L. B. Young, *JBC* **239,** 1415 (1964).

139. S. V. Konev, "Fluorescence and Phosphorescence of Proteins and Nucleic Acids." Plenum Press, New York, 1967.

140. S. Udenfriend, "Fluorecence Assay in Biology and Medicine." Academic Press, New York, 1965.

141. L. Brand, "Methods in Enzymology," Vol. 11, p. 776, 1967.

142. L. Stryer, *Science* **162,** 526 (1968).

143. R. F. Chen, *in* "Fluorescence" (G. G. Guilbault, ed.), p. 443. Marcel Dekker, New York, 1967.

143a. R. F. Chen, H. Edelhoch, and R. F. Steiner, *in* "Physical Principles and Techniques of Protein Chemistry" (S. J. Leach, ed.), Part A, p. 171. Academic Press, New York, 1969.

residues of the amino acid sequence; cofactors such as some coenzymes; and, extrinsic chromophores. The last type includes a number of compounds which are artificially incorporated (either by covalent or by non-covalent bonds) into specific binding sites on the enzyme. While the first two types of chromophoric groups are very useful as probes of structural features (*133*, *139*, *141*), they are not always found in the three-dimensional structure in locations or orientations which are favorable for the scrutiny of particular structural details. To overcome this problem, Weber (*135*) introduced the technique of binding polycyclic aromatic compounds to the protein in question and using their fluorescence properties as structural probes. If such a group is part of an enzyme, substrate or inhibitor, it can be incorporated specifically into the active site and used to great advantage to obtain information on the local structure and interactions within the site.

The determination of the polarity of a locus on an enzyme is based on the knowledge that a change in polarity of the immediate environment of the emitting group results in a displacement of the position of the fluorescence emission spectral band. For example, Stryer (*144*) has shown that the position of the emission band of 1-anilino-8-naphthalene sulfonate

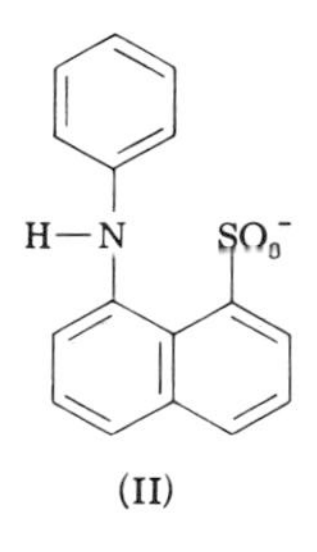

(II)

(ANS) was shifted to longer wavelengths as the polarity of solvent increased from that of *n*-octanol ($\lambda_{max} = 464$ nm) to that of ethylene glycol ($\lambda_{max} = 484$ nm). This red shift in band position is accompanied by a decrease in the quantum yield. The proximity of groups within an enzyme is measured by the efficiency of energy transfer between them. It is found that when the transfer is from the singlet state of one group to the singlet state of another group, the distances that may be measured are between 10 and 65 Å. Förster (*145*) treated this case quantitatively. Regarding singlet–singlet transfer as occurring via the interaction of dipoles on the donor and acceptor, he defined a distance R_0 (at which

144. L. Stryer, *JMB* **13**, 482 (1965).
145. T. Förster, *Ann. Physik,* [6]**2**, 55 (1948).

transfer is 50% efficient) in terms of spectroscopic and geometric parameters. Thus,

$$R_0 = 9.79 \times 10^3 (J n^{-4} \kappa^2 Q)^{1/6} \tag{1}$$

where J is the integral of overlap of the emission spectrum of the donor group and the absorption spectrum of the acceptor group, n is the refractive index of the medium, κ^2 is a factor defined by the mutual orientation of the two dipoles (it may vary between 0 and 4; for random orientation, its value is 2/3), and Q is the quantum yield of the donor. If the observed efficiency of transfer between two groups is e, then the distance between them, r, is $r = (e^{-1} - 1)^{1/6} R_0$. Recently, these relations have been examined critically (*146, 147*). In a very elegant study, Stryer and Haugland (*148*) tested Förster's theory on a system which consisted of a donor (α-naphthyl) and an acceptor (dansyl) group coupled to two ends of polyproline chains of known different lengths and were able to confirm fully the r^{-6} dependence. Systems in which the energy transfer is triplet–singlet, act over similar distances (*149*). In the case of triplet–triplet energy transfer, the effective distance is much shorter, less than about 12 Å, since the mechanism is by an electron–exchange interaction. Thus, with the proper donor and acceptor groups, fluorescence spectroscopy provides a set of rulers for measuring specific distances on enzyme molecules (*142*).

Information on the rigidity of structure, molecular shape, and presence of mobile structural segments of an enzyme molecule results from measurements of the decay of fluorescence polarization (*134–136, 150, 151*). The enzyme solution is illuminated with polarized radiation, exciting those molecules in which the absorbing groups are oriented preferentially in a direction parallel to the plane of polarization of the light, and the intensities of fluorescence parallel and perpendicular to the incident light are measured as a function of time. This gives the anisotropy parameter $A(t)$ at any moment t:

$$A(t) = \frac{I_{\parallel}(t) + 2I_{\perp}(t)}{I_{\parallel}(t) - I_{\perp}(t)} \tag{2}$$

where $I_{\parallel}(t)$ and $I_{\perp}(t)$ are the components of fluorescence parallel and perpendicular to the plane of polarization of the incident energy at any

146. J. Eisinger, *Biochemistry* **8,** 3902 (1969).
147. J. Eisinger, B. Feuer, and A. A. Lamola, *Biochemistry* **8,** 3908 (1969).
148. L. Stryer and R. P. Haugland, *Proc. Natl. Acad. Sci. U. S.* **58,** 719 (1967).
149. R. G. Bennett, R. P. Schwenker, and R. E. Kellogg, *J. Chem. Phys.* **41,** 3040 (1964).
150. A. Jablonski, *Z. Naturforsch.* **16a,** 1 (1961).
151. R. P. Haugland and L. Stryer, *in* "Conformation of Biopolymers" (G. N. Ramachandran, ed.), Vol. 1, p. 321. Academic Press, New York, 1967.

time t. This is related to the relaxation times ρ_i of various mobile components of the molecule by

$$A(t) = A_0 \sum_i \alpha_i \exp(-3t/\rho_i) \quad (3)$$

where A_0 is the anisotropy at time zero, t is the time, and α_i is the fractional contribution of relaxation i. For a spherical molecule, the relaxation time ρ_0 is related to molecular volume V since $\rho_0 = 3\eta V/kT$, where η is the solution viscosity, k is Boltzmann's constant, and T is the thermodynamic temperature. If the exciting radiation consists of an almost instantaneous flash (of duration of the order of a few nanoseconds), the anisotropy $A(t)$ can be measured directly since the duration of the fluorescence decay is usually of the order of 10–100 nanoseconds (nsec, 10^{-9} sec) (*142, 151–156*).

Not only is fluorescence a highly versatile technique, capable of yielding a wealth of structural information, but also it has the further advantages that measurements can be made at very low protein concentration. Difference spectroscopy normally requires working at total optical densities of the order of two; in the case of fluorescence, it is possible to work at enzyme concentrations an order of magnitude lower. Furthermore, the fluorescence spectrum reveals only the state of the excited molecules so that one sees the net effect of interaction of the molecules with the light. There is no need for any corrections resulting from overlap with incident radiation nor with spectra of molecules in the ground state. Let us examine now with the aid of specific examples the type of structural information that may be obtained.

A. Typical Cases

1. *Chymotrypsin*

The structure of chymotrypsin has been probed by McClure and Edelman (*157–159*) and Stryer and co-workers (*142, 151, 160, 161*). McClure

152. L. Hundley, T. Coburn, E. Garwin, and L. Stryer, *Rev. Sci. Instr.* **38,** 488 (1967).
153. H. Lami, G. Pfeffer, and G. Laustriat, *J. Phys.* (*Paris*) **27,** 398 (1966).
154. P. Wahl, *BBA* **175,** 55 (1969).
155. P. Wahl, *Compt. Rend.* **260,** 6891 (1965); **263,** 1525 (1966).
156. P. Wahl and S. N. Timasheff, *Biochemistry* **8,** 2945 (1969).
157. W. O. McClure and G. M. Edelman, *Biochemistry* **5,** 1908 (1966).
158. W. O. McClure and G. M. Edelman, *Biochemistry* **6,** 559 (1967).
159. W. O. McClure and G. M. Edelman, *Biochemistry* **6,** 567 (1967).

and Edelman examined the activation of chymotrypsinogen to chymotrypsin using as probe 2-*p*-toluidinylnaphthalene-6-sulfonate (TNS)

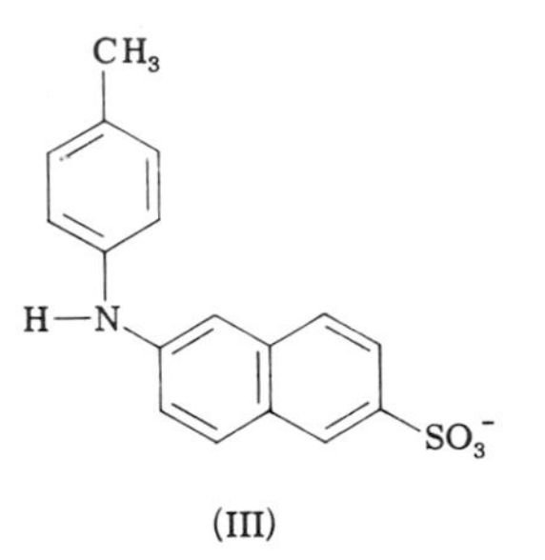

(III)

which is a specific fluorescent marker of hydrophobic regions (*157*). When solutions of chymotrypsinogen containing TNS were activated by the addition of trypsin ("fast" activation leading to δ-chymotrypsin), the fluorescence intensity increased 20-fold at a rate parallel with the appearance of chymotrypsin activity (*159*). It is apparent that on activation a structural change occurs, in which the proper hydrophobic environment for TNS is generated. Fluorescence of TNS-chymotrypsin was inhibited by the binding of substrate analogs, while binding of TNS to the enzyme inhibited noncompetitively the hydrolysis of acetyl-L-tyrosine ethyl ester, suggesting the presence of a hydrophobic binding site which is not part of the active site of the enzyme (*158*). When the pH was varied between 7 and 9, the fluorescence intensity decreased sharply, suggesting that the ionizing groups which influence fluorescence may be the same as those which control enzyme activity (*158*). It is interesting to recall that circular dichroism studies reveal a conformational change related to the activation of the enzyme, as well as to the state of ionization of key groups such as Ile 16 and Asp 194 (*72, 77, 79, 80*).

Haugland and Stryer (*151*) probed the polarity and flexibility of the active site of α-chymotrypsin. By reacting *p*-nitrophenyl anthranilate

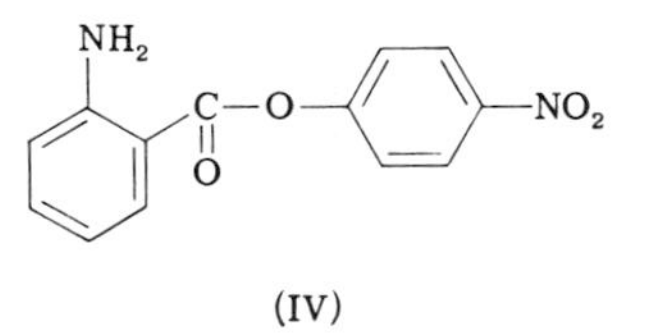

(IV)

160. W. C. Galley and L. Stryer, *Biochemistry* **8**, 1831 (1969).
161. W. C. Galley and L. Stryer, *Proc. Natl. Acad. Sci. U. S.* **60**, 108 (1968).

with the enzyme, they formed a fluorescent anthraniloyl derivative, in which the probing group was attached to the active site at Ser 195, and which was highly stable at neutral pH. The anthraniloyl group is particularly favorable for this kind of study. Its absorption and emission maxima at 342 and 422 nm are remote from those of the protein aromatic residues, which makes it possible to excite the probing group exclusively. By applying a nanosecond pulse (*142, 151*), the decay curves of the parallel and perpendicular components of the fluorescence were established, resulting in the dependence of the anisotropy $A(t)$ on time, as shown in Fig. 17. When the shape and finite duration of the exciting flash

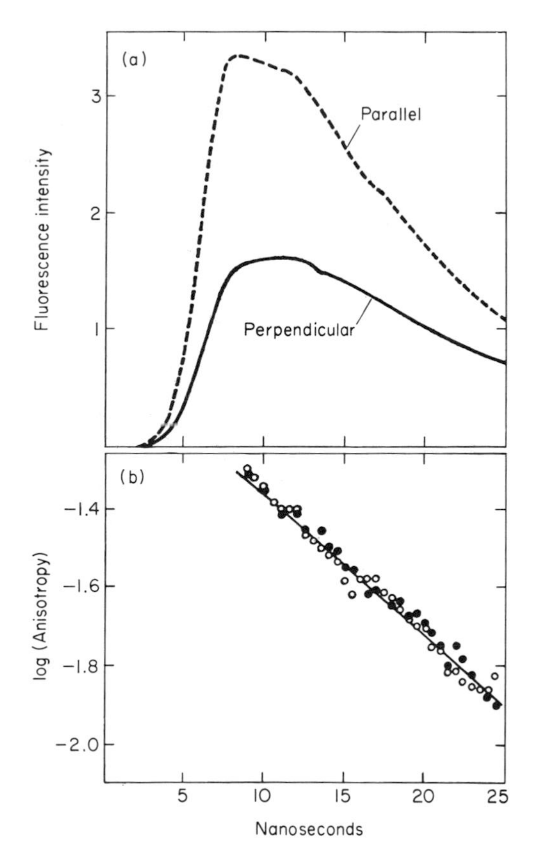

FIG. 17. Nanosecond fluorescence polarization of anthraniloyl chymotrypsin in 0.1 M phosphate buffer, pH 6.8, at 22°C. (a) Intensity of the parallel and perpendicularly polarized components of the fluorescence as a function of time. (b) Logarithm of the emission anisotropy as a function of time. The slope (when corrected for the finite duration of the light pulse) yields a rotational relaxation time of 52 nsec for the anthraniloyl chromophore [from Stryer (*142*)].

were accounted for by means of a convolution analysis (*152*, *154*, *156*), a rotational relaxation time of 52 nsec was obtained for the anthraniloyl group when it is attached to the enzyme. This value of ρ permitted the immediate elimination of molecular models in which the active site region of the enzyme is highly flexible. Indeed the expected value of ρ_0 at 20°C for an anhydrous spherical protein molecule with a molecular weight of 25,000 daltons is 22 nsec. Since the experimental value is more than twice that calculated for a rigid sphere, it must be concluded that the anthraniloyl chromophore is rigidly fixed to the active site of the enzyme and has only that rotational mobility which is part of the mobility of the whole enzyme. Furthermore, the high value of the relaxation time indicates that α-chymotrypsin is a hydrated nonspherical molecule, which is perfectly consistent with the known three-dimensional structure (*16*). The emission maximum of the anthraniloyl derivative in 1.0 *M* phosphate buffer of pH 6.8 is at 422 nm, i.e., very close to the position of methyl anthranilate in water (420 nm). Since the enclosure of this group in nonpolar environment results in a blue shift of its emission maximum (for methyl anthranilate the positions are 405 nm in methanol, 390 nm in dioxane, and 380 nm in cyclohexane), Haugland and Stryer (*151*) concluded that the environment of this group, when it is coupled to α-chymotrypsin, is highly polar; furthermore, the observed shifts of the absorption maximum indicate that water is excluded from the binding site. These conclusions are entirely reasonable since the structural region of α-chymotrypsin in the vicinity of the active site contains a number of polar, or even ionized, residues (e.g., Ile 16, Asp 194, Ser 195, and His 57), and it is possible that the anthraniloyl group is located in their vicinity rather than in a hydrophobic part of the cavity.

The same fluorescent probe was used to measure the distance between tryptophan residues and the active site. The excitation spectrum of the anthraniloyl derivative overlaps the emission spectrum of tryptophan residues. By applying the Förster theory (*145*) to this singlet–singlet energy transfer, Haugland and Stryer found a characteristic distance R_0 of 20 Å which, however, cannot be decomposed into distances between individual tryptophans and the active site. A further elaboration of this problem was provided by Galley and Stryer (*161*) who probed this distance by triplet–triplet energy transfer from *m*-acetylbenzenesulfonamide, covalently bonded to Ser 195 of α-chymotrypsin, to tryptophan residues of the enzyme. This system is amenable to such studies because the triplet level of the sulfonamide is higher than that of tryptophan, while the opposite is true of the singlet levels. Excitation of the probing chromophore resulted in no tryptophan emission, showing that none of the tryptophans are in the vicinity of the active site of the enzyme, a conclusion in agreement with the X-ray diffraction results (*16*).

2. *Carbonic Anhydrase*

Another interesting example of the fluorescence probing of an enzyme active site is found in the studies on bovine carbonic anhydrase. Using the same triplet–triplet energy transfer system as with α-chymotrypsin, Galley and Stryer (*161*) attached to the active site *m*-acetylbenzenesulfonamide, a known specific inhibitor. Excitation of the marker at 330 nm, where tryptophan does not absorb, resulted only in tryptophan phosphorescence with an energy transfer efficiency of close to 100%. This means that a tryptophan residue is present at the active site of carbonic anhydrase, or close to it.

In a similar study, Chen and Kernohan (*104*) probed the active site

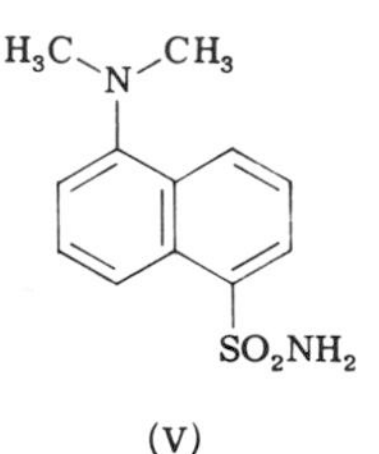

(V)

of bovine carbonic anhydrase B with the inhibitor, 5-dimethylaminonaphthalene-1-sulfonamide (DNSA). In water DNSA has an emission maximum at 580 nm and a quantum yield of 0.055. Incorporation into the enzyme resulted in a blue shift to 468 nm and a dramatic increase of the quantum yield to 0.84. Both changes show that contrary to the case of α-chymotrypsin the binding site is highly nonpolar, while the high quantum yield indicates in addition that the site is shielded from solvent interactions. The efficiency of energy transfer from the seven tryptophans to the single DNSA group of 85% suggests that all the tryptophans are within the critical transfer distance R_0 (*145*) from the probe and probably in favorable mutual orientation. The observed quenching of 73% of tryptophan fluorescence by the DNSA led Chen and Kernohan to conclude that the fluorescence efficiencies of the seven tryptophans are different and that the relative coordinates of the bound DNSA and the tryptophans are such that energy transfer is more probable from the less fluorescent residues. Measurements of fluorescence decay times gave a relaxation time of 30 nsec; therefore, the inhibitor is rigidly attached to the enzyme, which has a symmetrical hydrodynamic structure. These conclusions on the nature of the inhibitor binding site are fully in accord with those drawn from circular dichroism experiments (*46*) and with the known features of the three-dimensional structure deduced from X-ray crystallographic analysis (*102*), which shows that the inhibitor lies within a deep crevice in the enzyme with the sulfonamide group bound to the Zn ion. This has led Chen and Kernohan (*104*) to conclude with the observation that ". . . if the active site and the sulfonamide-binding site are indeed identical, we are left with the mild paradox that water reacts at or near a hydrophobic site." That this "paradox" may be true is further substantiated in the literature (*103*).

3. *Lysozyme*

Details of the interaction of lysozyme with substrate molecules have been examined by Lehrer and Fasman (*162, 163*) who used direct and differential fluorescence spectroscopy methods (*163a*). It is known that three of the six tryptophans of this enzyme are located in the region of the active site (*164, 165*). Binding of substrates or inhibitors results in alterations of the circular dichroism spectrum (*166*), a red shift in the ultraviolet absorption spectrum (*167, 168*) and changes in tryptophan fluorescence (*162, 163, 169–171*). Lehrer and Fasman (*162, 163*) examined the tryptophan fluorescence of lysozyme as a function of pH in the presence of ligands of increasing size. In the native enzyme, the quantum yield is very low, indicating considerable quenching. As the size of the ligand is increased, the intensities of the emission spectra are progressively enhanced and their maxima are shifted to lower wavelengths. The fluorescence intensities of this enzyme, as well as of its complexes with di-*N*-acetyl-D-glucosamine (diNAG) and tri-*N*-acetyl-D-glucosamine (triNAG), are shown in Fig. 18 as a function of pH. It is evident that binding of substrate greatly increases the fluorescence at neutral pH. A decrease in pH leads to a two-step quenching, between pH 7 and 5.5 and between pH 4 and 2, while an increase in pH above 8 also causes a sharp drop in fluorescence intensity. Analysis of difference spectra between lysozyme-triNAG complexes at various pH values (shown in Fig. 19) per-

162. S. S. Lehrer and G. D. Fasman, *BBRC* **23,** 133 (1966).
163. S. S. Lehrer and G. D. Fasman, *JBC* **242,** 4644 (1967).
163a. B. Bablouzian, M. Grourke, and G. D. Fasman, *JBC* **245,** 2081 (1970).
164. L. N. Johnson and D. C. Phillips, *Nature* **206,** 761 (1965).
165. D. C. Phillips, *Sci. Am.* **215,** 78 (1966).
166. A. N. Glazer and N. S. Simmons, *JACS* **88,** 2335 (1966).
167. K. T. Hayashi, G. Imoto, and M. Funatsu, *J. Biochem.* (*Tokyo*) **54,** 381 (1963); **55,** 516 (1964).
168. F. W. Dahlquist, L. Jao, and M. A. Raftery, *Proc. Natl. Acad. Sci. U. S.* **56,** 26 (1966).
169. R. F. Steiner and H. Edelhoch, *Nature* **192,** 873 (1961).
170. M. Shinitzky, V. Grisaro, D. M. Chipman, and N. Sharon, *ABB* **115,** 232 (1966).
171. R. F. Steiner and H. Edelhoch, *BBA* **66,** 341 (1963).

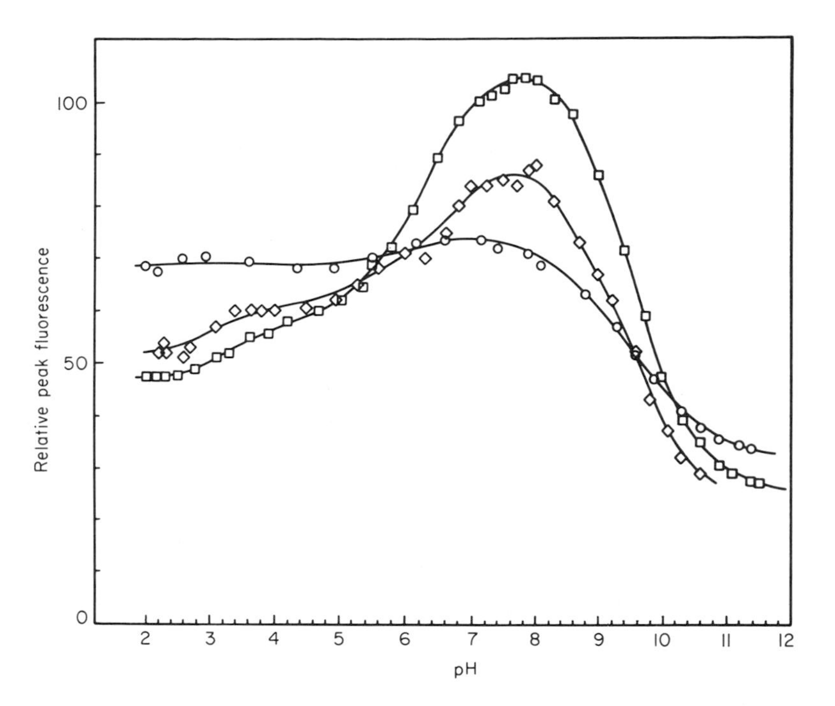

FIG. 18. Fluorescence plotted against pH: (○) lysozyme; (◇) lysozyme + 0.1% di-*N*-acetyl-D-glucosamine; and (□) lysozyme + 0.1% tri-*N*-acetyl-D-glucosamine. Enzyme emission at 345 nm; enzyme + di-*N*-acetyl-D-glucosamine emission at 337 nm; enzyme + tri-*N*-acetyl-D-glucosamine emission at 335 nm. In 0.2 *M* NaCl at 25°. Lysozyme concentration, 0.005%, $OD_{280} \approx 0.10$. Excitation at 280 nm, 1 cm path length cell. Relative fluorescence for same concentration of enzyme [from Lehrer and Fasman (*163*)].

mitted the conclusion that in complex with trimer two tryptophan residues are quenched independently by two abnormal carboxylic groups, one with an unusually high p*K* (~6.3) and the other with a low p*K* (~3.0) (*172, 173*); the high pH quenching is most probably due to energy transfer to a tyrosine residue (*174*) with a p*K* of 9.95. It is known that protonated carboxyls can quench tryptophan fluorescence (*175, 176*). The conclusion on the interaction of two different tryptophans with two carboxyls is based on the formation of different difference spectra for the two pH intervals. The difference spectra were analyzed in the following way. The

172. J. W. Donovan, M. Laskowski, Jr., and H. A. Scheraga, *JACS* **82**, 2154 (1960); **83**, 2686 (1961).
173. C. Tanford, *Advan. Protein Chem.* **17**, 69 (1962).
174. R. Cowgill, *BBA* **94**, 81 (1965).
175. G. D. Fasman, E. Bodenheimer, and A. Pesce, *JBC* **241**, 916 (1966).
176. A. White, *BJ* **71**, 214 (1959).

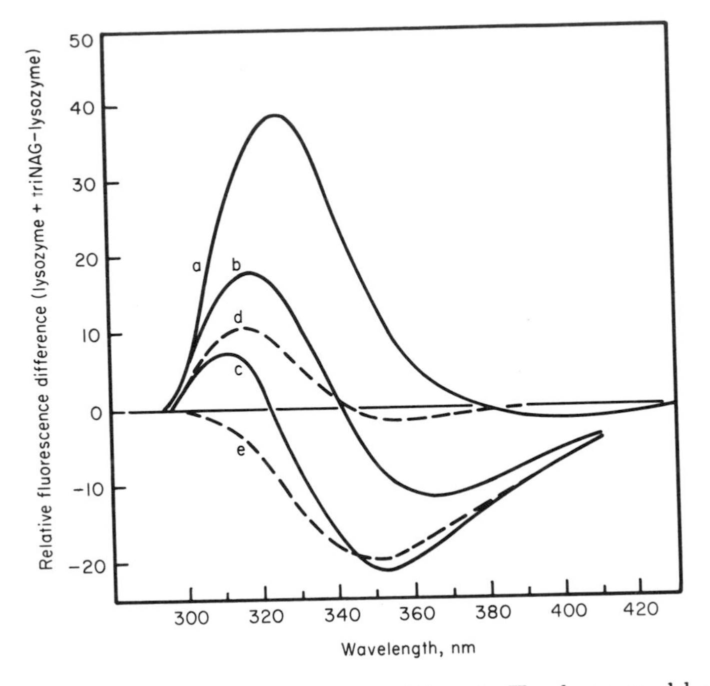

FIG. 19. Fluorescence difference spectra at different pH values caused by binding tri-*N*-acetyl-D-glucosamine (triNAG) (F[lysozyme + tri-*N*-acetyl-D-glucosamine] − F[lysozyme]). Curves a, pH 7.5; b, pH 5.5; c, pH 2.0; and d and e decomposition of c into two contributions [from Lehrer and Fasman (*163*)].

negative contribution to the pH 2.0 difference spectrum (curve e, Fig. 19) is the fluorescence of two residues in the free enzyme which are quenched on formation of the complex; the positive contribution (curve d, Fig. 19) corresponds to the spectral shift of the third residue. The peak of the quenched fluorescence (curve e) is at 345 nm, indicating presence of the corresponding tryptophan in an aqueous environment when it is not complexed with the substrate. Furthermore, the tryptophan residues which are preferentially quenched by ionized tyrosines are located in a nonaqueous environment since the energy transfer occurs in the wavelength region below 320 nm, which is characteristic of the fluorescence of nonpolar-type tryptophans. Comparison of the difference fluorescence data (*163*) with X-ray structural information (*165*) has led Lehrer and Fasman to postulate that the interactions observed in the quenching experiments are between Trp 108 and Glu 35 (high p*K*) and Trp 63 and Asp 101 (low p*K*), although interaction between Trp 63 and Asp 52 is also possible. Imoto and Rupley (*177*) have shown, by a combination of

177. T. Omoto and J. A. Rupley, *Federation Proc.* **27,** 392 (1968).

difference spectroscopy with chemical modification experiments, that the difference spectrum of lysozyme near 300 nm is a function of the interaction between Trp 108 and Glu 35. That the difference spectra of tryptophan-containing proteins are perturbed near 300 nm by the interaction of charges with tryptophan residues has been demonstrated recently by Ananthanarayan and Bigelow (*178*, *179*).

4. *Staphylococcal Nuclease*

An interesting study of active-site probing by various physical techniques has been carried out by Anfinsen and co-workers on the nuclease from *Staphylococcus aureus* (*126*, *180*). Here we shall limit our discussion to the fluorescence analysis (*180*, *181*). It was found that addition of the inhibitor, deoxythymidine 3′,5′-diphosphate (pdTp) greatly suppresses the tyrosine fluorescence without affecting the region of tryptophan emission (*171*). Following their solvent perturbation spectroscopy observation that binding of the nucelotide leads to the burial of all but two tyrosine residues (*126*), the authors carried out solvent perturbation experiments using fluorescence as the criterion of exposure. The fluorescence of the model compound, acetyltyrosinamide, is strongly enhanced by solvents of low polarity (*180*). When the fluorescence intensity of the enzyme and of the enzyme–inhibitor complex was measured in similar fashion as a function of ethanol concentration, the change observed with the free enzyme was almost identical to that found with the model compound, while for the complex the effect was much smaller, confirming the conclusion that complexing buries some tyrosine residues. Again, there was no effect on the single buried tryptophan. Binding of pdTp also resulted in a decrease of the relaxation time from 17.7 to 15.6 nsec, pointing to a small conformational change which increases the hydrodynamic symmetry of the enzyme molecule.

B. Conclusion

As is evident from the specific examples cited, fluorescence is a powerful probe of enzyme structure and enzyme–ligand interactions. This method can be used as a ruler to measure distances between specific

178. V. S. Ananthanarayan and C. C. Bigelow, *Biochemistry* **8**, 3717 (1969).
179. V. S. Ananthanarayan and C. C. Bigelow, *Biochemistry* **8**, 3723 (1969).
180. P. Cuatrecasas, H. Taniuchi, and C. B. Anfinsen, *Brookhaven Symp. Biol.* **21**, 172 (1969).
181. P. Cuatrecasas, H. Edelhoch, and C. B. Anfinsen, *Proc. Natl. Acad. Sci. U. S.* **58**, 2043 (1967).

groups as was done with α-chymotrypsin and carbonic anhydrase: It can detect specific interactions between amino acid residues, as in lysozyme, and it can measure molecular dimensions and rigidity and probe for the exposure to solvent of aromatic residues on an enzyme. Other possible applications are to the specific detection of given secondary structures (*182*), to the quarternary structures of subunit systems (*156*), and to the analysis of substrate and subunit interactions (*183*). As is true of all other solution physical techniques, however, fluorescence in all its ramifications becomes a most powerful probe when used in conjunction with other methods.

VIII. Ionizable Groups

The environment of ionizable groups of enzymes and changes in their state is probed most easily by measurements of their interactions with protons as the concentration of the latter is varied, i.e., by a determination of their titration curves. General acid–base titration curves have been widely used to characterize enzymes, and their detailed discussion is available in a number of reviews (*173, 184–189*); therefore, this presentation will be limited to the examination of some specific groups, with particular emphasis on the identification of ionizable residues involved in interactions.

While the determination of the proton binding capacity of an enzyme as a function of pH gives information on the overall electrostatic state of the molecule and its changes resulting from structural transitions, the ionization of particular groups and their structural environment may be examined quite readily with the help of spectroscopic differences between

182. J. Lynn and G. D. Fasman, *BBRC* **33**, 327 (1968).
183. M. E. Goldberg, S. York, and L. Stryer, *Biochemistry* **7**, 3662 (1968).
184. E. J. Cohn and J. T. Edsall, "Proteins, Amino Acids and Peptides," Chapter 20, p. 444. Reinhold, New York, 1943.
185. J. T. Edsall and J. Wyman, "Biophysical Chemistry," Vol. 1, Chapters 8 and 9. Academic Press, New York, 1958.
186. C. Tanford, "Physical Chemistry of Macromolecules," Chapters 7 and 8. Wiley, New York, 1961.
187. S. A. Rice and M. Nagasawa, "Polyelectrolyte Solutions." Academic Press, New York, 1961.
188. J. Steinhardt and S. Beychok, *in* "The Proteins" (H. Neurath, ed.), 2nd ed., Vol. 2, Chapter 8. Academic Press, New York, 1964.
189. S. N. Timasheff, *in* "Biological Polyelectrolytes" (A. Veis, ed.), p. 1. Marcel Dekker, New York, 1970.

their ionized and unionized states. The titration curves of tyrosine (*190*), histidine (*191*), and cysteine (*192*) residues may be obtained by ultraviolet difference spectroscopy, ionization of carboxylic groups can be followed in the infrared region (*193*), and histidine imidazoles have been titrated as well with nuclear magnetic resonance (NMR) as detector (*194–196*). Nonchromophoric ionizable residues may be titrated spectroscopically if their state of ionization perturbs the absorption or emission behavior of chromophores with which they interact (*197*); a typical example of such a titration has been described above (see Section VII,A,3) in the case of two abnormal carboxyls in lysozyme (one with a high pK, the other with a low pK) which perturb the fluorescence of specific tryptophan residues (*162, 163*).

The titration of chromophoric residues is rendered possible by the displacement of the absorption maxima to higher wavelengths when the pertinent groups become electrostatically charged. The positions of the bands are listed in Table V. Further details may be found in the review literature (*8, 107, 109a, 198*). The normal procedure consists in the determination of absorption spectra of the enzyme measured at various pH values using as reference an enzyme solution of identical concentration at a pH at which the groups in question are either all ionized or

TABLE V
ABSORPTION MAXIMA OF IONIZING CHROMOPHORES

Residue	Chromophore	Uncharged (nm)	Charged (nm)
Tyrosine	Phenolic	193	200
		222	235
		274	295
Histidine	Imidazole	211	237
Cysteine	Sulfhydryl	195	235
Asp, Glu	Carboxyl	1710 cm^{-1} [a]	1560 cm^{-1} [a]

[a] Infrared.

190. J. L. Crammer and A. Neuberger, *BJ* **37,** 302 (1943).
191. J. W. Donovan, *Biochemistry* **4,** 823 (1965).
192. J. W. Donovan, *Biochemistry* **3,** 67 (1964).
193. H. Susi, T. Zell, and S. N. Timasheff, *ABB* **85,** 437 (1959).
194. J. H. Bradbury and H. A. Scheraga, *JACS* **88,** 4240 (1966).
195. D. H. Meadows, J. L. Markley, J. S. Cohen, and O. Jardetzky, *Proc. Natl. Acad. Sci. U. S.* **58,** 1307 (1967).
196. D. H. Meadows, O. Jardetzky, R. M. Epand, H. H. Ruterjans, and H. A. Scheraga, *Proc. Natl. Acad. Sci. U. S.* **60,** 766 (1968).
197. J. W. Donovan, *Biochemistry* **6,** 3918 (1967).
198. G. H. Beaven and E. R. Holiday, *Advan. Protein Chem.* **7,** 319 (1952).

all unionized. The difference in absorption $\Delta\epsilon(\lambda)$ at a given wavelength λ is then directly related to the degree of ionization α since

$$\Delta\epsilon(\lambda) = [\epsilon_i(\lambda) - \epsilon_u(\lambda)]\alpha \tag{4}$$

where $\epsilon_i(\lambda)$ and $\epsilon_u(\lambda)$ are the molar absorption coefficients of the group in question in the ionized and unionized states, respectively.

Spectrophotometric titrations have been used extensively in the identification of normally and abnormally ionizing residues. Normal groups are usually exposed to solvent; they ionize with the intrinsic pK values obtained with the same residues when included in small peptides. Abnormally titrating groups are those which, because of interactions, ionize with apparent pK values which are either too high or too low; their ionization properties are normalized only by conformational changes. The interactions may be of various natures. Ionizable groups may be buried within the nonpolar folds of the protein molecule; such groups must remain neutral, even though the pH of the surrounding solvent corresponds to their ionized state, since the burial of a charge within the interior of a protein introduces a structure destabilizing free energy contribution of 50–100 kcal/mole (*84*), i.e., more than the net free energy of stabilization of a globular protein. If a group is buried in such manner, its pK is raised if it is anionic; the pK is lowered if the group is cationic. Typical examples are the abnormal tyrosines found in many enzymes, e.g., in bovine and human carbonic anhydrase (*24, 199*), as well as the buried histidines of the same enzyme (*199*). A pair of charged groups of opposite sign may be buried as an ion pair; in this case, the pK of the cationic group is raised and that of the anionic group is lowered. A typical example is found in the ion pair formed between Ile 16 and Asp 195 of chymotrypsin and DIP-chymotrypsin (*80, 85*). Ionizable groups may also interact via the formation of hydrogen bonds; the variation of the pK values in this case depends on whether the group is hydrogen bonded in the charged or neutral state and on whether it is the donor or acceptor; a typical example is found in the tyrosine–carboxyl interactions of ribonuclease (*200*). Furthermore, the pK of a group may be shifted significantly if it is located in the immediate vicinity of another charged group while fully in contact with solvent.

Typical titration spectra are shown in Fig. 20 for the ionization of pepsinogen tyrosines (*201*). In Fig. 20a the direct spectra of the zymogen

199. L. M. Riddiford, R. H. Stellwagen, S. Mehta, and J. T. Edsall, *JBC* **240,** 3305 (1965).

200. J. P. Riehm, C. A. Broomfield, and H. A. Scheraga, *Biochemistry* **4,** 760 (1965).

201. G. E. Perlmann, *JBC* **239,** 3762 (1964).

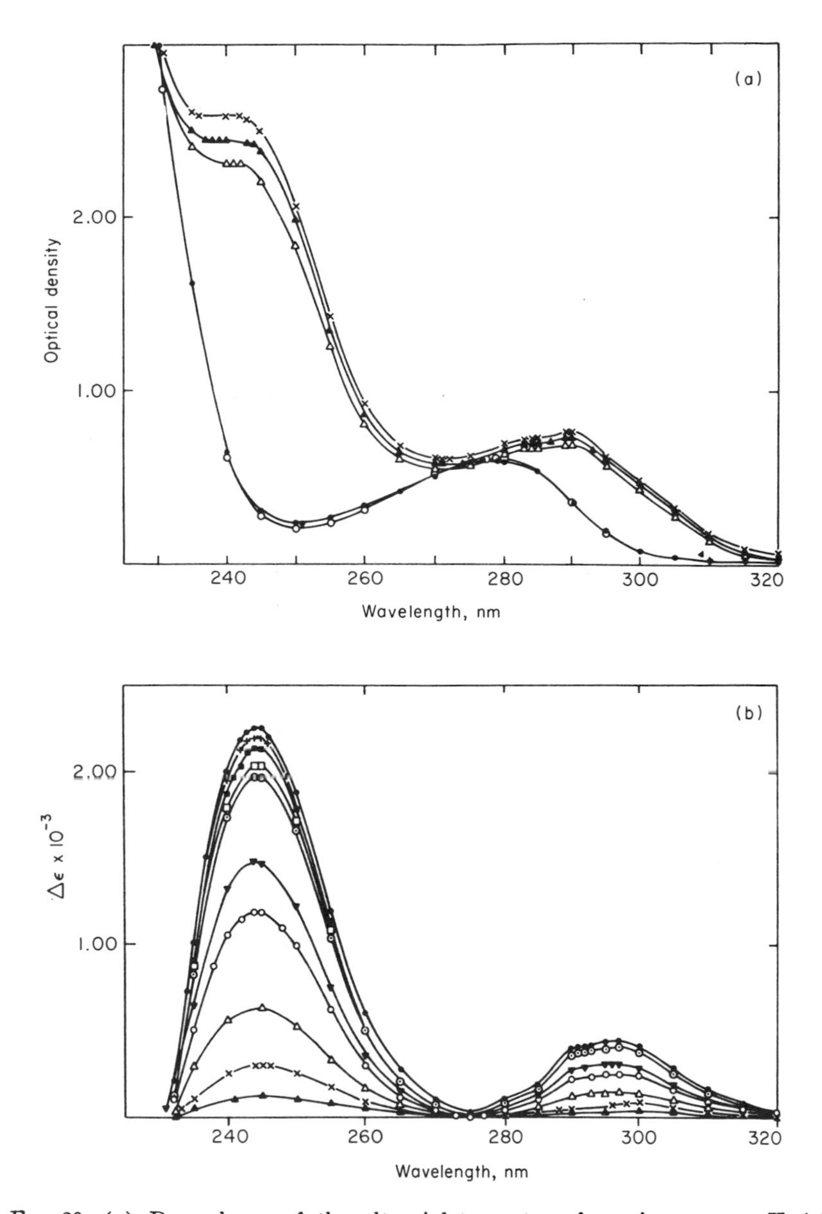

FIG. 20. (a) Dependence of the ultraviolet spectra of pepsinogen on pH (pH values as follows): (○) 8.36, (●) 9.25, (△) 11.75, (▲) 12.12, and (×) 12.56. (b) Dependence of the ultraviolet difference spectra of pepsinogen on pH (pH values as follows): (▲) 9.62, (×) 10.03, (△) 10.29, (○) 10.85, (▼) 11.29, (⊙) 11.75, (□) 11.93, (■) 12.12, (+) 12.38, and (●) 12.60. Both (a) and (b) from Perlmann (*201*).

show that as pH is raised the UV spectrum displays a general red shift and enhancement of absorption above 230 nm: new absorption maxima appear near 243 and 292 nm. The difference between these spectra and that of a standard pepsinogen solution at pH 7.7 is shown in Fig. 20b: Difference bands of increasing intensity with increasing pH are observed at 245 and 295 nm. Application of Eq. (4) to these data results in the titration curve of pepsinogen tyrosines. Comparison of the titration curves of this enzyme precursor obtained in dilute buffers and in the presence of urea revealed that while 15 tyrosine residues are normally available to solvent, two are embedded in the interior of the molecule. This conclusion is consistent both with the analysis of chemical modification studies (*202–204*), which permit a gradation in the state of exposure of these groups, and with observations on conformational transitions which occur in the alkaline range (*205–207*).

In making differential spectroscopic measurements, it is very important to choose for reference a solution in the proper state (*192, 208*). When abnormal groups are titrated, their spectra change because of (1) the ionization of the group and (2) transfer of the group from a state of interaction either with other groups or with the nonpolar interior of the protein to contact with solvent; the second effect frequently results in a perturbation difference spectrum without a change in the state of ionization. The change in pH may also result in the perturbation of the absorption spectra of other chromophores which absorb in the wavelength region of interest; these spectral changes evidently also contribute to the observed overall difference spectrum, and if not properly corrected for they lead to erroneous conclusions on the ionization of the groups of interest. Typical examples are found in the perturbation of tryptophan absorption when the tyrosines of aldolase are titrated (*192*) or the perturbation of tyrosine absorption during the titration of ribonuclease histidines (*191*).

Donovan (*192*) discussed this problem in his study of the ionization of tyrosines and cysteines in aldolase. This enzyme becomes unfolded during the course of the titration; thus, a contribution to the difference spectrum is made by the change in the environment of the chromophores

202. M. J. Gorbunoff, *Biochemistry* **7,** 2547 (1968).

203. G. E. Perlmann, *JBC* **241,** 153 (1966).

204. M. Sokolovsky, J. F. Riordan, and B. L. Vallee, *Biochemistry* **5,** 3582 (1966).

205. G. E. Perlmann, *JMB* **6,** 452 (1961).

206. R. M. Herriot, *J. Gen. Physiol.* **45,** 57 (1962).

207. V. Frattali, R. F. Steiner, and H. Edelhoch, *JBC* **240,** 112 (1965).

208. M. J. Kronman, *in* "Fine Structure of Proteins and Nucleic Acids" (G. D. Fasman and S. N. Timasheff, eds.), p. 271. Marcel Dekker, New York, 1970.

when the pH is raised from neutrality. In this case, it was found that use of a reference solution at acid pH eliminated this complication since this enzyme is highly unfolded in that pH region as well and the unionized chromophores are in contact with solvent. A similar perturbing effect was found in the titration of the ribonuclease histidines (*191*). This system is a particularly good example of the manner in which progressively more detailed information may be obtained on the state of ionizing groups as different probes are applied.

Ribonuclease contains four histidines, two of which, His 12 and His 119, form part of the active site (*17*, *18*, *209–211*). Analysis of its acid binding titration curve showed that all four imidazoles ionize normally with an average intrinsic pK of 6.5 (*212*). Donovan (*191*) examined directly the histidine ionization, using difference ultraviolet spectroscopy which, in ribonuclease, gives a maximum at 237 nm for imidazole ionization. The data could be fitted either with a curve calculated for four groups ionizing with an intrinsic pK of 6.5 or with a curve for two sets of two groups each ionizing with apparent pK values of 6.2 and 6.8, as shown in Fig. 21a. The imidazole difference spectra, however, are weak and they could not be obtained directly. The pH dependence of difference spectra between 250 and 300 nm showed that the ionization of imidazole or α-amino groups perturbed the absorption properties of one or more of the abnormal tyrosines. Since tyrosine absorbs close to 235 nm, a tyrosine perturbation spectrum could be expected in that region as well, i.e., overlapping with the histidine ionization difference spectrum. Using the extent of perturbation above 250 nm, Donovan (*191*) calculated the expected tyrosine difference spectrum at 235 nm; the resulting negative band is shown in Fig. 22. Subtraction of this negative absorption from the experimental curve resulted in the sought imidazole difference spectrum shown by the dashed line of Fig. 22. Since imidazole ionization also perturbed some phenylalanine residues, Donovan (*191*) could conclude that these residues, as well as buried tyrosines, are located in the vicinity of one or more histidines.

Finally, the ribonuclease histidine ionization has been the subject of an NMR study by Meadows *et al.* (*195*, *196*). Using this method, it was possible to follow the ionization of the four groups individually since the chemical shifts of their C_2 and C_4 protons are different and each group gives an independent peak. The results are shown in Fig. 21b.

209. H. A. Scheraga and J. A. Rupley, *Advan. Enzymol.* **24,** 161 (1962).

210. R. Heindrickson, W. H. Stein, A. M. Crestfield, and S. Moore, *JBC* **240,** 2921 (1965).

211. R. E. Cathou and G. G. Hammes, *JACS* **87,** 4674 (1965).

212. C. Tanford and J. D. Hauenstein, *JACS* **78,** 5287 (1956).

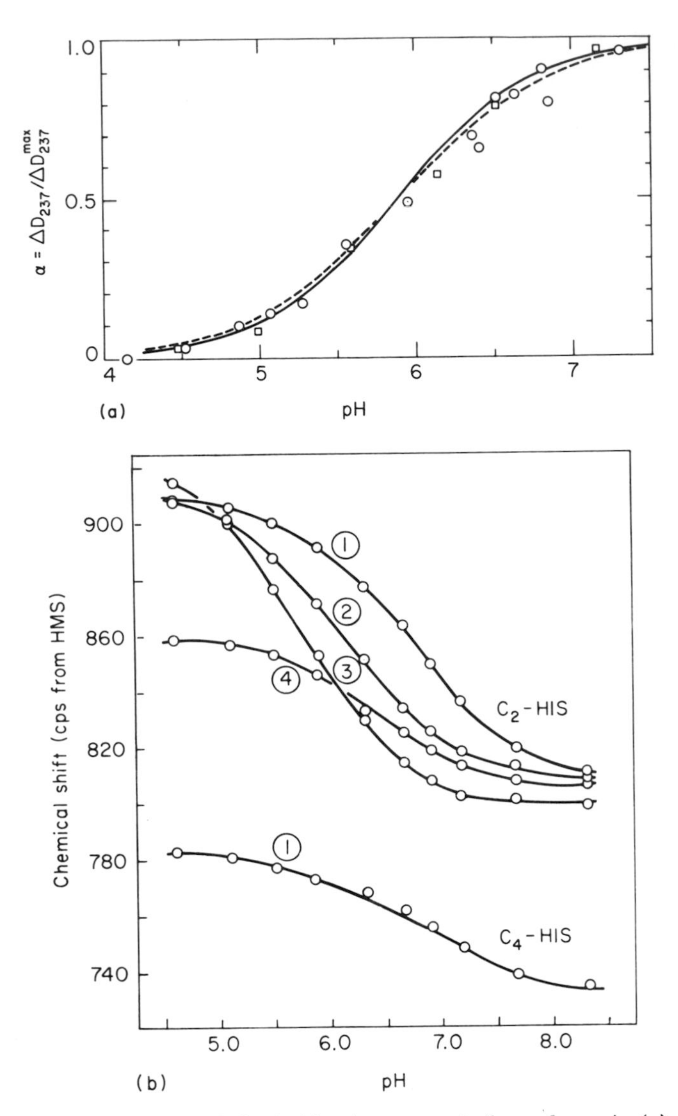

FIG. 21. Titration curves of the imidazole group of ribonuclease A. (a) Measured by ultraviolet difference spectra at 237 nm with a pH 4.1 reference uncorrected for superimposed perturbation of phenolic groups: (○) forward titration and (□) reverse titration from pH 6.5. The solid curve is calculated for four groups with an intrinsic pK of 6.5; the dashed curve is calculated for two groups with an intrinsic pK of 6.8 and two groups with an intrinsic pK of 6.2 [from Donovan (*191*)]. (b) Measured by NMR. The curves represent the variation of the chemical shift [expressed as cycles per second (cps) from a hexamethyldisiloxane (HMS) reference] as a function of pH for four C_2 hydrogen peaks and one C_4 hydrogen peak. The apparent pK's are for 32°C in 0.2 M deuteroacetate buffer [from Meadows *et al.* (*196*)].

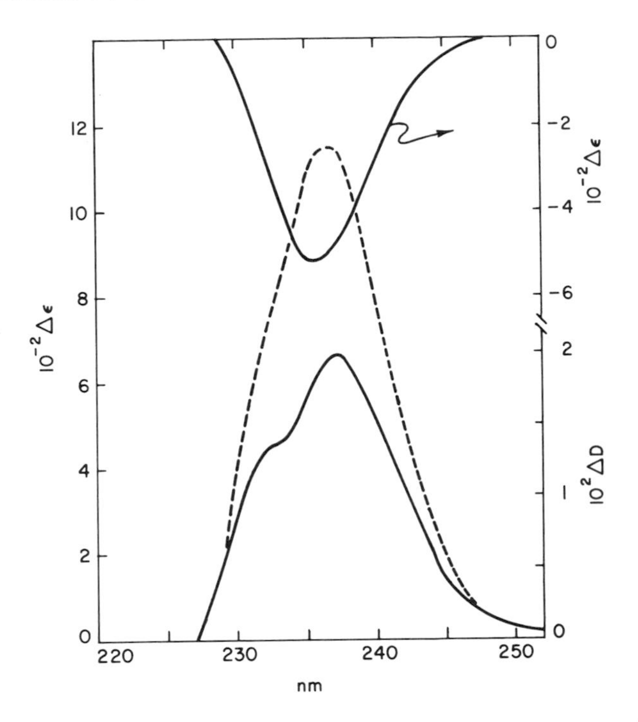

FIG. 22. Calculation of the corrected ultraviolet difference spectrum for the dissociation of protons from the four ribonuclease histidine residues. Lower solid curve: experimental difference spectrum for a pH 7.0 solution against a pH 4.1 reference; upper solid curve: calculated perturbation difference spectrum for affected tyrosines; dashed curve: correct imidazole difference spectrum [from Donovan (*191*)].

Four titration curves are obtained with pK values of (1) 6.7, (2) 6.2, (3) 5.8, and (4) 6.4. Carboxylmethylation of either His 12 or His 119 shifts peaks (2) and (3), indicating the mutual proximity in space of these two residues as can be expected from their presence within the active site; exchange of the His 12 C_2 proton for deuterium eliminated peak (2) permitting its assignment to residue 12; peak (4) was assigned to His 48 because of its anomalous chemical shift and line width, which is consistent with the buried state of this residue. This analysis established that the four histidines of ribonuclease have essentially normal pK values which, however, vary over nearly one pH unit. The pK values are 6.2 (His 12), 6.4 (His 48), 6.7 (His 105), and 5.8 (His 119). Similar histidine titrations have been performed with staphylococcal nuclease and lysozyme (*195*).

The histidine NMR spectra were further used by Roberts *et al.* (*213*)

213. G. C.-K. Roberts, J. Hannah, and O. Jardetzky, *Science* **165,** 504 (1969).

to probe the active site of ribonuclease with the spin-labeled inhibitor, 2,2,6,6-tetramethyl-4-hydroxypiperidin-1-oxyl monophosphate. Increasing amounts of the inhibitor resulted in the broadening of the C_2 peaks of His 12 and of His 119 (to a somewhat smaller extent), while His 105 was not affected, nor probably was His 48. Furthermore, the main aromatic region of the NMR spectrum remained unchanged. These results indicate that the unpaired electron of the inhibitor is relatively close to His 12, somewhat farther from His 119, and distant from the other two histidines as well as from the tyrosines and phenylalanines. Since addition of cytidine nucleotides to ribonuclease affects the NMR spectrum of phenylalanine 120 (*214*), it is evident that the free radical portion of the inhibitor does not lie on the cytidine binding site.

The abnormally titrating carboxyls of lysozyme have been the object of several studies. Donovan *et al.* (*172*) concluded from the titration curve that this enzyme contains at least one carboxyl with an abnormally high pK and one with an abnormally low pK. These groups have also been detected by the quenching of trytophan fluorescence (*162, 163*). Carboxyl ionization may be examined directly by difference infrared spectroscopy (*193*). Protonated carboxyls give a difference band at 1710 cm^{-1}; ionized groups give a band at 1560 cm^{-1}. Difference infrared spectra, determined on lysozyme over a wide pH range, fully confirmed the conclusion on the presence of both types of abnormal carboxyls (*215*). Sulfhydryl ionization gives rise to a difference spectrum, maximal at 235 nm. Using this peak, Donovan (*192*) was able to establish that aldolase contains about six available and 21 buried cysteine residues.

Conformational transitions in enzymes are frequently reflected in a steepening of the dependence of their proton uptake on pH. Typical examples are the titration curves of ribonuclease (*173, 212*) and carbonic anhydrase (*24, 199*) in the alkaline region, where abnormal tyrosines become exposed to solvent, or the titration curve of carbonic anhydrase in the pH region between 5 and 4 (*199*) where seven previously buried histidines are liberated to ionization. Structural transitions of an enzyme may be characterized in detail by examining the pH dependence of the derivative with respect to pH of the proton binding h, i.e., of the buffering capacity of the protein, $\beta' = (\partial h/\partial \text{pH})_{T,p}$ (*216, 217*). At the point of conformational transition a maximum is obtained in this

214. D. H. Meadows and O. Jardetzky, *Proc. Natl. Acad. Sci. U. S.* **61**, 406 (1968).

215. J. A. Rupley and S. N. Timasheff, in preparation.

216. G. E. Perlmann, A. Oplatka, and A. Katchalsky, *JBC* **242**, 5163 (1967).

217. T. M. Birshtein and O. B. Ptitsyn, "Conformation of Macromolecules," Chapter 10. Wiley (Interscience), New York, 1966.

curve. Integration of the area under the curve gives Δh, i.e., the number of groups exposed during the conformational change. The temperature dependence of this displacement along the pH scale of the maximum position of β' results in the entropy and enthalpy of the transition. In this way, Perlmann *et al.* (*216*) probed the conformational transition of pepsinogen. Their results, shown in Fig. 23, indicate that a pH- and temperature-dependent transition is occurring. Integration under the individual curves showed that the three histidine residues of pepsinogen were being liberated to contact with solvent during the transition. The upward shift of the apparent pK of these groups was attributed to ion-pair formation between imidazolium and carboxylate ions in the zymogen.

The above examples have shown specifically how the structural environment of various types of ionizable groups may be probed by making use either of their spectra or of secondary effects which ionization of these groups generate in the enzyme, for example, the perturbation of tyrosine absorption by ionizing histidines in ribonuclease. Generally, probing is most successful when the state of a group is altered between "normal" and "abnormal" in the course of the process. In some specific cases, like the histidines of ribonuclease, the detected small differences in ionization

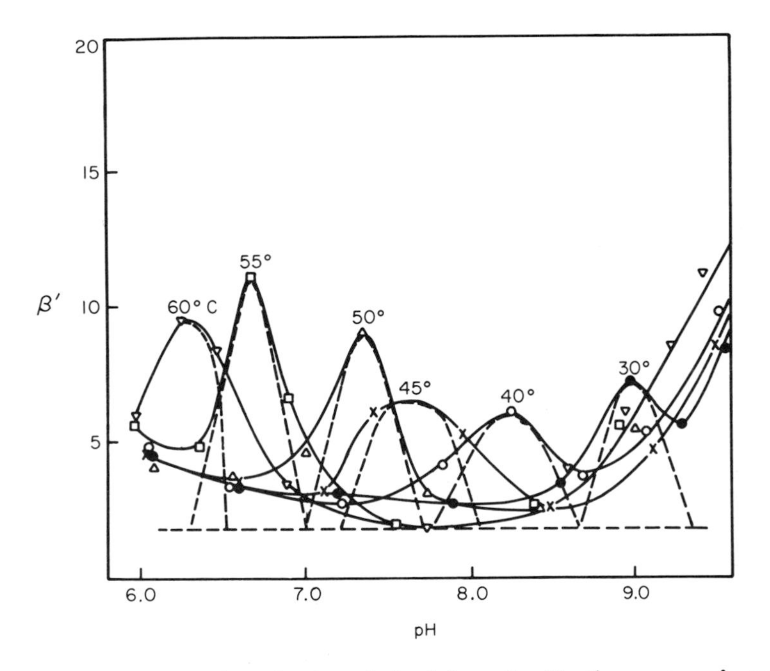

FIG. 23. Buffering capacity β' values derived from the titration curves of pepsinogen at various temperatures. The dashed lines under each curve indicate the pH limits used for the integration [from Perlmann (*216*)].

pK of individual "normal" groups reflect subtle differences in the environment of these groups and of their interactions with the environment. As the structures of enzymes become known in greater detail, it should become even more possible to analyze quantitatively these effects and, conversely, to use their occurrence as probes of detailed structural features.

IX. Geometry and Quarternary Structure

The overall geometry of enzyme molecules in solution may be probed with a variety of techniques, which measure either the hydrodynamic properties of these macromolecules or specific structural parameters. Hydrodynamic methods such as viscosity, flow birefringence, dielectric dispersion, and sedimentation velocity give little structural information beyond a very general description of the flexibility and degree of asymmetry of the molecules. For example, most globular proteins have intrinsic viscosities close to 3–4 cc/g, indicating that they are compact highly symmetrical molecules (*218*). Changes in this quantity are useful probes of conformational alterations without, however, giving any detailed structural information. For example, Conway and Koshland (*219*) have found that the process of binding a third molecule of DPN to tetrameric rabbit glyceraldehyde-3-phosphate dehydrogenase is accompanied by a change in intrinsic viscosity from 2.2 to 3.7 cc/g. Thus, while the structure remains globular, it becomes either more asymmetric or more flexible. Similarly, polarization of fluorescence has been used successfully to examine the general flexibility of some enzymes and the rigidity of quarternary structures formed of subunits (see Section VII); this technique, however, permits the drawing of conclusions concerning the motility or flexibility of specific regions of an enzyme by examining directly the rotational diffusion of the emitting chromophores within those regions.

The best method available at present for probing the geometry of macromolecules in solution is small-angle X-ray scattering. The development of absolute intensity equipment (*220–222*) has made it possible to

218. C. Tanford, "Physical Chemistry of Macromolecules," p. 394. Wiley, New York, 1961.

219. A. Conway and D. E. Koshland, Jr., *Biochemistry* **7**, 4011 (1968).

220. V. Luzzati, J. Witz, and R. Baro, *J. Phys. Phys. Appl.* **24**, 141A (1963).

221. O. Kratky, *Progr. Biophys. Mol. Biol.* **13**, 105 (1963).

222. H. Pessen, T. F. Kumosinski, S. N. Timasheff, R. R. Calhoun, Jr., and

determine directly a number of characteristic structural parameters (*223–227*). These are the molecular weight, the radius of gyration, the surface to volume ratio, the hydrated volume, and the degree of hydration. A combination of these parameters imposes strict limits on the geometry of the enzyme molecule. At higher angles, small-angle X-ray scattering curves display secondary maxima, which are highly sensitive to specific geometric details of the structure and enable further restriction of allowable structures of the macromolecules. Application of this technique to the absolute intensity scale has resulted in reasonable values for molecular parameters. For example, for ribonuclease a radius of gyration of 14.6 Å was measured (*222*); this should be compared with 15.0 Å calculated from the crystallographic data (*17, 18*). For lysozyme a molecular weight of 14,100 daltons was measured (*228*); the chemical value is 14,308 daltons.

Studies devoted to the probing of particular enzyme processes have been extremely limited up to now. Krigbaum and Godwin (*229*) examined the activation of chymotrypsinogen to chymotrypsin and concluded that this reaction does not result in gross structural changes. Their results gave almost identical radii of gyration and surface to volume ratios (S/V) for the zymogen ($R_g = 18.1$ Å, $S/V = 0.160$ Å^{-1}) and the product of slow activation, α-chymotrypsin ($R_g = 18.0$ Å, $S/V = 0.157$ Å^{-1}). The product of rapid activation, however, δ-chymotrypsin was found to be somewhat larger and more symmetrical than the precursor ($R_g = 19.0$ Å, $S/V = 0.146$ Å^{-1})

The technique of small-angle X-ray scattering seems to be particularly well suited to the characterization of the assembly geometry of subunit proteins (*230, 231*). Such a study has been carried out by Sund *et al.* (*232*) on beef liver glutamate dehydrogenase as a function of

---

J. A. Connelly, *in* "Advances in X-Ray Analysis" (B. L. Henke, J. B. Newkirk, and G. R. Mallett, eds.), Vol. 13, p. 618. Plenum Press, New York, 1970.

223. A. Guinier and G. Fournet, "Small Angle Scattering of X-rays." Wiley, New York, 1955.

224. W. W. Beeman, P. Kaesberg, J. W. Andergg, and M. B. Webb, *in* "Handbuch der Physik" (S. Flügge, ed.), Vol. 32, p. 321. Springer, Berlin, 1957.

225. V. Luzzati, *Acta Cryst.* **13,** 939 (1960).

226. S. N. Timasheff, *in* "Electromagnetic Scattering" (M. Kerker, ed.), p. 337. Pergamon Press, Oxford, 1963.

227. S. N. Timasheff, *J. Chem. Educ.* **41,** 314 (1964).

228. V. Luzzati, J. Witz, and A. Nicolaieff, *JMB* **3,** 367 (1961).

229. W. R. Krigbaum and R. W. Godwin, *Biochemistry* **7,** 3126 (1968).

230. J. Witz, S. N. Timasheff, and V. Luzzati, *JACS* **86,** 168 (1964).

231. S. N. Timasheff and R. Townend, *Nature* **203,** 517 (1964).

232. H. Sund, I. Pilz, and M. Herbst, *European J. Biochem.* **7,** 517 (1969)

enzyme concentration. It was found that this enzyme forms end-to-end aggregates which can be described best by the model of elliptical cylinders with long and short cross-section axes of 95 and 76 Å and a mass per unit length of 2340 daltons/Å. The cross section is independent of protein concentration, the length of the aggregates being linearly proportional to the molecular weight. The presence of a secondary maximum at an angle corresponding to a Bragg value of 56 Å indicates that the enzyme molecule is loosely built and possesses voids. With this geometric model it became possible to interpret quantitatively the hydrodynamic parameters of this enzyme (*233*). Furthermore, the knowledge from light scattering that the 2×10^6 daltons molecular species contains eight subunits (*234*) has permitted arrival at a geometry of the monomer which is best described as a globular structure with dimensions of 110, 95, and 76 Å along the three axes of the molecule.

X. Conclusions

In this chapter an attempt has been made to summarize by specific examples the type of structural information that may be obtained when enzymes are probed in solution by physical measurements and to stress the pitfalls that must be avoided. No attempt at an exhaustive coverage of available techniques was made. For example, the use of nuclear magnetic resonance as a conformational tool has been only mentioned (this technique, however, is discussed in some of its ramifications in Chapter 9 by Mildvan, this volume). Methods based on kinetic measurements have been totally excluded. Examination of the degree of success that has been achieved with various primarily spectroscopic techniques leads to the conclusion that in the present state of the art it is possible to gain a considerable amount of structural information by probing specific groups in various ways, as they are involved in interactions, either with other structural elements of the enzyme macromolecule or with ligands such as substrates, inhibitors, or chromophores artificially attached at selected positions. In general, interpretation of data obtained with any single technique is hazardous and extreme caution must be exercised in interpreting spectral and other results in terms of particular structures. The greatest amount of success has been achieved when several techniques were applied to the same question. Normally, the multitechnique approach gives as final result more information than the sum of the

233. H. Sund, *Acta Chem. Scand.* **17,** Suppl. 102 (1963).
234. H. Sund and W. Burchard, *European J. Biochem.* **6,** 202 (1968).

conclusions permitted by each technique individually. This approach allows and requires constant integration of all the pieces of information gleaned from individual experiments. It can be expected that the continuation of this approach with even more sensitive probes will bring its reward in terms of a progressively better and more detailed understanding of the manner in which enzymes and other biological macromolecules perform their various functions.

9

Metals in Enzyme Catalsis

ALBERT S. MILDVAN

I. Introduction

Of the 840 enzymes known in 1964 (*1*), 27% have metals built into their structures, require added metals for activity, or are further activated by metal ions. Since this value must be considered a lower limit, the activation of enzymes by metals is indeed widespread, and a detailed description of the roles of metals in enzyme catalysis would appear to be impossible. Clearly a conceptual or mechanistic approach to this subject, which might condense the vast amount of available information, is required. Such an approach is now possible for two major reasons. First, in the past decade the application of standard enzymological and chemical techniques as well as the specialized physical methods of X-ray crystallography, nuclear magnetic resonance, electron spin resonance, and chemical relaxation, to metal-activated and metalloenzymes has provided detailed insight into the role of metals in certain enzymes. Second, during this same period the application of classical and modern techniques has brought about a renaissance in the field of coordination chemistry, which has amplified our understanding of the mechanisms of formation and decomposition of metal complexes (*2–6*) and of reactions within the coordination spheres of metals (*7*, *8*). Generalities on the mechanisms of inorganic reactions have been formulated (*2–4*) analogous to the principles of organic reaction mechanisms developed earlier (*9*). These principles provide the enzymologist with a conceptual framework upon which to consider the roles of metals in enzymes. Conversely, the study of enzyme mechanisms may lead to new principles of coordination chemistry and thus will enlarge or modify the conceptual framework itself.

1. M. Dixon and E. C. Webb, "Enzymes," 2nd ed., p. 672. Academic Press, New York, 1964.
2. D. D. Brown, C. K. Ingold, and R. S. Nyholm, *JCS* p. 2674 (1953).
3. F. Basolo and R. G. Pearson, "Mechanisms of Inorganic Reactions," pp. 37, 173, and 232. Wiley, New York, 1967.
4. C. Langford and H. Gray, "Ligand Substitution Processes," p. 7. Benjamin, New York, 1965.
5. F. A. Cotton and G. Wilkinson, "Advanced Inorganic Chemisty," p. 172. Wiley (Interscience), New York, 1966.
6. M. Eigen, *Ber. Bunsenges. Physik. Chem.* **67**, 753 (1963).
7. D. H. Busch, *Advan. Chem. Ser.* **37**, 1 (1963).
8. M. M. Jones, "Ligand Reactivity and Catalysis," p. 116. Academic Press, New York, 1968.
9. C. K. Ingold, "Structure and Mechanism in Organic Chemistry," p. 310. Bell, London, 1953; see, also, J. Hine, "Physical Organic Chemistry," p. 123. McGraw-Hill, New York, 1962.

The interaction of metal ions with proteins and enzymes has been treated in many reviews. In 1950, Lehninger reviewed the entire field of metal-activated and metalloenzymes and pointed out the need for model studies of metal-catalyzed organic reactions (*10*).

In 1955, Vallee clearly defined metalloenzymes as enzymes with tightly bound metals that are retained in purification and set forth the criteria for their identification (*11*).

In an incisive review (*12*), Malmström and Rosenberg correctly pointed out that the major difference between metalloenzymes (which have built-in metals) and metal-activated enzymes (to which metals must be added for activity) is a quantitative one. In the latter enzymes, the affinity for the metal was relatively low (*13*), but the role of the metal might be the same as in the metalloenzymes. Although these authors considered only E—M—S and M—E—S types of complexes, they made the important point that kinetic studies alone were inadequate to distinguish between these two coordination schemes and that other studies, such as direct binding studies of metals to enzymes and to substrates, might assist in identifying the structure of the ternary complex. This and more recent reviews (*14–16*) summarize the structural properties of known metalloenzymes and metal-activated enzymes and consider their mechanisms of action.

Thermodynamic aspects of the interaction of metal ions with amino acids, peptides, proteins (*17*), and with enzymes and substrates (*18, 19*), have been reviewed and have formed the basis for mechanistic suggestions (*18, 19*).

Several authors have considered those enzymes and enzyme systems

10. A. L. Lehninger, *Physiol. Rev.* **30,** 393 (1950).
11. B. L. Vallee, *Advan. Protein Chem.* **10,** 317 (1955).
12. B. G. Malmström and A. Rosenberg, *Advan. Enzymol.* **21,** 131 (1959).
13. A stability constant of $10^8\,M^{-1}$ might serve as a rough dividing line between metalloenzymes and metal-activated enzymes. Thus, the stability constant of Zn-carboxypeptidase is $10^{10.5}\,M^{-1}$ (a metalloenzyme) and of Zn-enolase is $10^{5.3}\,M^{-1}$ (a metal-activated enzyme) [J. E. Coleman and B. L. Vallee, *JBC* **236,** 2244 (1961)]; also see Malmström *et al.* (*199*).

Since metals can combine with enzymes rapidly, a more fundamental distinction between metalloenzymes and metal-activated enzymes may be a kinetic one: In metalloenzymes, the E–M complex dissociates more slowly.

14. B. G. Malmström and J. B. Nielands, *Ann. Rev. Biochem.* **33,** 331 (1964).
15. B. L. Vallee, "The Enzymes," 2nd ed., Vol. 3, p. 225, 1960.
16. B. L. Vallee and J. E. Coleman, *Comprehensive Biochem.* **12,** 165 (1964).
17. F. R. Gurd and P. E. Wilcox, *Advan. Protein Chem.* **11,** 311 (1956).
18. A. E. Martell and M. Calvin, "Chemistry of the Metal Chelate Compounds," pp. 149 and 378. Prentice-Hall, Englewood Cliffs, New Jersey, 1952.
19. R. J. P. Williams, "The Enzymes," 2nd ed., Vol. 1, p. 391, 1959.

which are activated by monovalent cations (*20*) and have generalized the proposal, originally applied to pyruvate kinase (*21*), that monovalent cations might adjust or stabilize an active conformation of the enzyme. Suelter (*21a*) has suggested a direct chemical mechanism for enzyme activation by monovalent cations, namely, by interaction of the enzyme-bound monovalent cation with a keto–enol tautomer of the substrate.

The role of metals in enzyme-catalyzed oxidation–reduction reactions, as studied by rapid kinetics and EPR spectroscopy, has been reviewed (*22-24*), as have the chemical mechanisms of redox reactions in general (*25*). Hamilton has recently provided a valuable summary of the mechanisms proposed for electron transfer between ligands in coordination complexes and has suggested their applicability to various oxidase reactions catalyzed by metalloenzymes (*26*).

The purpose of this chapter is to examine the mechanisms of those coordination reactions and of those reactions catalyzed by metal-requiring enzymes which appear to be understood, and to seek generalities which might then be applied to other metal-activated enzymes. No attempt will be made for completeness of coverage; rather, this review will stress those principles that appear to have been established and others that seem attractive enough to merit further testing.

II. Properties of Metal Ions Relevant to Catalysis

A. General Properties of Metals and Ligands

Like protons, metal ions are Lewis acids or electrophiles, i.e., they can accept a share in an electron pair to form a σ bond. Moreover, metal ions may be considered "super acids" since they exist in neutral solution,

20. F. C. Happold and R. B. Beechey, *Biochem. Soc. Symp.* (*Cambridge, Engl.*) **15,** 52 (1958); H. J. Evans and G. J. Sorger, *Ann. Rev. Plant Physiol.* **17,** 47 (1966).
21. J. F. Kachmar and P. D. Boyer, *JBC* **200,** 669 (1953).
21a. C. H. Suelter, *Science* **168,** 789 (1970).
22. G. Palmer and H. Brintzinger, *in* "A Treatise on Electron and Coupled Energy Transfer in Biological Systems" (M. Klingenberg and T. E. King, eds.) (in press).
23. H. Beinert and G. Palmer, *Advan. Enzymol.* **27,** 105 (1965).
24. T. E. King, H. S. Mason, and M. Morrison, eds., "Oxidases and Related Redox Systems," Vols. 1 and 2. Wiley, New York, 1964.
25. F. H. Westheimer, *in* "The Mechanism of Enzyme Action" (W. D. McElroy and B. Glass, eds.), p. 301. Johns Hopkins Press, Baltimore, Maryland, 1954; F. H. Westheimer, "The Enzymes," 2nd ed., Vol. 1, p. 259, 1959.
26. G. A. Hamilton, *Advan. Enzymol.* **32,** 55 (1969).

often have a charge greater than +1, and in some cases can accept electrons into their low-lying vacant orbitals to form π bonds. Hence, metals can often serve as effective general acid catalysts for reactions that are catalyzed by protons (*27*). In contrast with protons, metal ions can function as three-dimensional templates for the binding and orientation of bases either independently or as chelates. Because of their filled orbitals, metal ions are much larger and more polarizable than protons and can therefore donate electrons to form π bonds as well as σ bonds.

The polarizability of metal ions, which correlates with their ability to donate electrons for π bonding, has been referred to as "softness" (*3, 28, 29*). Thus, soft or polarizable metal ions tend to be large, have several unshared and easily excited valence electrons, and have low positive charge. Hard metal ions are small, lack easily excited unshared valence electrons, and may have a higher positive oxidation state (Table I). The same terminology may be applied to ligands and correlates with their ability to accept π bonds. Thus, soft or polarizable ligands tend to have low electronegativities and are either easily oxidized or have low-lying vacant orbitals as in double bonds (Table I). Hard ligands have the opposite properties: low polarizability, high electronegativity, and inaccessibly high vacant orbitals. Hard, soft, and borderline ligands are listed in Table I.

From a survey of the affinities of metals for ligands, Pearson has set forth the general principle that ". . . hard acids prefer to associate with

TABLE I

CLASSIFICATION OF LEWIS ACIDS AND BASES OF BIOCHEMICAL INTEREST[a]

Hard	Borderline	Soft
	Lewis Acids	
H^+, Li^+, Na^+, K^+, Be^{2+}, Mg^{2+}, Ca^{2+}, Sr^{2+}, Mn^{2+}, Cr^{3+}, Co^{3+}, Fe^{3+}, As^{3+}, hydrogen bond donors	Fe^{2+}, Co^{2+}, Ni^{2+}, Cu^{2+}, Zn^{2+}, Pb^{2+}, R_3C^+ (carbonium ions)	Cu^+, Ag^+, Cs^+, Hg^+, Pd^{2+}, Cd^{2+}, Pt^{2+}, Hg^{2+}, RS^+, Br^+, HO^+, I_2, Br_2, M^0 (metal atoms), CH_2 (carbenes)
	Lewis Bases	
H_2O, OH^-, F^-, $CH_3CO_2^-$, PO_4^{3-}, SO_4^{2-}, CO_3^{2-}, ClO_4^-, NO_3^-, NH_3, RNH_2	$C_6H_5NH_2$, pyridine, imidazole, N_3^-, N_2, Cl^-, NO_2^-, SO_3^{2-}	R_2S, RSH, RS^-, Br^-, I^-, SCN^-, $S_2O_3^{2-}$, CN^-, RNC, CO, C_2H_4, C_6H_6, H^-, R^- (carbanions)

[a] From Pearson (*28, 29*).

27. M. L. Bender, *Advan. Chem. Ser.* **37,** 19 (1963).
28. R. G. Pearson, *Science* **151,** 172 (1966).
29. R. G. Pearson, *J. Chem. Educ.* **45,** 581 and 643 (1968).

hard bases and soft acids prefer to associate with soft bases" (*30*). Borderline cases behave in an intermediate way. From an examination of Table I it is seen that most metals and ligands of biochemical interest are hard with a few important exceptions: Cuprous, thiols, and hydride are soft; and Zn^{2+}, Fe^{2+}, Cu^{2+}, Co^{2+}, and imidazole are borderline.

B. Reactions of Metal Complexes

Metal complexes undergo the following classes of reaction (*7*):

1. Ligand substitution or addition at the metal atom.

$$[M(A)_m] + B \rightarrow [M(A)_{m-1}B] + A \quad (1)$$

$$[M(A)_m] + nB \rightarrow [M(A)_m(B)_n] \quad (2)$$

2. Reactions of the coordinated ligand(s).

$$[M(A)_m] + C \rightarrow [M(A)_{m-1}D] + E, \text{ etc.} \quad (3)$$

3. Reactions of the complex as a whole.

$$\text{D-}[M(A\text{–}A)_3] \xrightleftharpoons{\text{isomerization}} \text{L-}[M(A\text{–}A)_3], \text{ etc.} \quad (4)$$

4. Oxidation–reduction of the metal atom.

$$[M(A)_m]^n \rightarrow [M(A)_m]^{n+1} + e^- \quad (5)$$

Substitution reactions will be discussed according to their various mechanisms. Reactions of Classes 2 and 3 are so diverse that their mechanisms are best considered from the viewpoint of those electronic and structural properties of metal ions that are utilized, i.e., in terms of what the metal does in the reaction (*7, 8*).

The mechanism of oxidation–reduction reactions involving metal ions will be considered in Section VIII, which concerns itself with mechanisms of metallo-oxidoreductases.

1. *Ligand Substitution Mechanisms*

The mechanisms of ligand substitution reactions on metals have been treated in several excellent reviews (*3–6, 31*) and can only be summarized here.

30. An equivalent statement of Pearson's rule is: "Cations that indulge in ionic bonding prefer ligands that so indulge; cations that indulge in covalent bonding prefer ligands that so indulge."

31. M. Eigen and R. G. Wilkins, *Advan. Chem. Ser.* **47**, 55 (1965); M. Eigen and K. Tamm, *Z. Elektrochem.* **66**, 107 (1962).

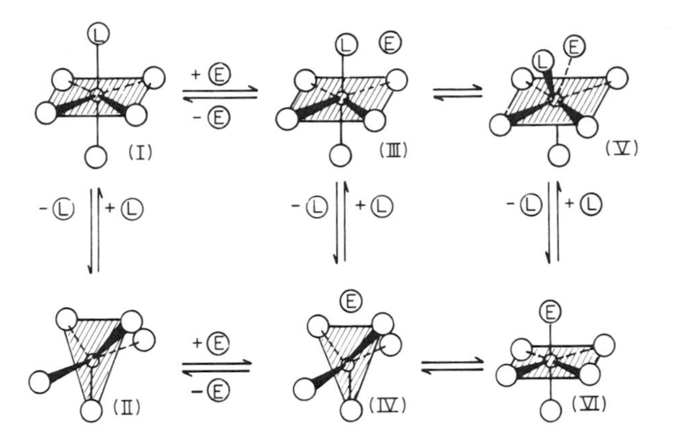

FIG. 1. Structures of intermediates in mechanisms of ligand substitution reactions of octahedral metal complexes: (L) represents the leaving ligand, and (E) the entering ligand. (See Table II for definitions of the mechanisms.)

Various mechanisms for ligand substitution reactions on octahedral metal complexes in an aqueous environment are contained in Fig. 1 and Table II. In Fig. 1, where the hypothetical structures of the intermediates and transition states are given, L refers to the ligand which is to leave and E refers to the entering or attacking ligand. For simplicity we will use the S_N1 and S_N2 nomenclature which Hughes, Ingold and co-workers introduced for substitution reactions on carbon (*9*), and which Brown *et al.* (*2*) have extended to substitution reactions on metal ions, although these terms refer to extreme or limiting cases.

In an extreme S_N1 mechanism, the leaving ligand dissociates in a rate-limiting step (I) → (II), which results in an intermediate (II) of reduced coordination number. This intermediate combines rapidly with the entering ligand (II) → [IV] → (VI) to form the product, (VI). Depending on

TABLE II
LIGAND SUBSTITUTION MECHANISMS IN OCTAHEDRAL COMPLEXES

Mechanism[a]	Reaction path[b]	Other names[c]
S_N1	(I) → (II) → [IV] → (VI)	S_N1-lim; dissociative
S_N2	(I) → (III) → [V] → (VI)	S_N2-lim; associative
S_N1–outer sphere	(I) → (III) → [IV] → (VI)	S_N1-ion pair; interchange-dissociative

[a] The nomenclature used here is that of Brown *et al.* (*2*) and Eigen (*6, 31*).

[b] See Fig. 1. The rate constants for steps between intermediates are numbered with arabic subscripts: (I) $\xrightarrow{k_{12}}$ (II).

[c] Basolo and Pearson (*3*) and Langford and Gray (*4*) have proposed these alternative systems of nomenclature.

its lifetime (τ), structure (IV) is either a transition state ($\tau \sim 10^{-13}$ sec) or a kinetic intermediate ($\tau > 10^{-13}$ sec). When the former is the case, the correct way to describe the second step is with a single rate constant k_{26}. The overall reaction then has two steps with rate constants k_{12} and k_{26}. In an extreme S_N2 mechanism the entering ligand displaces the leaving ligand (I) → (III) → [V] → (VI) by way of a single transition state [V], and with a single rate constant k_{16}. As in carbon chemistry (*9*), mechanisms intermediate between S_N1 and S_N2 are to be expected, depending upon the relative magnitudes of k_{34} and k_{35}, and depending upon whether structures (IV) and (V) are transition states or true kinetic intermediates.

An additional mechanistic category is required (S_N1—outer sphere) owing largely to the work of Eigen (*6*, *31*, *32*) who emphasized that a metal ion orders the structure of water not only in the inner coordination sphere but also in the second (outer) and possibly even the third layer. The complexes in which the entering ligand is in the second sphere, (III) and (IV), are referred to as outer sphere complexes or, when charged, ion pairs. Of course, all entering ligands must traverse the second sphere regardless of the mechanism. If a thermodynamic or kinetic property of the outer sphere complex contributes to the rate-controlling step in the formation of the product, i.e., if the outer sphere complex is a detectable intermediate, then an outer sphere or ion pair mechanism is operative.

In principle, an outer sphere complex of structure (III) can be converted to the product (VI) by either an S_N1 mechanism (III) → [IV] → (VI) in which the departure of water is rate limiting or by an S_N2 mechanism (III) → [V] → (VI) in which the coordination of the attacking ligand is rate limiting.

Experimentally, most complex-forming reactions between hydrated monovalent or divalent cations of biochemical interest and negatively charged ligands proceed by way of an S_N1 outer sphere mechanism (I) → (III) → [IV] → (VI), as demonstrated by the observations that the rates of complex formation are characteristic for a given metal and decrease in the order

$$Cs^+ > Rb^+ > K^+ > Na^+ > Li^+ > Cu^{2+} \approx Ca^{2+} > Zn^{2+} > Mn^{2+} > Fe^{2+} > Co^{2+} > Mg^{2+} > Ni^{2+}$$

from a value of 10^9 sec^{-1} to 10^4 sec^{-1} but are independent of the nature of the ligand (*6*, *31–33*). Moreover, the rates of formation of (VI) are generally equal to $k_{12}K[L]/(1 + K[L])$, where k_{12} is the rate constant

32. G. G. Hammes and J. I. Steinfeld, *JACS* **84**, 4639 (1962).
33. M. Eigen and G. G. Hammes, *Advan. Enzymol.* **25**, 1 (1963).

for dissociation of water from the given metal ion as determined independently and K is the stability constant (k_{13}/k_{31}) of the outer sphere complex.

The former (k_{12}) is characteristic for a given metal ion varying directly with its ion radius and bond lengths and inversely with its charge and ligand field stabilization parameters (*6, 31*). The latter (K) is dominated by electrostatic attraction, depending primarily on the ionic strength, the charge of the metal ion and ligand, and the distance of closest approach between them and may be calculated from electrostatic theory or measured independently (*3, 33–36*). (References, *3, 6,* and *33* give the characteristic water exchange rates which would equal k_{36} for various cations.)

Exceptions to the generality of the S_N1–outer sphere mechanism are found with uncharged ligands, where S_N1 mechanisms predominate; on metals with slowly exchanging ligands (Be^{2+}, Al^{3+}, and Co^{3+}), where hydrolysis of a coordinated water molecule may precede and catalyze ligand exchange (*3*); on square planar complexes of Ni^{II}, Pt^{II}, Pd^{II} where S_N2 mechanisms predominate, owing to the facile entrance of the attacking ligand onto an axial position (*4*); and in certain cases involving phosphate ligands (fluorophosphate and methyl phosphate) which enter the coordination spheres of Mn and Ni at rates significantly slower than water exchange (*37, 38*). Unusual stability of second sphere complexes have been suggested for these last cases, but an alternative explanation is that orthophosphate derivatives may be bidentate ligands and may slowly form a strained chelate (*vide infra*).

2. *Chelation Mechanisms*

As with most monodentate complexes, the rate of formation of most complexes of metal ions with flexible bidentate (and probably multidentate) ligands is limited by the rate of departure of the first water molecule from the aquo-metal ion (*31*) in conformity with an S_N1–outer sphere mechanism (VII) → (VIII) → (IX), followed by a rapid closure of the chelate ring, probably by an S_N2 mechanism (IX) → (X). The

34. R. M. Fuoss, *JACS* **79,** 3301 (1957); **80,** 5059 (1958).
35. J. Bjerrum, *Advan. Chem. Ser.* **62,** 178 (1967).
36. Interestingly, substitution reactions on carbon have recently been suggested to take place by outer sphere or ion-pair mechanisms [H. Weiner and R. A. Sneen, *JACS* **87,** 292 (1965); R. A. Sneen and J. W. Larsen, *ibid.* **91,** 362 (1969)]. Thus, concepts in organic chemistry and coordination chemistry have come full circle.
37. A. S. Mildvan, J. S. Leigh, Jr., and M. Cohn, *Biochemistry* **6,** 1805 (1967).
38. H. Brintzinger and G. G. Hammes, *Inorg. Chem.* **5,** 1286 (1966).

$$\underset{\text{(VII)}}{M(H_2O)_6} \underset{-L-L}{\overset{+L-L}{\rightleftharpoons}} \underset{\text{(VIII)}}{M(H_2O)_6\ L-L} \underset{-H_2O}{\overset{+H_2O}{\rightleftharpoons}} \underset{\text{(IX)}}{(H_2O)_5M-L-L} \underset{+H_2O}{\overset{\text{fast}\ -H_2O}{\rightleftharpoons}} \underset{\text{(X)}}{(H_2O)_4M\langle{}^{L}_{L}\rangle} \quad (6)$$

unusually high rate of (IX) → (X), as compared to the rate of combination of a second ligand (XI) → (XII), results from the statistical advantage of a unimolecular reaction [Eq. (6)] over a bimolecular reaction, [Eq. (7)] i.e., a larger effective concentration of the second ligand when

$$\underset{\text{(XI)}}{(H_2O)_5ML} \underset{-L}{\overset{+L}{\rightleftharpoons}} \underset{\text{(XII)}}{(H_2O)_4M(L)_2} + H_2O \quad (7)$$

it is chelated, and, as expected, manifests itself as a lower entropy barrier ($-T\Delta S^{\ddagger}$) to reaction (*3*). The effect is strictly analogous to the neighboring group effects found for substitution reactions on carbon. Basolo and Pearson (*3*) have pointed out that this single kinetic effect underlies the greater stability of chelate complexes, also an entropy effect (*18*), since other effects tend to cancel out. The most stable sizes of chelates involve five- and, to a lesser extent, six-membered rings.

An exception to the S_N1–outer sphere–chelation mechanism has been found (*5*) in the substitution of oxalate for water as in Eq. (8), for

$$[Cr(C_2O_4)_2(H_2O)_2]^- + C_2O_4^{2-} \rightarrow Cr(C_2O_4)_3^{3-} + 2H_2O \quad (8)$$

which the rate law is consistent with rapid addition of $C_2O_4^{2-}$, followed by slow rate-controlling closure of the chelate ring, presumably owing to steric hindrance. Such effects might be important in sterically crowded environments on enzymes. The formation of a strained chelate may contribute to the relatively slow rate of formation of $MnFPO_3$ (*37*) and Ni–PO_3OMe (*38*). Evidence for the existence of such structures has been found in the crystalline state (*39, 40*).

An opposite effect has been observed in the rate of complexation by Co^{2+} of pyrophosphate and tripolyphosphate which occur at 1.8 and 2.4 times faster than the rate of dissociation of water from Co^{2+}. This minor deviation from the general chelation mechanism has been attributed to a loosening of coordinated water ligands by the polyvalent ligand, while it is in the second coordination sphere rather than to a greater stability of the second sphere complex (*41*).

The dissociation of chelates takes place with stepwise, reversible dissociation of each ligand (*3, 42*) and may therefore be accelerated by

39. J. M. Mays, *Phys. Rev.* **131,** 38 (1963).
40. C. Calvo, *Inorg. Chem.* **7,** 1345 (1968).
41. G. G. Hammes and M. L. Morrell, *JACS* **86,** 1497 (1964).

metal ions, which could coordinate with an exposed ligand (*42*), or by other ligands, which could coordinate to the metal and prevent reclosure of the chelate ring (*42*).

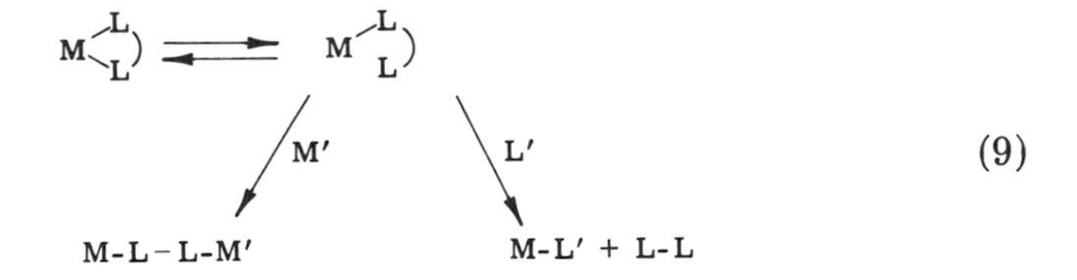

(9)

3. *Reactions of Coordinated Ligands*

Coordinated ligands can undergo a wide variety of reactions including hydrolysis, transamination, aldol condensations, bromination, insertions, carboxylations, and decarboxylation. Valuable reviews of the mechanisms of these reactions are available (*7, 8*). For brevity we will summarize only what the metal does in these reactions and the relevance to enzyme catalysis (Table III).

Jencks (*43*) has discussed four mechanisms by which an enzyme might bring about a rate acceleration: general acid-base catalysis, covalent catalysis, approximation of the reactants, and the induction of distortion or strain in the substrate, the enzyme, or both. As suggested by the model reactions summarized in Table III, metal ions can participate in each of these mechanisms.

Thus, metal ions can accept electrons from ligands via σ or π bonds and can thereby activate electrophiles and nucleophiles (general acid–base catalysis). Metals can donate electrons via σ or π bonds and thereby activate nucleophiles or act as nucleophiles themselves (covalent catalysis). The coordination sphere of a metal may be utilized to bring reactants together as separate ligands (approximation) or to form stable or unstable chelates (inducing distortion or strain). Chelation permits simultaneous holding and bonding effects of the type exemplified in Table III. By closing rings, chelation may also enhance reactivity by cyclic resonance (covalent catalysis). A metal ion may also mask a nucleophile and thereby prevent an unwanted side reaction from occurring.

The specific geometry of a coordination sphere may serve as a template to hold reactive groups in the proper position and thereby to control the stereochemical course of a reaction. Stereochemical control may result

42. D. W. Margerum, *R. Chem. Progr.* **24,** 237 (1963); *J. Phys. Chem.* **63,** 336 (1959).

43. W. P. Jencks, "Catalysis in Chemistry and Enzymology," pp. 8, 42, 163, and 282. McGraw-Hill, New York, 1969.

TABLE III

EFFECTS OF METAL IONS ON REACTIONS OF THEIR COORDINATED LIGANDS[a]

Effect	Example of effect in coordination chemistry	Ref.	Enzymes which may utilize the effect	Ref.
I. One site effects A. σ-Bond effects 1. Electron withdrawal a. Masking a nucleophile	$CH_3I \xrightarrow{CN^-} CH_3{-}CN + I^-$ $CH_3I \xrightarrow{Ag_4[Fe(CN)_6]} Fe(CN)_2(CN{-}CH_3)_4 + 4\,AgI$	*43a*	Histidine deaminase	*160, 202*
	$Co(NH_3)_6^{3+} + H^+ \longrightarrow$ No reaction	*8*		
b. Activation or promotion of an electrophile	$M^{+n}\cdots O{=}C(OR)R' \xrightarrow{\bar{O}H} M^{+n}{-}O{-}C(OR)(OH)R' \longrightarrow ROH + M^{+n}{-}O{-}C({=}O){-}R'$	*43b*	Kinases Lyases Pyruvate carboxylase	*37, 44* *60, 160, 218* *65, 206a, 207*
	$H{-}O\cdots H\cdots O{=}C(R){-}O{-}Co(NH_3)_5^{2+} \longrightarrow HO{-}C(R)(OH){-}O{-}Co(NH_3)_5^{2+} \longrightarrow R{-}C({=}O){-}OH + HO{-}Co(NH_3)_5^{2+}$	*3*	Carboxypeptidase Synthetases Xylose isomerase Aldolase	*134* *182, 183* *161* *162, 216*
c. Activation or promotion of a nucleophile	$(NH_3)_5Co{-}OH_2^{3+} + N_2O_3 \underset{+H^+}{\overset{-H^+}{\rightleftharpoons}} (NH_3)_5Co{-}O(H)\cdots O{=}N{-}O{-}N{=}O \longrightarrow (NH_3)_5Co{-}O{-}N{=}O + HNO_2$	*3*	Carbonic anhydrase ? Carboxypeptidase	*78, 239–241* *134*

2. Electron donation a. Metal acts as nucleophile	$\mathrm{Co^{I}(N_4)(py)} \xrightarrow{R-CH_2-Cl} [\mathrm{R-CH_2(Cl)\cdots Co(N_4)(py)}] \rightarrow \mathrm{R-CH_2-Co(N_4)(py)} + Cl^-$	*265*	Cobamide enzymes	*261–263*
B. π-Bond effects 1. π-Electron withdrawal	$M^{+n}-N\equiv C-R \;(+\,H-O-H) \longrightarrow M^{+n}-N(H)=C(OH)-R \longrightarrow M^{+n}-NH_2-C(=O)-R \xrightarrow[-H^+]{H_2O} M^{+n}-OH + R-C(=O)-NH_2$	*43c*	Pyruvate carboxylase Carboxypeptidase Alcohol dehydrogenase Polyphenol oxidase Uricase Galactose oxidase	*65*, *206a*, *207* *134* *64*, *258* *26* *26* *26*
	$C_6H_5-C(=O)OH \rightleftharpoons C_6H_5-C(=O)O^- + H^+ \quad pK = 5.7$ $(CO)_3Cr-C_6H_4-C(=O)OH \rightleftharpoons (CO)_3Cr-C_6H_4-C(=O)O^- + H^+ \quad pK = 4.8$	*43d*		

TABLE III (*Continued*)

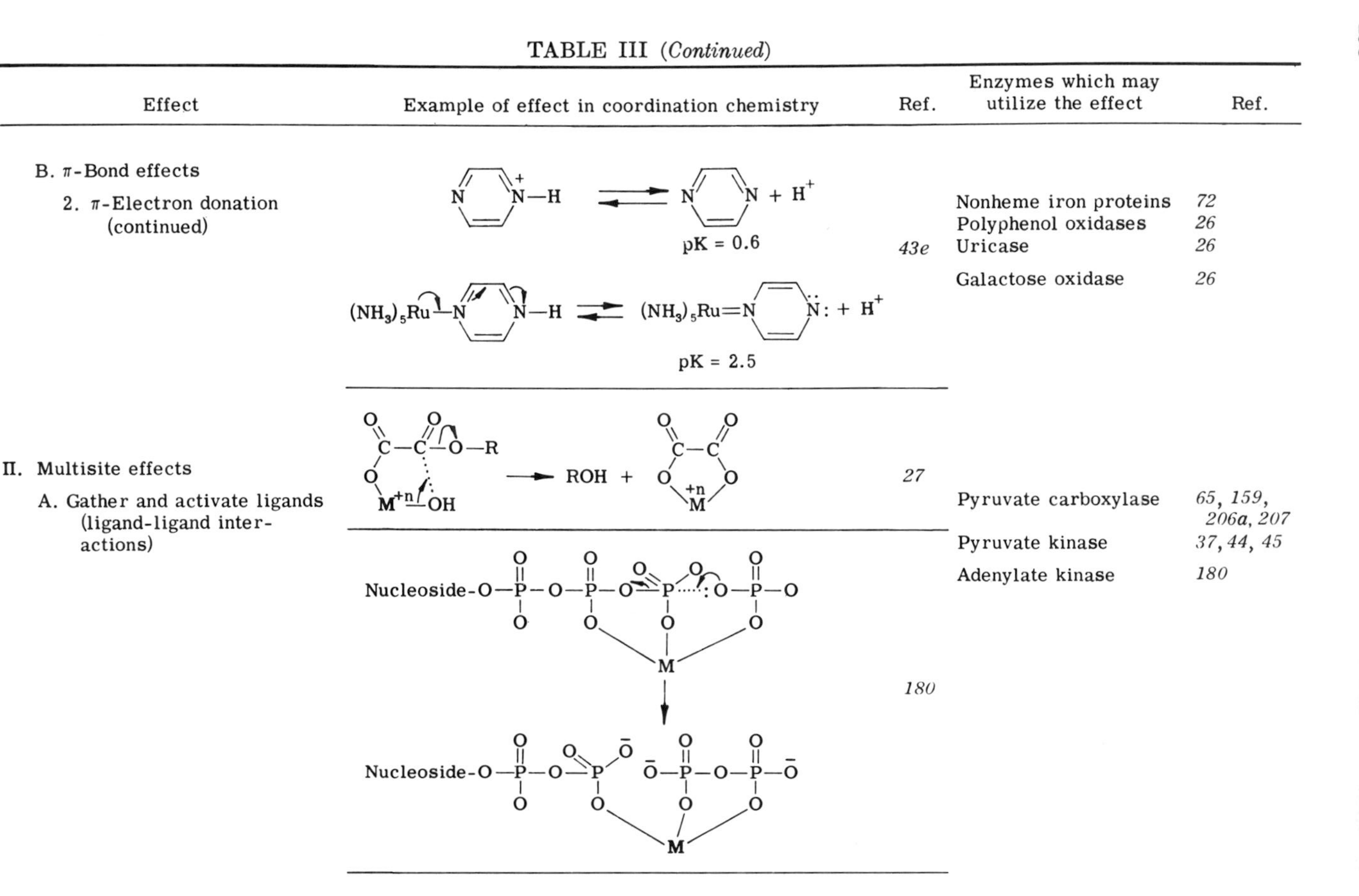

Effect	Example of effect in coordination chemistry	Ref.	Enzymes which may utilize the effect	Ref.
B. π-Bond effects				
2. π-Electron donation (continued)	pK = 0.6; pK = 2.5	*43e*	Nonheme iron proteins Polyphenol oxidases Uricase Galactose oxidase	*72* *26* *26* *26*
II. Multisite effects				
A. Gather and activate ligands (ligand-ligand interactions)	→ ROH +	*27*	Pyruvate carboxylase	*65, 159, 206a, 207*
	Nucleoside-O… → Nucleoside-O…	*180*	Pyruvate kinase Adenylate kinase	*37, 44, 45* *180*

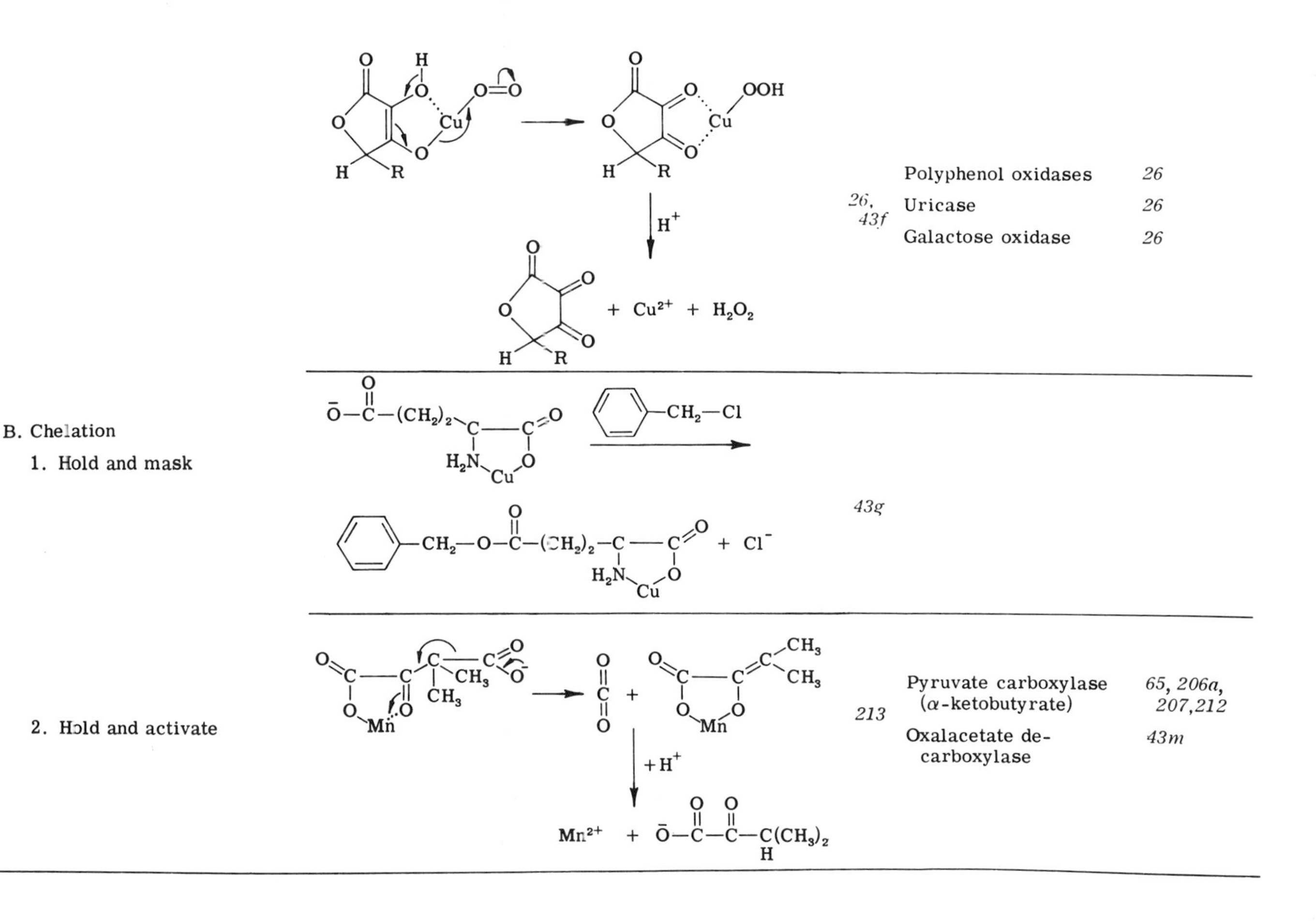

	Reaction	Reference	Enzyme	Reference
		26, 43f	Polyphenol oxidases	26
			Uricase	26
			Galactose oxidase	26
B. Chelation 1. Hold and mask		43g		
2. Hold and activate		213	Pyruvate carboxylase (α-ketobutyrate)	65, 206a, 207,212
			Oxalacetate de-carboxylase	43m

TABLE III (*Continued*)

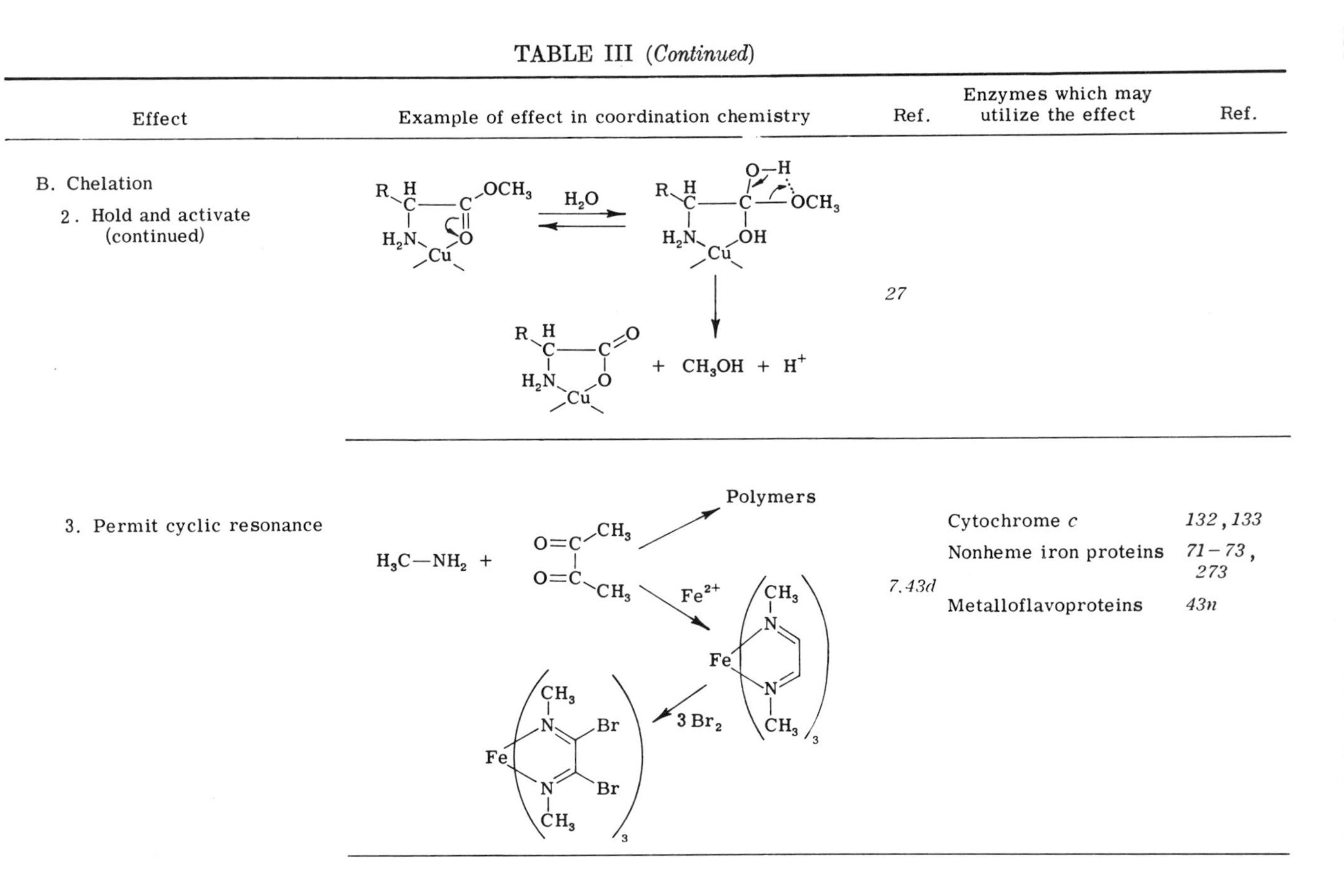

Effect	Example of effect in coordination chemistry	Ref.	Enzymes which may utilize the effect	Ref.
B. Chelation				
2. Hold and activate (continued)		27		
3. Permit cyclic resonance		7, 43d	Cytochrome *c*	*132, 133*
			Nonheme iron proteins	*71–73, 273*
			Metalloflavoproteins	*43n*

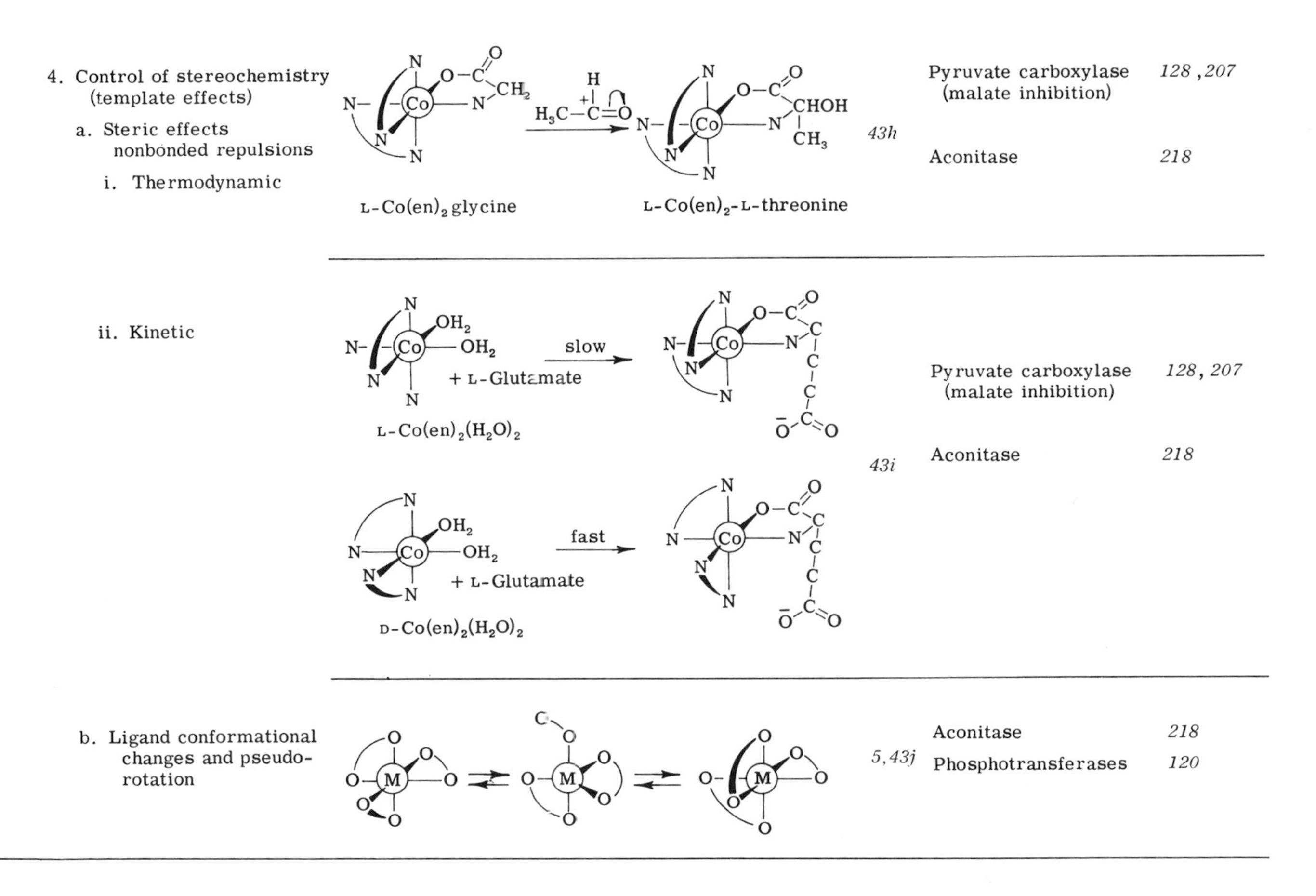

4. Control of stereochemistry (template effects) a. Steric effects nonbonded repulsions i. Thermodynamic	L-Co(en)$_2$glycine + $H_3C-C(H)=O$ → L-Co(en)$_2$-L-threonine	43h	Pyruvate carboxylase (malate inhibition)	128, 207
			Aconitase	218
ii. Kinetic	L-Co(en)$_2$(H$_2$O)$_2$ + L-Glutamate (slow); D-Co(en)$_2$(H$_2$O)$_2$ + L-Glutamate (fast)	43i	Pyruvate carboxylase (malate inhibition)	128, 207
			Aconitase	218
b. Ligand conformational changes and pseudo-rotation		5, 43j	Aconitase	218
			Phosphotransferases	120

TABLE III (*Continued*)

Effect	Example of effect in coordination chemistry	Ref.	Enzymes which may utilize the effect	Ref.
4. Control of stereochemistry (template effects) c. Equilibrium template reactions (continued)	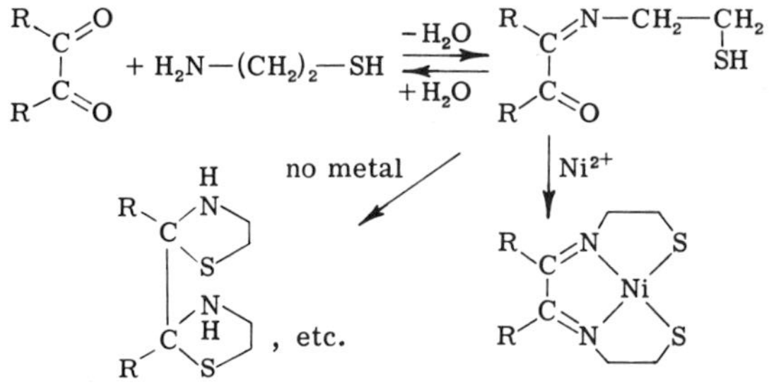 $R-C(=O)-C(=O)-R + H_2N-(CH_2)_2-SH \underset{+H_2O}{\overset{-H_2O}{\rightleftharpoons}} R-C(=N-CH_2-CH_2-SH)-C(=O)-R$; no metal → R–C(H–N, S ring)–C(N–H, S ring)–R, etc.; Ni^{2+} → R–C(=N)–C(=N)–R with S, Ni, S (Ni complex)	*43k*	Folding of taka-amylase and of deoxyribonuclease	*246*, *247*, *248*
d. Kinetic template reactions	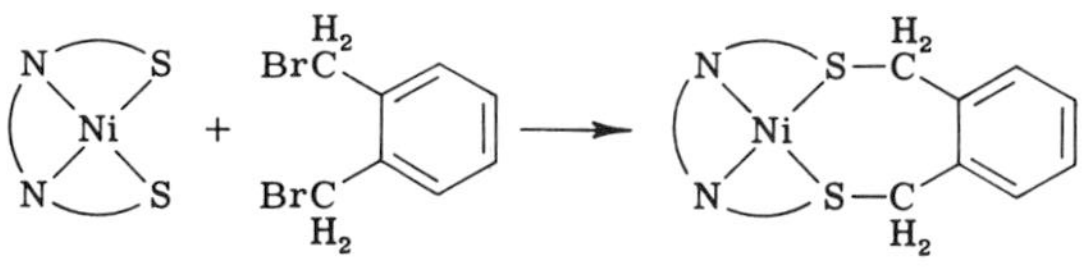 N, S, Ni, N, S complex + $BrCH_2$–C_6H_4–CH_2Br → N, Ni, N with S–CH_2–C_6H_4–CH_2–S macrocycle	*43k*	Folding of taka-amylase and of deoxyribonuclease	*246*, *247*, *248*
5. Strain	Co^{3+}(–O–)(–O–)P(=O)–OR + H_2O ⇌ Co(–O–P(=O)(–OR)–O^-)(OH_2) but Co^{3+}(–O–)(–O–)P(=O)–OH + H_2O ⇌ Co(–O–P(=O)(–OH)–O^-)(OH_2)	*39*, *40*, *120*, *431*	Phosphotransferases	Sec. VII, A
			D-Xylose isomerase	Sec. VII, C, 3
			Hemoproteins	*43o*

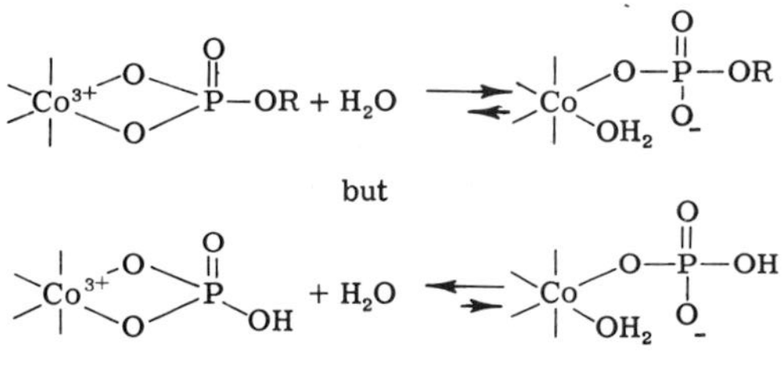

[a] Modified from a classification scheme devised by D. H. Busch.

from steric hindrance, restricting the possible conformations a ligand may assume (inducing distortion), from selective affinity for one member of an equilibrium mixture, or by a kinetic selection of the path taken by an irreversible process. These effects will be discussed in detail with reference to individual enzyme mechanisms.

III. Coordination Schemes of Enzymes, Metals, and Substrates

As previously defined (*11–13, 15, 16*) metalloenzymes are enzymes which retain stoichiometric, tightly bound, functional metal ions upon purification in contrast to metal-activated enzymes which require the addition of metal ions for activity. Since the major difference between these classes is usually the affinity of the active complex for the metal or the rate of dissociation of E—M (*13*), borderline cases exist in which partial retention of metal activators occur on purification, and an increase in residual activity is observed upon adding metal ions to a purified enzyme. When further purification is carried out with the help of chelating agents the residual activity may be reduced to a minimum and an absolute dependence on added metal ions may be established.

There is now considerable evidence (*12, 44–47*) which indicates that

---

References for Table III:

43a. F. Höltzl, *Monatsh. Chem.* **48,** 71 (1927).

43b. E. Bamann and H. Trappman, *Advan. Enzymol.* **21,** 169 (1959).

43c. D. A. Buckingham, F. P. Dwyer, and A. M. Sargeson, *Australian J. Chem.* **17,** 622 (1964).

43d. G. E. Coates, "Organometallic Compounds," 2nd ed., p. 273. Methuen, London, 1965.

43e. P. Ford, DeF. P. Rudd, R. Gaunder, and H. Taube, *JACS* **90,** 1187 (1968).

43f. M. M. T. Kahn and A. E. Martell, *JACS* **89,** 4176 (1967).

43g. W. E. Hanby, S. G. Wales, and J. Watson, *JCS* p. 3241 (1950).

43h. M. Murakami and K. Takahashi, *Bull. Chem. Soc. Japan* **32,** 308 (1959).

43i. J. H. Dunlop, R. D. Gillard, and N. C. Payne, *JCS* p. 1469 (1967).

43j. R. C. Fay and T. S. Piper, *Inorg. Chem.* **3,** 348 (1964).

43k. M. C. Thompson and D. H. Busch, *JACS* **86,** 213 and 3651 (1964).

43l. S. F. Lincoln and D. R. Stranks, *Australian J. Chem.* **21,** 57 (1968).

43m. G. W. Kosicki and F. H. Westheimer, *Biochemistry* **7,** 4303 (1968); G. W. Kosicki, *ibid.* p. 4310.

43n. P. Hemmerich, F. Müller, and A. Ehrenberg, *in* "Oxidases and Related Redox Systems" (T. E. King, H. S. Mason, and M. Morrison, eds.), Vol. 1, p. 157. Wiley, New York, 1964.

43o. R. Lumry, "The Enzymes," 2nd ed., Vol. 1, p. 157, 1959.

44. A. S. Mildvan and M. Cohn, *JBC* **240,** 238 (1965); **241,** 1178 (1966).

metal-activated enzymes have a simple stoichiometric relationship in their functional ternary complexes among enzyme sites, bound metal, and bound substrate which is often 1:1:1. For metal-activated enzymes, which form 1:1:1 complexes of enzyme, metal, and substrate, four coordination schemes are possible:

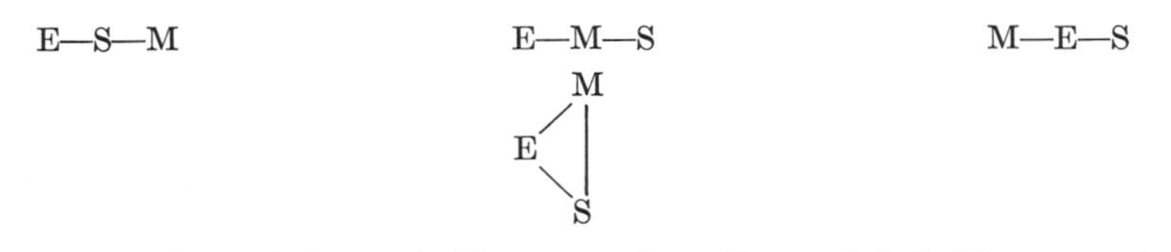

These are the substrate bridge complex, the metal bridge complexes [simple and cyclic (*48*)], and the enzyme bridge complex. For metallo-enzymes the substrate bridge complex is not possible.

A. Techniques for Determining the Coordination Scheme

Table IV summarizes the techniques that have been found useful for determining the coordination schemes of ternary complexes of enzyme, metal, and substrate. The techniques have been divided into two groups: provisional techniques are widely applicable and convenient, but fail to provide an unequivocal diagnosis of the coordination scheme; confirmatory techniques are less general, but are capable of providing an unequivocal answer.

1. *Provisional Techniques*

A useful way of distinguishing among coordination schemes is by binding studies. Thus, enzymes which form E–S–M complexes will not bind metals tightly or specifically in the absence of substrates. Substrates which participate in E–S–M complexes are generally nucleoside di- and triphosphates which have a high affinity ($K_D < 1$ mM) for the metal ion. Enzymes which form simple metal bridge complexes will not bind substrates specifically in absence of the metal, and those which form cyclic metal bridge complexes bind substrates in absence of the metal, but, typically, more strongly in the presence of the metal. The enzyme

45. A. M. Reynard, L. F. Hass, D. D. Jacobsen, and P. D. Boyer, *JBC* **236,** 2277 (1961).
46. G. L. Cottam and R. L. Ward, *ABB* **132,** 308 (1969).
47. R. S. Miller, A. S. Mildvan, H. Chang, R. Easterday, H. Maruyama, and M. D. Lane, *JBC* **243,** 6030 (1968).
48. The term *ménage à trois* has been suggested for this complex.

TABLE IV

EXPERIMENTAL METHODS FOR DETERMINING THE COORDINATION SCHEMES OF TERNARY COMPLEXES OF ENZYME, METAL, AND SUBSTRATE IN SOLUTION

		Substrate bridge	Metal bridge	Enzyme bridge
	Coordination scheme	E–S–M	E–M–S; E–M–S triangle (E bonded to M and S, M bonded to S)	M–E–S
Provisional techniques	Binding studies[a]	$K_1 < 10^{-3}\,M$ $K_D = \infty$ $K_2 \sim K_S$	$K_S = \infty$ (E–M–S); $K_S < \infty$ (triangle) $K_D < \infty$ $K_1 \geq K_3$	$K_S < \infty$ $K_D < \infty$ $K_3 \sim K_S$
	Enhancement (ϵ) behavior of H_2O	Type I ($\epsilon_b < \epsilon_T$)	Type II ($\epsilon_b > \epsilon_T$)	Type III ($\epsilon_b = \epsilon_T$)
	Effect of Ca^{2+}	Usually activates	Usually inhibits	Activates or inhibits
Confirmatory techniques	NMR of substrates with E–Mn	—	Enhanced relaxation of substrate nuclei	Deenhanced relaxation of substrate nuclei
	EPR spectra of Mn in M–S and E–M–S	Spectra of M–S and E–S–M identical	Spectrum of M–S taller and narrower than that of E–M–S or E–M–S triangle (E bonded to M and S, M bonded to S)	—
	EPR spectrum of Fe or Cu in E–M–S	—	Hyperfine splitting by magnetic nuclei of substrate	No hyperfine splitting

[a] Definitions of dissociation constants: $K_1 = \frac{(M)(S)}{(MS)}$; $K_2 = \frac{(E)(MS)}{(EMS)}$; $K_3 = \frac{(EM)(S)}{(EMS)}$; $K_D = \frac{(E)(M)}{(EM)}$, $K_S = \frac{(E)(S)}{(ES)}$.

bridge complex, in the simplest case, should bind metal and substrate independently i.e., there is no mutual effect of the metal on the affinity for the substrate and conversely. Because of the occurrence of site–site interaction (*49*) and changes in protein conformation on binding of substrates to enzymes (*50*), these generalities are only approximate and exceptions to them are found; these will be considered later (Section III,B). Kinetic studies cannot generally distinguish among these coordination schemes but are a necessary supplement to the binding studies. If the dissociation constants from the binding studies are inconsistent with those deduced from the kinetic studies, then the binding studies are not detecting the kinetically active complexes. Several techniques are available for measuring the affinity and stoichiometry of complexes of enzymes, metals, and substrates such as equilibrium dialysis (*51*), the ultracentrifuge (*52*), ultraviolet spectroscopy (*53*), optical rotatory dispersion (*16*), extrinsic or intrinsic fluorescence (*54*), column chromatography (*55*), the kinetic protection method (*56*), and the dialysis rate method (*57*). If the divalent cation used is Mn^{2+}, the techniques of electron spin resonance (ESR) (*58*), which measures free Mn, and the enhancement of the proton relaxation rate of water (PRR) (*59*), an NMR technique which measures an effect of bound Mn may be used.

The enhancement technique, introduced into enzymology by Cohn (*60*, *61*), is an exceedingly powerful tool for detecting binary and ternary complexes of Mn with proteins and other macromolecules (*62*). A detailed review of the applications of this technique to enzymology is in press (*63*). Here, only a summary can be given.

49. J. Monod, J. P. Changeux, and F. Jacob, *JMB* **6**, 306 (1963).
50. D. E. Koshland, Jr., *Proc. Natl. Acad. Sci. U. S.* **44**, 98 (1958); D. E. Koshland, Jr. and K. E. Neet, *Ann. Rev. Biochem.* **37**, 359 (1968).
51. I. M. Klotz, F. M. Walker, and R. B. Pivan, *BJ* **68**, 1486 (1946); see, also, Englund *et al.* (*184*).
52. S. F. Velick, J. Hayes, and J. Harting, *JBC* **203**, 527 (1953).
53. C. H. Suelter and W. Melander, *JBC* **238**, PC4108 (1963).
54. P. D. Boyer and H. Theorell, *Acta Chem. Scand.* **10**, 447 (1956); H. Theorell and J. S. McKinley McKee, *ibid.* **15**, 1811 (1961); see also, Zewe *et al.* (*80b*).
55. J. P. Hummel and W. J. Dryer, *BBA* **63**, 530 (1962).
56. F. Labeyrie and E. Stachiewicz, *BBA* **52**, 136 (1961); A. S. Mildvan and R. A. Leigh, *ibid.* **89**, 393 (1964).
57. S. P. Colowick and F. C. Womack, *JBC* **244**, 774 (1969).
58. M. Cohn and J. Townsend, *Nature* **173**, 1090 (1954).
59. A. S. Mildvan and M. Cohn, *Biochemistry* **2**, 910 (1963).
60. M. Cohn and J. S. Leigh, Jr., *Nature* **193**, 1037 (1962).
61. M. Cohn, *Biochemistry* **2**, 623 (1963).
62. J. Eisinger, R. G. Shulman, and B. M. Szymanski, *J. Chem. Phys.* **36**, 1721 (1962).
63. A. S. Mildvan and M. Cohn, *Advan. Enzymol.* (1970) (in press).

Briefly, paramagnetic ions [or radicals (*64*)], which contain unpaired electrons profoundly increase the longitudinal ($1/T_1$) and transverse ($1/T_2$) relaxation rates of magnetic nuclei, such as water protons, which are located in their immediate environment. For a paramagnetic ion such as $Mn(H_2O)_6^{2+}$ this effect occurs predominantly on the protons which are present in, or exchangeable into, the first coordination sphere, i.e., the protons of the water ligands. At high frequencies, the magnitude of the effect of Mn on the relaxation rate of the ligand protons is inversely related to the sixth power of the distance between Mn and the ligand protons, and to the residence time of the ligand in the coordination sphere of the Mn. The effect of Mn is directly related to the coordination number for the ligand and to the correlation time for the Mn–proton interaction. The correlation time (τ_c) is the time constant of that process which modulates the interaction between the Mn and the ligand protons and is usually dominated by the time constant for the relative rotational motion of Mn and its coordinated water. For $Mn(H_2O)_6^{2+}$, τ_c is the tumbling time of the complex which is of the order of 3×10^{-11} sec. Upon binding of Mn to the surface of a macromolecule, the coordination number for water may halve, the water exchange rate may increase or decrease by an order of magnitude, but the tumbling will slow (i.e., the correlation time will increase) by as much as three orders of magnitude. Hence the last-mentioned effect usually outweighs the former and an enhancement of the relaxation rate will be observed. Thus an enhancement of the effect of Mn on the relaxation rate of water protons in the presence of an enzyme indicates that Mn has bound to a site which is tumbling slowly, i.e., to the surface of the enzyme. For substrate bridge complexes, little or no enhancement would be observed unless all three components, enzyme, metal, and substrate, were present, i.e. the enhancement of the ternary complex (ϵ_T) exceeds that of the binary mixture of Mn and enzyme (ϵ_b) (type I behavior). For metal bridge complexes, an enhanced effect is observed on forming a binary complex of Mn and enzyme. Upon adding a substrate, which would replace one or more coordinated water molecules, the effect of Mn on water would generally decrease, i.e., the enhancement of the ternary complex (ϵ_T) is less than the enhancement of the binary E–Mn complex (ϵ_b) (type II behavior). For enzyme bridge complexes, an enhanced binary complex can form, but in the absence of interaction the addition of substrate should not change the enhancement (i.e., $\epsilon_T = \epsilon_b$; type III behavior).

Because of the fact that multiple parameters determine the relaxation rate, exceptions to these generalities have occasionally been observed

64. A. S. Mildvan and H. Weiner, *Biochemistry* **8**, 552 (1969); *JBC* **244**, 2465 (1969).

and independent techniques are required to confirm the coordination scheme (*63*). Measurement of the proton relaxation rate of water is, however, a rapid exploratory technique and also permits the determination of the stoichiometry and the dissociation constants of ternary complexes of enzyme, metal, and substrate (*63*). Table V lists the coordination schemes of ternary complexes of thirty-one enzymes determined by their enhancement behavior or by the other techniques indicated.

2. *Confirmatory Techniques*

The most direct and general way of demonstrating enzyme–metal–substrate bridge complexes in solution is by examination of the effect of an enzyme-bound paramagnetic metal (such as Mn^{2+}) on the relaxation rates of the nuclei of the substrate. This method was first used to detect a pyruvate kinase–Mn–FPO_3 bridge complex by fluorine NMR (*37*), and has since been applied to a number of other metal bridge complexes (*63*, *65*, *66*). Under favorable conditions, such experiments may provide information on the rate of exchange of the substrate into the coordination sphere of the metal, the dipolar distance between the substrate nuclei and the metal, and the nature of the bonds connecting a substrate nucleus with the paramagnetic center (*65*, *66*). In most cases, where a metal bridge complex is demonstrated, limiting values for these parameters can be estimated (*63*, *66*). Hence, unlike static binding techniques (including X-ray crystallography), the kinetic properties of the ternary complex are examined in the same experiment permitting one to determine whether the complex detected is kinetically competent to function in the catalysis.

Another powerful, but less general, technique for detecting E–M–S complexes is to demonstrate hyperfine splitting in the EPR spectrum of a paramagnetic metal by a magnetic nucleus of the substrate, since hyperfine coupling between an unpaired electron and a nucleus operates only through chemical bonds (*67*). The first application of this technique to a protein–metal–ligand system was by Kotani (*68*), who demonstrated direct coordination of fluoride by iron in crystalline metmyoglobin fluoride at liquid nitrogen temperature, which has been confirmed in

65. A. S. Mildvan and M. C. Scrutton, *Biochemistry* **6**, 2978 (1967).

66. A. S. Mildvan, *in* "Johnson Foundation Symposium on Probes for Macromolecular Structure and Function" (B. Chance, T. Yonetani, and M. Cohn, eds.). Academic Press, New York (in press).

67. M. Barfield and M. Karplus, *JACS* **91**, 1 (1969).

68. M. Kotani and H. Morimoto, *in* "Magnetic Resonance in Biological Systems" (A. Ehrenberg, B. G. Malmström, and T. Vänngård, eds.), p. 135. Pergamon Press, Oxford, 1967.

solution at 25° by the effect of metmyoglobin on $1/T_1$ and $1/T_2$ of F^- (*69*). The EPR method has been employed by several workers to elucidate the ligands of iron in nonheme iron proteins (*70–73*). Because this approach requires well-resolved EPR spectra, it is probably limited to complexes of Fe in the frozen state and to complexes of copper. A further limitation is introduced by the possibility of structural changes in metalloproteins upon freezing (*74*).

A useful confirmatory technique for substrate bridge complexes involving nucleoside di- and triphosphates is to demonstrate an identity of the EPR spectrum of Mn in the binary Mn–substrate complex with that in the ternary complex. Since the EPR spectrum of Mn^{2+} is very sensitive to the ligand field experienced by the metal (*75*), the absence of a change in the EPR spectrum in the presence of independent evidence for formation of a ternary complex (e.g., an increased enhancement) strongly supports the existence of an E–S–M complex. This technique was first used by Cohn and Leigh to demonstrate the creatine kinase–ADP–Mn complex (*60*). Conversely, a reduction in the amplitude of the EPR spectrum of Mn-ATP, consistent with the amount of ternary complex formed, upon the addition of pyruvate kinase, has been used to confirm a metal bridge complex (*76*).

The lack of effect of a bound substrate on the EPR spectrum of Cu^{2+} enzymes in the presence of independent evidence of substrate binding may be used to confirm an M–E–S complex (*77*). An M–E–S complex (Mn-pyruvate kinase-F) was deduced from the lack of effect of the enzyme–Mn complex on the relaxation rates of fluoride despite independent evidence (water relaxation rate) for the binding of fluoride (*37*).

69. A. S. Mildvan, N. M. Rumen, and B. Chance, *Abstr. 156th ACS Meeting, Atlantic City, New Jersey, 1968* Biol. 32; also *in* "Johnson Foundation Symposium on Probes for Macromolecular Structure and Function" (B. Chance, T. Yonetani, and M. Cohn, eds.). Academic Press, New York (in press).

70. T. C. Hollocher, F. Solomon, and T. E. Ragland, *JBC* **241,** 3452 (1966).

71. D. V. DerVartanian, W. H. Orme-Johnson, R. E. Hansen, H. Beinert, R. L. Tsai, J. C. M. Tsibris, R. C. Bartholomaus, and I. C. Gunsalus, *BBRC* **26,** 569 (1967).

72. J. C. M. Tsibris, R. C. Tsai, I. C. Gunsalus, W. H. Orme-Johnson, R. E. Hansen, and H. Beinert, *Proc. Natl. Acad. Sci. U. S.* **59,** 959 (1968).

73. W. H. Orme-Johnson, R. E. Hansen, H. Beinert, J. C. M. Tsibris, R. C. Bartholomaus, and I. C. Gunsalus, *Proc. Natl. Acad. Sci. U. S.* **60,** 368 (1968).

74. T. Yonetani and A. Ehrenberg, *in* "Magnetic Resonance in Biological Systems" (A. Ehrenberg, B. G. Malmström, and T. Vänngård, eds.), p. 155. Pergamon Press, Oxford, 1967.

75. A. Hudson and G. R. Luckhurst, *Mol. Phys.* **16,** 395 (1969).

76. J. Donaldson and A. S. Mildvan, unpublished observations (1965).

77. W. E. Blumberg, M. Goldstein, E. Lauber, and J. Peisach, *BBA* **99,** 187 (1965).

Infrared spectroscopy has recently been used to detect a Zn–carbonic anhydrase–CO_2 complex in which the CO_2 is not directly coordinated to the metal (*78*).

The technique of X-ray crystallography has been used to determine the total structure of several ternary complexes of enzyme, metal, and substrate or inhibitor (*79*). However, a fundamental limitation of this method, as with all physical methods for detecting complexes (*80*), is that there is no assurance that the complex observed in the Fourier synthesis is kinetically active or kinetically competent to participate in the catalysis. Techniques are needed for determining the kinetic and thermodynamic properties of those enzyme–substrate complexes which are observed crystallographically. Initial developments in this important field have recently been reviewed (*79*) and are discussed by Eisenberg in Chapter 1 of Volume I. Another limitation of X-ray crystallography in the detection of enzyme–metal–substrate bridge complexes arises from the fact that a water ligand is replaced by a substrate ligand. Hence the difference Fourier synthesis (E–M–S minus E–M–OH_2) reveals a minimum in the electron density (i.e., in the signal-to-noise ratio) at the position of the coordinated atom of the substrate.

B. Experimentally Determined Coordination Schemes

1. *Ternary Complexes*

Table V lists the complexes of thirty-one enzymes according to their tentative coordination schemes as determined by the methods outlined in Table IV and discussed above. Those ten enzymes and two hemoproteins for which coordination schemes have been established by a confirmatory method are so indicated.

From the studies summarized in Table V, several generalities emerge. Most, but not all, kinase reactions form enzyme–nucleotide–metal complexes. The important exceptions are those kinases and other phosphotransferase enzymes that utilize pyruvate or phosphoenolpyruvate. These and other enzymes which catalyze reactions of phosphoenolpyruvate form enzyme–metal–substrate bridge complexes as do carboxylating enzymes. Calcium is often an inhibitor of enzymes that form metal bridge complexes but an alternative activator of enzymes which form substrate bridge complexes. Because of its unusually rapid ligand exchange rates (*33*), Ca may be unsuitable to function in certain metal

78. M. C. Riepe and J. H. Wang, *JBC* **243,** 2779 (1968).
79. L. Stryer, *Ann. Rev. Biochem.* **37,** 25 (1968).
80. B. Chance, *Advan. Enzymol.* **12,** 153 (1951).

bridge complexes. With polyphosphates, the conformation of the chelate ring formed by CaATP and CaADP may be inadequate to function in E–M–S complexes (Section V,A). An exception to this generality may occur with staphylococcal nuclease where an enzyme–Ca–nucleotide bridge complex, suggested by binding studies (*81*, *82*), has been detected crystallographically (*83*).

2. *Higher Substrate Complexes*

Most enzymic reactions involve two substrates, and metal-requiring enzymes are no exception. Hence, complexes of the type EMS_1S_2 are to be expected. The second substrate may form complexes with coordination schemes of each of the types previously discussed. Using the proton relaxation rate method, binding studies have been carried out for pyruvate kinase (*44*) and creatine kinase (*84*) involving the formation of abortive quaternary complexes of enzyme, metal, substrate, and a product to prevent a reaction from occurring. Their probable coordination schemes are:

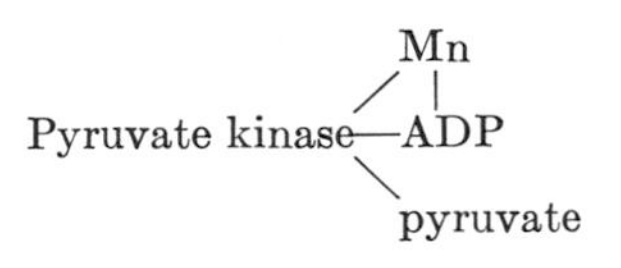

and

$$\text{Creatine kinase}\begin{matrix}\diagup\ \text{ADP-Mn}\\ \diagdown\ \text{creatine}\end{matrix}$$

As indicated, the substrate ADP forms a metal bridge complex with pyruvate kinase and a substrate bridge complex with creatine kinase (*44*, *60*, *84*). The products pyruvate and creatine are believed to form enzyme bridge complexes with respect to the Mn, as determined by the

81. P. Cuatrecasas, S. Fuchs, and C. B. Anfinsen, *JBC* **242,** 3063 (1967).

82. P. Cuatrecasas, H. Taniuchi, and C. B. Anfinsen, *Brookhaven Symp. Biol.* **21,** 172 (1968).

83. F. A. Cotton, E. E. Hazen, Jr., D. C. Richardson, J. S. Richardson, A. Arnone, and C. J. Bier, *Acta Cryst.* **A25,** S188 (1969); A. Arnone, C. J. Bier, F. A. Cotton, E. E. Hazen, Jr., D. C. Richardson, and J. S. Richardson, *Proc. Natl. Acad. Sci. U. S.* **64,** 420 (1969). A preliminary estimate of the distance between the Ca and the phosphorus of the phosphate [personal communication from E. E. Hazen (1969)] of 4.7–4.8 Å would permit an intervening water molecule, suggesting as an alternative possibility that the metal might be promoting the nucleophilicity of water in the catalysis. In either case an $E\begin{matrix}\diagup M\\ \quad|\\ \diagdown S\end{matrix}$ complex would ultimately be formed.

84. W. J. O'Sullivan and M. Cohn, *JBC* **241,** 3104 and 3116 (1966).

TABLE V
COORDINATION SCHEMES OF TERNARY COMPLEXES OF ENZYME, METAL, AND SUBSTRATE[a]

Substrate bridge		Metal bridge		Enzyme bridge	
E–S–M	Ref.	E–M–S or E(–M–S, cyclic)	Ref.	S–E–M	Ref.
Muscle creatine kinase EPR*, ε, K, B, Ca	*(60, 84, 110, 173)*	Muscle pyruvate kinase NMR*, EPR*; ε, B, Ca	*(37, 44, 45, 76)*	Citrate lyase ε, (Ca?)	*(80j, k)*
Brain creatine kinase ε	*(111)*	Pyruvate carboxylase NMR*; ε, B	*(65, 159, 207)*	Dopamine hydroxylase EPR*, ε	*(77)*
Muscle adenylate kinase EPR*; ε, B, (Ca ?)	*(80a, 112, 181)*	Histidine deaminase NMR*; ε	*(160)*	UDPG pyrophosphorylase ε, B S = PP_i	*(182)*
Arginine kinase ε	*(113)*	D-Xylose isomerase NMR*; ε	*(161)*	Tryptophan RNA synthetase ε, B S = PP_i	*(183a)*
Tetrahydrofolate synthetase ε	*(86)*	Yeast aldolase NMR*; (ε?)	*(162)*	Valine RNA synthetase ε, B S = PP_i	*(183)*
3-PGA kinase ε	*(61)*	Carboxypeptidase X-ray*, NMR*; ε, B	*(134, 141, 147)*	*E. coli* glutamine synthetase B (structural site)	*(95–97)*
Yeast hexokinase ε, B	*(61, 80b)*	Yeast pyruvate kinase ε, B	*(91)*	Carbonic anhydrase IR* S = CO_2	*(78)*
UDPG pyrophosphorylase ε, B S = UTP	*(182)*	PEP carboxykinase ε, B	*(47)*	Pyruvate kinase NMR*, ε S = fluoride	*(37)*

Enzyme / evidence	Reference
Tryptophan RNA synthetase	*(183)*
ε, B S = ATP	
Valine RNA synthetase	*(183a)*
ε S = ATP	
PEP carboxylase	*(47)*
ε	
PEP synthase	*(80c)*
ε, B	
Enolase	*(60, 156, 199)*
ε, B	
Phosphoglucomutase	*(80d-f, 157)*
B, K, ε	
Inorganic pyrophosphatase	*(61)*
ε	
Ribulosediphosphate carboxylase	*(198)*
ε	
DNA polymerase	*(184, 185, 185a)*
B, Ca, ε	
Staphylococcal nuclease	*(82, 83)*
B, X-ray	
Carbonic anhydrase	*(163)*
Cl^-, NMR*	
β-Methylaspartase	*(80g, 165)*
B, ε	
Metmyoglobin azide	*(80h)*
X-ray*	
Metmyoglobin fluoride	*(68, 69, 80i)*
EPR*, NMR*, X-ray*	

[a] Definition of symbols: EPR*, confirmed by electron paramagnetic resonance; NMR*, confirmed by nuclear relaxation of substrate; X-ray*, confirmed in the crystalline state by X-ray diffraction; IR*, confirmed by infrared spectroscopy of bound substrate; ε, consistent with enhancement data on the relaxation rate of water; B, consistent with binding studies; Ca, consistent with the effect of Ca; K, consistent with the results of transient state or temperature jump kinetic studies; ?, inconsistent with the results of the indicated technique.

References for Table V:

80a. S. A. Kuby, T. A. Mahowald, and E. A. Noltmann, *Biochemistry* **1,** 748 (1962).

80b. V. Zewe, H. J. Fromm, and R. Fabiano, *JBC* **239,** 1625 (1964).

80c. K. Berman and M. Cohn, *Federation Proc.* **28,** 602 (1969); *BBA* **141,** 214 (1967), and references therein.

80d. W. J. Ray, Jr. and G. A. Roscelli, *JBC* **241,** 2596 and 3499 (1966).

80e. W. J. Ray, Jr., *JBC* **242,** 3737 (1967).

80f. W. J. Ray, Jr. and A. S. Mildvan, unpublished observations (1969).

80g. G. A. Fields and H. J. Bright, unpublished observations (1968).

80h. L. Stryer, J. C. Kendrew, and H. C. Watson, *JMB* **8,** 96 (1964).

80i. H. C. Watson and B. Chance, *in* "Hemes and Hemoproteins" (B. Chance, R. W. Estabrook, and T. Yonetani, eds.), p. 149. Academic Press, New York, 1966.

80j. R. L. Ward and P. A. Srere, *BBA* **99,** 270 (1965).

80k. S. Dagley and E. A. Dawes, *BBA* **17,** 177 (1955).

enhancement behavior. In the case of the fluorokinase reaction of pyruvate kinase, the latter type of complex has been confirmed for the fluoride ion (*37*) which, like pyruvate, can accept a phosphoryl group from ATP (*85*). In the case of formyl tetrahydrofolate synthetase, a three substrate enzyme, an active quaternary complex (E–ATP–Mn, tetrahydrofolate), and an abortive quaternary complex (E–ADP–Mn, tetrahydrofolate) have been detected by proton relaxation (*86*). The large reduction in enhancement suggests that tetrahydrofolate may donate a ligand to the Mn in these quaternary complexes (*86*).

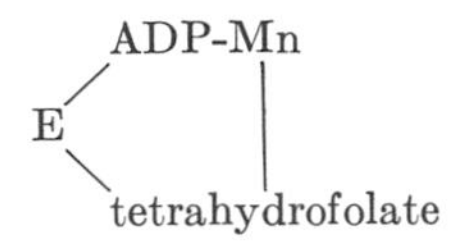

In all of the higher substrate complexes studied, the dissociation constants obtained in the binding studies agreed with those calculated from kinetic data.

3. *Higher Metal Complexes*

Nucleoside polyphosphates can bind more than one metal ion to form higher metal complexes (*87, 88*). With enzyme complexes of certain nucleotide substrates, indirect evidence has been obtained for the formation of higher complexes of the type:

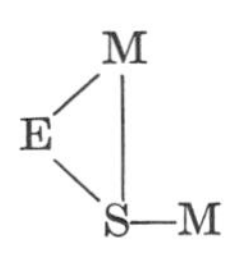

containing two metal ions. Webster (*89*) has obtained evidence for the functioning of a higher metal complex involving two divalent cations (M_1 and M_2) in the first step of the acetyl-CoA synthetase reaction. A reasonable coordination scheme for these metals (*89*) is as shown:

$$\mathrm{E}\genfrac{}{}{0pt}{}{\diagup\,\mathrm{ATP\text{-}M_2}}{\diagdown\,\mathrm{M_1}} + \mathrm{Acetate} \rightleftharpoons \mathrm{E}\genfrac{}{}{0pt}{}{\diagup\,\mathrm{Acetyl\text{-}AMP}}{\diagdown\,\mathrm{M_1}} + \mathrm{M_2\text{-}Pyrophosphate} \tag{10}$$

$$\mathrm{E}\genfrac{}{}{0pt}{}{\diagup\,\mathrm{Acetyl\text{-}AMP}}{\diagdown\,\mathrm{M_1}} + \mathrm{CoA} \rightleftharpoons \mathrm{Acetyl\text{-}CoA} + \mathrm{E\text{-}M_1} \tag{11}$$

85. M. Flavin, H. Castro-Mendoza, and S. Ochoa, *JBC* **229**, 981 (1957); A. Tietz and S. Ochoa, *ABB* **78**, 477 (1958).
86. R. H. Himes and M. Cohn, *JBC* **242**, 3628 (1967).
87. J. Lowenstein, *Nature* **187**, 570 (1960).
88. R. L. Ward and J. A. Happe, *BBRC* **28**, 785 (1968).
89. L. Webster, *JBC* **240**, 4164 (1965); **242**, 1232 (1967).

The requirement for M_1, which is bound to the enzyme and which is necessary for the tight binding of acetyl-AMP to the enzyme, may be met by Ni^{2+}, Fe^{2+}, Ca^{2+}, or Cu^{2+} while the requirement for M_2 may be met by Mg^{2+}, Mn^{2+}, Fe^{2+}, Co^{2+}, or Ca^{2+}. Both steps also require a monovalent cation which stabilizes the enzyme-bound acetyl adenylate but which is not required for the binding of acetyl adenylate (*89*). Higher metal complexes have also been detected by binding studies with phosphoenolpyruvate carboxykinase, GDP, GTP, and Mn (*47*). The kinetic role of these complexes is not known at present.

The existence of such a complex of muscle pyruvate kinase

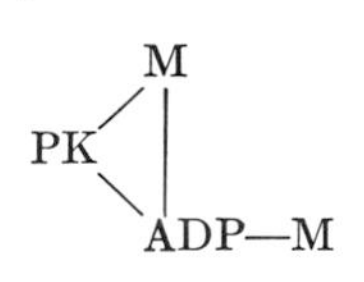

with the same enzymic activity as

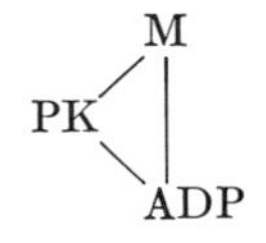

would resolve an anomaly in the kinetic data on this enzyme (*44*, *90*, *91*). Such a complex of pyruvate kinase is suggested by the observation that the stable 1:1 chromic complex of ADP (Cr–ADP) which cannot exchange water ligands, and therefore cannot form a metal bridge complex, is a potent competitive inhibitor of pyruvate kinase with respect to MnADP in the presence of enough Mn to saturate the enzyme sites (*92*). As will be discussed in the next section, a complex of the type M—E—S (with M bridging E and S) may function in glutamine synthetase (*93*). A higher metal complex involving two or three Ca ions in combination with the enzyme, transglutaminase, appears to be necessary for catalysis of amide hydrolysis (*94*).

90. W. W. Cleland, *Ann. Rev. Biochem.* **36**, 77 (1967).

91. A. S. Mildvan, J. Hunsley, and C. H. Suelter, *in* "Johnson Foundation Symposium on Probes for Macromolecular Structure and Function" (B. Chance, T. Yonetani, and M Cohn, eds.), Academic Press, New York (in press).

92. K. Hirsch, A. S. Mildvan, and A. Kowalsky, *Abstr., 158th ACS Meeting, New York, 1969* Biol. 52.

93. E. R. Stadtman, B. M. Shapiro, A. Ginsburg, H. S. Kingdon, and M. D. Denton, *Brookhaven Symp. Biol.* **21**, 378 (1968).

94. J. E. Folk, P. W. Cole, and J. P. Mullooly, *JBC* **242**, 2615 (1967).

IV. Enzyme Bridge Complexes (M—E—S)

Little is known about the detailed role of metals in M–E–S complexes. Presumably it is structural, to stabilize a catalytically active protein conformation, as has been suggested for monovalent cations (*20*).

A purely structural role for divalent cations in glutamine synthetase from *E. coli* is suggested by the observations that preincubation with Ca^{2+}, Mg^{2+}, or Mn^{2+}, but not Co^{2+}, slowly reactivates the enzyme which has been inactivated by EDTA (*95, 96*), but that Mg^{2+}, Mn^{2+}, or Co^{2+}, but not Ca^{2+}, are necessary for maximal activity in the enzyme assay (*95, 97*). The binding of the metal ions in the preincubation is followed by a slow first-order change in protein conformation to a more compact and symmetric structure (as judged by physical studies), in which aromatic amino acids are buried in a nonpolar environment and in which the twelve subunits are more difficult to dissociate (*96*). Twelve tight binding sites for Mn on the enzyme (as well as two classes of weaker binding sites) have been found by equilibrium dialysis (*93*). Under comparable conditions, the affinity of the tight binding sites for Mn ($K_D = 4\ \mu M$) is consistent with the concentration of Mn which produces half-maximal change in the UV spectrum of the protein ($\sim 15\ \mu M$), suggesting that the tight Mn binding sites influence the protein conformation (*90*).

The structural role of divalent cations in enzymes is not limited to M–E–S complexes. As will be discussed later, in the case of pyruvate kinase, the activating divalent cation alters the conformation of the protein (*53, 98, 99*) and also functions as a metal bridge to the substrate (*37, 44*). Complexes of the type M–E–S_2 exist in two substrate enzymes of the type:

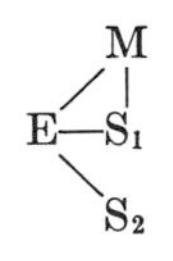

The role of the metal is to hold and activate the other substrate (S_1), as

95. H. S. Kingdon, J. S. Hubbard, and E. R. Stadtman, *Biochemistry* **7**, 2136 (1968).

96. B. M. Shapiro and A. Ginsburg, *Biochemistry* **7**, 2153 (1968).

97. H. S. Kingdon and E. R. Stadtman, *J. Bacteriol.* **94**, 949 (1967).

98. F. J. Kayne and C. H. Suelter, *JACS* **87**, 897 (1965).

99. C. H. Suelter, R. Singleton, Jr., F. J. Kayne, S. Arrington, J. Glass, and A. S. Mildvan, *Biochemistry* **5**, 131 (1966).

appears to be the case in the fluorokinase reaction of pyruvate kinase (*37*):

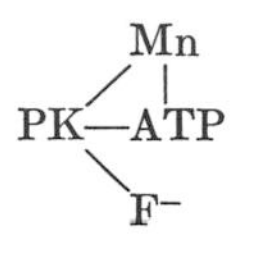

More is known about the role of metals in E–S–M and E–M–S complexes where there is direct coordination of the substrate by the metal in the active complex.

V. Substrate Bridge Complexes (E—S—M)

Substrate bridge complexes appear to be restricted to ternary complexes of nucleoside di- and triphosphates. The structure of nucleoside triphosphates and their metal complexes will be discussed in detail.

A. Electronic Structure of ATP and Its Metal Complexes

Cohn had pointed out in 1959 that most enzymic reactions of ATP involve either phosphoryl transfer (kinases, ATPases, etc.) or nucleotidyl transfer (synthetases and DNA polymerase) (Fig. 2) (*100*).

This point may be explicable in part in terms of the electronic structure of the polyphosphate chain.

A crystallographic study of sodium tripolyphosphate (*101*) has per-

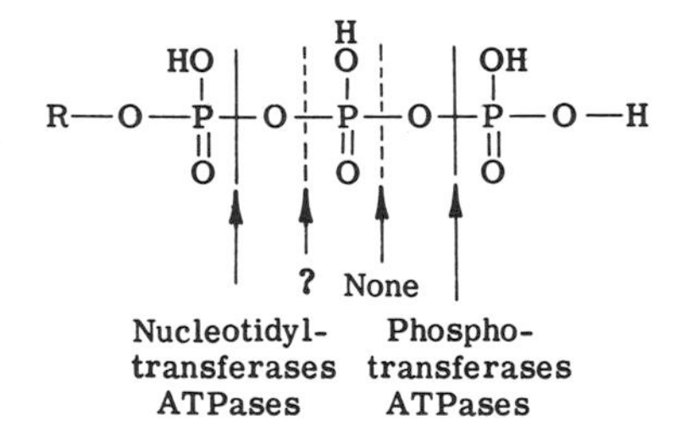

Fig. 2. Types of bond cleavage of ATP [from Cohn (*100*)]. The bond in question is cleaved in the reaction catalyzed by phosphoenolpyruvate synthase (*80c*).

100. M. Cohn, *J. Cellular Comp. Physiol.* **54**, Suppl. 1, 17 (1959).
101. D. R. Davies and D. E. C. Corbridge, *Acta Cryst.* **11**, 315 (1958).
102. L. Pauling, "The Nature of the Chemical Bond," 3rd ed., p. 323. Cornell Univ. Press, Ithaca, New York, 1960.

mitted the estimation by Pauling (*102*) of the bond number of the various P—O bonds as follows:

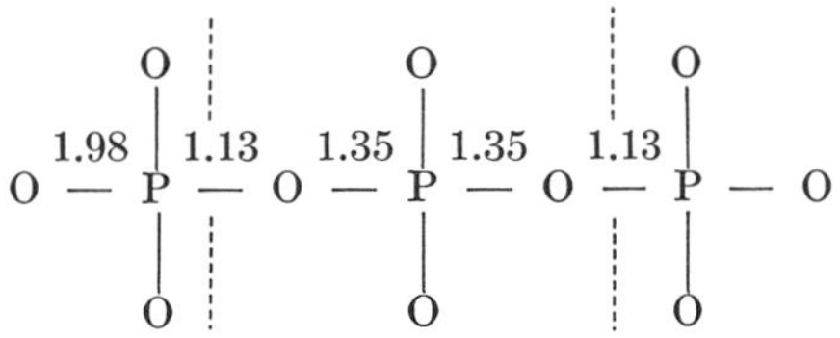

```
          O ¦      O      ¦ O
     1.98 | 1.13  1.35 | 1.35  1.13 |
       O — P — O — P — O — P — O
          | ¦      |      ¦ |
          O ¦      O      ¦ O
```

The bonds connecting the bridging oxygens to the outer phosphorus atoms are longer and weaker (13% double bond character) than the bonds to the central phosphorus atom (35% double bond character). Thus the bonds which are usually broken enzymically are inherently weaker than those which are preserved. In solution, the ^{31}P NMR spectrum of ATP indicates that the chemical shift of the central (β) phosphorus is 11 ppm upfield from the inner (α) phosphorus atom and 14 ppm upfield from the outer (γ) phosphorus atom (*103*), which may be interpreted in terms of greater electronic shielding of the β phosphorus atom as compared to the α and γ ones. Hence, the phosphorus atoms, which are commonly attacked in enzymic reactions, are inherently (10–14 ppm) less shielded and therefore more susceptible to nucleophilic attack. The divalent cations Mn, Ca, and Zn slightly deshield the β and γ phosphorus atoms by 2–3 ppm at all pH values studied (*103*), which would increase their susceptibility to nucleophilic attack.

Crystallographic evidence for P—O bond lengthening on coordination of a metal ion to pyrophosphate is equivocal, but the coordination of a second metal to the same oxygen in the crystal produces significant P—O bond lengthening consistent with partial loss of double bond character (*104*).

The binary metal complexes of ATP (*103, 105*) and of ADP (*106*) in solution are chelate structures in which the phosphate oxygens as well as the adenine nitrogens may contribute ligands to the metal. With Mn, Zn, Co, and Cu, unequivocal evidence for adenine coordination in solution has been obtained by NMR (*103, 105*). The interaction of Mg with the ring has been detected by ORD (*107*) but not by NMR (*103*) (Fig. 3). The conformation, in solution, of the six-membered chelate ring involving pyrophosphates and cations is unknown. Studies using both skeletal and

103. M. Cohn and T. R. Hughes, Jr., *JBC* **235,** 3250 (1960); **237,** 176 (1962).
104. C. Calvo, *Inorg. Chem.* **7,** 1345 (1968).
105. H. Sternlicht, R. G. Shulman, and E. W. Anderson, *J. Chem. Phys.* **43,** 3123 and 3133 (1965).
106. J. S. Leigh, Jr., Ph.D. Dissertation, University of Pennsylvania, 1970.
107. W. G. McCormick and B. H. Levedahl, *BBA* **34,** 303 (1959).

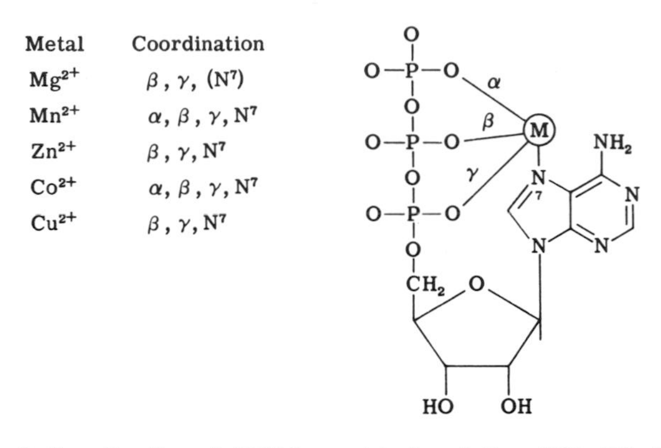

Metal	Coordination
Mg^{2+}	$\beta, \gamma, (N^7)$
Mn^{2+}	$\alpha, \beta, \gamma, N^7$
Zn^{2+}	β, γ, N^7
Co^{2+}	$\alpha, \beta, \gamma, N^7$
Cu^{2+}	β, γ, N^7

FIG. 3. Coordination of ATP by metals in solution (*103, 105, 107*).

space-filling models suggest that a chair conformation, which is rigid, eclipses the charged oxygens of pyrophosphate and might therefore be unstable (Fig. 4). A boat conformation, which has high internal mobility and can interconvert among six orientations, can either eclipse or stagger the oxygens depending on the orientation of the boat. The staggered conformation could therefore minimize repulsions between oxygens and provide greater stability (Fig. 4).

Crystallographic studies of a number of metal–pyrophosphate complexes reveal boat conformations. A predominantly staggered orientation of the charged oxygens is observed for metals with ionic radii < 1.0 Å (Mg, Co, Ni, Cu, Zn, and Na), but a predominantly eclipsed orientation is observed for larger cations with ionic radii ≥ 1.0 Å (Ca, Sr, Ba, and Cd) (*104, 108*).

The relevance of these findings to the dissolved state and to enzyme catalysis remains to be explored. It is of interest that Ca, which forms an eclipsed pyrophosphate complex, is a potent activator of E–S–M en-

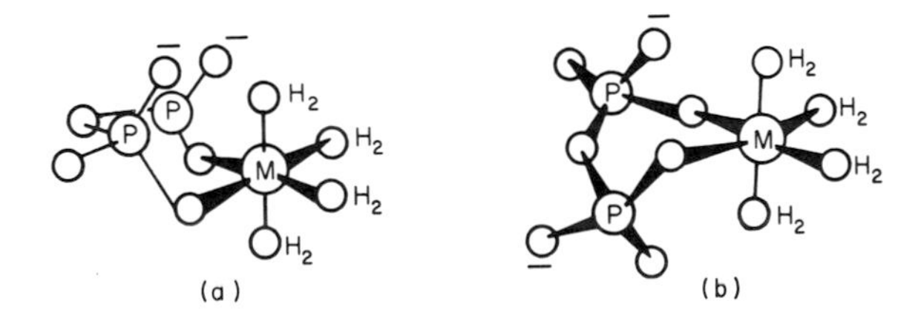

FIG. 4. Conformations of metal–pyrophosphate chelates: (a) chair and (b) boat. The boat conformation permits better separation of the negatively charged oxygens.

108. C. Calvo, *Can. J. Chem.* **43,** 1147 (1965); N. Webb, *Acta Cryst.* **21,** 942 (1966); J. C. Grenier and R. Masse, *Bull. Soc. Franc. Mineral. Crist.* **90,** 285 (1967).

zymes but a potent competitive inhibitor of most E–M–S enzymes; Mg, which forms a staggered pyrophosphate complex, activates both types of enzymes.

The mechanism of formation of these metal nucleotide complexes [Eq. (12)] involves ligand substitution reactions on the metal in which coordinated water molecules are replaced by phosphate and nitrogen ligands. The important kinetic studies of Hammes, Eigen, and co-workers (*109*) indicate that the rates of combination of ADP^{3-} with Mg^{2+} (and probably Ca^{2+}) and of ATP^{4-} with Ni^{2+}, Mg^{2+}, Co^{2+} (and probably Mn^{2+} and Ca^{2+}) are limited by the rate of dissociation of the first water molecule from the coordination sphere of the hydrated metal ion. These results are consistent with the most general chelation mechanism described above [Eq. (6)] in which an S_N1–outer sphere substitution of the first water ligand is followed by rapid closure of the chelate ring.

$$M(H_2O)_6 + ATP \underset{\text{fast}}{\overset{\text{fast}}{\rightleftharpoons}} M(H_2O)_6\cdot ATP \underset{+H_2O}{\overset{-H_2O}{\rightleftharpoons}} (H_2O)_5M{-}O{-}\underset{O}{\overset{O}{P}}{-}O{-}\underset{O}{\overset{O}{P}}{-}O{-}\underset{O}{\overset{O}{P}}{-}O{-}Ad$$

$$\text{fast } -n\,H_2O \;\swarrow\!\nearrow\; \text{slow } +n\,H_2O \qquad (12)$$

$$O{-}\underset{O}{\overset{O}{P}}{-}O{-}\underset{O}{\overset{O}{P}}{-}O{-}\underset{O}{\overset{O}{P}}{-}O{-}Ad \quad (\text{three phosphate O and Ad coordinated to } M(H_2O)_{5-n})$$

B. Mechanisms of Formation of Substrate Bridge Complexes

The ternary E–S–M complexes of nucleoside triphosphates are formed predominantly by the reactions:

$$ATP^{4-} + M(H_2O)_6^{2+} \rightleftharpoons ATP\text{–}M(H_2O)_3^{2-} + 3H_2O \qquad (13)$$
$$E + ATP\text{–}M(H_2O)_3^{2-} \rightleftharpoons E\text{–}ATP\text{–}M(H_2O)_3 \qquad (14)$$

The alternative pathway,

$$E + ATP^{4-} \rightleftharpoons E\text{–}ATP \qquad (15)$$
$$E\text{–}ATP + M(H_2O)_6^{2+} \rightleftharpoons E\text{–}ATP\text{–}M(H_2O)_3 + 3H_2O \qquad (16)$$

is not excluded, but it probably contributes little to the reaction at low concentrations of enzyme, although it may become important at high

109. G. G. Hammes and S. A. Levison, *Biochemistry* **3,** 1504 (1964); H. Diebler, M. Eigen, and G. G. Hammes, *Z. Naturforsch.* **15b,** 554 (1960); M. Eigen and G. G. Hammes, *JACS* **82,** 5951 (1960); **83,** 2786 (1961).

enzyme concentrations where it has been found to be faster, under equivalent conditions, than pathways (13) and (14) (*110*). These observations cast doubt on the generality (*90*) that nuceloside di- and triphosphates always react in the form of their metal complexes, and that the uncomplexed nucleotides invariably act as inhibitors.

When the Mn–ATP and Mn–ADP complexes combine with creatine kinase (*60, 111*), or when the Mn–ATP complex combines with adenylate kinase (*112*), there are no changes in the ligand field about the Mn as detected by EPR, but a large increase in the enhancement occurs because of an increase in the rigidity (i.e., correlation time) of the environment of the coordinated water (*112*). That this increased rigidity is important to the catalysis is suggested by the parallel relationship between the correlation time and the V_{max} for a series of nucleotides with these enzymes and with arginine kinase. This parallel behavior of enhancement and maximal catalytic rate has been interpreted in terms of a graded conformation change in the enzyme induced, as suggested by Koshland (*50*), by the binding of the various metal–nucleotide substrates (*84, 112, 113*). Powerful independent evidence for substrate bridge complexes and for substrate-induced conformation changes in creatine kinase have come from a temperature-jump kinetic study of the binding of ADP, ATP, and their Mg, Mn, and Ca complexes (*110*). The rate constants for all of the metal complexes and for the free nucleotides (at 11°) are of the following order:

$$\text{E} + \text{M–ADP} \underset{10^3\text{–}10^4\ \text{sec}^{-1}}{\overset{10^6\text{–}10^7\ M^{-1}\ \text{sec}^{-1}}{\rightleftharpoons}} (\text{E–ADP–M})_1 \underset{10^3\ \text{sec}^{-1}}{\overset{10^4\ \text{sec}^{-1}}{\rightleftharpoons}} (\text{E–ADP–M})_2 \qquad (17)$$

$$\text{E} + \text{M–ATP} \underset{>10^4\ \text{sec}^{-1}}{\overset{>10^7\ M^{-1}\ \text{sec}^{-1}}{\rightleftharpoons}} (\text{E–ATP–M})_1 \overset{\sim 10^4\ \text{sec}^{-1}}{\longleftrightarrow} (\text{E–ATP–M})_2 \qquad (18)$$

The high magnitude and approximate equality of the second-order rate constants with the three cations and with the free nucleotides argue strongly against a metal bridge complex which would be limited by water exchange rates of the order of 10^5, 10^7, and 10^8 sec^{-1} for Mg, Mn, and Ca, respectively (*31, 33*). In addition to the second-order combination of the substrate with the enzyme, a rapid first-order process, presumably a conformation change, was detected in each case. The rate of this process was three times greater than the maximum rate of ATP formation and twenty times greater than the maximum rate of ADP formation in the overall

110. G. G. Hammes and J. K. Hurst, *Biochemistry* **8**, 1083 (1969).
111. J. S. Taylor, Ph.D. Dissertation, University of Pennsylvania, 1969.
112. W. J. O'Sullivan and L. Noda, *JBC* **243**, 1424 (1968).
113. W. J. O'Sullivan, R. Virden, and S. Blethen, *European J. Biochem.* **8**, 562 (1969).

reaction (*114*), consistent with its role in the catalysis. The nature of the conformation changes are unknown in detail, but they appear to cause a release of protons (*110*) and to render a thiol group more susceptible to alkylation (*115*). The extent of the conformation change in the ternary complex is independent of the activating ability of the cation (*110, 116*).

C. Mechanism of Reactions within Substrate Bridge Complexes

For phosphoryl transfer to occur, the phosphoryl acceptor must be present. For most kinases this requires prior or subsequent combination with the phosphoryl acceptor (S_2) to form a quaternary $E\langle^{ATP\text{–}M}_{S_2}$ complex, although in some cases such as nucleoside diphosphokinase, direct phosphorylation of the enzyme occurs in the ternary E–ATP–M complex (*117*). The direct evidence on pyruvate kinase (*37*) (an E–M–S system to be discussed later), and in model reactions (*118*), strongly suggests that the atom which accepts the phosphoryl group does not enter the coordination sphere of the metal ion but that the phosphoryl group which is transferred does enter the coordination sphere.

Water relaxation data on creatine kinase are consistent with, but do not prove, this view. Hence, the role of the metal in phosphoryl transfer reactions in E–S–M complexes may be solely to activate the phosphorus atom which is to be attacked, by σ and π electron withdrawal, which could lower the $\Delta H^{\ddagger}$ barrier, and to form a rigid polyphosphate (and adenine) complex of the proper conformation in the quaternary complex: a template effect which would lower the statistical ($-T\Delta S^{\ddagger}$) barrier to reaction. The electrophilic effect of a metal ion on ATP, which is weak in free solution (*103*), may be enhanced on the surface of an enzyme where the effective dielectric constant is lower and is estimated to be comparable to that of ethanol (*119*).

Bidentate coordination of the phosphoryl group to be transferred might further facilitate electron withdrawal from phosphorus (*37*), but it may also permit the operation of another effect. The phosphoryl transfer

114. S. A. Kuby, L. Noda, and H. A. Lardy, *JBC* **210,** 65 (1954).

115. W. J. O'Sullivan, H. Diefenbach, and M. Cohn, *Biochemistry* **5,** 2666 (1966).

116. J. S. Leigh, Jr., and M. Cohn, unpublished observations (1969).

117. N. Mourad and R. E. Parks, Jr., *JBC* **241,** 3838 (1966); O. Wålinder, *ibid.* **243,** 3947 (1968).

118. B. S. Cooperman, *Abstr. 158th ACS Meeting New York, 1969* Biol.-59; *Biochemistry* **8,** 5005 (1969).

119. D. C. Turner and L. Brand, *Biochemistry* **7,** 3381 (1968); G. M. Edelman and W. O. McClure, *Accounts Chem. Res.* **1,** 65 (1968).

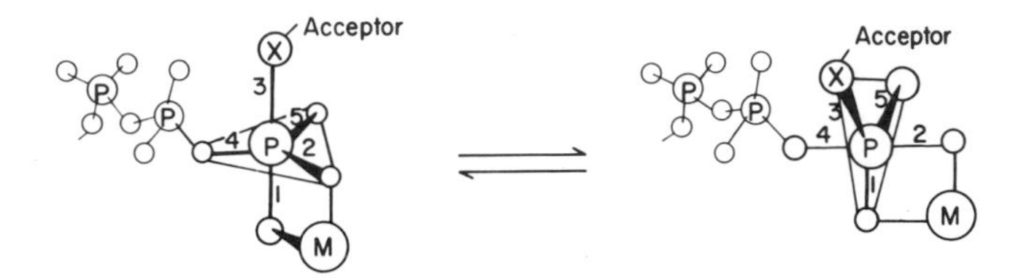

FIG. 5. Pseudorotation mechanism for a phosphoryl transfer reaction which requires a metal ion. The phosphoryl acceptor has entered along the axial bond 3. The metal, M, chelates the pentacoordinate phosphorus intermediate. As the pseudorotation proceeds from left to right, the metal and bond 2 move away from the reader, while bonds 3 and 5 move toward the reader. The acceptor (bond 3) becomes equatorial and the leaving pyrophosphate becomes axial (bond 4), from which it departs.

reaction has long been considered a nucleophilic displacement on phosphorus of the S_N2 variety with pentacoordinate phosphorus as a transition state or intermediate. Bidentate coordination of such a pentacoordinate intermediate by a metal ion might permit a pseudorotation mechanism to operate (*120*) (Fig. 5), as has been proposed to explain the 10^6-fold acceleration of the hydrolysis of ethylene methyl phosphate as compared to trimethyl phosphate (*121*). The divalent cation in the bidentate complex might serve the same role as ethylene (*120*). By constraining the O—P—O bond angle of the pentacoordinate phosphorus to a value close to 90°, the metal ion would insure that one P—O bond (1) would be a long axial bond and the other (2) would be a shorter equatorial bond (Fig. 5). The nucleophilic acceptor would have entered the complex along an axial bond 3. Pseudorotation would rapidly interconvert the P—O bonds 1 and 2 to equatorial and axial, respectively. At the same time the acceptor would become firmly attached to the phosphorus via an equatorial bond, 3, while the leaving ADP would be shifted to a longer axial bond, 4, from which it would rapidly depart.

In support of this hypothesis, the hydrolyses of methyl phosphates coordinated to Co^{III} in complexes in which bidentate phosphate coordination was possible were accelerated by two orders of magnitude over the hydrolysis of comparable monodentate complexes (*120*). However, alternative mechanisms for the model reaction are possible such as attack by a neighboring ligand (*120*). Pseudorotation has also been suggested in the phosphoryl transfer reaction catalyzed by ribonuclease, which requires no metal but involves a cyclic phosphate intermediate similar to ethylene phosphate (*122*).

120. F. J. Farrell, W. A. Kjellstrom, and T. G. Spiro, *Science* **164,** 320 (1969).
121. F. H. Westheimer, *Accounts Chem. Res.* **1,** 70 (1968).
122. J. H. Wang, *Science* **161,** 328 (1968).

Since an activating metal such as Mg is coordinated to the β and γ phosphorus in ATP and to the pyrophosphate of ADP (Fig. 3) (*103*, *106*), a substitution of an outer (γ) phosphate for an inner (α) phosphate must occur some time during the course of the reaction in which ADP is converted to ATP. This migration of the metal might occur on or off the enzyme. That this occurs on the enzyme is suggested by two observations. First, the stable chromic complex of ADP $[(Cr(ADP)(OH)(OH_2)_{1 \text{ or } 2}]^{-1}$, which cannot exchange ligands, is not a substrate but a potent general inhibitor of phosphoryl transfer reactions that occur with creatine kinase, pyruvate kinase, and oxidative phosphorylation (*92*). Second, EPR (*60*) and water relaxation data on creatine kinase (*84*) are consistent with an E–ADP–Mn complex in which both phosphoryl groups are coordinated to Mn, and an E–ATP–Mn complex in which three phosphoryl groups are coordinated to Mn, as in the binary metal nucleotide complexes (Fig. 3) (*103*, *106*).

As indicated in Eqs. (13) to (16), all pathways may operate in the dissociation of E–S–M complexes. By comparing the rates of dissociation of metal nucleotide complexes (*109*) with those of creatine kinase–nucleotide–metal complexes (*110*), it may be concluded that Mg leaves this enzyme predominantly in combination with the nucleotide product but that the alternative pathway in which Mn and Ca may dissociate from their ternary complexes to form free cations may also operate.

VI. Metal Bridge Complexes (E—M—S and $E\langle{}^{M}_{S}$)

A. Development of the Concept

In a classic review published in 1937 (*123*), Hellerman summarized his work on arginase which led to his original proposal of the concept of an enzyme–metal–substrate bridge complex (Fig. 6). The introduction of the metal bridge concept into enzymology, even though it remains to be proven for arginase, was a valuable contribution since it has stimulated much research. Smith and co-workers (*124*) extended this concept to the

123. L. Hellerman, *Physiol. Rev.* **17**, 454 (1937).

124. E. L. Smith, N. C. Davis, E. Adams, and D. H. Spackman, *in* "The Mechanism of Enzyme Action" (W. D. McElroy and B. Glass, eds.), p. 291. Johns Hopkins Press, Baltimore, Maryland, 1954; E. L. Smith, *Advan. Enzymol.* **12**, 191 (1951).

TABLE VI

THERMODYNAMIC AND KINETIC PROPERTIES OF ESTABLISHED ENZYME–METAL–SUBSTRATE, ENZYME–METAL–INHIBITOR, AND RELATED BRIDGE COMPLEXES IN SOLUTION[a]

Protein	Metal	Ligand	$-\log (K_3)$ (M)	$\log (k_{off})$ (sec^{-1})	E_a (kcal/mole)	$\log (k_{on})$ ($M^{-1} sec^{-1}$)	$\log (k_{36})$ (sec^{-1})
Pyruvate kinase[b] (muscle)	Mn^{2+}	FPO_3^{2-}	2.59	≥ 4.53	≥ 2.6	≥ 7.12	≥ 5.52
		H_2O	—	7.93	19.9	—	—
Pyruvate carboxylase[c]	Mn^{2+}	Pyruvate	2.33	4.32	3.1	6.65	6.17
		α-Ketobutyrate	2.45	4.0	—	6.45	5.97
		Oxalacetate	3.38	4.11	15.9	7.49	5.89
		D-Malate	0.23	4.30	2.4	4.53	2.94
		L-Malate	2.06	< 3.30	—	< 5.36	< 3.77
		H_2O	—	6.18	2.3	—	—
Carboxypeptidase A[d]	Mn^{2+}	Indole-Ac	3.15	3.68	—	6.83	6.35
		t-Butyl-Ac	2.27	4.11	—	6.38	5.90
		Br-Ac	1.01	4.64	—	5.65	5.17
		Methoxy-Ac	0.15	> 5.34	—	> 5.49	> 5.01
		H_2O	—	> 6.42	—	—	—
Histidine deaminase[e]	Mn^{2+}	Urocanate	2.70	≥ 4.49	—	≥ 7.19	≥ 6.71
		Imidazole	1.46	≥ 4.72	—	≥ 6.18	≥ 6.18
		H_2O	—	≥ 6.33	—	—	—

D-Xylose isomerase[f]	Mn^{2+}	α-D Xylose	2.66	≥4.63	—	≥7.29	≥7.29
		H_2O	—	≥5.77	—	—	—
Yeast aldolase[g]	Mn^{2+}	FDP	2.71	≥4.68	—	≥7.39	≥5.79
		H_2O	—	5.85	1.8	—	—
Carbonic anhydrase[h] (bovine)	Zn^{2+}	Cl^-	0.72	≥6.60	—	≥7.32	≥6.84
Metmyoglobin[i] (seal)	Fe^{3+}	F^-	1.96	—	—	4.54[j]	—

[a] Definition of symbols: $E\text{–}M\text{–}L \underset{k_{on}}{\overset{k_{off}}{\rightleftharpoons}} E\text{–}M + L$, $K_3 = (E\text{–}M)(L)/(E\text{–}M\text{–}L)$ as defined in Table IV; k_{36} is the specific rate of formation of a coordinated complex from an outer sphere complex as defined in Table II and is calculated by dividing k_{on} by the stability constant of the outer sphere complex. For a divalent cation interacting with an uncharged ligand, a monoanion, a dianion, a trianion, and a tetra-anion, the values are approximately 1, 3, 40, 120, and 400 M^{-1}, respectively, as calculated from electrostatic theory (*31, 32, 34, 65*) or as measured experimentally (*3, 35, 109*).

[b] Mildvan *et al.* (*37*).

[c] Mildvan and Scrutton (*65, 128, 159, 212*).

[d] Shulman *et al.* (*141, 147*).

[e] Givot *et al.* (*160*).

[f] Mildvan and Rose (*161*).

[g] Kobes *et al.* (*162*).

[h] Ward (*163*).

[i] Mildvan *et al.* (*69*).

[j] Rate constant for S_N2 displacement of F^- on metmyoglobin-F by F^-.

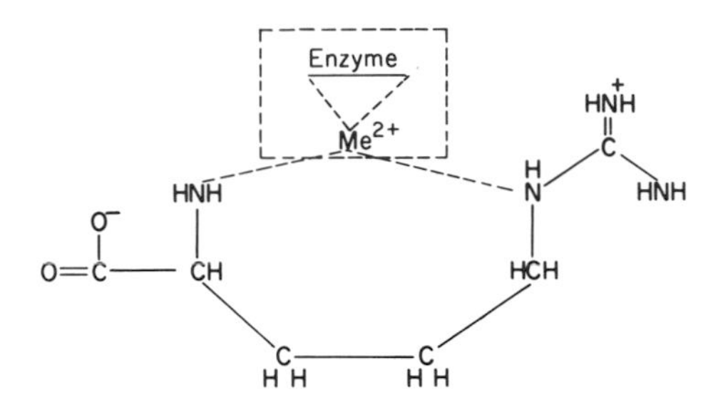

FIG. 6. Hellerman's original proposal of a metal bridge complex in arginase, the first in enzymology [from Hellerman (*123*)].

metal-activated peptidases, prolidase and leucine aminopeptidase; and Klotz, in a review of his binding studies (*125*), suggested that the role of the metal bridge was to stabilize the transition state of the peptide substrate. Metal bridge complexes have since been suggested as functional intermediates in a wide variety of metal-activated and metalloenzymes (*14, 16, 26, 44, 47, 60, 126, 127*) and in other biological processes (*127*). In all such cases the evidence has come from provisional techniques at best (Table IV) and was therefore subject to alternative interpretation. The first direct demonstration of an E–M–S complex was in the case of the fluorokinase reaction of pyruvate kinase where a short-lived E–Mn–O_3PF complex was detected in solution by ^{19}F nuclear relaxation (*37*). Subsequently, the techniques of nuclear relaxation have been used to detect labile E–M–S and E–M–I complexes of six other enzymes (Table V) and to measure their kinetic (Table VI) and structural properties (Fig. 7). X-ray crystallography has been used to detect metal bridge complexes of two enzymes (Table V). With any physical technique (Section III) the mere detection of an enzyme–metal–substrate complex is no assurance that it represents the kinetically active species (*80*). Thus with pyruvate carboxylase, a labile E–Mn–tris base complex has been demonstrated by NMR which is of no catalytic consequence (*128*). For consideration as a functional intermediate in the catalysis, the complex must be demonstrated to possess those kinetic and thermodynamic properties demanded by the rate equation of the enzyme.

The discussion which follows will include both metalloenzymes and metal–enzyme complexes.

125. I. M. Klotz, *in* "The Mechanism of Enzyme Action" (W. D. McElroy and B. Glass, eds.), p. 257. Johns Hopkins Press, Baltimore, Maryland, 1954.
126. A. S. Mildvan, M. C. Scrutton, and M. F. Utter, *JBC* **241,** 3488 (1966).
127. G. C. Eichorn, *Advan. Chem. Ser.* **37,** 37 (1963).
128. M. C. Scrutton and A. S. Mildvan, unpublished observations (1967).

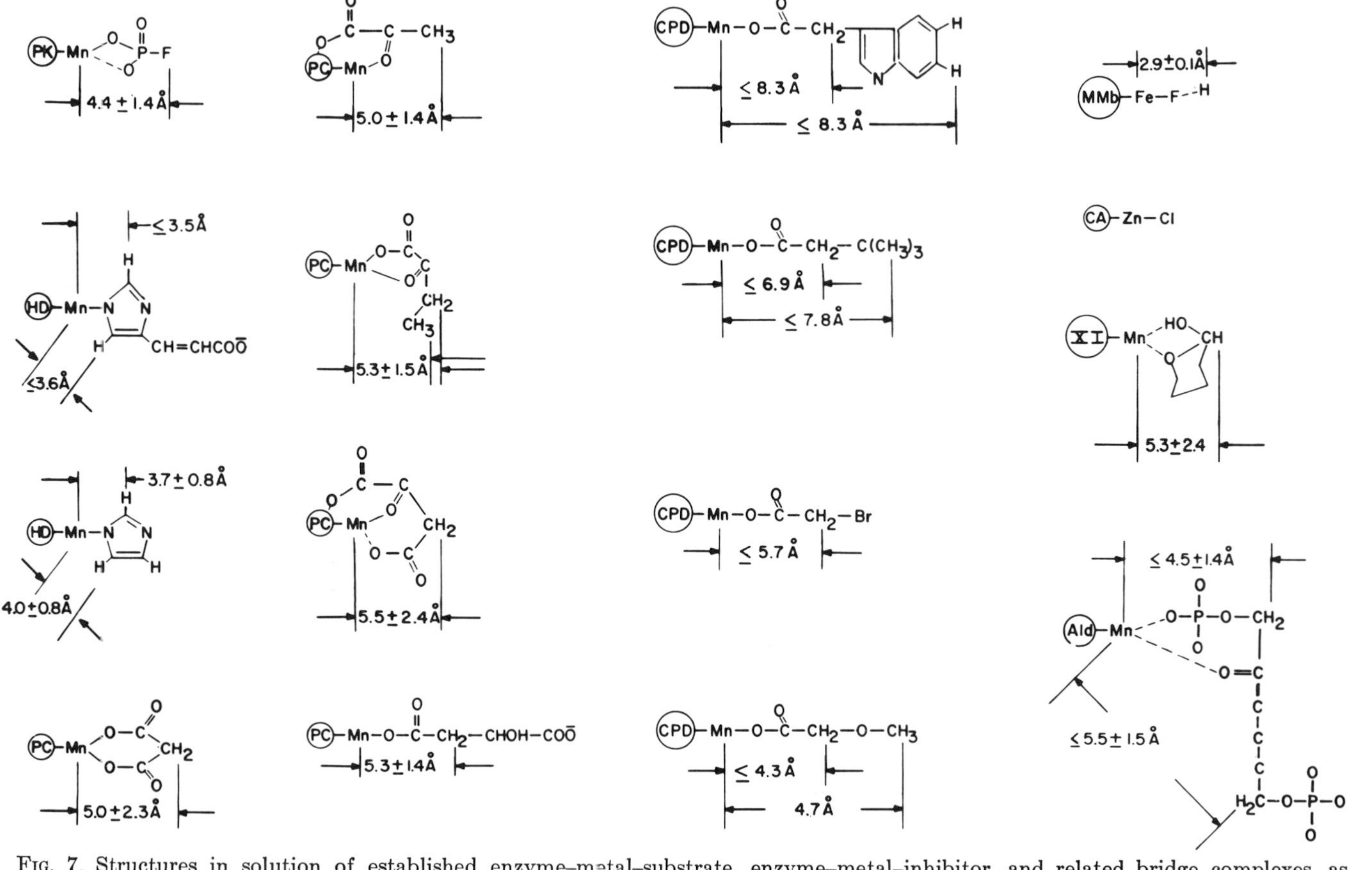

FIG. 7. Structures in solution of established enzyme–metal–substrate, enzyme–metal–inhibitor, and related bridge complexes, as determined by nuclear magnetic relaxation. Abbreviations and references are: PK, pyruvate kinase (*37*); HD, histidine deaminase (*160*); PC, pyruvate carboxylase (*65, 128, 159, 208*); CPD, carboxypeptidase A (*147*); MMb, metmyoglobin (*69*); CA, carbonic anhydrase (bovine) (*163*) XI, D-xylose isomerase (*161*); and Ald, aldolase (yeast) (*162*).

B. The Enzyme–Metal Linkage

1. *The Nature of the Ligands*

The combination of X-ray crystallography and the analysis of amino acid sequences has thus far proved the only reliable means of identifying the ligands donated by a protein to a metal. The limited information available on metalloproteins and metal–protein complexes from these techniques is summarized in Table VII, where the role of imidazole is apparent. The generality of the imidazole ligand for metals had been suggested (*44*) from indirect studies of the effect of pH on the stability constants of the Mn complexes of pyruvate kinase (*44*) and enolase (*129*). However, all thermodynamic techniques for identifying ligands must be viewed with suspicion since the affinity of bases for protons and

TABLE VII
Nature and Geometry of Ligands Donated by Proteins to Metals Established by Crystallography and Sequence Studies

Protein	Metal	Ligands	Coordination geometry	References
Metmyoglobin	Fe	Histidine F8; heme	~Octahedral	(*129a*)
	Cu	Histidine A10; lysine A14; (asparagine GH4)[a]	—	(*129b*)
	Zn	Histidine GH1; lysine A14; (asparagine GH4)[a]	—	(*129b*)
	Au	Histidine GH1; histidine B5	—	(*129a*)
	Hg	Histidine GH1	—	(*129a*)
Methemoglobin	Fe	Histidine F8; heme	~Octahedral	(*129c*)
Cytochrome c	Fe	Histidine 18; (methionine)[b]; heme	Octahedral	(*283–285*)
Carboxypeptidase A	Zn	Histidine 69; histidine 196; glutamate-72	~Tetrahedral	(*129d, 134*)
Rubredoxin	Fe	4 Cysteines	Tetrahedral	(*277*)

[a] Uncertain if directly coordinated.
[b] Not established crystallographically [see Harbury (*286*)].

129a. J. C. Kendrew, *Brookhaven Symp. Biol.* **15,** 216 (1962).
129b. L. J. Banaszak, H. C. Watson, and J. C. Kendrew, *JMB* **12,** 130 (1965).
129c. M. F. Perutz and F. S. Mathews, *JMB* **21,** 199 (1966).
129d. The X-ray study (*134*) identified residue 196 as a ligand for zinc. The amino acid sequence of H. Neurath and co-workers identified residue 196 as histidine [W. N. Lipscomb, personal communication (1969)].
129. B. G. Malmström and L. E. Westlund, *ABB* **61,** 186 (1956).

metal ions may be very different on the surface of an enzyme than in free solution. Similarly, chemical modification of groups is an unreliable way to identify ligands since modification of a distant, nonliganding group may cause a structural change in the protein, which would lower the affinity of the liganding groups of the protein for the metal. The literature contains several examples of incorrect ligand assignments based on indirect thermodynamic studies and chemical modification experiments. Clearly, reliable techniques for determining the ligands for metals on proteins in solution are needed.

Promising beginnings have been made by the use of EPR and NMR techniques. Hyperfine coupling in the EPR spectra of Cu^{II}–protein complexes (*130*) may be used to detect nitrogen ligands. The combination of EPR with specific labeling of protein ligands using magnetic nuclei (e.g., ^{33}S) has been used to identify sulfur coordination in nonheme iron proteins (*70–73*). Nuclear magnetic resonance contact shifts may prove useful for the identification of ligands in complexes of paramagnetic ions such as Ni^{II}, Co^{II}, and Fe^{III} (*131–133*).

Table VII indicates that hard ligands (carboxyl), borderline ligands (imidazole and pyrrole), or soft ligands (thiol) may be donated by a protein to a metal, depending on the function of the protein. In the case of carboxypeptidase, the two imidazole ligands for Zn suggest that the Zn is functioning as a borderline-soft acid rather than a hard acid. If so, then the Zn would have a higher affinity for the softer carbonyl group of the substrate than for a hard hydroxyl group or water molecule in the ternary complex, which would argue in support of the Zn-carbonyl mechanism rather than the Zn-hydroxide mechanism (*134*) (see Section VII). Jørgensen (*135*) has referred to this tendency for ligands of like softness or hardness (Table I) to flock together around a given metal as *symbiotic behavior.*

130. B. G. Malmström and T. Vänngård, *JMB* **2,** 118 (1960); D. C. Gould and H. S. Mason, *in* "Biochemistry of Copper" (J. Peisach, P. Aisen, and W. E. Blumberg, eds.), p. 35. Academic Press, New York, 1966.

131. D. R. Eaton, A. D. Josey, W. D. Phillips, and R. E. Benson, *Discussions Faraday Soc.* **34,** 77 (1962).

132. A. Kowalsky, *Biochemistry* **4,** 2382 (1965).

133. K. Wüthrich, *Proc. Natl. Acad. Sci. U. S.* **63,** 1071 (1969); also *in* "Johnson Foundation Symposium on Probes for Macromolecular Structure and Function" (B. Chance, T. Yonetani, and M. Cohn, eds.). Academic Press, New York. (in press).

134. W. N. Lipscomb, J. A. Hartsuck, G. N. Reeke, Jr., F. A. Quiocho, P. H. Bethge, M. L. Ludwig, T. A. Steitz, H. Muirhead, and J. C. Coppola, *Brookhaven Symp. Biol.* **21,** 24 (1968); W. N. Lipscomb, J. A. Hartsuck, F. A. Quiocho, and G. N. Reeke, Jr., *Proc. Natl. Acad. Sci. U. S.* **64,** 28 (1969).

135. C. R. Jørgensen, *Inorg. Chem.* **3,** 1207 (1964).

Other examples of symbiotic behavior are found in hemoproteins. In hemoglobin and myoglobin the Fe^{II}, which is already coordinated to borderline-soft pyrrole and imidazole ligands, prefers to coordinate the soft CO and O_2 ligands, but it does not coordinate the hard H_2O ligand (*136*). In metmyoglobin, the harder Fe^{III} can coordinate H_2O and ^-OH ligands, but it prefers the softer cyanide and azide ligands (*137*).

2. *The Coordination Geometry*

The coordination geometry (octahedral, tetrahedral, square, etc.) of enzyme–metal complexes, as well as of small coordination complexes, may be determined by X-ray crystallography (Table VII) or in principle by optical (*5*), ORD (*138*), and EPR spectroscopy (*139, 140*).

Thus, the Co^{II}–pyruvate kinase complex has a typical octahedral spectrum (*44*). Octahedral coordination is consistent with the high enhancement of the proton relaxation rate of the Mn–pyruvate kinase complex suggesting the presence of several coordinated water molecules and with the high stability of this complex suggesting several ligands from the protein (*44*). The coordination geometries of five metalloproteins established by X-ray crystallography are given in Table VII.

The crystallographic results on carboxypeptidase (*134*) (Table VII) indicate Zn to be in a tetrahedral environment with three ligands from the protein, which leaves only one hydrated and easily substituted site. Proton relaxation studies of the Mn-substituted carboxypeptidase complex are consistent with one water ligand (*141*). The visible spectrum of the Co^{II} complex of carboxypeptidase (*142*) differs from that of a symmetrical tetrahedral complex such as $(CoCl_4)^{-2}$ (*143*), which may be due to the nature of the ligands and to deviations from perfect tetrahedral symmetry. Analogous spectral observations on Co^{II} carbonic anhydrase (*144*) and on Co^{II} alkaline phosphatase (*145*), as well as the EPR parameters of various nonheme iron and Cu^{II} proteins, have been interpreted in terms of a strained coordination with unusually high re-

136. C. L. Nobbs, H. C. Watson, and J. C. Kendrew, *Nature* **209,** 339 (1966).
137. R. Lemberg and J. W. Legge, "Hematin Compounds and Bile Pigments," pp. 219 and 221. Wiley (Interscience), New York, 1949.
138. R. G. Wilkins and M. J. G. Williams, *in* "Modern Coordination Chemistry" (J. Lewis and R. G. Wilkins, eds.), p. 174. Wiley (Interscience), New York, 1960.
139. S. I. Chan, B. M. Fung, and H. Lütje, *J. Chem. Phys.* **47,** 2121 (1967).
140. H. Levanon and Z. Luz, *J. Chem. Phys.* **49,** 2031 (1968).
141. R. G. Shulman, G. Navon, B. J. Wyluda, D. C. Douglass, and T. Yamane, *Proc. Natl. Acad. Sci. U. S.* **56,** 39 (1966).
142. J. E. Coleman and B. L. Vallee, *JBC* **235,** 390 (1960).
143. N. S. Gill and R. S. Nyholm, *JCS* p. 3997 (1959).
144. S. Linskog and P. O. Nyman, *BBA* **85,** 462 (1964).
145. R. T. Simpson and B. L. Vallee, *Federation Proc.* **27,** 291 (1968).

activity; an *entatic state,* ". . . with energy closer to that of a unimolecular transition state than to that of a conventional stable molecule, thereby constituting an energy poised domain" (*146*).

The entatic concept thus constitutes a kinetic theory of the ground state and transition state for coordination reactions which is based solely on static observations of the ground state. Direct examination of the kinetic and thermodynamic properties of metals in enzymes (*63, 66,* Table VI) offers little support for this theory. Rates of ligand substitution, ligand deprotonation, and electron transfer, which are found in simple model systems, are often as high or higher than those found for metalloenzymes or metal-activated enzymes. Thus, Mn^{2+} exchanges water ligands at a rate of 10^7 sec^{-1} in the $Mn(H_2O)_6$ complex and at a rate of 10^6–10^8 sec^{-1} in various enzyme environments (*63, 66*). Similarly, Mn^{2+} exchanges substrate or product ligands at the same rate or at slower rates on enzymes, including carboxypeptidase (*147*), as in simple complexes (*31, 33*). Fe^{III} loses a proton at the rate of 10^6 sec^{-1} in the aquo complex but at 10^4 sec^{-1} in various methemoproteins (*69, 148*). The electron exchange between ferricyanide and ferrocyanide takes place at a specific rate of 10^5 M^{-1} sec^{-1} (*149*), which is ten times faster than the electron exchange between reduced and oxidized cytochrome c (*132*). The electron transfer between ferrocyanide and Fe^{III} $(phenanthroline)_3$ ($> 10^8$ M^{-1} sec^{-1}) (*150*) is at least an order of magnitude faster than the electron transfer between Fe^{II} cytochrome c and ferricyanide (*151*). In each of these examples, the small symmetric metal complex was as reactive or more reactive than the corresponding, asymmetric metal–protein complex, indicating that the protein has not conferred kinetic reactivity upon the metals which are inherently more reactive than the limting rates of enzyme catalysis (*33*). Moreover, the high affinity of carboxypeptidase for Zn (*13*) and of pyruvate kinase for Mn (*44*), compared with the stability of chelates which use the same ligands (*17, 152*), argues against a thermodynamically strained coordination.

The coordination of a metal by an enzyme does, in certain cases (Sec-

146. B. L. Vallee and R. J. P. Williams, *Proc. Natl. Acad. Sci. U. S.* **59,** 498 (1968).

147. G. Navon, R. G. Shulman, B. J. Wyluda, and T. Yamane, *Proc. Natl. Acad. Sci. U. S.* **60,** 86 (1968).

148. S. Maricic, A. Ravilly, and A. S. Mildvan, *in* "Hemes and Hemoproteins" (B. Chance, R. W. Estabrook, and T. Yonetani, eds.), p. 157. Academic Press, New York, 1966.

149. A. Loewenstein and G. Ron, *Inorg. Chem.* **6,** 1604 (1967).

150. B. M. Gordon, L. L. Williams, and N. Sutin, *JACS* **83,** 2061 (1961).

151. N. Sutin and D. R. Christman, *JACS* **83,** 1773 (1961).

152. L. G. Sillen and A. E. Martell, *Chem. Soc.* (London), *Spec. Publ.* **17** (1964); C. N. C. Drey and J. S. Fruton, *Biochemistry* **4,** 1258 (1965).

tion VII), confer the appropriate geometry for catalysis among groups which are inherently highly reactive, or the metal may function by any of the other mechanisms listed in Table III.

C. Mechanisms of Formation of Binary Enzyme–Metal Complexes

Little data are available on the rates of combination of metal ions with metal-activated enzymes. Preliminary investigations of the combination of Ni with pyruvate kinase (*153*) suggest that the rate is limited by the rate of departure of water from the coordination sphere of Ni (S_N1 outer sphere chelation) as has been found for simple Ni amino acid chelates (*32*). The combination of pyruvate kinase with monovalent cations (and with Mn^{2+}), which lose water more rapidly than Ni (Section II,B,1), appears to be followed by slower conformational changes in the protein ($\leq 10^4$ sec^{-1}) (*154, 155*) analogous to the rate-limiting ring closure found for reaction (8). The rate of the fastest conformational change is rapid compared to the maximal turnover number of pyruvate kinase ($\sim$300 sec^{-1}) (*44*).

The activation of a number of metal-requiring peptidases such as prolidase and leucine aminopeptidase by Mn (*124*) has long been known to be a slow reaction requiring several hours. Since Mn can exchange ligands in less than a microsecond (Table VI), this slow activation is undoubtedly owing to conformation changes in the protein which are required either to close chelate rings to the metal (*124*) or to complete the structure of the active site in some other way. It is clear that these slow conformation changes do not recur with each turnover of the enzyme, even though the metal may bind and dissociate rapidly. Thus, the conformation change is a step which is kinetically separable from the metal-binding step.

$$E + M(H_2O)_6 \rightleftharpoons E\text{–}M(H_2O)_{6-n} + nH_2O \quad (19)$$

$$E\text{–}M(H_2O)_{6-n} \rightleftharpoons E^*\text{–}M(H_2O)_{6-n} \quad (20)$$

Protein conformation changes on combination of metal activators with enzymes have been directly detected by UV difference spectroscopy for pyruvate kinase (*98, 99*), enolase (*156*), phosphoglucomutase (*157*), and

153. F. J. Kayne, unpublished observations (1968).
154. G. C. Czerlinski and A. S. Mildvan, unpublished observations (1967).
155. F. J. Kayne, *in* "Johnson Foundation Symposium on Probes for Macromolecular Structure and Function" (B. Chance, T. Yonetani, and M. Cohn, eds.). Academic Press, New York (in press).
156. D. P. Hanlon and E. W. Westhead, *BBA* **96**, 537 (1965); *Biochemistry* **11**, 4247, 4255 (1969).

glutamine synthetase (*96*). The correlation of these conformation changes with catalysis may be positive, negative, or absent. A positive correlation is found with pyruvate kinase which binds the inhibitory Ca^{2+} at the same site as the activating Mn^{2+} (*44*), without undergoing a conformation change detectable by UV spectroscopy (*158*).

Conversely, with phosphoglucomutase (*157*), cations which differ in activating ability by two orders of magnitude (e.g., Mg and Zn) induce the same protein difference spectrum in the binary complex (*157*) (no correlation); and, in the ternary E–M–S complex, the most activating cation (Mg^{2+}) induces the least spectral change (negative correlation) (*157*).

D. Mechanisms of Formation of Ternary Metal Bridge Complexes

For metalloenzymes the metal bridge complex must form by a combination of the substrate with the metal enzyme complex, E–M + S $\rightleftharpoons$ E$<$(M | S), etc. The kinetics of this process have been investigated by nuclear relaxation for the Mn metalloenzyme, pyruvate carboxylase (*65*, *128*, *159*). The rate constants ($k_{3,6}$) for the coordination of the substrates, pyruvate ($10^{6.2}$ sec^{-1}), α-ketobutyrate ($10^{6.0}$ sec^{-1}), and oxalacetate ($10^{5.9}$ sec^{-1}), by the enzyme-bound Mn, from an outer sphere complex, agree with the rate of dissociation of water protons from the coordination sphere of enzyme-bound Mn ($10^{6.2}$ sec^{-1}), strongly suggesting an S_N1 outer sphere mechanism, as has been found for simple complexes of Mn (*31*).

$$\begin{array}{ccccc} \text{E–Mn(OH}_2) + (\text{OH}_2)\text{S} & \overset{\text{fast}}{\rightleftharpoons} & \text{E–Mn(OH}_2)\text{S} & \underset{k_{\text{off}}}{\overset{k_{3,6}}{\rightleftharpoons}} & \text{E–Mn–S} \\ 10^{6.2}\ \text{sec}^{-1} \updownarrow & & +\text{H}_2\text{O} & & +\text{H}_2\text{O} \\ \text{E–Mn} + \text{H}_2\text{O} & & & & \end{array} \tag{21}$$

Analogous kinetic data consistent with an S_N1–outer sphere mechanism have been obtained for the coordination of substrate or inhibitor ligands by the Mn complexes of histidine deaminase (*160*), D-xylose isomerase (*161*), yeast aldolase (*162*) and Zn-carbonic anhydrase (*163*) (see Table

157. W. J. Ray, Jr., *JBC* **244,** 3740 (1969); E. J. Peck, Jr. and W. J. Ray, Jr., *ibid.* pp. 3748 and 3754.

158. C. H. Suelter, unpublished observations (cited in Mildvan and Cohn, *44*).

159. M. C. Scrutton, *Federation Proc.* **28,** 534 (1969); M. C. Scrutton and A. S. Mildvan, *ABB* (in press).

160. I. Givot, A. S. Mildvan, and R. H. Abeles, *Federation Proc.* **29,** 531 (1970).

161. A. S. Mildvan and I. A. Rose, *Federation Proc.* **28,** 534 (1969); also K. Schray, A. S. Mildvan, and I. A. Rose, unpublished observations (1969).

VI). As with nonenzymic coordination reactions, the S_N1–outer sphere mechanism is therefore widespread but not general. Thus, studies of the rate of binding of FPO_3^{2-} to the pyruvate kinase–Mn complex by ^{19}F nuclear relaxation suggest that ~500 water ligands may change before one fluorophosphate enters the coordination sphere of Mn (*37*); and with pyruvate carboxylase, 1100 water ligands on Mn will exchange before the inhibitor, D-malate, will coordinate (*128*). The reason for these "slow" coordination reactions is not clear. Delayed closure of a strained chelate ring may occur in the former case and steric interference with coordination of the first ligand may occur in the latter case.

The coordination of various inhibitors by Mn-carboxypeptidase may be fit by an S_N1–outer sphere mechanism (*147*) only if one assumes that the stability constants of the outer sphere complexes decrease with decreasing size of the inhibitor, from a typical value of $\leq 2.6\ M^{-1}$ for indoleacetic acid (*34, 35, 65*) to an exceedingly low value of $0.17\ M^{-1}$ for bromoacetate.

The kinetic data for fluoride exchange into metmyoglobin fluoride are best fit by an S_N2 mechanism (*69*) (Table VI), which would involve a septa coordinate transition state for fluoride [Fig. 1, structure (V)] analogous of the structure predicted by Griffith for oxyhemoproteins (*164*).

For metal-activated enzymes, three pathways can operate for the formation of the metal bridge complex:

(1) $$E \underset{-M}{\overset{+M}{\rightleftharpoons}} E\text{–}M \underset{-S}{\overset{+S}{\rightleftharpoons}} E\langle{}^{M}_{S} \qquad (22)$$

(2) $$E \underset{-S}{\overset{+S}{\rightleftharpoons}} E\text{–}S \underset{-M}{\overset{+M}{\rightleftharpoons}} E\langle{}^{M}_{S} \qquad (23)$$

(3) $$M \underset{-S}{\overset{+S}{\rightleftharpoons}} M\text{–}S \underset{-E}{\overset{+E}{\rightleftharpoons}} E\langle{}^{M}_{S} \qquad (24)$$

Certain of these pathways may predominate, depending on the individual rate constants and the concentrations of E, M, and S.

Thus, for pyruvate kinase–Mn–FPO_3, Eq. (22) has been shown to

162. R. Kobes, A. S. Mildvan, and W. J. Rutter, *Abstr. 158th ACS Meeting, New York, 1969* Biol.-58; also unpublished observations (1969).
163. R. L. Ward, *Biochemistry* **8,** 1879 (1969).
164. J. S. Griffith, *Proc. Roy, Soc.* **A235,** 23 (1956).

operate by direct kinetic measurements of this process (*37*), though pathways (23) and (24) have not been excluded. For pyruvate kinase-Mn-phosphoenolpyruvate, the operation of pathways (22), (23), and (24) are consistent with binding studies and with a steady state kinetic analysis (*44*). For pyruvate kinase–Mn–ADP, the random operation of pathways (22), (23), and (24) provide a reasonable fit to substrate and metal binding studies and to kinetic studies with Mn^{2+} and Mg^{2+} (*44*). An alternative and simpler scheme for ADP, involving only Eq. (24), has been proposed for pyruvate kinase and for all other enzymes which utilize nucleotide substrates (*90*). However, this more restrictive scheme is inconsistent with the binding data and provides no better fit to the kinetic data on pyruvate kinase (*91*). Moreover, recent rapid quenching experiments (*91*) indicate that the pathways yield the same amount of

$$\text{Pyruvate kinase–Ni} + \text{ADP} \rightarrow \rightarrow \text{products} \qquad (25)$$

and

$$\text{Pyruvate kinase} + \text{NiADP} \rightarrow \rightarrow \text{products} \qquad (26)$$

product (pyruvate) in one turnover, before the Ni has had time to dissociate from the enzyme. For histidine deaminase (*160*) and β-methylaspartase (*165*), kinetic and binding data are consistent with the operation of all pathways to form the respective ternary complexes. With D-xylose isomerase, the kinetic data fail to detect an E–S complex (*166, 167*) and the stability of M–S is low. Hence, the enzyme–Mn–D-xylose bridge complex (*101*) forms predominantly by Eq. (22) under the conditions of the kinetic experiment. Similar kinetic results were obtained with the related enzyme, L-arabinose isomerase (*168*).

E. Reactions within the Coordination Sphere of Metal Bridge Complexes

The structures in solution, as found by NMR, for various enzyme–metal–substrate bridge complexes, which possess thermodynamic and kinetic properties consistent with their participation in catalysis, are summarized in Fig. 7. From these structures the reaction mechanisms and the role of the metal may be inferred and will be discussed in detail in Section VII. With pyruvate kinase, pyruvate carboxylase, histidine deaminase, and probably also with carboxypeptidase, D-xylose isomerase, and aldolase, the primary role of the metal appears to be the withdrawal

165. H. J. Bright, *JBC* **240,** 1198 (1965).
166. K. Yamanaka, *BBA* **151,** 670 (1968).
167. K. Yamanaka, *ABB* **131,** 502 (1969).
168. T. Nakamatu and K. Yamanaka, *BBA* **178,** 156 (1969).

of electrons from the substrate, predominantly via σ bonding (Table III, 1A-1) to activate electrophiles (Table III, 1A-1b), or to create nucleophiles (Table III, 1A-1c). The gathering of ligands (Table III, IIA) may be important in pyruvate carboxylase and pyruvate kinase, and the masking of a nucleophile may prevent side reactions in histidine deaminase (Table III, IA-1a). Additional roles of metals which vary with the details of individual enzyme mechanisms are summarized in Table III and will be discussed in the next section.

VII. The Role of Metals in the Mechanisms of Specific Enzymes

As pointed out elsewhere (*63*), a complete description of the mechanism of action of an enzyme requires a thorough knowledge of the following points:

(1) The geometric and electronic structure at the active site of the enzyme and of its complexes with substrates and products.
(2) The affinity of substrates and specificity of substrates.
(3) The kinetic scheme of the reaction, i.e., the number and arrangement of chemical steps in the catalytic process.
(4) The chemical mechanism of each step, i.e., a description of the structures of the intermediate complexes, transition states, and electronic and atomic rearrangements involving the enzymes and substrates.
(5) The rate constants of the individual steps.
(6) A rationale for the magnitude of the rate constants in terms of structure.

Obviously, no single technique can provide information on all of these points. The presence of a metal at or near the active site can often serve as a probe (*63*, *169*) which can increase the number of techniques that can be applied to the problem of working out an enzyme mechanism. Nuclear magnetic relaxation with paramagnetic metals can supply some key data, particularly relevant to points (1), (2), (4), and in some cases, (5); other forms of spectroscopy (EPR, optical, ORD, and CD) can provide information relevant to point (1) and, in some cases, (2); X-ray crystallography provides a powerful approach toward points (1) and (2); for points (2), (3), and (5), binding studies, steady state and transient state kinetics are needed; the study of appropriate model reactions may elucidate point (6). Hence, the following discussion of

169. B. L. Vallee and J. F. Riordan, *Ann. Rev. Biochem.* **38**, 733 (1969).

the role of metals in enzyme mechanisms will necessarily include the results of several types of studies relevant to the mechanisms.

A. Phosphoryl and Nucleotidyl Transferring Enzymes

As seen in Table V, although most kinases form E–S–M complexes with nucleotides, some kinase enzymes, notably those with pyruvate or phosphoenolpyruvate as substrates, form E—M—S complexes where the metal ion interacts with the enzyme, as well as the nucleotides.

1. *Creatine Kinase*

Much has already been said about the mechanism of creatine kinase in this review since it is the prototype of the kinases which utilize a substrate bridge (E–S–M) complex. As mentioned above, the active ternary E–nucleotide–M complexes form predominantly by a combination of the enzyme with the metal–nucleotide complex without altering the nature of the ligands in the coordination sphere of Mn (*60*, *84*, *111*, *115*) and at a rate which is independent of the nature of the metal (*110*). The "rigidity" of the coordination sphere, as measured by the correlation time, increases in the ternary complex. The enhancements of the PRR for the ternary complexes of five nucleoside diphosphates, ADP, deoxy-2′-ADP, deoxy-3′-ADP, IDP, and UDP revealed a 1:1 correspondence between the correlation time (rigidity), the increased reactivity of the essential sulfhydryl group, and the maximal velocity of the enzymic reaction for these substrates (*84*). A similar enhancement pattern was observed for the triphosphates. Thus, the more active nucleotide substrates form a ternary complex in which the hydration sphere of Mn may be more immobilized as a result of a conformational change at the active site, in support of Koshland's induced fit hypothesis of substrate specificity (*50*). The rate of the conformation change is consistent with its role in catalysis (*110*).

The effect of adding creatine to form an abortive quaternary complex $\left(E\begin{smallmatrix}\text{ADP-Mn}\\ \text{creatine}\end{smallmatrix}\right)$ is to lower the rate of water ligand exchange on Mn and to further increase the correlation time (rigidity) of the hydration sphere sphere of Mn. Thus, the presence of M–ADP and creatine on the enzyme together cause a profound change in conformation at the active site reflected both in the immobilization of water in the environment of the metal ion and the rate of water ligand exchange (*84*). The change in structure between the ternary and quaternary complexes is reflected in the reactivity of the essential sulfhydryl group toward iodoacetic acid (*115*) and the susceptibility of the enzyme toward tryptic digestion (*170*).

The geometry of the active site of creatine kinase has recently been further investigated with another paramagnetic probe by reacting the essential sulfhydryl group (one per site) with the iodoacetamide derivative of a nitroxide spin label, *N*-(1-oxyl-2,2,5,5-tetramethyl-3-pyrrolidinyl) iodoacetamide (*171*). In the ternary E–ADP–M complex, the extent of the conformational change (as measured by changes in the EPR spectrum of the spin label) is independent of the cations for Mg, Ca, Sr, and Ba, which vary by two orders of magnitude in activating ability (*171*), [as also found in the kinetic study (*110*)]. In the quaternary complex, however, the metal specificity manifests itself by a graded change in the EPR spectrum of the spin label (*172*). The calculated distance between the spin label and various nuclei of the substrates as determined by nuclear relaxation as well as all other known facts about creatine kinase (*173, 174*) is accommodated by the mechanism shown in Fig. 8, which has been modified in several respects from the mechanisms suggested earlier by Rabin and Watts (*175*) and by Crane (*176*). The principal modification from the Rabin and Watts structure is the shift from an E–M–S complex to an E–nucleotide–M complex. The mechanism suggested by Crane has been expanded to two separate steps: one involving phosphoryl transfer (a → b), and the second involving a ligand exchange on the metal coupled to a conformational change of the E–M–nucleotide complex (b → c).

As discussed above, the phosphoryl transfer reaction (a → b) may involve pseudorotation of a pentacoordinate phosphorus intermediate (Fig. 5), although there is no direct evidence for it. The role of the metal in this reaction is thus to promote the electrophilicity of phosphorus by σ electron withdrawal (Table III; IA-1b), to serve as a template for the conformation change in the ternary and quaternary complexes (b → c) (Table III; IB-4b), and, possibly, to serve as a template for pseudorotation in the phosphoryl transfer reaction (Fig. 5).

170. L. W. Cunningham and L. W. Jacobs, *Biochemistry* **7**, 143 (1968).

171. J. S. Taylor, J. S. Leigh, Jr., and M. Cohn, *Proc. Natl. Acad. Sci. U. S.* **64**, 219 (1969); also unpublished observations (1969).

172. M. Cohn, *Quart. Rev. Biophys.* **3**, 61 (1970).

173. S. A. Kuby and E. A. Noltmann, "The Enzymes," 2nd ed., Vol. 5, p. 515, 1962.

174. J. F. Morrison and E. James, *BJ* **97**, 37 (1965).

175. B. R. Rabin and D. C. Watts, *BJ* **85**, 507 (1962).

176. R. K. Crane, *Comp. Biochem.* **15**, 200 (1964).

176a. R. Kassab, C. Roustan, and L. Pradel, *BBA* **167**, 308 (1968); T. A. Mahowald, *Federation Proc.* **28**, 601 (1969).

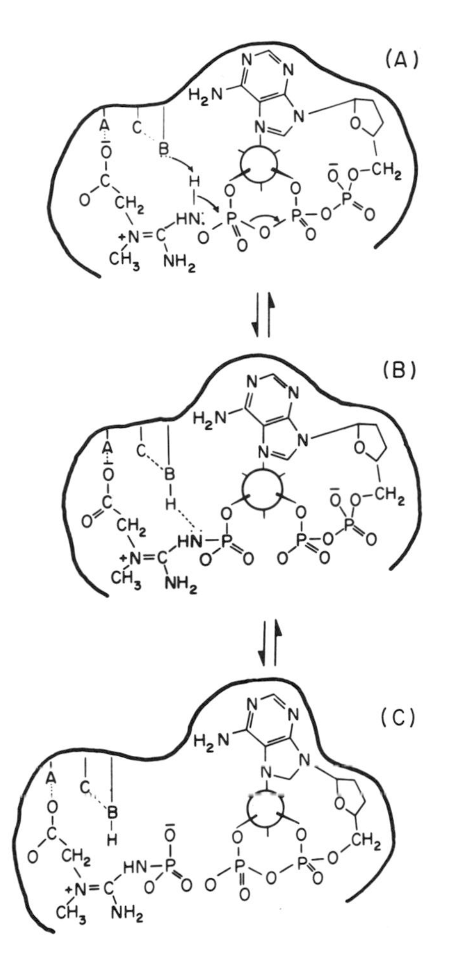

FIG. 8. Mechanism of creatine kinase [from Mildvan and Cohn (*63*) and Cohn (*172*)]. The phosphoryl transfer (a $\rightleftarrows$ b) is separated from the ligand exchange and coupled conformational change (b $\rightleftarrows$ c): (A) cationic group, possibly an ϵ-amino group of lysine (*176a*), (B) probably the essential sulfhydryl group (*175*), and (C) a basic group, either carboxylate or imidazole, which shields the sulfhydryl group (*175*).

2. *Arginine Kinase*

The water relaxation data for arginine kinase (*113*) is entirely analogous to that of creatine kinase. The stereospecificity of L-arginine as a substrate is clearly correlated with its large effect in lowering the enhancement of the E–ADP–Mn complex, in contrast to the relatively small effect of the inhibitor D-arginine.

A suggestion that this enzyme might function in a ping-pong kinetic scheme with a phosphorylated protein intermediate (*177*) is inconsistent with the low rate of the required isotopic exchange reactions (*177*), with the direct detection of quaternary complexes (*113*), and with the approximate agreement between the K_m values of the substrates MnADP, MnATP, Mn-2′-dADP, and their respective dissociation constants from the enzyme (*113*). The latter observations suggest a rapid random equilibrium kinetic scheme (*113, 178*), as had been found for creatine kinase (*174*). Hence the detailed mechanism and the roles of the metal suggested for creatine kinase may also apply to arginine kinase.

3. *Adenylate Kinase*

The data on the role of Mn in adenylate kinase is more limited since only ternary complexes with ATP could be investigated by magnetic resonance techniques. An E–ATP–Mn complex, suggested by the enhancement data, has been established by EPR (Table V). The ternary ATP complex exhibits a unique feature in that the observed enhancement at high temperature is higher than any other enhancement observed. In fact, it exceeds the predicted theoretical maximum (*63, 179*), suggesting a highly immobilized hydration sphere in the ternary complex.

There is a fairly good correlation between enhancement and reaction velocity for a series of nucleotides at room temperature but not at other temperatures. These findings and model studies support a mechanism of the type proposed by Lowenstein (*180*), with the modifications indicated in Fig. 9. The role of the metal may be to gather and activate ligands, promote the electrophilicity of phosphorus, and serve as a template for conformation changes, possibly pseudorotation (Fig. 5). The findings of maximal activity at a nucleotide-metal ratio of 2:1 is in accord with this mechanism (*181*). The failure of Ca^{2+} to activate this enzyme (*181*) is, however, more typical of E–M–S complexes (Table IV).

4. *Nucleotidyl Transferring Enzymes*

The reaction catalyzed by UDPG pyrophosphorylase, UTP + glucose-1-phosphate $\rightleftarrows$ UDPG + PP_i, is a nucleotidyl rather than a phosphoryl group transfer from a nucleoside triphosphate. As with creatine kinase, the enhancement behavior is consistent with a substrate bridge complex (Table V) (*182*). However, unlike creatine kinase, this enzyme does not

177. M. L. Uhr, F. Marcus, and J. F. Morrison, *JBC* **241,** 5428 (1966).
178. R. Virden, D. C. Watts, and E. Baldwin, *BJ* **94,** 536 (1965); E. Smith and J. F. Morrison, *JBC* **244,** 4224 (1969).
179. B. Sheard and E. M. Bradbury, *Progr. Biophys. Mol. Biol.* (1970) (in press).
180. J. Lowenstein, *BJ* **70,** 222 (1958).
181. L. Noda, "The Enzymes," 2nd ed., Vol. 6, p. 139, 1962.
182. G. H. Reed and M. Cohn, unpublished observations (1969).

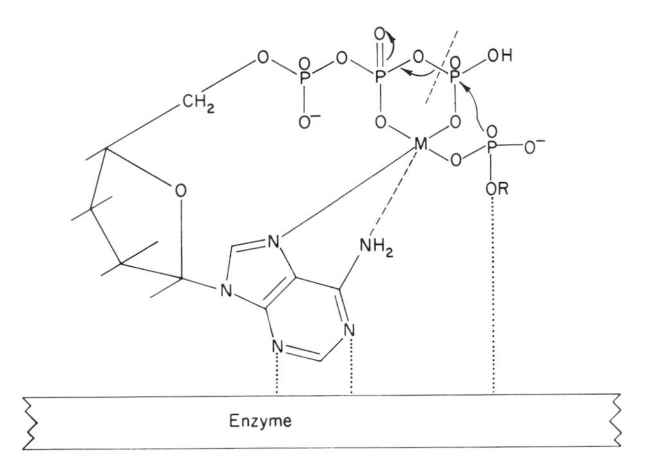

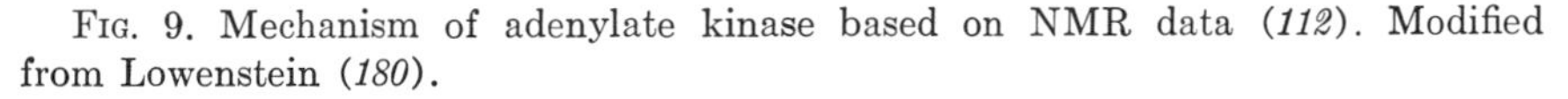
FIG. 9. Mechanism of adenylate kinase based on NMR data (*112*). Modified from Lowenstein (*180*).

form an enhanced E–pyrophosphate–Mn complex (Table V). Another notable difference between the ternary nucleoside triphosphate complex with this enzyme and the kinases is in the binding of Mn. Upon addition of either creatine kinase or pyruvate kinase to a solution containing Mn and nucleotide, the amount of free Mn decreases as would be anticipated in the formation of a metal bridge complex or from the shift in equilibria in a nucleotide bridge complex. However, upon formation of the ternary E–M–S complex with UDPG pyrophosphorylase, as evidenced by a PRR enhancement for water, free Mn is increased, as detected by EPR. Analogous observations have been made with tryptophan (*183*) and valine synthetase (*183a*), which also transfer nucleotidyl groups (Table V).

Several alternative explanations for these observations have been offered (*63*). The most attractive (but as yet unproved) explanation is that the binding of MnUTP to the enzyme (in an E–S–M complex) causes the Mn to shift its coordination of the triphosphate from tridentate α,β,γ coordination (high affinity) to α coordination (low affinity):

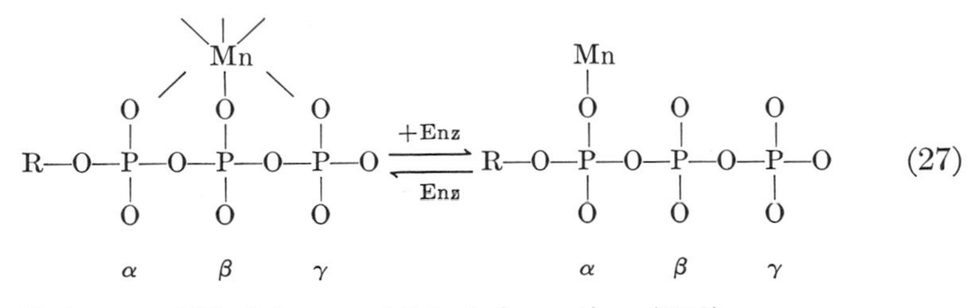

(27)

183. B. K. Joyce and M. Cohn, unpublished observations (1968).
183a. M. Yaniv and M. Cohn, unpublished observations (1967).

Such a shift of the Mn (possibly facilitated by a monovalent cation) (*180*) would concentrate its σ electron withdrawing ability on the phosphoryl group to be attacked. This mechanism may be relevant to a wide class of biosynthetic enzymes that catalyze group transfer such as DNA polymerase. However, the latter enzyme has recently been shown to require metal for the tight binding of deoxynuclcoside triphosphate substrates (*184*) and to be inhibited by Ca^{2+} (*185*), suggesting an E–M–S complex (Table V). The enhancement behavior of this enzyme (Type II) (*185a*) is consistent with this view.

5. *Muscle Pyruvate Kinase*

This enzyme, the prototype of the metal bridge kinases, has already been discussed in Section VI, where it was concluded that a random kinetic scheme for the formation of all ternary complexes provides the best fit to the available data. The large enhancement of the water relaxation rate in the presence of the binary E–Mn complex (*186*) is greater than the enhancements of all ternary complexes (*44*). The enhancements of the ternary complexes (ϵ_T) of the "phosphorylated" substrates (PEP and ATP) were found to be less than ϵ_T of the corresponding "nonphosphorylated" substrates (pyruvate and ADP) (*44*) over a wide range of temperatures (*187*), in conformity with the suggestion (*188*) that the phosphoryl group undergoing transfer was directly coordinated to the enzyme-bound Mn. Moreover, the relative values of ϵ_b, ϵ_T(ADP), and ϵ_T(ATP) were quantitatively consistent with a relative coordination number for water (q) in E–Mn:E–MnADP:E–MnATP of 4 or 5:3:2, respectively (*44*), over a range of temperatures (*187*).

An enzyme–Mn–phosphoryl bridge complex was verified by the enhanced effect of E–Mn on $1/T_1$, and $1/T_2$ of the fluorine nucleus of FPO_3^{2-} (*37*), the product of a side reaction catalyzed at the same site of the enzyme (*85*):

$$\text{ATP} + \text{F}^- \xrightarrow[\text{HCO}_3^-]{\text{Mn}^{2+}\text{K}^+} \text{ADP} + \text{FPO}_3 \tag{28}$$

184. P. T. Englund, J. T. Huberman, T. M. Jovin, and A. Kornberg, *JBC* **244,** 3038 (1969).

185. H. M. Keir, *Prog. Nucleic Acid Res. Mol. Biol.* **4,** 81 (1965); L. A. Loeb, *JBC* **244,** 1672 (1969).

185a. J. P. Slater, L. A. Loeb, and A. S. Mildvan, unpublished observations (1970).

186. Recent experiments (J. Reuben, M. Cohn, G. L. Cottam, and A. S. Mildvan) with aged preparations of pyruvate kinase with high specific activity reveal approximately four binding sites for Mn rather than two as previously reported for fresh preparations (*44*). The significance of these findings is as yet unknown. The

The distance between the enzyme-bound Mn and the fluorine (Fig. 7) (*37*) was consistent with coordination through oxygen rather than fluorine. The exchange rate ($1/\tau_M$) of FPO_3^{2-} into the coordination sphere of Mn (Table VI) (*37*) was estimated from $1/T_2$ to be at least three orders of magnitude faster than the turnover number of the fluorokinase reaction, indicating that the metal bridge complex was a kinetically competent intermediate. The substrate PEP displaced FPO_3^{2-} from the E–Mn–FPO_3 complex, as determined by fluorine NMR, at a concentration consistent with the relative affinity of the two substrates for E–Mn established by substrate kinetics and by enhancement of water relaxation. The nearly equal enhancement values for water protons of E–Mn–FPO_3 and E–Mn–PEP suggested a similar metal bridge complex for PEP (*37*).

The large decrease of ϵ of water on forming the ternary E–Mn–PEP complex probably results from a conformational change in the ternary complex, as detected independently by UV difference spectroscopy (*99*). The relaxation data suggest an order of magnitude decrease in the water exchange rate on Mn (to a value of 0.6×10^6 sec^{-1}) on the binding of pyruvate to E–Mn–ADP (*187*). This may be due to steric hindrance of water escape, to a further conformation change as detected independently by UV difference spectroscopy for E–Mn–pyruvate (*99*), or to the direct coordination of pyruvate by enzyme-bound Mn. Direct coordination of fluoride, however, was not detected (*37*). These observations together with the evidence by Boyer and co-workers for direct phosphoryl transfer in a quaternary complex (*45, 188*) are accommodated by the mechanism given in Fig. 10. As with other kinases, the role of the Mn is to promote an electrophile (the phosphoryl group which is attacked and transferred), to gather and activate ligands, and possibly to serve as a template for conformation changes, including pseudorotation (Fig. 5). The monovalent cation which is required for this reaction (*21*) may be necessary for the proper conformation of the binary (*98, 99, 189*) and ternary (*190*) complexes, or it may be more directly involved in binding the carboxyl group of PEP to the enzyme (*91*).

---

enzyme is composed of four apparently identical subunits [G. L. Cottam, P. F. Hollenberg, and M. J. Coon, *JBC* **244**, 1481 (1969)].

187. M. Cohn, *in* "Magnetic Resonance in Biological Systems" (A. Ehrenberg, B. G. Malmström, and T. Vänngård, eds.), p. 101. Pergamon Press, Oxford, 1967.

188. P. D. Boyer, "The Enzymes," 2nd ed., Vol. 6, p. 95, 1962.

189. G. J. Sorger, R. E. Ford, and H. J. Evans, *Proc. Natl. Acad. Sci. U. S.* **54**, 1614 (1965).

190. A. S. Mildvan and M. Cohn, *Abstr. 6th Intern. Congr. Biochem., New York, 1964* IUB, Vol. IV, p. 111.

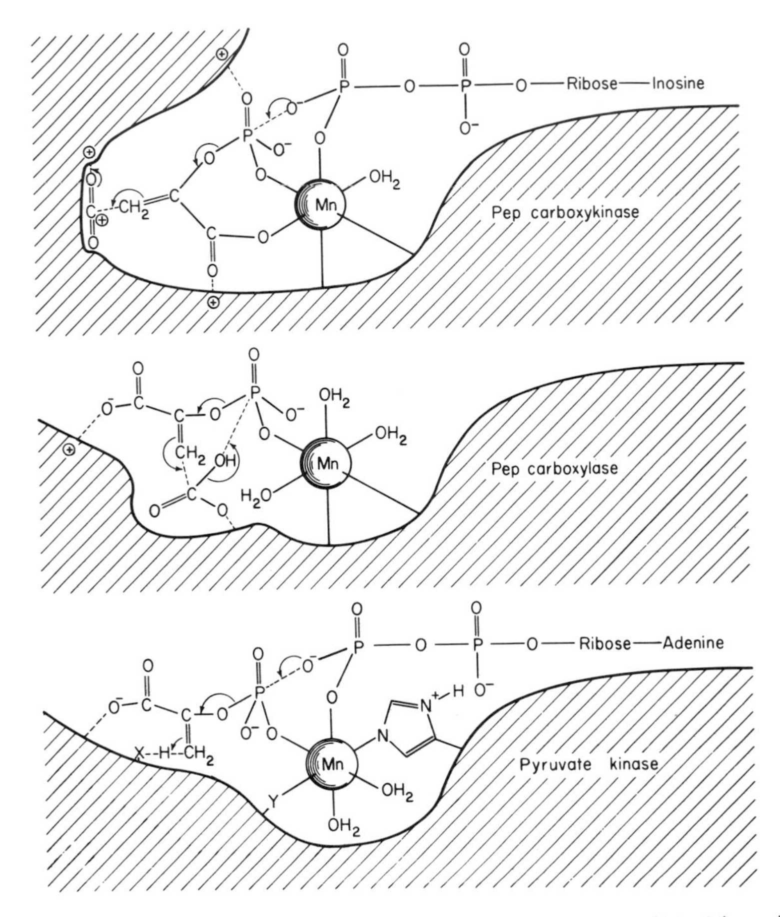

FIG. 10. Homology of the mechanisms of pyruvate kinase (*37*, *44*) and the phosphoenolpyruvate carboxylating enzymes (*47*) [from Miller *et al.* (*47*)].

6. *Yeast Pyruvate Kinase*

Unlike the muscle enzyme, which shows simple hyperbolic substrate kinetics (*44*, *45*) yeast pyruvate kinase shows a sigmoidal relationship between PEP concentration and reaction velocity (*191*). Fructosediphosphate activates the yeast enzyme by raising the apparent affinity for PEP, but not the maximal velocity, and converts the sigmoidal rela-

191. B. Hess, R. Haeckel, and K. Brand, *BBRC* **24,** 824 (1966); R. Haeckel, B. Hess, W. Lauterborn, and K. H. Wüster, *Z. Physiol. Chem.* **249,** 699 (1968); J. R. Hunsley and C. H. Suelter, *Federation Proc.* **26,** 559 (1967); *JBC* **244,** 4815, 4819 (1969).

tionship to a hyperbolic one (*191*). Thus, the yeast enzyme is "allosteric" in its kinetic behavior although the muscle enzyme is not (*44, 45*).

Like the muscle enzyme, yeast pyruvate kinase shows type II enhancement behavior (Table V) (*91*). However, unlike the muscle enzyme which binds up to four Mn ions/mole with nearly equal affinity (*186*), the yeast enzyme binds up to six Mn ions with dissociation constants which differ by nearly three orders of magnitude (*91*). The substrate PEP decreases the enhancement (from a value of ~15) to a value of ~2 (*192*), similar to that of ϵ_T (PEP) for the muscle enzyme (*44*), suggesting similar structures for the ternary PEP complexes. In binding studies, FDP (which also decreases ϵ) appears to compete with PEP when only one Mn is present per enzyme molecule, but raises the affinity for PEP (in qualitative and quantitative agreement with the kinetic data) when three Mn/enzyme are present.

It therefore appears that the role of the Mn in the mechanism of yeast pyruvate kinase may be similar to that in the muscle enzyme (Fig. 10) with the additional factor of site–site interaction among the Mn and PEP binding sites, which raises the affinity of the enzyme for PEP. A mechanism for this site–site interaction has been proposed (*91*). As depicted in Fig. 10, the binding site for PEP on pyruvate kinase consists of two subsites: the Mn, which binds the phosphoryl group, and another group [possibly a hydrogen bond donor or a monovalent cation (*192a*)], which binds the carboxylate group. The affinity of pyruvate kinase for PEP in the proposed mechanism would depend on the distance between these subsites, reaching a maximum at the optimum distance (~6 Å) and diminishing when this distance becomes too large or too small. On the nonallosteric muscle enzyme, this distance might be fixed, but on the allosteric yeast enzyme this distance might be adjustable. The binding of the first molecule of PEP (or of the activator FDP) at the first PEP site optimizes the distance between subsites at the next site for PEP. If the two subsites were on different subunits, this effect could manifest itself by changes in subunit interaction.

B. Phosphoenolpyruvate Carboxylating Enzymes

The enzymes, PEP-carboxykinase (*193*), PEP-carboxylase (*194*), and pyruvate kinase, catalyze homologous reactions of PEP, all involving

192. G. L. Cottam, A. S. Mildvan, and C. H. Suelter, unpublished observations (1969).

192a. F. J. Kayne and J. Reuben, *JACS* **92,** 220 (1970).

193. M. F. Utter and K. Kurahashi, *JBC* **207,** 787 (1954).

194. R. S. Bandurski and C. M. Greiner, *JBC* **204,** 781 (1953).

phosphoryl transfer to a nucleophile, a tautomeric shift of the double bond, and the addition of a positively charged atom at C-3 of PEP.

$$\text{PEP} + \text{IDP} + \text{HCO}_3^- \underset{\text{PEP-carboxykinase}}{\overset{\text{Mn}^{2+}}{\rightleftharpoons}} \text{OAA} + \text{ITP} + \text{H}^+ \quad (29)$$

$$\text{PEP} + \text{CO}_2 \underset{\text{PEP-carboxylase}}{\overset{\text{Mn}^{2+}}{\rightleftharpoons}} \text{OAA} + \text{P}_i \quad (30)$$

$$\text{H}^+ + \text{PEP} + \text{ADP} \underset{\text{pyruvate kinase}}{\overset{\text{Mn}^{2+}\text{K}^+}{\rightleftharpoons}} \text{pyruvate} + \text{ATP} \quad (31)$$

All of these enzymes have been found to show the same type of enhancement behavior (Table V) (*47*), and agreement was observed between dissociation constants determined by kinetics (*195*) and by binding studies using nuclear relaxation, equilibrium dialysis, and gel filtration (*47*). Hence, mechanisms homologous to that for pyruvate kinase (Fig. 10) have been proposed for the carboxylating enzymes (*47*). A similar mechanism has been suggested for the related enzyme, carboxytransphosphorylase (*196*), although Mn binding studies have not been reported. Another carboxylating enzyme, ribulosediphosphate carboxylase (*197*), shows the same type of enhancement behavior (*198*), suggesting an analogous role for Mn^{2+}.

C. Lyases and Related Enzymes

1. *Enolase*

Enolase catalyzes a reaction of PEP very different from those discussed above, namely, the addition of water, in which the double bond of PEP is polarized toward carbon atom 2; a proton adds to C-2 and a hydroxyl adds to C-3. The shift of the double bond in the enolase reaction is thus opposite to the tautomeric shift which occurs with the other PEP enzymes.

$$\bar{\text{O}}-\overset{\text{O}}{\overset{\|}{\text{C}}}-\underset{\underset{+\,\text{CH}_2}{\|}}{\text{C}}-\text{O}-\underset{\underset{\bar{}}{\text{O}}}{\overset{\overset{\text{O}}{\|}}{\text{P}}}-\bar{\text{O}} + \text{H}^+-\bar{\text{O}}\text{H} \underset{\text{enolase}}{\overset{\text{Mn}}{\rightleftharpoons}} \text{H}-\overset{\overset{\text{O}\diagdown\ \diagup\bar{\text{O}}}{\text{C}}}{\overset{|}{\text{C}}}-\text{O}-\underset{\underset{\bar{}}{\text{O}}}{\overset{\overset{\text{O}}{\|}}{\text{P}}}-\bar{\text{O}} \quad (32)$$
$$\qquad\qquad\qquad\qquad\qquad\qquad\qquad\qquad\quad \text{CH}_2\text{OH}$$

195. R. S. Miller and M. D. Lane, *JBC* **243**, 6041 (1968).
196. H. Lochmüller, H. G. Wood, and J. J. Davis, *JBC* **241**, 5678 (1966).
197. J. M. Paulsen and M. D. Lane, *Biochemistry* **5**, 2530 (1966).
198. M. S. Wishnick, A. S. Mildvan, and M. D. Lane, unpublished observations (1969).

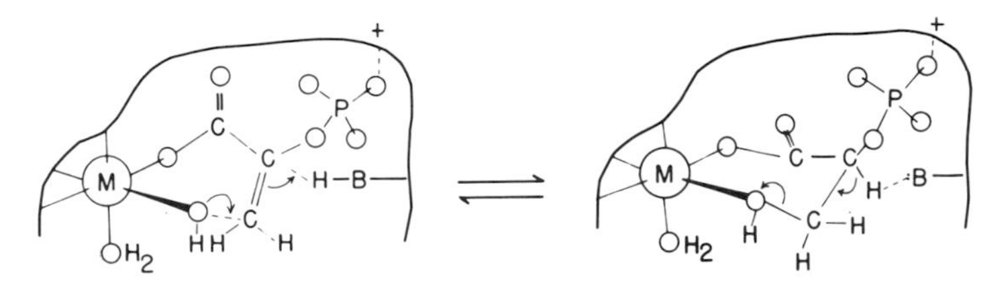

FIG. 11. Tentative mechanism for the enolase reaction consistent with kinetic, NMR, and other binding data (*60, 156, 199*).

The metal and substrate binding studies of Malmström and co-workers (*199*) and Wold and Ballou (*200*) were interpreted in terms of an enzyme–metal–substrate bridge complex (*199*). In conformity with this view, yeast enolase shows type II enhancement behavior (*60*). Because of the opposite polarity of the electromeric shift, the ternary E–M–S complex of enolase must differ in structure from that formed in pyruvate kinase. In pyruvate kinase the phosphoryl group is probably coordinated to the Mn (Fig. 10). In enolase, the Mn might coordinate the carboxyl group of PEP (activation of an electrophile) or, more likely, the OH which is to attack the C-3 of PEP (activation of a nucleophile), or both (Fig. 11).

2. *Histidine Deaminase*

The enzyme, histidine ammonia-lyase (*201*), requires Mn for activity and catalyzes the following elimination reaction:

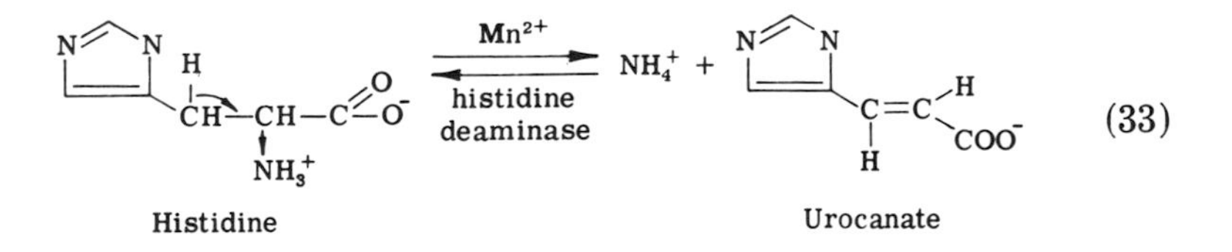

The mechanism of this reaction has already been referred to in Section VI, since E–M–S complexes have been established (Table VI and Fig. 7) (*160*). This enzyme shows type II enhancement behavior with the substrates histidine, urocanate, and the competitive inhibitors imidazole, histamine, and cysteine (*160*). Addition of alanine or glycine caused very

199. B. G. Malmström, T. Vänngård, and M. Larsson, *BBA* **30,** 1 (1958); B. G. Malmström, "The Enzymes," 2nd ed., Vol. 5, p. 471, 1961.
200. F. Wold and C. E. Ballou, *JBC* **227,** 301 and 313 (1957).
201. H. Tober and A. H. Mehler, "Methods in Enzymology," Vol. 2, p. 228, 1955.

little reduction of ϵ_b, although the latter is a competitive inhibitor ($K_I = 6$ mM), suggesting that only the imidazole moiety of the substrate provides a ligand to the enzyme-bound Mn. Direct evidence for such complexes was obtained by the effects of the E–Mn complex on $1/T_1$ and $1/T_2$ of the carbon-bound protons of imidazole and of urocanate (*160*). The exchange rates of urocanate and of imidazole (Table VI) are more than two orders of magnitude faster than the maximal velocity of the enzymatic reaction ($\sim$70 sec^{-1}) (*202*). The calculated distances (Fig. 7) between the manganese and the imidazole protons are consistent with direct coordination of imidazole. Smith *et al.* (*202*) have obtained evidence suggesting that the ammonium group of the substrate histidine interacts covalently with an enzyme-bound carbonyl group or a group of similar reactivity. A mechanism which accommodates all the data is given in Fig. 12. The role of the metal is σ electron withdrawal at the imidazole ligand to increase the acidity of a methylene proton, facilitating its removal by a base. The resulting carbanion eliminates ammonia and forms a double bond. Another role of the metal may be to mask the nucleophilic imidazole of the substrate histidine (Table III; IA-1a) to prevent its attacking the carbonyl group of the enzyme. In this connection the related enzyme, phenylalanine ammonia-lyase, which catalyzes the deamination of phenylalanine by the same carbonyl-type mechanism (*203*), does not require a metal ion for activity, presumably because the substrate phenylalanine lacks a liganding nucleophile on its side chain.

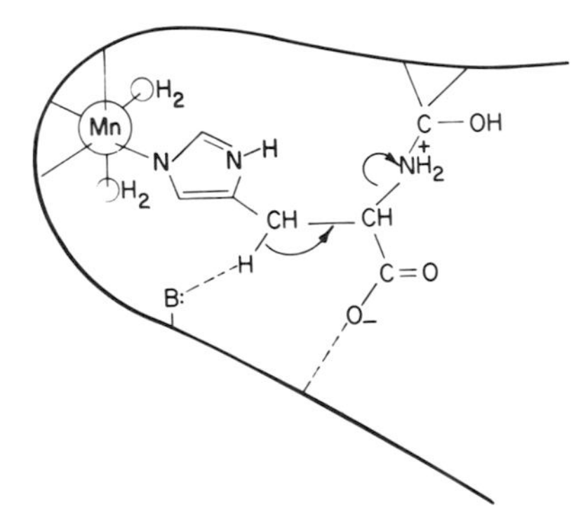

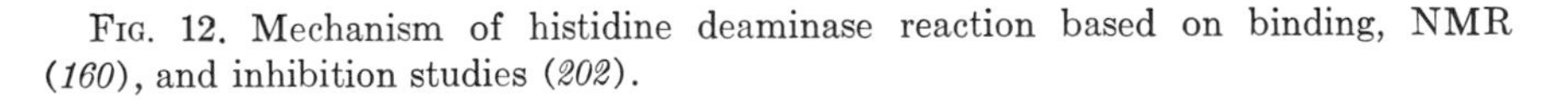
FIG. 12. Mechanism of histidine deaminase reaction based on binding, NMR (*160*), and inhibition studies (*202*).

202. T. A. Smith, F. H. Cordelle, and R. H. Abeles, *ABB* **120,** 724 (1967).
203. K. R. Hanson and E. A. Havir, *Federation Proc.* **28,** 602 (1969).

3. D-*Xylose Isomerase*

As shown by isotopic studies (*204*), the reaction catalyzed by this enzyme may be considered to be a sequence of an elimination reaction (to form a *cis*-enediol intermediate) followed by the addition of water or of the alcohol at carbon C-5 of the *cis*-enediol to form the product, D-xylulose. The enzyme shows type II enhancement behavior with the

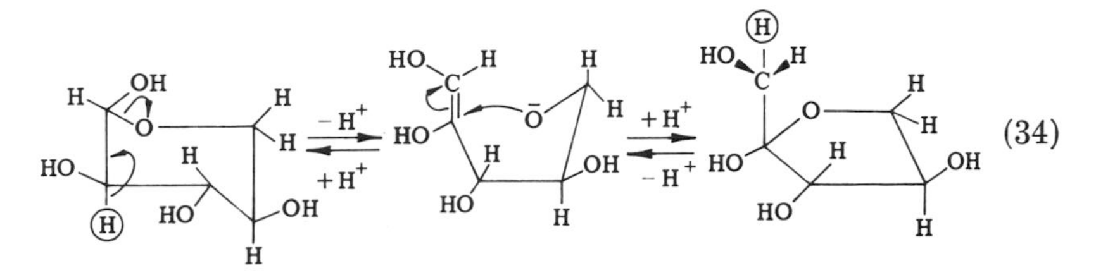

substrates D-xylose, D-glucose, and the inhibitors D-xylitol and D-sorbitol, suggesting the formation of metal bridge complexes (*161*). The dissociation constants of these complexes determined by enhancement agreed with those obtained by kinetics (*166, 167*). An enzyme–Mn–xylose bridge complex was confirmed since the enzyme enhanced the effects of Mn on the relaxation rates of the C_1 proton of α-D-xylose but not β-D-xylose (*161*). The exchange rate of α-D-xylose into this complex (Table VI) was more than two orders of magnitude faster than V_{max}. The distances between the enzyme-bound Mn and the C-1 proton of D-xylose (Fig. 7) were consistent with direct coordination by Mn of an oxygen or hydroxyl on C_1 of the sugar, but the precise structure of the complex is unknown. In the tentative mechanism presented in Fig. 13, the role of the Mn is to promote the electrophilicity of the proton which is removed in forming the *cis*-enediol and to facilitate opening of the glycosidic ring by inducing strain via a four-membered chelate ring. Indirect evidence for strain is the 5- to 11-fold lower affinity of the E–Mn complex for the substrate D-xylose, as compared to the inhibitor D-xylitol (*161*).

Metal bridge complexes may function in the related reactions catalyzed by L-arabinose isomerase, which requires Mn for activity (*168*), and mannose-6-phosphate isomerase, which is a Zn metalloenzyme (*205*).

4. *Pyruvate Carboxylase*

Pyruvate carboxylase from chicken liver mitochondria is the first naturally occurring Mn-metalloenzyme to be found (*206*). As summarized in

204. I. A. Rose, E. L. O'Connell, and R. P. Mortlock, *BBA* **178,** 376 (1969).
205. R. W. Gracy and E. A. Noltmann, *JBC* **243,** 4109 and 5410 (1968).
206. M C. Scrutton, M. F. Utter, and A. S. Mildvan, *JBC* **241,** 3480 (1966).

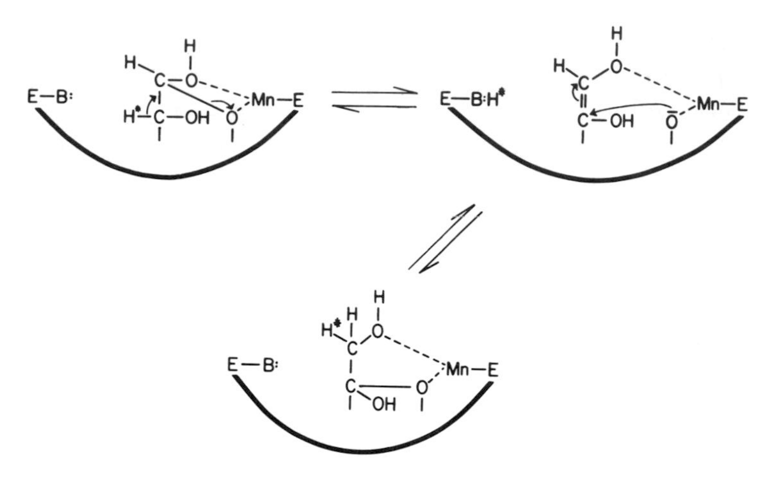

FIG. 13. Tentative mechanism of the D-xylose isomerase reaction based on NMR and stereochemical studies (*161, 204*).

a recent review (*206a*), the enzyme contains four subunits, four Mn, and four biotins per mole, and catalyzes the following reactions:

$$\text{E–biotin} + \text{ATP} + \text{HCO}_3^- \xrightleftharpoons{\text{Mg}^{2+},\ \text{acetyl CoA}} \text{ADP} + \text{P}_i + \text{E–biotin–CO}_2 \quad (35)$$

$$\text{E–biotin–CO}_2 + \text{pyruvate} \rightleftharpoons \text{E–biotin} + \text{oxalacetate} \quad (36)$$

The E–biotin–CO_2 complex carboxylates α-ketobutyrate, an analog of the normal substrate at ~3% of the rate of the reaction with pyruvate.

$$\text{E–biotin–CO}_2 + \alpha\text{-ketobutyrate} \overset{\text{slow}}{\rightleftharpoons} \text{E–biotin} + \beta\text{-methyloxalacetate} \quad (36a)$$

Only the substrates of steps (36) and (36a) (pyruvate, α-ketobutyrate, oxalacetate, β-methyloxalacetate) interact with enzyme-bound Mn as indicated by their effect on the enhancement of the water relaxation rate (*207*). The direct interaction of enzyme-bound Mn with pyruvate (*65*), α-ketobutyrate (*65*), and oxalacetate (*159*) has been demonstrated by studies of the relaxation rates of the carbon-bound protons of these substrates. As mentioned in Section VI, the kinetics of the exchange of H_2O, pyruvate, α-ketobutyrate, and oxalacetate into the coordination sphere of Mn provides strong support for an S_N1–outer sphere mechanism for the formation of the E–Mn–S complexes (Table VI).

206a. M. C. Scrutton and A. S. Mildvan, *in* "Symposium on CO_2: Chemical, Biochemical, and Physiological Aspects" (J. T. Edsall *et al.*, eds.), p. 207. N.A.S.A. Publication No. SP-188, 1969.

207. A. S. Mildvan, M. C. Scrutton, and M. F. Utter, *JBC* **241,** 3488 (1966).

The ranges of distances in solution (Fig. 7) (*208*) calculated between enzyme-bound Mn and the protons of bound pyruvate, α-ketobutyrate, and oxalacetate are consistent with direct coordination to Mn. However, in the case of pyruvate and oxalacetate they fail to distinguish between carboxyl or carbonyl ligands. Carbonyl coordination, as suggested by the high value of the hyperfine coupling constant (*65*), defines the role of Mn in the catalysis—to polarize the carbonyl group by σ (and possibly π) electron withdrawal, increasing the acidity of a methyl proton and facilitating its removal by a base.

Because of the exceedingly weak basicity of the biotin carbonyl group (*209*), a base from the enzyme may be necessary to deprotonate the pyruvate, although no direct evidence for such a base exists. The resulting nucleophilic carbanion may then attack the carboxyl group of carboxybiotin in a concerted (Fig. 14) or stepwise mechanism. Because carboxybiotin analogs are inert to nucleophilic attack in model reactions (*210*), an additional mode of activation is required. As suggested by the high activation energy for oxalacetate dissociation (Table VI) (*159*), the Mn may also coordinate the carboxyl group of carboxybiotin (gathering ligands) and promote its electrophilicity by electron withdrawal. The mechanism (*65, 206a, 207*) is depicted in Fig. 14. The suggested nucleophilic role of the sulfur in biotin to assist the polarization of the biotin carbonyl (*207*) appears unlikely because of the inability to detect transannular interaction by several independent techniques (*209*). However, a purely kinetic effect solely in the transition state has not been excluded. A kinetic difference has been found in the rate of carboxylation of biotin analogs (biotin $>$ oxybiotin $>$ desthiobiotin) by acetyl-CoA carboxylase (*211*).

From the ϵ_T values of water (*207, 212*), the relatively inactive substrate, α-ketobutyrate, appears to be a bidentate ligand in conformity with the model reaction (*213*), while the most active substrate, pyruvate, is probably a monodentate, carbonyl ligand, an important difference in detailed structure. The reduced rate of carboxylation of the α-keto-

208. The distances calculated for pyruvate carboxylase in Fig. 7 have been revised but still overlap with those given in Mildvan and Scrutton (*65*), as a result of an improved estimate of the correlation time [see Mildvan and Cohn (*63*)].

209. C. E. Bowen, E. Rauscher, and L. L. Ingraham, *ABB* **125,** 865 (1968); M. Caplow, *Biochemistry* **8,** 2656 (1969).

210. M. Caplow, *JACS* **87,** 5774 (1965).

211. E. Stoll, E. Ryder, J. B. Edwards, and M. D. Lane, *Proc. Natl. Acad. Sci. U. S.* **60,** 986 (1968).

212. M. C. Scrutton and A. S. Mildvan, *Biochemistry* **7,** 1490 (1968).

213. A. Kornberg, S. Ochoa, and A. H. Mehler, *JBC* **174,** 159 (1948); R. Steinberger and F. H. Westheimer, *JACS* **73,** 429 (1951).

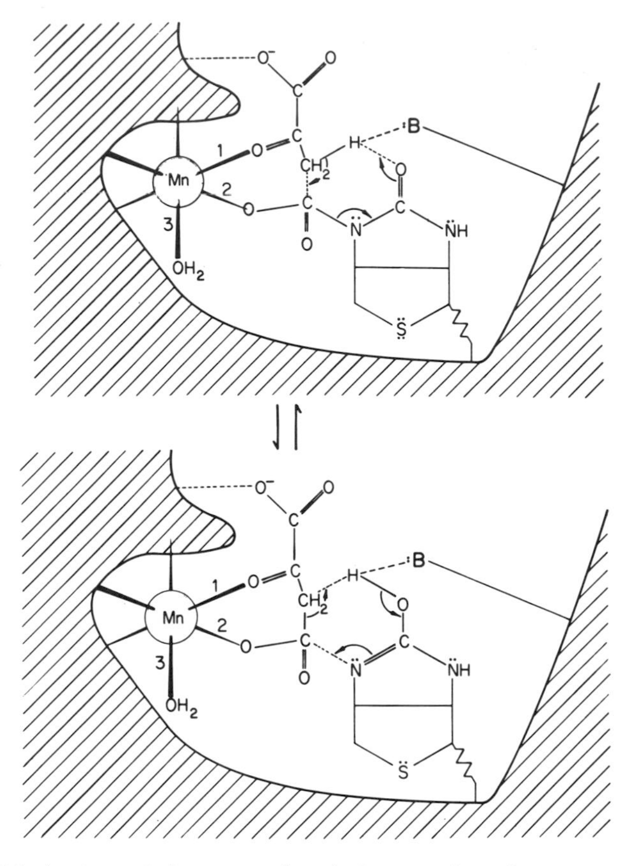

FIG. 14. Mechanism of the transcarboxylation reaction of pyruvate carboxylase modified from references *65* and *207* by the addition of a basic group (B) to assist the proton transfer as discussed in text.

butyrate complex may therefore result from a reduction in the electrophilicity of the Mn at the carbonyl ligand owing to coordination of the negatively charged carboxylate ligand or to an unfavorable conformation of the α-ketobutyrate chelate complex. The stability constants of the enzyme complexes of pyruvate and α-ketobutyrate are approximately equal (*207, 212*).

The large reduction in ϵ, caused by the inhibitors oxalate and malonate suggest that these compounds also function as bidentate ligands on the enzyme-bound Mn (*207, 212*). As with simple coordination complexes, the stability constant of the five-membered chelate ring formed by the enzyme–Mn–oxalate complex (10^6 M^{-1}) is much greater than that of the six-membered chelate ring formed by the enzyme–Mn–malonate complex

($10^{2.1}\ M^{-1}$). Steric factors also contribute to this large effect on ligand binding as indicated by the recent observation of stereospecific inhibition by L-malate which binds to the enzyme-bound Mn with a 30-fold greater affinity than does D-malate (*128*). Water and malate relaxation data are consistent with L-malate functioning as a bidentate ligand and D-malate as a monodentate ligand on the Mn. These findings represent enzymic examples of a thermodynamic template effect (Table III; IIB-4a). The slow binding of D- and L-malate, as compared to water dissociation (Table VI), indicates a kinetic template effect as well. The rate of exchange of L-malate was too slow ($k_{off} < 10^{3.3}\ sec^{-1}$) to be measured in the NMR experiment (*128*).

5. *Yeast Aldolase*

Aldolase catalyzes the reaction:

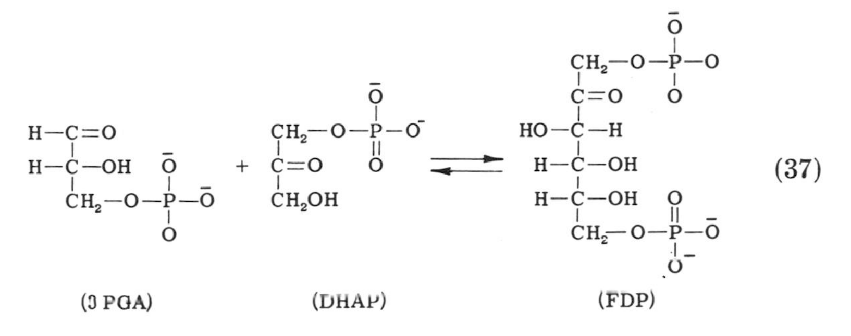

The enzyme from yeast is a Zn metalloenzyme (*214*) in which the tightly bound Zn may be replaced by Mn and Co to yield an active preparation (*215*). The enhancement factor ϵ_b of the stable binary E–Mn complex was 5.0 (*162*). Surprisingly, the addition of the substrates, fructose diphosphate, or dihydroxyacetone phosphate, produced small (10%) increases in the enhancement at room temperature (type I behavior), as found with M–E–S complexes. The addition of 3-phosphoglyceraldehyde decreased the enhancement by 10% (type II behavior). The dissociation constants of these complexes from enhancement titrations agreed with those determined kinetically (*162*). A more direct NMR experiment for establishing the coordination scheme by comparing the effects of the Zn-enzyme, the Co-enzyme, and the Mn-enzyme on the relaxation rates of the C-1 protons of fructose diphosphate and the C-3 protons of dihydroxyacetone phosphate, and acetol phosphate revealed interaction be-

214. W. J. Rutter, "The Enzymes," 2nd ed., Vol. 5, p. 341, 1961.

215. R. D. Kobes, R. T. Simpson, B. L. Vallee, and W. J. Rutter, *Biochemistry* **8**, 585 (1969).

tween the paramagnetic Mn^{II} ions and these protons, consistent with direct coordination (Fig. 7 and Table VI) (*162*). The latter findings suggest a metal bridge structure for type II (metallo)-aldolases as previously proposed (*216*), in which the metal functions to withdraw electrons from the bond to be broken. In type I aldolases, a Schiff base functions in place of a metal as the electrophile (*217*).

6. *Aconitase*

This enzyme catalyzes the reactions shown in Fig. 15, which also summarizes the stereochemistry of the reaction.

An ingenious "ferrous wheel" mechanism for aconitase has been proposed by Glusker (*218*), which is based on the conformations of citrate and isocitrate found in the crystalline state and on established principles of coordination chemistry. The ferrous wheel mechanism (Fig. 16) provides reasonable chemical explanations for the substrate specificity, the unusual stereochemistry of the intramolecular proton transfer (*219, 220*), the obligatory exchange of the hydroxyl of the substrate with the solvent (*220*), and the requirement for iron and cysteine (*221*). As indicated, the mechanism involves an E–Fe^{II}–citrate bridge complex in which the Fe^{II} by σ (and π) electron withdrawal at the C-3 hydroxyl promotes the acidity of the C-2 proton of citrate. This proton is transferred to a base and the OH remains (as water) in the coordination sphere of Fe^{II}, as citrate eliminates water to form *cis*-aconitate. The *cis*-aconitate has a

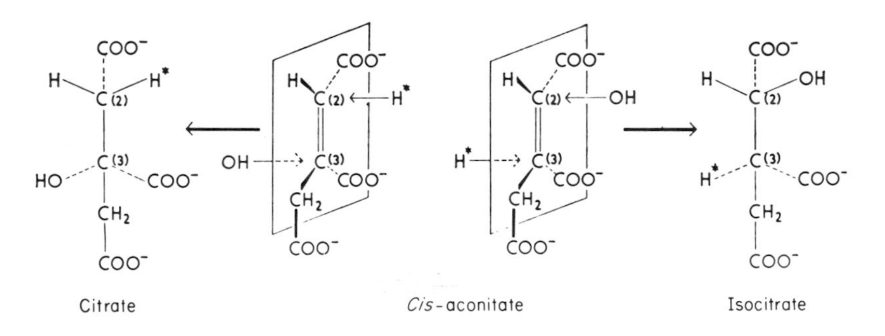

FIG. 15. Stereochemistry and proton retention in the aconitase reaction based on isotopic studies (*219, 220*), as discussed by Glusker (*218*).

216. W. J. Rutter, *Federation Proc.* **23**, 1248 (1964).
217. E. Grazi, T. Cheng, and B. L. Horecker, *BBRC* **7**, 250 (1962).
218. J. P. Glusker, *JMB* **38**, 149 (1968).
219. O. Gawron, A. J. Glaid, III, and T. P. Fondy, *JACS* **83**, 3634 (1961).
220. I. A. Rose and E. L. O'Connell, *JBC* **242**, 1870 (1967).
221. J. F. Morrison, *BJ* **58**, 685 (1954).

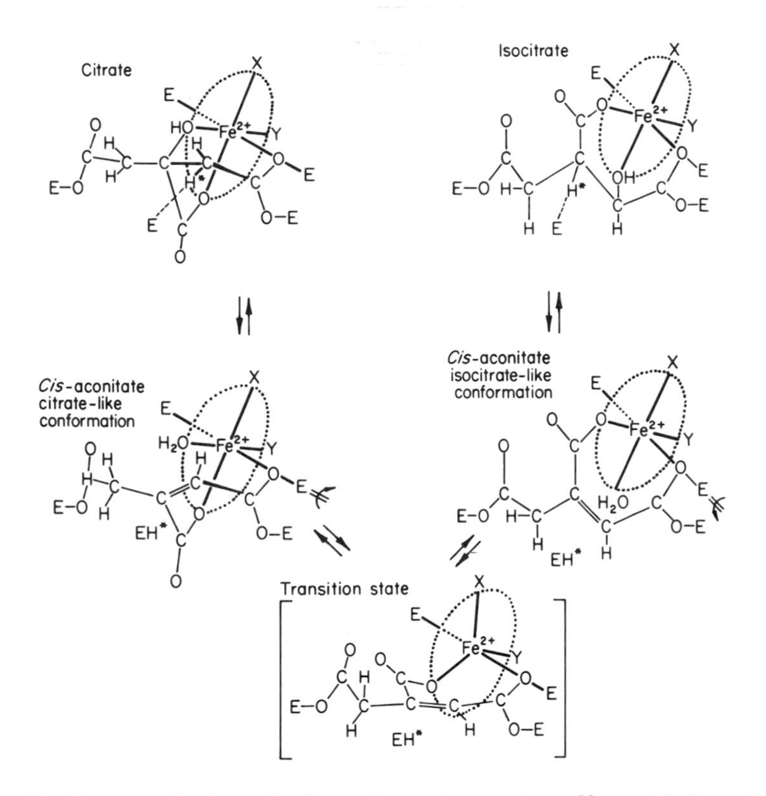

FIG. 16. Ferrous wheel mechanism of aconitase action [from Glusker (*218*)]. For discussion, see text.

"citrate-like" conformation. In order for the enzyme to form isocitrate of the proper stereochemistry, the bound intermediate, *cis*-aconitate must change its conformation from citrate-like to isocitrate-like. The driving force for the conformational change of bound *cis*-aconitate is simply a ligand (water) substitution reaction on the iron in which the entering water molecule approaches trans to the leaving water molecule. (The substitution is depicted as an S_N1 mechanism with a trigonal bipyramid transition state.) The net effect of the trans-substitution of water is to produce a 90° rotation of the ferrous wheel, which pulls the coordinated *cis*-aconitate into an isocitrate-like conformation. This conformation change of enzyme-bound *cis*-aconitate (as shown by models) places the C-3 carbon of *cis*-aconitate near the proton which was originally removed from the C-2 carbon of citrate and places the C-2 carbon of *cis*-aconitate near the new water ligand of the iron. Addition of these components then completes the formation of isocitrate of the proper stereochemistry, which has retained the citrate proton but has obligatorily exchanged its OH with the solvent. The Fe^{II} acts as a template for this conformation change and also as an electrophile. The spin state of Fe^{II} is unknown. The cysteine-Fe combination may be preferred because π-bonding ligands such as sulfur increase the tendency of octahedral complexes of high spin d^6 cations to distort to form the trigonal bipyramidal intermediate (*3*, *222*) and labilize low spin d^6 cations to substitution reactions (*3*).

The only deficiency in this elegant mechanism is direct experimental proof. Recent measurements of the effects of the inactive $(Mn)_2$ aconitase complex on the relaxation rates of H_2O and citrate protons demonstrate the existence of an aconitase–Mn–citrate bridge complex (*222a*).

7. *General Comment on Lyases and Related Enzymes*

The reactions catalyzed by aconitase, enolase, D-xylose isomerase, and histidine deaminase involve eliminations. A general role of the metal in these eliminations, as in the related pyruvate carboxylase and aldolase reactions, is to assist the departure of a proton from a carbon atom to form a full or incipient carbanion. In each case the metal may do so by coordinating an electronegative atom attached to a carbon atom which is β or γ to the proton to be removed and by σ electron withdrawal to increase its acidity and assist its removal by a base:

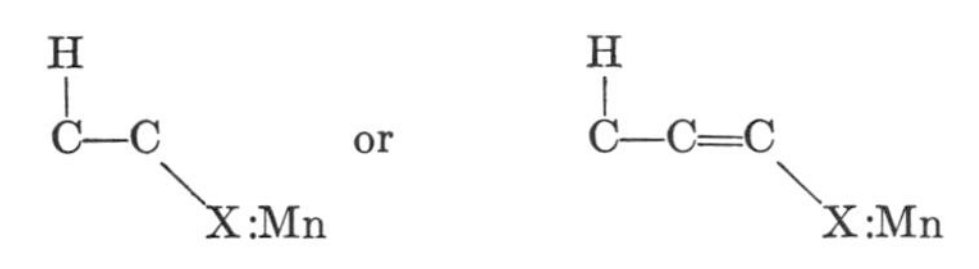

A similar role has been suggested for the metal in the elimination reaction catalyzed by β-methylaspartase (*223*) where a carbanion has been demonstrated by isotope exchange experiments (*223*). Indirect evidence for an electrophilic role of the metal is provided by the effect of the ionic radius of the activator on the maximal rate of the elimination reaction catalyzed by enolase (*199*), β-methylaspartase (*223*), and on the phosphoryl transfer reactions catalyzed by phosphoglucomutase (*157*) and pyruvate kinase (*44*, *188*). In these cases, where the substrate would donate an oxygen ligand, small divalent cations with high charge density and "hardness" such as Mg and Co are better activators than the larger and somewhat softer Mn^{2+}, Zn^{2+}, and Cd^{2+}. In the histidine deaminase

222. L. E. Orgel, "Introduction to Transition Metal Chemistry," p. 62. Methuen, London, 1960.

222a. J. J. Villafranca and A. S. Mildvan, *Abstr. Middle Atlantic Regional ACS Meeting Newark, Delaware, April 1970*, p. 22.

223. H. J. Bright, L. L. Ingraham, and R. E. Lundin, *BBA* **81**, 576 (1964); H. J. Bright, *Biochemistry* **6**, 1191 (1967).

reaction (*224*) and in the hydroxylamine kinase reaction of pyruvate kinase (*225*), the order appears to be reversed ($Mn^{2+} > Mg^{2+}$, Co^{2+}), possibly owing to the coordination of the "softer" imidazole ligand of the substrate in the former, and the nitrogen ligand of NH_2OH in the latter.

Irregularities are found in all correlations of ionic radius with activating ability which may result from changes in the pH optima of a reaction upon changing the metal activator (*226*), from variable π donation by the metals, and from variable crystal field effects on the coordination geometry (*16*, *19*, *222*).

D. Hydrolases and Related Enzymes

1. *Carboxypeptidase*

As a result of the incisive crystallographic studies of bovine carboxypeptidase A (*134*), the progress on its amino acid sequence (*227*), and the extensive modification and kinetic studies in solution (*169*, *228*) carried out in the laboratories of Lipscomb, Neurath, and Vallee, respectively, more is known about the structure of this than of any other metalloenzyme. Yet the precise role of the Zn remains to be established. The ligands for the Zn have been discussed in Section VI,B (Table VII). We will now consider the mechanism. The crystal structure of the cross-linked enzyme complex of glycyl tyrosine at 2 Å reveals that the terminal carboxyl group of the substrate is hydrogen bonded to arginine-145, the tyrosine ring is held in a hydrophobic pocket, and the peptide bond to be cleaved is positioned near three functional groups of the enzyme (Fig. 17): (a) The Zn, which is approximately tetrahedral (Table VII). In absence of substrate it coordinates a single water or hydroxyl ligand, depending on the pH: (b) The OH group of tyrosine-248, which has moved 12 Å to a position near the peptide bond upon binding of substrate. An intact tyrosine had previously been found essential for peptidase activity (*229*): (c) glutamate-270, which is very near the peptide bond to be cleaved.

224. I. Givot and R. H. Abeles, unpublished observations (1969).

225. F. P. Kupiecki and M. J. Coon, *JBC* **234,** 2428 (1959); G. L. Cottam, F. P. Kupiecki, and M. J. Coon, *ibid.* **243,** 1630 (1968).

226. C. Monder, *Biochemistry* **4,** 2677 (1965).

227. H. Neurath, R. A. Bradshaw, L. H. Ericssen, D. R. Babin, P. H. Petra, and K. A. Walsh, *Brookhaven Symp. Biol.* **21,** 1 (1968); M. Nomoto, N. G. Srinivasan, R. A. Bradshaw, R. D. Wade, and H. Neurath, *Biochemistry* **8,** 2755 (1969); P. H. Petra, R. A. Bradshaw, K. A. Walsh, and H. Neurath, *ibid.* p. 2762.

228. B. L. Vallee and J. F. Riordan, *Brookhaven Symp. Biol.* **21,** 91 (1968).

229. J. F. Riordan, M. Sokolovsky, and B. L. Vallee, *Biochemistry* **6,** 358 (1967).

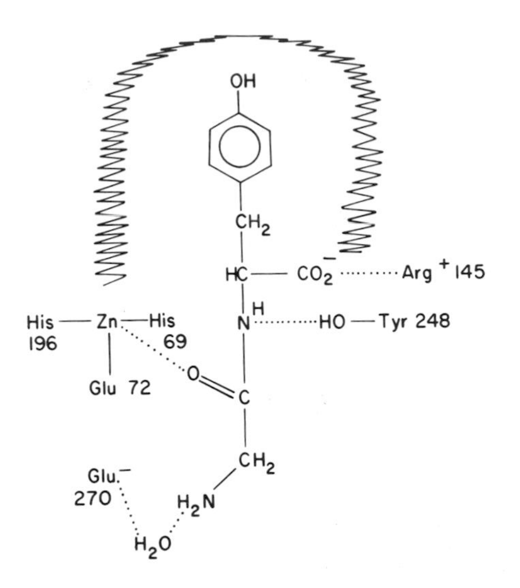

FIG. 17. Environment of the bound substrate in the crystalline complex of carboxypeptidase A with glycyl tyrosine. Modified from Lipscomb *et al.* (*134*) according to the amino acid sequence (*129d*).

Although the crystals have significant activity (*230*), the kinetic and thermodynamic properties of the complex observed in the Fourier synthesis are as yet unknown. Assuming it to be the active complex, Lipscomb and co-workers have considered several mechanisms and have proposed two which are consistent with the X-ray data, as well as with the relevant data in the literature.

(1) The Zn-carbonyl mechanism (Fig. 18A) proposes the displacement

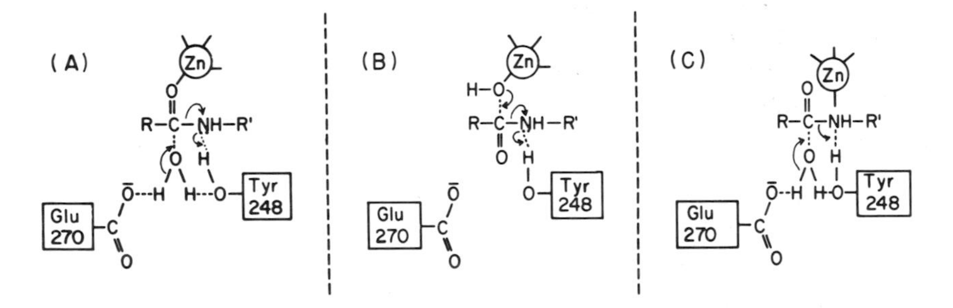

FIG. 18. Summary of mechanisms for carboxypeptidase-catalyzed hydrolysis of peptides: (A) Zn-carbonyl mechanism (*134*), (B) Zn-hydroxide mechanism (*134*), and (C) Zn-peptide nitrogen mechanism [figure is from Mildvan and Cohn (*63*)].

230. F. A. Quiocho and F. M. Richards, *Proc. Natl. Acad. Sci. U. S.* **52,** 833 (1964); *Biochemistry* **5,** 4062 (1966).

of a water on the Zn by the carbonyl oxygen of the peptide, resulting in polarization of the carbonyl group by σ (and possibly π) electron withdrawal, as suggested by Klotz (*125*).

(2) The Zn-hydroxide mechanism (Fig. 18B), which appears considerably less likely from the crystallographic studies because of steric interferences with substrate binding, can nevertheless not be excluded. In this mechanism the Zn increases the nucleophilicity of the attacking water or hydroxyl group. This mechanism is analogous to one suggested for carbonic anhydrase (see next section).

(3) A third mechanism, which might be called the Zn-peptide nitrogen mechanism, utilizes the Zn as an electrophilic catalyst on the peptide nitrogen rather than on the carbonyl oxygen. This mechanism has steric problems more serious than mechanism (2) and is, therefore, the least likely on crystallographic grounds.

Vallee and Coleman have proposed a complex which involves coordination of both the carbonyl oxygen and the peptide nitrogen (*16*, *228*). Mechanisms (1) and (2) [but not (3)] might ultimately involve coordination of the newly formed carboxylate anion of the hydrolyzed product by the Zn. Evidence suggesting the existence of such complexes in solution has been obtained by nuclear relaxation studies of the effect of the Mn-substituted carboxypeptidase on the relaxation rate of water (*141*). The enhanced effect is reduced to a low value in the presence of saturating amounts of the inhibitor phenylpropionate (type II behavior), suggesting the replacement of a single coordinated water by the monodentate carboxylate ligand. A number of weaker carboxylate inhibitors have been studied directly (*147*) by measurements of the relaxation rates of their carbon-bound protons, permitting the estimation of their limiting exchange rates (Table VI) and their dipolar distances from the bound Mn (Fig. 7). As discussed above (Section VI), the exchange rates are roughly consistent with an S_N1–outer sphere mechanism. The calculated distances are consistent with carboxylate coordination to Mn. However, the relationship between the binding sites detected by nuclear relaxation and the catalytic site utilized by the substrates is not clear. The binding sites for substrates and inhibitors are large, consisting of several partially overlapping subsites (*134*, *231*). Three binding sites for *p*-iodophenylpropionate near the Zn have been detected crystallographically (*232*). Further, the type of inhibition caused by phenylpropionate is mixed rather

231. N. Abramowitz, I. Schechter, and A. Berger, *BBRC* **29**, 862 (1967); B. L. Vallee, J. F. Riordan, J. L. Bethune, T. L. Coombs, D. S. Auld, and M. Sokolovsky, *Biochemistry* **7**, 3547 (1968).

232. T. A. Steitz, M. L. Ludwig, F. A. Quiocho, and W. N. Lipscomb, *JBC* **242**, 4662 (1967).

than strictly competitive (*233*), suggesting partial or no overlap of the inhibitor and the substrate sites.

At present it must be concluded that although mechanism (1) is most likely from steric considerations, is consistent with Jorgensen's symbiotic rule (Section VI,B,1), and would probably be the most effective means of promoting the hydrolysis of peptides, mechanism (2) has not been excluded by the kinetic, magnetic resonance, or the crystallographic data. Mechanism (3) has the least support from the magnetic resonance and the crystallographic data.

2. *Carbonic Anhydrase*

This enzyme, the first Zn-metalloenzyme to have been discovered (*234*), effectively catalyzes the hydration of carbonyl compounds such as CO_2 (*235*), and aldehydes (*236*), and the slower hydrolysis of esters (*237*). An informative review of the kinetic and physical properties of the isozymes from erythrocytes has been written by Edsall (*238*). The potent noncompetitive inhibition by sulfide ($K_I = 0.3\ \mu M$) and the failure of chelating agents to inhibit the enzyme led Davis to suggest that CO_2 was not directly coordinated to the Zn (*239*) but that the metal might function to convert a coordinated water to the more nucleophilic hydroxyl in a "Zn-hydroxide" mechanism (*78, 239–241*):

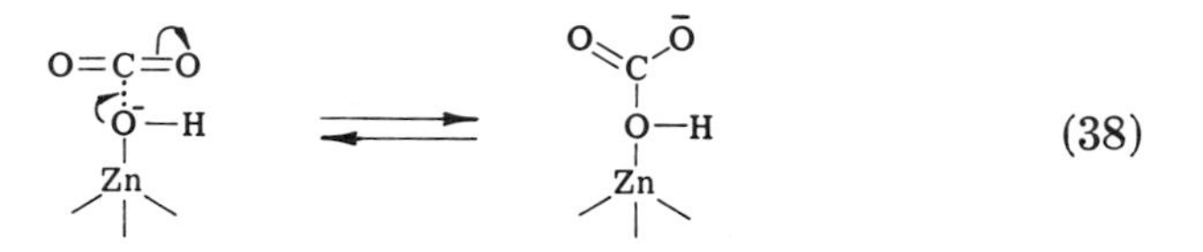

(38)

In this mechanism the substrate CO_2 is in the second coordination sphere of Zn and the product HCO_3^- is directly coordinated. More direct evidence for a substrate bridge complex (CO_2–E–Zn) has been obtained by infrared difference spectroscopy (*78*), which revealed only a small decrease

233. R. Lumry and E. L. Smith, *Discussions Faraday Soc.* **20,** 105 (1955).
234. D. Keilin and T. Mann, *BJ* **34,** 1163 (1940).
235. N. Meldrum and F. J. W. Roughton, *J. Physiol.* (*London*) **80,** 113 (1933).
236. Y. Pocker and J. E. Meany, *Biochemistry* **4,** 2535 (1965).
237. R. E. Tashian, D. P. Douglas, and Y. L. Yu, *BBRC* **14,** 256 (1964); B. G. Malmström, P. O. Nyman, B. Strandberg, and B. Tilander, *in* "Structure and Activity of Enzymes" (T. W. Goodwin, J. I. Harris, and B. S. Hartley, eds.), p. 121. Academic Press, New York, 1964.
238. J. T. Edsall, *Harvey Lectures* **62,** 191 (1967).
239. R. P. Davis, "The Enzymes," 2nd ed., Vol. 5, p. 545, 1961.
240. Y. Pocker and J. T. Stone, *Biochemistry* **6,** 668 (1967).
241. J. E. Coleman, *JBC* **242,** 5212 (1967).

(—2.5 cm^{-1}) in the asymmetric stretching frequency of enzyme-bound CO_2 consistent with a hydrophobic environment for CO_2 but inconsistent with direct coordination or distortion. However, the dissociation constant for CO_2 determined spectroscopically was eight times greater than that obtained kinetically (*242*), leaving open the possibility that the complex detected by IR may not be the kinetically active complex. The binding of azide to the enzyme caused a large increase (+48 cm^{-1}) in its stretching frequency which is consistent with the direct coordination of azide to Zn. This finding does not prove direct coordination since protonation of azide also causes large positive spectral shifts (*78*). The dissociation constant of azide from the spectral data agreed within a factor of three with its inhibitor constant (*240*). In contrast with the noncompetitive kinetic data (*239, 240, 243*) the binding of CO_2, HCO_3^-, ^-OH, and azide, as detected by IR, could be fit by assuming mutual competition for a single region of the enzyme (*78*). Hence, the CO_2, which is not directly coordinated to the Zn, may bind near the Zn, possibly in the second coordination sphere.

Direct coordination of the inhibitory chloride ion to Zn was demonstrated by the highly enhanced effect of the Zn-enzyme on the transverse nuclear quadropolar-relaxation rate of Cl^- (*163*). The mechanism of this enhancement is analogous to that of magnetic relaxation enhancement (see Section III,A) and results from hindered rotation of the Zn–Cl complex on the enzyme (*163, 244*). An S_N1–outer sphere mechanism for the binding of Cl^- to the enzyme-bound Zn may be inferred by calculation of the limiting rate constants for Cl binding from the NMR data (*163*) and the K_I value (*240*) (see Table VI). The rate constant k_{36} for the formation of the inner sphere Cl^- complex from a second sphere complex on enzyme-bound Zn ($\geq 10^{6.84}$ sec^{-1}) is in rough agreement with that found for the binding of Cl^- to inorganic Zn ($10^{7.4}$ sec^{-1}) which occurs by an S_N1 outer sphere mechanism (*31*). This finding suggests that the second coordination sphere of Zn is available for occupancy by Cl^- and presumably by CO_2 in the reaction mechanism. Cyanide and sulfonamide inhibitors displace chloride from the Zn (*163*). Crystallographic studies at 5.5 Å indicate that the sulfonamide group binds very near the Zn (*245*). The apoenzyme and the inactive Mn, Ni, Cu, Cd, and Hg enzymes fail to bind acetazoleamide (*241*). All of these findings, as well as the effect of pH on the enzymic activity (*238*), on the Zn–Cl interaction (*163*),

242. J. C. Kernohan, *BBA* **96**, 304 (1965).
243. J. A. Verpoorte, S. Mehta, and J. T. Edsall, *JBC* **242**, 4221 (1967).
244. T. R. Stengle and J. D. Baldeschwieler, *JACS* **89**, 3045 (1967).
245. K. Fridborg, K. K. Kannan, A. Liljas, J. Lundin, B. Strandberg, R. Strandberg, B. Tilander, and G. Wiren, *JMB* **25**, 505 (1967).

and on the proton release upon binding of H_2S and HCN (*241*) are consistent with the Zn-hydroxide mechanism of Eq. (38).

3. *Calcium-Activated Hydrolases*

A diverse group of enzymes which hydrolyze macromolecules is activated by Ca which functions to adjust thc protein conformation and, in at least one case, may participate directly in the catalysis by way of an E–S–M bridge complex (E bonded to S and to M, with S and M also bonded to each other).

Cogent examples of the template effect of metals in controlling the stereochemistry of protein folding are afforded by the requirement for a single divalent cation (preferably Ca^{2+}) for the reactivation of denatured, reduced taka-amylase A (*246*, *247*) and of reduced pancreatic deoxyribonuclease (*248*). In the former case a tightly bound Ca (*246*) is needed for restoring the proper conformation to permit closing of the fourth disulfide bond but not the first three. Conformation changes in the protein paralleling the reactivation have been detected by UV and fluorescence spectroscopy (*247*). In the latter enzyme (*248*), a weakly bound Ca, Mn, Cu, or Mg is necessary for closing the first of two disulfide linkages which is required for activity. The presence of Ca during the reduction of deoxyribonuclease preserves one of the two disulfides and the activity of the enzyme (*248*).

Staphylococcal nuclease requires Ca for activity and for the tight binding of nucleotides (*81*). Conversely, the binding of Ca is detectable by gel filtration only in the presence of nucleotides. The number of Ca ions in the ternary complex parallels the number of phosphates in the nucleotide, but the affinity for Ca is low. These findings together with a number of physical and chemical properties of the ternary complex (*82*) have been interpreted in terms of a cyclic enzyme–metal–nucleotide bridge complex. Evidence suggesting a metal bridge complex has recently been obtained in a crystallographic study at 4 Å resolution (*83*) in which the 5′-phosphate of the inhibitor iodothymidine 3′5′-diphosphate was found to bind close to a single Ca site on the enzyme. The other binding site for Ca in this ternary complex was not detected in the crystalline state. As with pyruvate kinase (*37*), the phosphoryl group which is to be transferred during the catalysis may enter the coordination sphere of the

246. T. Takagi and T. Isemura, *J. Biochem.* (*Tokyo*) **57,** 89 (1965).
247. T. Friedmann and C. J. Epstein, *JBC* **242,** 5131 (1967).
248. P. A. Price, W. H. Stein, and S. Moore, *JBC* **244,** 929 (1969).

enzyme-bound metal to permit σ electron withdrawal and facilitation of nucleophilic attack on phosphorus.

VIII. Oxidation–Reduction Reactions Involving Metals

The mechanisms of oxidation–reduction reactions of coordination complexes are best considered in two parts: (A) oxidation–reduction reactions within individual metal complexes, and (B) electron transfer between pairs of metal complexes.

A. Oxidation–Reduction Reactions within Individual Metal Complexes

1. *Mechanistic Principles*

Oxidation–reduction reactions within individual metal complexes may involve electron transfer between a ligand and the metal, between pairs of ligands on the same metal, or between a ligand and a nonliganding molecule.

The electrons transferred may be even or odd in number, may be accompanied by atom or group transfer, and may involve ionic or radical intermediates. The roles of the metal in such reactions are analogous to those outlined in Table III for nonredox reactions. Thus, metals may oxidize or reduce ligands directly by σ or π electron withdrawal or donation (IA and B). Transition metals with several stable valence states are required for the direct oxidation or reduction of ligands. Transition or nontransition metals may indirectly facilitate electron transfer reactions by electron withdrawal or donation (Table III), by masking side reactions of the ligand, or by gathering the oxidant and reductant as separate ligands, and by completing a conjugated system between them in which the filled *d* orbitals of the metal may function in a manner analogous to a double bond (see Table III, reaction IIA) (*26*).

The following examples of enzyme mechanisms have been selected because they illustrate the above principles. Many detailed reviews of the subject of oxidoreductases involving individual metal complexes are available (*22–26*).

2. *NAD-Linked Dehydrogenases*

Many, but not all, dehydrogenases contain Zn, a nontransition metal ion which does not undergo oxidation–reduction reactions. The role of the Zn in dehydrogenases has long been a subject of discussion and remains unknown. Several proposals have been made suggesting that the Zn might serve as a bridge to coordinate the NADH via (a) the adenine (*249*), (b) the pyrophosphate (*250*), (c) the nicotinamide carbonyl (*251*), (d) the pyridone ring nitrogen (*22*), or (e) the nicotinamide C^5 carbon (*252*). (Apparently, the amide nitrogen and the ribose oxygens remain unclaimed.) Simultaneous coordination of the substrate (*22, 249, 251, 252*) was also proposed in some mechanisms.

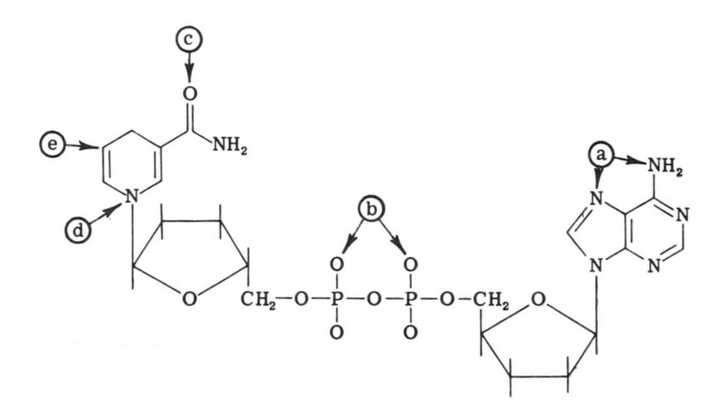

The finding that the two competitive inhibitors for NAD, ADPribose and *o*-phenanthroline, failed to compete against each other (*253*), together with the assumed coordination of the enzyme-bound Zn by *o*-phenanthroline (*254*), argued against (a) and (b) (*254*). However, the direct and rapid coordination of *o*-phenanthroline by Zn has been questioned (*255*). The recent finding that ADP-R·, a paramagnetic analog of NAD (*256*), can bind with the same stoichiometry and affinity to both native alcohol dehydrogenase and to the Zn-free apoenzyme casts further doubt on (a) and (b). Moreover, it is not apparent how coordination at (c) and (e) would facilitate hydride transfer, although coordination at (d) remains a possibility (*64*).

249. K. Wallenfalls and H. Sund, *Biochem. Z.* **329**, 59 (1957).

250. E. M. Kosower, *BBA* **56**, 474 (1962).

251. N. Evans and B. R. Rabin, *European J. Biochem.* **4**, 548 (1968).

252. H. Mahler and J. Douglas, *JACS* **79**, 1159 (1957).

253. T. Yonetani and H. Theorell, *ABB* **106**, 243 (1964).

254. T. Yonetani, *Biochem. Z.* **338**, 300 (1963); D. D. Ulmer and B. L. Vallee, *Advan. Enzymol.* **27**, 37 (1965).

255. B. M. Anderson, M. L. Reynolds, and C. D. Anderson, *BBA* **113**, 235 (1966); L. H. Piette and G. P. Rabold, *in* "Magnetic Resonance in Biological Systems" (A. Ehrenberg, B. G. Malmström, and T. Vänngård, eds.), p. 351. Pergamon Press, Oxford, 1967.

256. H. Weiner, *Biochemistry* **8**, 526 (1969). ADP-R· stands for ADP-4(2,2,6,6-tetramethylpiperidine-1-oxyl).

The threefold smaller enhancement of the proton relaxation rate (*64*) with E-ADP-R·, as compared to (E(Zn)$_4$)(ADP-R·), suggests that Zn immobilizes water near the unpaired electron of ADP-R·. The unpaired electron is localized in a region corresponding to the ribotide bond connecting the pyridine nitrogen and the C_1 ribose carbon atom of NAD. Similar reductions in ϵ to those found in the absence of Zn can also be found by forming the ternary substrate complexes (E(Zn)$_4$)(ADP-R·)(S) with the substrates ethanol and acetaldehyde or with the inhibitor isobutyramide, suggesting that the substrates replace some of the water immobilized by the Zn. Moreover, from nuclear relaxation studies of the substrates and inhibitors, it is known that they bind to the enzyme, at a position which is very near ($\leqslant$5.2 Å) the pyridine nitrogen of NAD (*64*). These findings, together with the relatively slow rate of binding of ethanol to (E(Zn)$_4$)ADP-R· ($\leqslant 10^{5.7}$ M^{-1} sec^{-1}) (*64*) and of acetaldehyde to (E(Zn)$_4$)NAD (10^5 M^{-1} sec^{-1}) (*257*), are best fit by assuming that only the substrates ethanol and acetaldehyde coordinate to the Zn in a rate-limiting ligand substitution reaction (*64, 258*), although no direct proof of an E–M–S complex has been provided. In certain sterically hindered complexes, Zn binds ligands with a rate constant as low as $10^{5.7}$ M^{-1} sec^{-1} (*259*).

A tentative mechanism for alcohol dehydrogenase based on these considerations and on the distances calculated to the unpaired electron of ADP-R· (*64*) is shown in Fig. 19.

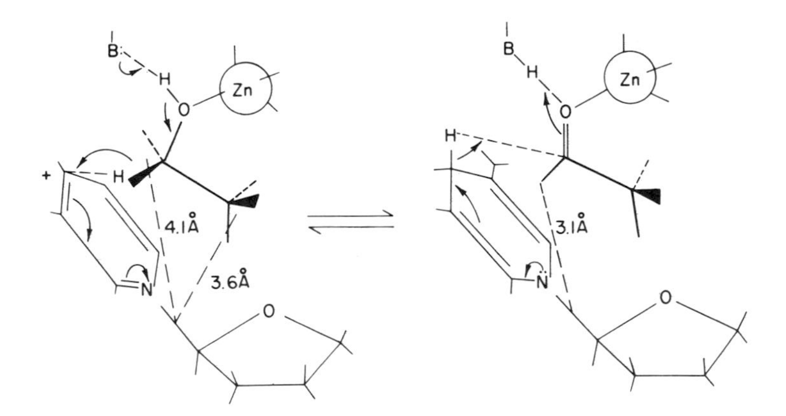

FIG. 19. Role of Zn in a tentative mechanism of the liver alcohol dehydrogenase reaction based on distances calculated between enzyme-bound substrates and enzyme-bound paramagnetic NAD analog (*64, 256*).

257. J. D. Shore and M. J. Gilleland, *Federation Proc.* **28**, 345 (1969).
258. H. Theorell and T. Yonetani, *Biochem. Z.* **338**, 537 (1963).
259. H. G. Hertz, *Z. Electrochem.* **65**, 36 (1961).

Detailed reviews of the relevant mechanisms from coordination chemistry have been published (*3, 266–268*).

1. *Mechanistic Principles*

Two mechanisms for electron transfer between pairs of metal complexes have been elucidated: the inner sphere mechanism (adjacent and remote) and the outer sphere mechanism.

In the inner sphere electron transfer mechanism, as demonstrated by Taube and co-workers (*269, 270*), a ligand is shared by the two metals in the transition state. In the adjacent inner sphere mechanism (*269*), both metals are coordinated to the same polar group of the ligand [Eq. (41)]. In the remote inner sphere mechanism (*270*), each metal is co-

$$(NH_3)_5Co^{III}Cl + Cr^{II}(H_2O)_6 \rightarrow [(NH_3)_5Co\text{—}Cl\text{—}Cr(H_2O)_5] + H_2O \xrightarrow{5H^+,\ 5H_2O} Co^{II}(H_2O)_6 + ClCr^{III}(H_2O)_5 + 5NH_4^+ \quad (41)$$

ordinated to a different polar group of the same ligand [Eq. (42)].

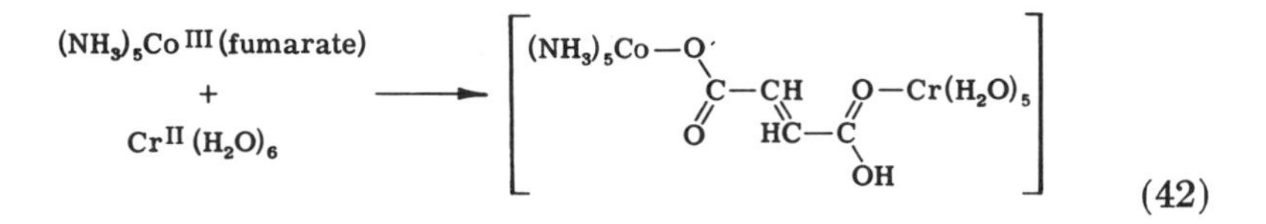

$$\xrightarrow{+5\,H^+} \xrightarrow{+5\,H_2O} Co^{II}(H_2O)_6 + (\text{fumarate})Cr^{III}(H_2O)_5 + 5NH_4^+$$

In each case the shared ligand serves to conduct electrons directly through its filled orbitals by resonance transfer (tunneling) (*266*), or the ligand may form a bridging radical during the transfer and function as a true chemical intermediate (*270*). The net transfer of the ligand which has occurred in reactions (41) and (42) is not an essential part of the oxidation–reduction reaction but serves to demonstrate ligand sharing in the transition state. The rates of such reactions may be limited by the rate of the ligand substitution reaction which is necessary to form the bridged transition state.

In the outer sphere electron transfer mechanism there is no common

266. P. George and J. S. Griffith, "The Enzymes," 2nd ed., Vol. 1, p. 347, 1959.
267. W. L. Reynolds and R. Lumry, "Mechanisms of Electron Transfer." Ronald Press, New York, 1966.
268. J. Halpern, *Quart. Rev.* (*London*) **15,** 207 (1961).
269. H. Taube, H. Myers, and R. J. Rich, *JACS* **75,** 4118 (1953).
270. H. Taube, *Advan. Chem. Ser.* **49,** 107 (1965).

ligand, and both coordination spheres remain essentially intact in the transition state. Electrons pass from the reductant to the oxidant either during a collision of the coordination spheres or while the coordination spheres are held together by weak bridging groups (*3, 267*). Reaction

$$Fe^{II}(DMP)_3 + Ir^{IV}Cl_6 \rightleftharpoons [Fe^{II}(DMP)_3 \cdots Ir^{IV}Cl_6] \rightleftharpoons$$
$$[Fe^{III}(DMP)_3 \cdots Ir^{III}Cl_6] \rightleftharpoons Fe^{III}(DMP)_3 + Ir^{III}Cl_6 \quad (43)$$

(43), in which the DMP (4,7 dimethyl-1,10-phenanthroline) complex of iron reduces the hexachloroiridate complex, is an exceedingly rapid outer sphere reaction (*271*).

The following metalloproteins and metalloenzymes provide biochemical examples of these mechanisms.

2. *Nonheme Iron Proteins*

Proposed theories of the structure of the coordination sphere of non-heme iron proteins involve pairs of sulfur bridges between pairs of iron atoms (*272, 273*). One theory (*273*) specifically postulates a distorted tetrahedral array of sulfur ligands about the iron (Fig. 21) and thus requires the absence of water molecules in the coordination sphere of iron. This theory is based in part on EPR spectroscopy of reduced non-heme iron proteins which must be carried out at liquid nitrogen tempera-

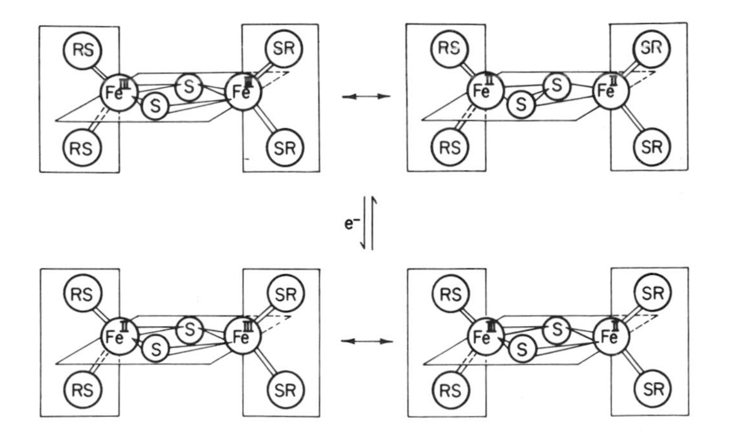

FIG. 21. Proposed structures of the coordination sphere of iron in oxidized (upper) and reduced (lower) ferredoxins [from Brintzinger *et al.* (*273*)].

271. J. Halpern, R. J. Legare, and R. Lumry, *JACS* **85,** 680 (1963).

272. D. C. Blomstrom, E. Knight, Jr., W. D. Phillips, and J. F. Weiher, *Proc. Natl. Acad. Sci. U. S.* **51,** 1085 (1964).

273. H. Brintzinger, G. Palmer, and R. Sands, *Proc. Natl. Acad. Sci. U. S.* **55,** 397 (1966).

ture. Freezing is known to cause structural alterations in other proteins (*74*). The electronic state of the iron atoms in the oxidized state may be low spin ferric, either diamagnetic (owing to spin pairing) or paramagnetic with exchange or dipolar broadening, to explain the absence of an EPR spectrum (*273*). A critical test of the hypothesis of Fig. 21 was to study the effect of spinach ferredoxin and adrenodoxin on the relaxation rates of water protons (*274*) which may be done at room temperature. Both proteins increased the relaxation rate of water equally at 25°, indicating paramagnetism in the oxidized forms, but the enhancement factors (compared to the ferric aquo cation) were very low ($\epsilon_b = 8.5 \times 10^{-3}$). The reduced forms were about half as effective in relaxing water ($\epsilon_b = 4.3 \times 10^{-3}$) as the oxidized forms. The low magnitude of the relaxivity, its negative temperature dependence, and the approximate equality of the longitudinal and transverse relaxation rates are consistent with outer sphere relaxation. Similar findings have been made with the oxidized ferredoxins from *Clostridium acidi urici* ($\epsilon_b = 2.4 \times 10^{-4}$) and *pasteurianium* ($\epsilon_b = 3.8 \times 10^{-4}$) (*275*), which is known to be paramagnetic at room temperature (*272*), and with oxidized ($\epsilon_b = 0.050$) and reduced ($\epsilon_b = 0.017$) rubredoxin from *Pseudomonas oleovorans* (*276*) in which the iron is probably high spin ferric and thus has a greater relaxivity. These results strongly suggest that water protons do not gain access to the coordination sphere of iron in nonheme iron proteins and thus support the tetrahedral sulfur theory. The crystallographic structure at 2.5 Å resolution for a rubredoxin from *Clostridium pasteurianium* confirmed a tetrahedral sulfur coordination and the absence of water ligands to iron (*277*). However, the high spin rubredoxins may not be typical of all nonheme iron proteins. Direct evidence, in the frozen state, for Fe–S interaction in nonheme iron proteins has been obtained by the observation of hyperfine splittings of the EPR resonance of Fe by substitution of ^{33}S for ^{32}S in *Azotobacter vinelandii* (*70*).

Purified nonheme iron proteins from this organism and from *Pseudomonas putida* gave Fe–S hyperfine splittings of a magnitude suggesting direct Fe–S bonding to both stable and labile sulfur (*71, 72*). By substitution of selenium for sulfur, the interaction of both of the labile selenium atoms with iron in Pseudomonas ferredoxin and adrenodoxin was demonstrated (*73*). Analogous evidence for Fe— — —Fe interaction,

274. A. S. Mildvan, R. W. Estabrook, and G. Palmer, *in* "Magnetic Resonance in Biological Systems" (A. Ehrenberg, B. G. Malmström, and T. Vänngård, eds.), p. 175. Pergamon Press, Oxford, 1967.

275. J. C. Rabinowitz and A. S. Mildvan, unpublished observations (1968).

276. J. Peterson and A. S. Mildvan, unpublished observations (1968).

277. J. R. Herriot, L. C. Sieker, and L. H. Jensen, *Acta Cryst.* **A25,** S-186 (1969).

using a reconstituted nonheme iron protein from *Pseudomonas putida*, has been obtained (*72*). The only aspect of the proposed structure (Fig. 21) which has not been directly demonstrated is S–S bonding, a resonance form in this theory and a required component of an alternative structure proposed by Rabinowitz (*278*).

Regardless of the detailed electronic structure of the iron, which is still under discussion (*279*), it is clear from the crystallography (*277*) and from the relaxivity studies (*274–276*) that the iron atoms are shielded from the solvent in both the oxidized and reduced states. Hence the mechanism of electron transfer to nonheme iron from external reductants appears to be of the outer sphere type in which electrons pass from reductant to iron during a collision, or in a complex, without significant alteration in the number or arrangement of the ligand atoms directly coordinated to iron. The migration of electrons between pairs of iron atoms within the same ferredoxin molecule (*72*) probably takes place by resonance transfer through bridging sulfur atoms in an adjacent inner sphere mechanism:

$$\mathrm{{}^{III}Fe\langle {}^{S}_{S} \rangle Fe^{II}} \longleftrightarrow \mathrm{{}^{II}Fe\langle {}^{S}_{S} \rangle Fe^{III}} \tag{44}$$

3. *Xanthine Oxidase and Aldehyde Oxidase*

Both of these metalloflavoproteins produce a two electron oxidation of their respective substrates and a two electron reduction of molecular oxygen to form hydrogen peroxide. Both contain two molybdenum atoms, two FAD molecules, and eight nonheme iron atoms per molecule (*280, 281*). Aldehyde oxidase also contains one or two molecules of coenzyme Q10 (*281*). The optical, EPR, and kinetic properties of these enzymes have recently been thoroughly reviewed (*22*). The available evidence on both enzymes is consistent with a remote inner sphere mechanism of electron transfer between reduced $\mathrm{Mo^{V}}$ and nonheme iron which may be mediated by a flavin radical functioning as a chemical intermediate. The chemical mechanism of electron transfer is supported by the time

278. J. C. Rabinowitz, private communication (1969), cited in Mildvan and Cohn (*63*).

279. J. F. Gibson, D. O. Hall, J. H. M. Thornley, and F. R. Whatley, *Proc. Natl. Acad. Sci. U. S.* **56**, 987 (1966).

280. R. C. Bray, "The Enzymes," 2nd ed., Vol. 8, p. 533, 1963; R. C. Bray, G. Palmer, and H. Beinert, *in* "Oxidases and Related Redox Systems" (T. E. King, H. S. Mason, and M. Morrison, eds.), Vol. 1, p. 359. Wiley, New York, 1964.

281. K. V. Rajogopalan, I. Fridovich, and P. Handler, *JBC* **237**, 922 (1962); K. V. Rajogopalan and P. Handler, *ibid.* **239**, 2022 (1964).

lags observed between the appearance of EPR spectra characteristic of Mo^V, a flavin semiquinone, and reduced nonheme iron. Linear sequences of electron transfer have therefore been suggested, but alternative path-

$$\text{Substrate} \rightarrow \text{Mo} \rightarrow \text{flavin} \rightarrow \text{Fe} \rightarrow O_2 \qquad (45)$$

ways have not been ruled out (*22, 280, 282*). The sites of substrate and oxygen binding remain to be established. Because of the general inaccessibility of nonheme iron to ligands, including water (*274–276*), iron seems an unlikely site for oxygen binding. A reasonable alternative to the linear pathway for electron transfer may therefore be

Substrate → [Mo ↗ FAD, Mo ↘ Fe, FAD ⇅ Fe] FAD → O_2 (46)

The pathway of Eq. (46) is supported by the recent finding that deflavo xanthine oxidase can no longer reduce oxygen but retains its ability to reduce cytochrome c while oxidizing xanthine (*282a*).

4. *Cytochrome c*

Of the electron transferring hemoproteins, the structure of cytochrome c is the most thoroughly understood. The X-ray crystallographic studies of horse heart ferricytochrome c at 4 Å resolution (*283–285*) reveal the iron to be coordinated by four pyrrole ligands from the heme, an imidazole, and a sixth ligand which is not imidazole but probably a methionine sulfur (*286*) (Table VII). The heme is embedded in a crevice with only one edge of the porphyrin ring exposed to the solvent (*283–285*).

Studies of the effect of oxidized cytochrome c on the relaxation rate of water protons reveal the relaxation mechanism to be outer sphere (*287*).

282. K. V. Rajogopalan, P. Handler, G. Palmer, and H. Beinert, *JBC* **243,** 3784 and 3797 (1968).

282a. H. Komac, V. Massey, and G. Palmer, *JBC* **244,** 1692 (1969).

283. R. E. Dickerson, M. L. Kopka, J. E. Weinzierl, J. Varnum, D. Eisenberg, and E. Margoliash, *JBC* **242,** 3015 (1967).

284. E. Margoliash, W. M. Fitch, and R. E. Dickerson, *Brookhaven Symp. Biol.* **21,** 259 (1968).

285. R. E. Dickerson, M. L. Kopka, C. L. Borders, Jr., J. Varnum, J. E. Weinzierl, and E. Margoliash, *JMB* **29,** 77 (1967).

286. H. Harbury, *in* "Hemes and Hemoproteins" (B. Chance, R. W. Estabrook, and T. Yonetani, eds.), p. 391. Academic Press, New York, 1966.

287. A. Kowalsky, *Federation Proc.* **28,** 603 (1969).

These observations indicate that the iron is shielded from the solvent in both the crystalline and the dissolved states. Nevertheless, ferrocytochrome c can be oxidized rapidly by $Fe(CN)_6^{3-}$ with a rate constant of 1.6×10^7 M^{-1} sec^{-1} (*151*) and even by ferricytochrome c at an appreciable rate (10^4 M^{-1} sec^{-1}) (*132*). The mechanisms of these electron transfer reactions are most certainly "outer sphere," as has been suggested above for the ferredoxins. Electron transfer to ferricytochrome c may be facilitated by the extensive delocalization of the unpaired electron (or hole) from the iron to the periphery of the heme, as detected by contact shifts in the high resolution NMR spectra (*132, 133*). Therefore, of the various mechanisms of reduction of cytochrome c previously considered (*266*), outer sphere electron transfer to the edge of the porphyrin followed by resonance transfer across the porphyrin ring to the iron would appear to provide the lowest kinetic barrier.

The c-type cytochrome of the bacterium, chromatium, which is photooxidized at the onset of photosynthesis in this organism, has been shown to transfer its electron to an acceptor [presumably oxidized chlorophyll (*288*)] by a resonance transfer or tunneling process below 100°K, as indicated by the high rate of oxidation at 4°K (300 sec^{-1}) and the negligible activation energy (<4 cal/mole) between 100°K and 4°K. Above 100°K the activation energy is 3.3 kcal/mole (*289, 290*).

These findings, which are of considerable mechanistic interest, suggest that those outer sphere processes which require energy, presumably to produce collisions and distortions, are frozen out at 100°K. The precise nature of the bridging groups, if any, remains to be elucidated.

A "remote inner sphere" reduction of cytochrome c by Cr^{II} has been discovered by Kowalsky (*291*). In this mechanism the binding of Cr^{II} to cytochrome c (hence the sharing of a ligand) precedes electron transfer. The Cr^{III} which is formed exchanges ligands very slowly and therefore remains bound to the protein molecule at the locus from which it has reduced the iron. This locus appears not to be on the heme (*291*).

The cytochrome c which has been reduced by Cr^{II} may be reoxidized by ferricyanide or cytochrome oxidase, and the oxidized product, which contains Cr^{III}, may be re-reduced by cytochrome c reductase but not by Cr^{II} (*291*). These findings further support a mechanistic difference between reduction by Cr^{II} (remote inner sphere) and reduction or oxidation by enzymes or ferricyanide (outer sphere).

288. W. W. Parson, *BBA* **153,** 248 (1968).
289. D. DeVault and B. Chance, *Biophys. J.* **6,** 825 (1966).
290. D. DeVault, J. H. Parkes, and B. Chance, *Nature* **215,** 642 (1967).
291. A. Kowalsky, *Federation Proc.* **26,** 674 (1967).

IX. Conclusions

This review has emphasized several milestones in the development of our present state of knowledge of the role of metals in enzyme catalysis. Such important concepts as the metal bridge complex, the substrate bridge complex, the S_N1–outer sphere mechanism, the reaction mechanisms of coordinated ligands and of electron transfer processes appear to have been established in certain enzymes. Other concepts such as the "ferrous wheel" mechanism of ligand conformation change, the pseudorotation of ligands, and the coordination of substrates by Zn in dehydrogenases are now amenable to direct experimental test. Direct and critical experiments have been made possible by such empirical advances as the introduction of chemical relaxation, nuclear relaxation, electron spin resonance, and X-ray crystallographic techniques into enzymology.

The applications of modern physical techniques to biochemical problems often raise more questions than they answer. However, we may console ourselves with the fact that the questions now being raised are more detailed and more penetrating than those deemed possible a decade ago.

ACKNOWLEDGMENTS

The author is grateful to Dr. D. H. Busch for his invaluable advice on the preparation of Table III and to Drs. M. Cohn, F. A. Cotton, E. E. Hazen, Jr., W. N. Lipscomb, G. Palmer, W. J. Ray, Jr., T. Spiro, and K. Wüthrich for sending preprints of their work.

Author Index

Numbers in parentheses are reference numbers and indicate that an author's work is referred to although his name is not cited in the text.

C

D

G

H

I

J

K

L

M

N

S

Y

Z

Subject Index

B

D

H

I

K

N

Q

R

U

V

W

X

Z